APPLIED MYCOLOGY AND BIOTECHNOLOGY

VOLUME 2

AGRICULTURE AND FOOD PRODUCTION

APPLIED MYCOLOGY AND BIOTECHNOLOGY

VOLUME 2

AGRICULTURE AND FOOD PRODUCTION

Edited by

George G. Khachatourians
Department of Applied Microbiology & Food Sciences
College of Agriculture
University of Saskatchewan
Saskatoon, SK, Canada

Dilip K. Arora
Department of Botany
Banaras Hindu University
Varanasi, India

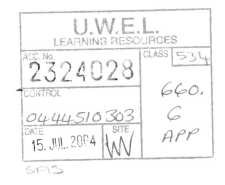

2002
ELSEVIER
Amsterdam - London - New York - Oxford - Paris - Shannon - Tokyo

ELSEVIER SCIENCE B.V.
Sara Burgerhartstraat 25
P.O. Box 211, 1000 AE Amsterdam, The Netherlands

First edition 2002

Library of Congress Cataloging in Publication Data
A catalog record from the Library of Congress has been applied for.

ISBN: 0 444 51030 3

♾ The paper used in this publication meets the requirements of ANSI/NISO Z39.48-1992 (Permanence of Paper).
Printed in The Netherlands.

Preface

The fungal kingdom consists of one of the most diverse groups of living organisms. They are numerous and ubiquitous, and undertake many roles, both independently, and in association with other organisms. In modern agriculture and food industry, fungi feature in a wide range of diverse processes and applications. In the food and drink arena role of fungi are historically important as mushrooms, in fermented foods, and as yeasts for baking and brewing. These roles are supplemented by the use of fungal food processing enzymes and additives, and more recently the development of protein based foodstuffs from fungi. On the detrimental side, fungi are important spoilage organisms of stored and processed foodstuffs. This balance of beneficial and detrimental effects is reflected in many other areas, in agriculture and horticulture such as certain mycorrhizal fungi may be necessary for seed germination and plant health, or may be used as biocontrol agents against weeds and invertebrates. The successful application of biotechnological processes in agriculture and food using fungi may therefore require the integration of a number of scientific disciplines and technologies. These may include subjects as diverse as agronomy, chemistry, genetic manipulation and process engineering. The practical use of newer techniques such as genetic recombination and robotics has revolutionized the modern agricultural biotechnology industry, and has created an enormous range of possible further applications of fungal products.

This volume of Applied Mycology and Biotechnology completes the set of two volumes dedicated to the coverage of recent developments on the theme "Agriculture and Food Production". The first volume provided overview on fungal physiology, metabolism, genetics, and biotechnology and highlighted their connection with particular applications to food production. The second volume examines various specific applications of mycology and fungal biotechnology to food production and processing. In the second volume, we present the coverage on two remaining areas of the theme, food crop production and applications in the foods and beverages sector. In our deliberations to examine content we asked several major questions related to agri-food production sector and applied mycology and biotechnology: (1) what were the most serious sources and causes of losses in production agriculture and food to involve fungi?; (2) what was the role and future potential for control strategies through fungal biotechnology?; (3) what benefits and values could have been added to the sector by fungal biotechnology and applied mycology? The editorial boards in selecting the coverage have assembled the best authors and select information available. We hope our readers will agree with our choices. The different aspects of the topics are organized in 12 chapters. In the first six chapters, we present the recent coverage of literature and work done in the area of genetics and biotechnology of brewer's yeasts, genetic diversity of yeasts in wine production, production of fungal carotenoids, recent biotechnological developments in the area of edible fungi, single cell protein, and fermentation of cereals. The next three chapters deal with the possibilities of applications of fungi to control stored grain mycotoxins, fruits and vegetables diseases. The last three chapters deal with agricultural applications of fungus plant interactions, whether harmful (weeds and plant pathogens) or beneficial (mycorrhizas). These chapters also examine the potential role of fungal biotechnology in changing our practice and the paradigm of food productivity by plants.

The interdisciplinary and complex nature of the subject area combined with the need to consider the sustainability of agri-food practices, its economics and industrial perspectives required a certain focus and selectivity of subjects. In this context where the turnover of literature is less than 2 years, we hope these chapters and its citations should help our readers arrive at comprehensive, in depth information on role of fungi in agricultural food and feed technology. As a professional reference, this book is targeted towards agri-food producer research establishments, government and academic units. Equally useful should this volume be for teachers and students, both in undergraduate and graduate studies, in departments of food science, food technology, food engineering, microbiology, applied molecular genetics and of course, biotechnology.

We are indebted to many authors for their up-to-date discussions on various topics. We thank Dr. Adriaan Klinkenberg and Ms. Anna Bela Sa-Dias at Elsevier Life Sciences for their encouragement, active support, cooperation and dedicated assistance in editorial structuring. We are looking forward to working together toward future volumes and enhancing the literature on the topics related to the potential upcoming areas of applied mycology and biotechnology.

George G. Khachatourians, Ph.D.
Dilip K. Arora, Ph. D.

Editorial Board for Volume 2

Editors

George G. Khachatourians
Department of Applied Microbiology and Food Science
College of Agriculture
University of Saskatchewan
Saskatoon, Canada
Tel: +1 306 966 5032
Fax: +1 306 966 8898
E-mail: khachatouria@sask.usask.ca

Dilip K. Arora
Department of Botany
Banaras Hindu University
Varanasi, India
Tel: +91 542 316770
Fax: +91 542 368141
E-mail: dkarora@banaras.ernet.in

Associate Editors

Deepak Bhatnagar	USDA/ARS, New Orleans, USA.
Christian P. Kubicek	Technical University of Vienna, Austria.
Helena Nevalainen	Macquarie University, Australia.
J. Ponton	Universidad del Pais Vasco, Spain.
C. A. Reddy	Michigan State University, USA.
Jose-Ruiz-Herrera	Centro de Investigacion y Estudios Avanzados del I.P.N., Mexico.
Anders Tunlid	Lund University, Sweden.
Gunther Winkelmann	University of Tubingen, Germany.

Contents

Contributors

Luciano Avio Dipartimento di Chimica e Biotecnologie Agrarie, University di Pisa, Via del Borghetto 80 56124 Pisa, Italy.

Tahia Benitez Department of Genetics, Faculty of Biology, University of Seville, Apartado 109, E-41080 Seville, Spain.

Deepak Bhatnagar Food and Feed Safety Research Unit, U. S. Department of Agriculture, Agricultural Research Service, Southern Regional Research Center, New Orleans, Louisiana 70124, USA.

S. M. Boyetchko Agriculture and Agri-Food Canada, Saskatoon Research Centre, 107, Science Place, Saskatoon SK S7N OX2, Canada.

Robert Brown Food and Feed Safety Research Unit, U. S. Department of Agriculture, Agricultural Research Service, Southern Regional Research Center, New Orleans, Louisiana 70124, USA.

A. J. Caesar USDA/ARS, 1500 N. Central Avenue, Sidney, Montana 59270, USA.

J. I. Castrillo School of Biological Sciences, Biochemistry Division, University of Manchester, 2.205, Stopford Building, Oxford Road, Manchester M13 9PT, U.K.

R. Charudattan University of Florida, Plant Pathology Department, 1453 Fifield Hall, Gainesville, Florida 32611, USA.

Thomas Cleveland Food and Feed Safety Research Unit, U. S. Department of Agriculture, Agricultural Research Service, Southern Regional Research Center, New Orleans, Louisiana 70124, USA.

Antonio C. Codon Department of Genetics, Faculty of Biology, University of Seville, Apartado 109, E-41080 Seville, Spain.

Samir Droby Dept of Postharvest Science, ARO, The Volcani Center, P.O. Box 6, Bet Dagan 5250, Israel.

Carlos Echavari-Erasun Department of Food Microbiology, Food Research Institute, University of Wisconsin, 1925 Willow Dr. 53706, Madison, WI, USA.

Kenneth Ehrlich Food and Feed Safety Research Unit, U. S. Department of Agriculture, Agricultural Research Service, Southern Regional Research Center, New Orleans, Louisiana 70124, USA.

Ahmed El Ghaouth MICRO FLO Company, Memphis, TN 38117, USA.

Manuela Giovannetti Dipartimento di Chimica e Biotecnologie Agrarie, University di Pisa, Via del Borghetto 80 56124 Pisa, Italy.

Eric A. Johnson Department of Food Microbiology, Toxicology and Bacteriology, Food Research Institute, University of Wisconsin, 1925 Willow Dr. 53706, Madison, WI, USA.

Lisa Korsten Department of Microbiology and Plant Pathology, University of Pretoria, Pretoria 0002, South Africa.

Cherl-Ho Lee Dept. of Food Engineering, CAFST, Korea University, Seoul 136-701, Korea.

Sang Sun Lee Department of Biology, Korea National University of Education, Chungbuk 363-791, Korea.

Koshlendra K. Pandey Indian Vegetable Research Institute, Gandhi Nagar (Naria), P. B. No. 5002, P.O. BHU, Varanasi 221 005, India.

Padam K. Pandey Indian Vegetable Research Institute, Gandhi Nagar (Naria), P. B. No. 5002, P.O. BHU, Varanasi 221 005, India.

Julio Polaina Instituto de Agroquimica y Tecnologia de Alimentos Consejo Superior de Investigaciones Cientificas Apartado de Correos 73, E46100-Burjasot (Valencia), Spain.
Raj D. Rai National Research Centre for Mushrooms, Chambaghat, Solan 173 213, H.P, India.
E. N. Rosskopf USDA/ARS, 2199 S. Rock Road, Fort Pierce, Florida 34945, USA.
Joseph L. Smilanick USDA-ARS, 2021 South Peach Avenue, Fresno, CA, 93727, USA.
U. O. Ugalde Department of Applied Chemistry, Faculty of Chemistry, University of Basque Country, P.O. Box 1072, 20080 San Sebastian, Spain.
Charles Wilson Appalachian Fruit Research Station, USDA/ARS, 45 Wiltshire Road, Kearneysville, WV 25430, USA.
Michael Wisniewski Appalachian Fruit Research Station, USDA/ARS, 45 Wiltshire Road, Kearneysville, WV 25430, USA.
L. J. C. Xavier Agriculture and Agri-Food Canada, Saskatoon Research Centre, 107, Science Place, Saskatoon, SK S7N OX2, Canada.

Applied Mycology and Biotechnology
Volume 2. Agriculture and Food Production

Brewer's Yeast: Genetics and Biotechnology

Julio Polaina
Instituto de Agroquímica y Tecnología de Alimentos, Consejo Superior de Investigaciones Científicas, Apartado de Correos 73, E46100-Burjasot (Valencia), Spain (E-Mail:jpolaina@iata.csic.es).

The advance of Science in the 19th century was a decisive force for the development and expansion of the modern brewing industry. Correspondingly, the brewing industry contributed important scientific achievements, such as Hansen's isolation of pure yeast cultures. Early studies on yeast were connected to the development of different scientific disciplines such as Microbiology, Biochemistry and Genetics. An example of this connection is Winge's discovery of Mendelian inheritance in yeast. However, genetic studies with the specific type of yeast used in brewing were hampered by the complex constitution of this organism. The emergence of Molecular Biology allowed a precise characterization of the brewer's yeast and the manipulation of its properties, aimed at the improvement of the brewing process and the quality of the beer.

1. INTRODUCTION

The progress of chemistry, physiology and microbiology during the 19th Century, allowed a scientific approach to brewing that caused a tremendous advancement on the production of beer. The precursor of such approach was the French microbiologist Louis Pasteur. At this time, the Danish brewer Jacob Christian Jacobsen, also founded the Carlsberg Brewery and the Carlsberg Laboratory. In Jacobsen's own words, the purpose of the Carlsberg Laboratory was: *"By independent investigation to test the doctrines already furnished by Science and by continued studies to develop them into as fully scientific a basis as possible for the operation of malting, brewing and fermentation".* Louis Pasteur (1822-1895) demonstrated that alcoholic fermentation is a process caused by living yeast cells. His conclusion was that fermentation is a physiological phenomenon by which sugars are converted in ethanol as a consequence of yeast metabolism. In 1876, Pasteur published *"Etudes sur la Bière"*, which followed the trend of his previous book *"Etudes sur le Vin"*, published ten years earlier. In *Etudes sur la Bière*, he dealt with the diseases of beer and described how the fermenting yeast was often contaminated by bacteria, filamentous fungi, and other yeasts. However, the importance of Pasteur in relation with brewing is due to his discovery of yeast as the agent of fermentation. His more specific contributions to this field are not to be considered among his greatest achievements. Probably, this had something to do with the fact that he did not like beer. Pasteur's work in connection with yeast and the brewing industry has been recently reviewed by Anderson [1] and Barnett [2]. A crucial achievement for the development of the brewing industry was accomplished by Emil Christian Hansen (1842-1909). Originally trained as a house painter and a primary school teacher, E. C. Hansen later became a botanist and a mycologist. In 1877, he was employed as a fermentation physiologist at the Carlsberg Brewery. Familiar with the work of Pasteur and facing the problems of microbial contamination that often caused serious troubles in breweries, Hansen pursued the idea of obtaining pure yeast cultures. To this end, he estimated the amount of yeast cells present in a beer sample. He made serial dilutions of the sample until he reached an estimated concentration of 0.5 cells per ml, and used 1 ml aliquots of the diluted suspension to inoculate many individual flasks containing wort. After about a week of incubation, roughly half of the cultures contained a single yeast colony, very few contained two or more colonies, and no growth was observed in the other half of the flasks. Hansen concluded from this experiment that it was possible to obtain a single colony consisting of the uncontaminated descendants of an individual cell. He performed additional experiments in which, starting with a mixture of two or more types of yeast, he was able to recover pure cultures of each different type. Another important contribution of Hansen to the work with yeast was the introduction of cultures on "solid medium". For this purpose he adapted the procedure devised by Robert Koch for bacteria. Yeast colonies were grown on glass plates, on the surface of a jellified medium prepared with gelatin. Hansen's new techniques allowed him to obtain pure cultures of different brewing strains and also to characterize contaminant strains that caused different beer diseases. In 1883, the Carlsberg Brewery started industrial production of lager beer with one of Hansen's pure cultures. This event became a milestone of the industrial revolution, since it meant the transition from small-scale, artisan brewing to large-scale, modern production. The path led by the Carlsberg Brewery was soon followed by other companies, and in the next few years the technique of brewing with pure yeast cultures became standard in Europe and North America and caused an exponential growth of beer production all over the world. An exciting account of the work of Hansen has been given by von Wettstein [3].

Øjvind Winge was born in Århus (Denmark) in 1886, shortly after the first industrial brewing with a pure yeast culture. Winge was a very capable biologist who mastered different disciplines, including botany, plant and animal genetics, and mycology. In 1921, he became Professor of Genetics, firstly at the Veterinary and Agricultural University of Copenhagen and several years later at University of Copenhagen. Winge took the position of Director of the Department of Physiology at the Carlsberg Laboratory in 1933. When established in his new position, he recovered the collection of natural and industrial yeast strains gathered by Hansen and Albert Klöcker, who both had preceded him at the Department of Physiology. Winge faced the problem that brewer's yeast strains were not able to sporulate, or did so very poorly, which made them unsuitable for genetic analysis. Therefore, he focused his attention on baker's yeast (*S. cerevisiae*), which had long been a favorite organism for biochemical studies, and different varieties of *Saccharomyces* capable of sporulation (*S. ludwigii, S. chevalieri, S. ellipsoideus*, and others). With the help of a micromanipulation system of his own design, Winge carried out dissection of the *asci* of sporulated yeast cultures and followed the germination of individual spores. He concluded that *Saccharomyces* has a normal alternance of unicellular haploid and diploid phases, i. e. it should behave genetically according to Mendel's laws. In collaboration with O. Laustsen, Winge reported the first results of tetrad analysis. After a lag period imposed by World War II, Winge started a very productive period that is marked by his collaboration with Catherine Roberts. Together, they discovered the gene that controls homothallism and many genes that control maltose and sucrose fermentation. They also found that haploid yeast strains might have several copies of the genes involved in the fermentation of these sugars. They coined the expression polymeric genes to designate a repeated set of genes that perform the same function. The beginning of fission yeast *(Schizosaccharomyces pombe)* genetics is also linked to Winge. Urs Leupold spent a research stay in Winge's Department of Physiology where he established the mating system and described the first cases of Mendelian inheritance for this yeast [4]. The work of Winge in connection with yeast has been reviewed by R. K. Mortimer [5]. The birth of yeast genetics had a strong Scandinavian clout since besides Winge, the other prominent figure was Carl C. Lindegren, born in 1896 in Wisconsin,

USA, in a family of Swedish immigrants. The most transcendent achievement of Lindegren in connection with yeast genetics was the discovery of the mating types. This led to development of stable haploid cultures of both mating types and served to start the cycle of mutant isolation and genetic crosses that made of *Saccharomyces* one of the most conspicuous organisms for genetic research. Other important achievements were the discovery of the phenomenon of gene conversion and the elaboration of the first genetic maps of the yeast. The work and the controversial personality of Lindegren have been the subject of an inspiring book chapter [6].

In 1847, the brewer J. C. Jacobsen started the production of bottom fermented (lager) beer at a brewery that he built in Valby, in the outskirts of Copenhagen. He named his brewery Carlsberg after his five years old son Carl, who later became a maecenas of arts in Denmark. J. C. Jacobsen was one of the pioneers of industrialization in Denmark. He introduced new procedures in the brewing process that soon became standard and gave Carlsberg a rapid success. In 1875-76, J. C. Jacobsen established the Carlsberg Foundation and the Carlsberg Laboratory. The Carlsberg Laboratory was divided in two Departments, Physiology and Chemistry. As a tradition, both Departments have focused their work mainly, albeit not exclusively, on processes and organisms of special significance for brewing, such as yeast and barley. The first director of the Department of Chemistry was Johan Kjeldahl, who invented the procedure for the determination of organic nitrogen that carries his name. Undoubtedly, the most popular contribution of the Department of Chemistry was the concept of pH, due to Søren P. L. Sørensen who was head of the Department from 1901 to 1938. Of outstanding scientific significance was the work of the following director, Kaj U. Lindestrøm-Lang, who devised the terms primary, secondary, and tertiary structure, to describe the structural hierarchy in proteins. The contributions of two former directors of the Department of Physiology, Hansen and Winge, have been summarized above. More recent work carried out with yeast will be dealt with in the following sections. Together with the work with yeast, the Department of Physiology has produced important contributions related to chlorophyll biosynthesis [7,8].

2. GENETIC CONSTITUTION OF BREWER'S YEAST

Saccharomyces cerevisiae is one of the best genetically characterized yeast as its genome is fully sequenced and analyzed exhaustively [9]. Procedures for genetic manipulation of *S. cerevisiae* are available on tap. Being a eukaryotic, the key of its success lies in the selection of a model strain with a perfect heterothallic life cycle [10]. In contrast, brewer's yeast is refractory to the genetic procedures used with laboratory strains. The main reason is its low sexual fertility. Like most other industrial yeast, brewing strains do not sporulate or do so with low efficiency. Even in those cases that they show a suitable sporulation frequency, most spores are not viable. The use of appropriate techniques and patient work, carried out mostly at the Carlsberg Laboratory during the last two decades, has lead to the elucidation of the genetic constitution of a representative strain of brewer's yeast. This work has been recently reviewed by Andersen et al. [11].

2.1. Strain Types

There are basically two kinds of yeast used in brewing that correspond to the ale and lager types of beer. Ale beer is produced by a top-fermenting yeast that works at about room temperature, ferments quickly, and produces beer with a characteristic fruity aroma. The bottom-fermenting lager yeast works at lower temperatures, about 10-14°C, ferments more slowly and produces beer with a distinct taste. The vast majority of beer production worldwide is lager. It is difficult to make generalizations concerning the yeast strains used for the industrial production of beer, since they are generally ill characterized and very few comparative studies have been reported. Bottom fermenting, lager strains are usually labeled *Saccharomyces carlsbergensis*. Although strains from different sources show differences regarding cell size, morphology and frequency of spore formation, it is unlikely that these differences reflect a significant genetic divergence. Only one strain, Carlsberg production strain 244, has been extensively analyzed and most of the studies described in the following sections have been conducted with this strain.

2.2. Genetic Crosses

Early attempts to carry out conventional genetic analysis with brewer's yeast faced the problems of poor sporulation and low viability [12]. To overcome this difficulty, several researchers hybridized brewing strains with laboratory strains of *S. cerevisiae* [13-16]. Notwithstanding the poor performance of brewing strains, viable spores were recovered from them. Some of the spores had mating capability and could be crossed with *S. cerevisiae* to generate hybrids easier to manipulate. The meiotic offspring of the hybrids was repeatedly backcrossed with laboratory strains of *S. cerevisiae* to bring particular traits of the brewing strain into an organism amenable to analysis. This procedure was followed to study flocculence, an important character in brewing [13,17]. Gjermansen and Sigsgaard [18] carried out a detailed analysis of the meiotic offspring of *S. carlsbergensis* strain 244. They obtained viable spore clones of both mating types. Cell lines with opposite mating type were crossed pairwise to generate a number of hybrids that were tested for brewing performance. One of them was as good as the original strain. Additionally, the clones derived from strain 244 with mating capability served as starting material for further genetic analysis which are described in the following section.

2.3 *kar* Mutants and Chromosome Transfer

Nuclear fusion (karyogamy), which takes place following gamete fusion (plasmogamy), is the event that instates the diploid phase in all organisms endowed with sexual reproduction. J. Conde and collaborators carried out a genetic analysis of nuclear fusion in *S. cerevisiae* by isolating mutations in different genes that control the process (*kar* mutations) [19,20]. The *kar* mutations served as a basis for a comprehensive study of the molecular mechanisms that control karyogamy, carried out by Rose and collaborators (see review by Rose) [21]. The *kar* mutations have been particularly useful tools to investigate cytoplasmic inheritance [22-24]. Additionally, the *kar* mutations supplied new genetic techniques. For instance, the chromosome number of virtually any *Saccharomyces* strain can be duplicated upon mating with a *kar2* partner [25]. These new tools and techniques opened a new way for the characterization of the brewer's yeast. Nilsson-Tillgren et al. [26] and Dutcher [27], described that when a normal *Saccharomyces* strain mates with a *kar1* mutant, transfer of genetic information occurs at a low frequency between nuclei (Fig. 1). Nuclear transfer events also occurs with *kar2* and *kar3* mutants [20]. Using strains with appropriate genetic markers, one can select the transfer of specific chromosomes. Nilsson-Tillgren et al. [28] used *kar1*-mediated chromosome transfer to obtain a *S. cerevisiae* strain that carried an extra copy of chromosome III from *S. carlsbergensis*. Since the brewing strain does not mate normally, the strain used in *kar* crosses was a meiotic derivative of strain 244 with mating capability [18]. When disomic strains for chromosome III (also referred to as chromosome addition strains) were crossed to haploid *S. cerevisiae* strains, normal spore viability was obtained, allowing tetrad analysis. In this process, one of the two copies of chromosome III can be lost. If the original *S. cerevisiae* copy is lost, the result is a "chromosome substitution strain" carrying a complete *S. cerevisiae* chromosome set, except chromosome III, which comes from *S. carlsbergensis*. Meiotic analysis of crosses between chromosome III addition strains and laboratory strains of *S. cerevisiae* revealed two important facts: (i) the functional equivalence of chromosome III for the brewing strain and *S. cerevisiae*, since ascospore viability and chromosome segregation were normal, and (ii) in spite of the functional equivalence, the two copies of chromosome III were different since the overall frequency of recombination between them was much lower than that expected for perfect homologues. The new procedure allowed the analysis of entire chromosomes from the brewing strain, placed into a laboratory yeast that could easily be manipulated genetically. The work with *S. carlsbergensis* chromosome III was followed by the analysis of chromosomes V, VII, X , XII and XIII [29-32].

2.4. Molecular Analysis

A clear picture of the genetic composition of *S. carlsbergensis* emerged from Southern hybridization experiments and from the first gene sequences from this yeast. The paper by Nilsson-Tillgren et al. [28], where the transfer of a chromosome III from the brewing strain to *S. cerevisiae* was reported, included a detailed Southern analysis of the *HIS4* gene contained in this chromosome. Five yeast strains were used in this analysis. Two were *S. cerevisiae* strains carrying mutant alleles of the *HIS4* gene, a point mutation and a deletion respectively. The other three strains were *S. carlsbergensis* 244, a chromosome III substitution strain and a chromosome addition strain. DNA samples from each one of the five strains were digested with restriction endonucleases, electrophoresed in an agarose gel and hybridized with a labeled probe that contained the *HIS4* gene. The pattern of bands obtained for the brewing strain and the chromosome addition strain were found to be composed by the bands characteristic of *S. cerevisiae*, plus other, extra bands, which showed weaker hybridization. This result indicated the presence in the brewing strain (and also in the addition strain) of two versions of chromosome III, one virtually indistinguishable from that of *S. cerevisiae*, and another with a reduced level of sequence homology. Therefore, the brewer's yeast must be an alloploid, or species hybrid, presumably arisen by hybridization between *S. cerevisiae* and another species of *Saccharomyces*. This conclusion was corroborated by similar analysis carried out for several other genes [29-36]. Determination of the nucleotide sequence of a number of *S. carlsbergensis* genes provided a precise characterization of the difference between the two types of homologous alleles present in the brewing yeast. This analysis has been carried out for *ILV1* and *ILV2* [37]; *URA3* [38]; *HIS4* [39]; *ACB1* [40]; *MET2* [41]; *MET10* [42] and *ATF1* [43]. Pooled data indicate a nucleotide sequence divergence of 10-20% within coding regions and higher outside.

2.5 Ploidy

Finding a sound answer for the long-standing question of how many chromosomes are contained in brewer's yeast, has taken a long time. The relative DNA content of *S. carlsbergensis* 244 has been recently determined by flow cytometry. Results obtained show that the genetic constitution of this strain must be close to tetraploidy [38]. Since it is known that *S. carlsbergensis* is an alloploid generated by the hybridization of two different *Saccharomyces spp.*, the question arises of what is the contribution of each parental species to the hybrid. Pooled data obtained from gene replacement experiments and meiotic analysis of genes located in chromosomes VI, XI, XIII and XIV, suggest that *S. carlsbergensis* contains four copies of each one of these chromosomes, two from each parental species [11]. However, this can not be generalized to all chromosomes. Results of experiments in which

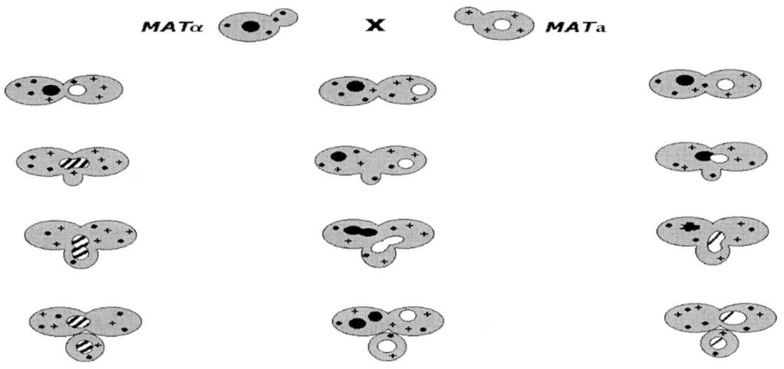

Fig. 1. Wild type and *kar* crosses of *Saccharomyces cerevisiae*. Two haploid cells with opposite mating types are shown on the upper part of the figure. Nuclei are represented either as black or white circles. Small dots and crosses represent cytoplasmic elements. The left column shows the evolution of a normal zygote, formed by the fusion of two wild type cells. Karyogamy occurs shortly after cell fusion, generating a diploid nucleus (represented as a black and white striped circle). The cytoplasmic elements from both parental cells get mixed. The diploid nucleus divides mitotically and the zygote buds off diploid cells. The central column represent the most frequent evolution of a zygote formed in a cross in which at least one of the parental cells has a *kar* mutation. Karyogamy does not take place. The unfused, haploid nuclei, divide mitotically, generating a heterokaryon. The zygote buds off haploid cells with cytoplasmic components from both parents. These cells are named heteroplasmons or cytoductants. The column on the right represents an instance of chromosome transfer. The haploid nuclei in the newly formed zygote undergo abortive karyogamy. Nuclear material from one nucleus is transferred to the other. This phenomenon originates an incomplete nucleus (represented in black) that degenerates, and an aneuploid nucleus (represented in white with a black stripe). The zygote buds off aneuploid cells (chromosome addition line) .

the segregation of different *in vitro* labeled alleles of the *HIS4* gene was analyzed [38], indicate the presence in the brewing yeast of five copies of chromosome III. Of these copies, four are *S. carlsbergensis*-specific, and only one corresponds to the *S. cerevisiae*.

2.6. Origin of Brewing Strains

The hybrid nature of the brewing yeast explains its poor sexual performance. Divergence between homeologous sequences impairs chromosome pairing and recombination, which are requisites for a proper meiotic function. Sexual reproduction appears in Evolution as a mechanism that recombines the genetic material of organisms to generate variability. It offers adaptive advantages to a changing environment through the random generation of new genotypes. On the contrary, abolition of sex is advantageous when the purpose is to keep unchanged a given property. The maintenance over the centuries of a brewing procedure to produce beer with particular organoleptic properties likely caused the selection of a particular type of yeast. The hybrid, vegetative vigor of this yeast assured a good fermentative capability, whereas its sexual infertility would keep fixed the genetic constitution responsible for the "good beer" phenotype. Sequence analysis shows that one of the two parental species that generated *S. carlsbergensis* was *S. cerevisiae*, but the precise identification of the other contributor is less clear. Several studies [43-46] point to *S. bayanus*. Other studies have pointed to *S. monacensis* as a better candidate [35,40,41]. However, recent analysis indicates that *S. monacensis* is itself a hybrid [40,47,48]. According to proteomic analysis, strain NRRL Y-1551 is the closest current candidate [47]. An interesting possibility is that *S.*

carlsbergensis has been generated by more than one event of hybridization. Thus, lager strains of different origin, labeled *S. carlsbergensis*, could be independently generated hybrids of slightly different genetic constitution.

3. GENETIC MANIPULATION

Yeast and barley play an active, primary role in the brewing process. The other two beer ingredients, water and hops, have secondary roles. Yeast is the fermenting agent, which transforms the carbohydrates stored in the grain of barley into ethanol. It produces a battery of compounds that ultimately result in the aroma and flavor of the beer. Barley is not solely a source of fermentable sugars. During the process of malting, cells in the germinating barley seeds secrete enzymes that are required to digest the starch into simpler sugars, mainly maltose and glucose, which can be assimilated by the yeast. Many properties of barley, in particular those affecting its carbohydrate content and composition, but also other characteristics, are very important for the quality of beer. Genetic engineering can be used to modify the properties of yeast and barley in ways that improve their performance in brewing. Different experimental approaches directed to the modification of the brewer's yeast, to produce beer with better properties or new characteristics. In most cases, technical advances allow the construction of new strains of yeast with the desired properties. Currently however, public concern about the use of genetically modified food poses a barrier to the industrial use of these strains.

3.1. Accelerated Maturation of Beer

The production of lager beer comprises two separate fermentation stages. The main fermentation, in which the fermentable sugars are converted in ethanol, is followed by a secondary fermentation, referred to as maturation or lagering. The most important function of maturation is the removal of diacetyl, a compound that causes an unwanted buttery flavor in beer. Diacetyl is formed by the spontaneous (non-enzymatic) oxidative decarboxylation of α-acetolactate, an intermediate in the biosynthesis of valine. In yeast, as in other organisms, the two branched-chain amino acids, isoleucine and valine, are synthesized in an unusual pathway in which a set of enzymes, acting in parallel reactions, lead to the formation of different end products. Like diacetyl is formed as a by-product of valine biosynthesis, a related compound, 2-3-pentanedione, is formed by decarboxylation of α-aceto-α-hydroxybutirate in the isoleucine biosynthesis. Both compounds, diacetyl and α-aceto-α-hydroxybutirate produce a similar undesirable effect in beer, although much more pronounced in the case of diacetyl. Together, they are referred to as vicinal diketones. Diacetyl is converted to acetoin by the action of diacetyl reductase, an enzyme from the yeast. The maturation period, which lasts several weeks, assures the conversion of the available α-acetolactate into diacetyl and the subsequent transformation of diacetyl into acetoin. The amounts formed of this last compound do not have a significant influence on beer flavor. Preventing diacetyl formation would reduce or even make unnecessary the lagering period. This would represent a considerable benefit for the brewing industry. Different approaches have been devised to eliminate diacetyl (Fig. 2). A first one requires the manipulation of the isoleucine-valine biosynthetic pathway, either by blocking the formation of the diacetyl precursor α-acetolactate, or by increasing the flux of the pathway at a later stage, channeling the available α-acetolactate into valine before it is converted into diacetyl. Masschelein and collaborators were first to suggest that a deleterious mutation of the brewer's yeast *ILV2* gene would solve the diacetyl problem. This gene encodes the enzyme acetohydroxyacid synthase, which catalyzes the synthesis of α-acetolactate, from which diacetyl is formed [49,50]. This or any alternative action on the valine pathway requires the manipulation of specific genes encoding enzymes of the pathway. These genes have been cloned from *S. cerevisiae* and characterized [51-54]. *S. carlsbergensis*–specific alleles of the *ILV* genes from the brewer's strain have also been cloned [32,37,55,56]. Because of the genetic complexity of the brewing strain (a hybrid with about four copies of each gene, two from each parent), the abolition of the *ILV2* function requires the very laborious task of eliminating each of the four copies of the gene present in the yeast. This result has not been reported so far. An alternative could be to boost the activity of the enzymes that direct the following steps in the conversion of α-acetolactate into valine: the reductoisomerase, encoded by *ILV5* and possibly the dehydrase, encoded by *ILV3* [57-60]. To achieve the desired effect, it could be sufficient to manipulate only one of the four copies of the *ILV* genes present in the brewer's yeast. A clever procedure to inhibit the *ILV2* function, by using an antisense RNA of the gene, has been reported [61]. However, a later note from the same laboratory stated that the reported results were incorrect [62]. Another approach makes use of an enzyme, α-acetolactate decarboxylase, which catalyzes the direct conversion of acetolactate into acetoin, bypassing the formation of dyacetyl. This enzyme is produced by different microorganisms [63]. Its use for the accelerated maturation of beer was suggested years ago [64,65], and currently is commercially available for this use. An obvious alternative is to express a gene encoding α-acetolactate decarboxylase in the brewing yeast. This has been carried out by different groups [66-68].

3.2. Beer Attenuation and the Production of Light Beer

Conversion of barley into wort that can be fermented requires two previous processes: malting and mashing. During malting, the barley grain is subjected to partial germination,

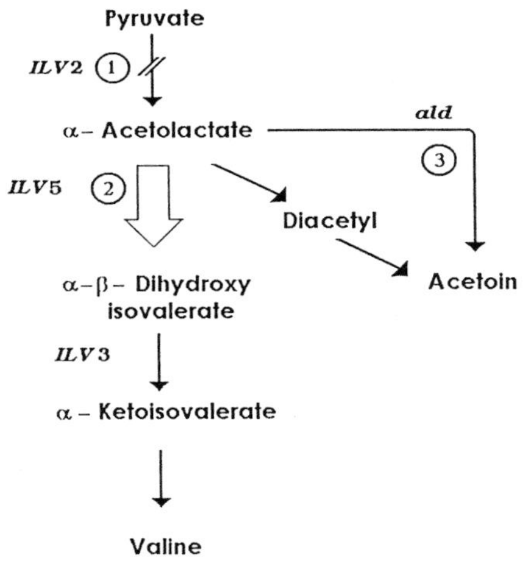

Fig. 2. Strategies designed to prevent the presence of diacetyl in beer. 1. Elimination of *ILV2*. This prevents the synthesis of the enzyme acetohydroxyacid synthase, required for the formation of the diacetyl precursor, acetolactate. 2. Overexpression of the *ILV5*. This increases the activity of the enzyme, which converts □-acetolactate into dihydroxy isovaleriate, the following intermediate of valine biosynthesis. As a consequence, the amount of □-acetolactate that can be transformed into diacetyl is reduced. 3. Expression in brewer's yeast of the *ald* gene encoding bacterial acetolactate decarboxylase. This enzyme avoids the formation of diacetyl, by converting the available acetolactate into acetoin. Commercial preparations of the enzyme are available as beer additive to accelerate maturation.

achieved by moistening, and subsequent drying. Germination induces the synthesis of amylase and other enzymes that allow the seed to mobilize its reserves. The dried malt is milled and the resulting powder is mixed with water and allowed to steep at warm temperatures. During mashing, amylases digest the seed's starch, liberating simpler sugars, chiefly maltose. This process is critical, since the brewing yeast is unable to hydrolyze starch. The enzymatic action of barley's amylases on starch yields fermentable sugars, but also oligosaccharides (dextrins) which remain unfermented during brewing. Dextrins represent an important fraction of the caloric content of beer. In current brewing practice, it is quite common to add exogenous enzymes. Thus glucoamylase can be added to the mash to improve the digestion of the starch. If the enzymatic treatment is carried out exhaustively, the dextrins are completely hydrolyzed, and the result is a light beer with substantially lower caloric content, for which there is a significant market demand in some parts of the world. A convenient alternative to the addition of exogenous glucoamylase is to endow the brewer's yeast with the genetic capability of synthesizing this enzyme. A variety of *S. cerevisiae*, formerly classified as a separate species (*S. diastaticus*), produces glucoamylase. Because of its close phylogenetic relationship with the brewing yeast, *S. diastaticus* is an obvious source of the glucoamylase gene.

The percentage of the sugar in the wort that is converted into ethanol and CO_2 by the yeast is called attenuation. Microbial contamination of beer is often associated with a pronounced increase in the attenuation value, which is known as superattenuation. This effect is due to the fermentation of dextrins, which are hydrolyzed by amylases produced by the contaminant microorganisms. *S. diastaticus* was characterized as a wild yeast that caused superattenuation [69]. Similarly to the synthesis of invertase or maltase by *Saccharomyces*, the synthesis of glucoamylase is controlled by a set of at least three polymeric genes, designated *STA1*, *STA2* and *STA3* [70]. This genetic system is complicated by the existence in normal *S. cerevisiae* strains of a gene, designated *STA10,* which inhibits the expression of the other *STA* genes [71]. Recently, the *STA10* gene has been identified with the absence of Flo8p, a transcriptional regulator of both glucoamylase and flocculation genes [72]. The sequence of the *STA1* gene was first determined by Yamashita et al. [73]. Different species of filamentous fungi, in particular some of the genus *Aspergillus*, produce powerful glucoamylases. The gene that encodes the enzyme of *A. awamori* has been expressed in *S. cerevisiae* [74]. Available information about the genetic control of glucoamylase production by *Saccharomyces* and current technology makes the construction of brewing strains with this capability relatively easy.

3.3. Beer Filterability and the Action of β-glucanases

Brewing with certain types or batches of barley, or using certain malting or brewing practices, can yield wort and beer with high viscosity, very difficult to filtrate. When this problem arises, the beer may also present hazes and gelatinous precipitates. Scott [75] pointed out that this problem was caused by a deficiency in β-glucanase activity. The substrate of this enzyme, β-glucan, is a major component of the endosperm cell walls of barley and other cereals. During the germination of the grain, β-glucanase degrades the endosperm cell walls, allowing the access of other hydrolytic enzymes to the starch and protein reserves of the seed. Insufficient β-glucanse activity during malting gives rise to an excess of β-glucan in the wort, which causes the problems. The addition of bacterial or fungal β-glucanases to the mash, or directly to the beer during the fermentation, is a common remedy. The construction of a brewing yeast with appropriate β-glucanase activity would make unnecessary the treatment with exogenous enzymes. Suitable organisms to be used as sources of the β-glucanase gene are *Bacillus subtilis* and *Thricoderma reesei*, from which the commercial enzyme preparations used in brewing are prepared. The genes from both have been characterized [76-79] and brewer's yeast expressing β-glucanase activity have been constructed [80]. An alternative is to make use of the gene encoding barley β-glucanase, the enzyme that naturally acts in malting. This gene has been characterized and expressed in *S. cerevisiae* [81-83]. However, the barley enzyme has lower thermal resistance than the microbial enzymes, which is a limitation for its use against the β-glucans present in wort. Consequently, the enzyme has been engineered to increase its thermal stability [84,85].

3.4. Control of Sulfite Production in Brewer's Yeast

Sulfite has an important, dual function in beer. It acts as an antioxidant and a stabilizing agent of flavor. Sulfite is formed by the yeast in the assimilation of inorganic sulfate, as an intermediate of the biosynthesis of sulfur-containing amino acids, but its physiological concentration is low. Hansen and Kielland-Brandt [86] have engineered a brewing strain to enhance sulfite level to a concentration that increases flavor stability. The formation of sulfite from sulfate is carried out in three consecutive enzymatic steps catalyzed by ATP sulfurylase, adenylsulfate kinase and phosphoadenylsulfate reductase. In *S. cerevisiae*, these enzymes are encoded by *MET3*, *MET14* and *MET16* [87-89]. In turn, sulfite is converted firstly into sulfide, by sulfite reductase, and then into homocysteine by homocysteine synthetase. This last compound leads to the synthesis of cysteine, methionine and S-adenosylmethionine. It has been proposed that S-adenosylmethionine plays a key regulatory role by repressing the genes of the pathway [90-92]. However, more recent evidence assigns this function to cysteine [93]. Anyhow, because of the regulation of the pathway, yeast growing in the presence of methionine contains very little sulfite. To increase its production in the brewing yeast, Hansen and Kielland-Brandt [86] planned to abolish sulfite reductase activity. This would increase sulfite concentration, as it cannot be reduced. At the same time, the disruption of the methionine pathway prevents the formation of cysteine and keeps free from repression the genes involved in sulfite formation. Sulphite reductase is a tetramer with an α_2 β_2 structure. The α and β subunits are encoded by the *MET10* and *MET5* genes, respectively [42,94]. Hansen and Kielland-Brandt undertook the construction of a brewing strain without *MET10* gene function. The allotetraploid constitution of *S. carlsbergensis* made it extremely difficult to perform the disruption of the four functional copies of the yeast. Therefore, they used allodiploid strains, obtained as meiotic derivatives of the brewer's yeast. These allodiploids contains two homeologous alleles of the *MET10* gene, one similar to the version normally found in *S. cerevisiae* and another which is *S. carlsbergensis*-specific. It is known that some allodiploids can be mated to each other to regenerate tetraploid strains with good brewing performance[18]. The functional *MET10* alleles present in the allodiploids were replaced by deletion-harboring, non-functional copies, by two successive steps of homologous recombination. New allotetraploid strains with reduced or abolished *MET10* activity were then generated by crossing the manipulated allodiploids. The brewing performance of one of these strains, in which the *MET10* function was totally abolished, met the expectations. Hansen and Kielland-Brandt [95] have used another strategy to increase the production of sulfite which relies in the inactivation of the *MET2* gene function. The *MET2* gene encodes *O*-acetyl transferase. This enzyme catalyzes the biosynthesis of *O*-acetyl homoserine, which binds hydrogen sulfide to form homocysteine [96]. Similarly to the inactivation of *MET10*, inactivation of *MET2* impedes the formation of cysteine, depressing the genes required for sulfite biosynthesis.

3.5. Yeast Flocculation

As beer fermentation proceeds, yeast cells start to flocculate. The flocs grow in size, and when they reach a certain mass start to settle. Eventually, the great majority of the yeast biomass sediments. This phenomenon is of great importance to the brewing process because it allows separation of the yeast biomass from the beer, once the primary fermentation is over. The small fraction of the yeast that is left in the green beer is sufficient to carry out the subsequent step, the lagering. Flocculation is a cell adhesion process mediated by the interaction between a lectin protein and mannose [97-99]. Stratford and Assinder [100] carried out an analysis of 42 flocculent strains of *Saccharomyces* and defined two different phenotypes. One was the known pattern observed in laboratory strains that carried the *FLO1* gene. They found, in some ale brewing strains, a new flocculation pattern characterized by being inhibited by the presence in the medium of a variety of sugars, including mannose, maltose, sucrose and glucose, whereas the *FLO1* type was sensitive only to mannose. The genetic analysis of flocculation has revealed the existence of a polymeric gene family analogous to the *SUC*, *MAL*, *STA* and *MEL* families [101,102]. The *FLO1* gene has been extensively characterized [103-107], which encodes a large, cell wall protein of 1,537 amino acids. The protein is highly glycosylated. It has a central domain harboring direct repeats rich in serine and threonine (putative sites for glycosylation). Kobayashi et al. [108] have isolated a flocculation gene homolog to *FLO1* that corresponds to the new pattern described by Stratford and Assinder [100]. This result is consistent with the hybrid nature of the brewing yeast. In addition to the structural genes encoding flocculins, other *FLO* genes play a regulatory role. For instance, the *FLO8* gene (alias *STA10*) encodes a transcriptional activator that in addition to flocculation regulates glucoamylase production, filamentous growth and mating [72,109-113].

3.6. Beer Spoilage Caused by Microorganisms

Microbial contamination of beer, caused by bacteria or wild yeast is a serious problem in brewing. To overcome the contamination, commonly sulfur dioxide and other chemicals are added, but this practice faces restrictive legal regulation and consumer rejection. An attractive alternative is to endow the brewing yeast with the capability of producing anti-microbial compounds. A specific example is the expression in *S. cerevisiae* of the genes required for the biosynthesis of pediocin, an antibacterial peptide from *Pediococcus acidilactici* [114]. Another example is the transfer to brewing strains of the killer character, conferred by the production of a toxin active against other yeasts [115,116].

3.7. Enhanced Synthesis of Organoleptic Compounds

The yeast metabolism during beer fermentation gives rise to the formation of higher alcohol, esters and other compounds which make an important contribution to the aroma and taste of beer. A first group of compounds important to beer flavor are isoamyl and isobutyl alcohol and their acetate esters. These compounds derive from the metabolism of valine and leucine [117]. Two genes, *ATF1* and *LEU4*, encoding enzymes involved in the formation of these compounds, have been successfully manipulated to increase theirs synthesis. *ATF1* encodes alcohol acetyl transferase. It has been shown that its over-expression causes increased production of isoamyl acetate [118]. *LEU4* encodes α-isopropylmalate synthase, an enzyme that controls a key step in the formation of isoamyl alcohol from leucine. This enzyme is inhibited by leucine [119,120]. Mutant strains resistant to a toxic analog of leucine are insensitive to leucine inhibition [119]. Mutants of this type, obtained from a lager strain, produce increased amounts of isoamyl alcohol and its ester [121].

4. CONCLUSIONS

Development of molecular biology in the 20th century has brought many new opportunities for technical improvements in the field of brewing industry. The basic scientific questions concerning the genetic nature of the brewer's yeast and different physiological problems related to brewing (secondary fermentation, flocculation, etc.) have been answered. Instruments to construct a new generation of brewer's yeast strains, designed to circumvent common problems of brewing, have been developed. A fine example is the work of Hansen and Kielland-Brandt [86] that led to the construction of a brewing yeast with increased sulfite production. Presently, the main obstacle for the development and industrial implementation of improved brewing yeast is not technical but psychological. Public concern about the safety of genetic engineering and pressure, often misguided, from various groups, force the brewing companies to refrain from innovation in these directions. Nevertheless, it is easy to forecast that in the future, genetic engineering will bring to the brewing industry, as well as to other food industries, a plethora of better and safer products.

Acknowledgment. I thank Professor Morten Kielland-Brandt for many useful suggestions and critical reading of the manuscript.

5. REFERENCES

1. Anderson, R. G. (1995). Louis Pasteur (1822-1895): An assessment of his impact on the brewing industry. Eur. Brew. Conv. Congr., 13-23.
2. Barnett, J. A. (2000). A history of research on yeast 2: Louis Pasteur and his contemporaries, 1850-1880. Yeast 16:755-771.
3. Wettstein, D. von (1983). Emil Christian Hansen Centennial Lecture: from pure yeast culture to genetic engineering of brewers yeast. Eur. Brew. Conv. Congr., 97-119.
4. Leupold, U. (1950). Die Vererbung von Homothallie und Heterothallie bei *Schizosaccharomyces pombe*. C. R. Trav. Lab. Carlsberg Ser. Physiol. 24:381-480.
5. Mortimer, R. K. (1993). Øjvind Winge: Founder of yeast genetics. In: The Early Days of Yeast Genetics. Ed. by M. N. Hall and P. Linder. Cold Spring Harbor Laboratory Press, Cold Spring Harbor, New York, pp. 3-16.
6. Mortimer, R. K. (1993). Carl C. Lindegren: Iconoclastic Father of *Neurospora* and Yeast Genetics. In: The Early Days of Yeast Genetics. Ed. by M. N. Hall and P. Linder. Cold Spring Harbor Laboratory Press, Cold Spring Harbor, New York, pp. 17-38.
7. Kannangara, C. G. , Gough, S. P., Oliver, R. P., and Rasmussen, S. K. (1984). Biosynthesis of aminolevulinate in greening barley leaves VI. Activation of glutamate by ligation to RNA. Carlsberg Res. Commun. 49:417-437.
8. Gough, S. P., Petersen, B. O., and Duus, J. Ø. (2000). Anaerobic chlorophyll isocyclic ring formation in *Rhodobacter capsulatus* requires a cobalamin cofactor. Proc. Natl. Acad. Sci. 97:6908-6913.
9. Goffeau, A. (2000). Four years of post-genomic life with 6,000 yeast genes. FEBS Lett. 480:37-41.
10. Mortimer, R. K. and Johnston, J. R. (1986). Genealogy of principal strains of the Yeast Genetics Stock Center. Genetics 113:35-43.
11. Andersen, T. H., Hoffmann, L., Grifone, R., Nilsson-Tillgren, T., and Kielland-Brandt, M. C. (2000). Brewing Yeast Genetics. EBC Monograph 28, Fachverlag Hans Carl, Nürnberg, pp.140-147.
12. Winge, Ø. (1944). On segregation and mutation in yeast. Compt. Rend. Trav. Lav. Carlsberg Ser. Physiol. 24:79-96.
13. Thorne, R. S. W. (1951). The genetic of flocculence in *Saccharomyces cerevisiae*. Compt. Rend. Trav. Lav. Carlsberg Ser. Physiol. 25:101-140.
14. Johnston, J. R. (1965). Breeding yeast for brewing, I. Isolation of breeding strains. J. Inst. Brew. 71:130-135.
15. Johnston, J. R. (1965). Breeding yeast for brewing, II. Production of hybrid strains. J. Inst. Brew. 71:135-1137.
16. Anderson, E., and Martin, P. A. (1975). The sporulation and mating of brewing yeast. J. Inst. Brew. 81:242-247.

17. Lewis, C. W., Johnston, J. R., and Martin, P. A. (1976). The genetics of yeast flocculation. J. Inst. Brew. 82:158-160.
18. Gjermansen, C., and Sigsgaard, P. (1981). Construction of a hybrid brewing strain of *Saccharomyces carlsbergensis* by mating of meiotic segregants. Carlsberg Res. Commun. 46:1-11.
19. Conde J., and Fink, G. R. (1976). A mutant of *Saccharomyces cerevisiae* defective for nuclear fusion. Proc. Natl. Acad. Sci. 73:3651-3655.
20. Polaina, J., and Conde, J. (1982). Genes involved in the control of nuclear fusion during the sexual cycle of *Saccharomyces cerevisiae*. Mol. Gen. Genet. 186:253-258.
21. Rose, M. D. (1996). Nuclear fusion in the yeast *Saccharomyces cerevisiae*. Annu. Rev. Cell Dev. Biol. 12:663-695.
22. Livingston, D. M. (1977). Inheritance of the 2 micrometer DNA plasmid from *Saccharomyces*. Genetics 86:73-84.
23. Lancanshire, W. E., and Mattoon, J. R. (1979). Cytoduction: a tool for mitochondrial studies in yeast. Utilization of the nuclear-fusion mutation *kar1*-1 for transfer of drug^r and *mit* genomes in *Saccharomyces cerecisiae*. Mol. Gen. Genet. 170:333-344.
24. Wickner, R. B. (1980). Plasmids controlling exclusion of the K_2 killer double-stranded RNA plasmid of yeast. Cell 21:217-226.
25. Polaina, J., Adam, A. C., and del Castillo, L. (1993). Self-diploidization in *Saccharomyces cerevisiae kar2* heterokaryons. Curr. Genet. 24:369-372.
26. Nilsson-Tillgren, T., Petersen, J. G. L., Holmberg, S., and Kielland-Brandt, M. C. (1980). Transfer of chromosome III during *kar* mediated cytoduction in yeast. Carlsberg Res. Commun. 45:113-117.
27. Dutcher, S. K. (1981). Internuclear transfer of genetic information in *kar1*-1/*KAR1* heterokaryons in *Saccharomyces cerevisiae*. Mol. Cell Biol. 1:245-253.
28. Nilsson-Tillgren, T., Gjermansen, C., Kielland-Brandt, M. C., Petersen, J. G. L., and Holmberg, S., and (1981). Genetic differences between *Saccharomyces carlsbergensis* and *S. cerevisiae*. Analysis of chromosome III by single chromosome transfer. Carlsberg Res. Commun. 46:65-76.
29. Nilsson-Tillgren, T., Gjermansen, C., Holmberg, S., Petersen, J. G. L., and Kielland-Brandt, M. C. (1986). Analysis of chromosome V and the *ILV1* gene from *Saccharomyces carlsbergensis*. Carlsberg Res. Commun. 51:309-326.
30. Kielland-Brandt; M. C., Nilsson-Tillgren, T., Gjermansen, C., Holmberg, S., and Pedersen, M. B. (1995). Genetic of brewing yeast. In: The Yeast, Second Edition, Vol. 6. A. H. Rose, A. E. Wheals, J. S. Harrison (eds). Academic Press, London, pp. 223-254.
31. Casey, G. P. (1986). Molecular and genetic analysis of chromosome X in *Saccharomyces carlsbergensis*. Carlsberg Res. Commun. 51:343-362.
32. Petersen, J. G. L., Nilsson-Tillgren, T., Kielland-Brandt, M. C. Gjermansen, C., and Holmberg, S. (1987). Structural heterozygosis at genes *ILV2* and *ILV5* in *Saccharomyces carlsbergensis*. Current Genet. 12:167-174.
33. Holmberg, S. (1982). Genetic differences between *Saccharomyces carlsbergensis* and *S. cerevisiae*. II. Restricition endonuclease analysis of genes in chromosome III. Carlsberg Res. Commun. 47:233-244.
34. Pedersen, M. B. (1985). DNA sequence polymorphisms in the genus *Saccharomyces*. II. Analysis of the genes *RDN1*, *HIS4*, *LEU2* and *Ty* transposable elements in Carlsberg, Tuborg and 22 Bavarian Brewing strains. Carlsberg Res. Commun. 50:263-272.
35. Pedersen, M. B. (1986). DNA sequence polymorphisms in the genus *Saccharomyces*. III. Restriction endonuclease fragment patterns of chromosomal regions in brewing and other yeast strains. Carlsberg Res. Commun. 51:163-183.
36. Pedersen, M. B. (1986) DNA sequence polymorphisms in the genus *Saccharomyces*. IV. Homeologous chromosomes III of *Saccharomyces bayanus*, *S. carlsbergensis*, and *S. uvarum*. Carlsberg Res. Commun. 51:185-202.
37. Gjermansen, C. (1991). Comparison of genes in *Saccharomyces cerevisiae* and *Saccharomyces carlsbergensis*. Ph. D. thesis, University of Copenhagen, Denmark.
38. Hoffmann, L. (1999). The defective sporulations of lager brewing yeast. Ph. D. thesis, University of Copenhagen, Denmark.
39. Porter, G., Westmoreland, J., Priebe, S., and Resnick, M. A. (1996). Homologous and homeologous intermolecular gene conversion are not differentially affected by mutations in the DNA damage or the mismatch repair genes *RAD1*, *RAD50*, *RAD51*, *RAD52*, *RAD54*, *PMS1*, and *MSH2*. Genetics 143:755-767.

40. Børsting, C., Hummel, R., Schultz, E. R., Rose, T. M., Pedersen, M. B., Knudsen, J., and Kristiansen, K. (1997). *Saccharomyces carlsbergensis* contains two functional genes encoding acyl-CoA binding protein, one similar to the *ACB1* gene from *S. cerevisiae* and one identical to the *ACB1* gene from *S. monacensis*. Yeast 13:1409-1421.
41. Hansen, J., and Kielland-Brandt, M. C. (1994). *Saccharomyces carlsbergensis* contains two functional *MET2* alleles similar to homologues from *S. cerevisiae* and *S. monacensis*. Gene 140:33-40.
42. Hansen, J., Cherest, H., and Kielland-Brandt, M. C. (1994). Two divergent *MET10* genes, one for *Saccharomyces cerevisiae* and one from *Saccharomyces carlsbergensis*, encode the □ subunit of sulfite reductase and specify potential binding sites for FAD and NADPH. J. Bacteriol. 176:6050-6058.
43. Fujii, T. H., Yoshimoto, H., Nagasawa, N., Bogaki, T., Tamai, Y., and Hamachi, M. (1996). Nucleotide sequences of alcohol acetyltransferase genes from lager brewing yeast, *Saccharomyces carlsbergensis*. Yeast 12:593-598.
44. Martini, A. V., and Kurtzman, C. P. (1985). Deoxyribonucleic acid relatedness among species of the genus *Saccharomyces sensu stricto*. Int. J. Syst. Bacteriol. 35.508-511.
45. Tamai, Y. T., Momma, T., Yoshimoto, H., and Kaneko, Y. (1998). Co-existence of two types of chromosome in the botton fermenting yeast *Saccharomyces pastorianus*. Yeast 14:923-933.
46. Yamagishi, H., and Ogata, T. (1999). Chromosomal structures of bottom fermenting yeasts. Syst. Appl. Microbiol. 22:341-353.
47. Joubert, R., Brignon, P., Lehmann, C., Monribot, C., Gendre, F., and Boucherie, H. (2000). Two-dimensional gel analysis of the proteome of lager brewing yeast. Yeast 16:511-522
48. Tamai, Y., Tanaka, K., Umemoto, N., Tomizuka, K., and Kaneko, Y. (2000). Diversity of the *HO* gene encoding an endonuclease for mating-type conversion in the bottom fermenting yeast *Saccharomyces pasterianus*. Yeast 16:1335-1343.
49. Cabane, B. Ramos-Jeunnehomme, C., Lapage, N., and Masschelein, C. A. (1974). Vicinal diketones - the problem and prospective solutions. Am. Soc. Brew. Chem. Proc.1973:94-99.
50. Ramos-Jeunehomme, C., and Masschelein, C. A. (1977). Controle genetique de la formation des dicetones vicinales chez *Saccharomyces cerevisiae*. Eur. Brew. Conv. Congr., Amsterdam, 267-283.
51. Polaina, J. (1984). Cloning of the *ILV2, ILV3* and *ILV5* genes of *Saccharomyces cerevisiae*. Carlsberg Res. Commun. 49:577-584.
52. Falco, S. C., Dumas, K. S., and Livak, K. J. (1986). Nucleotide sequence of the yeast *ILV2* gene which encodes acetolactate synthase. Nucleic Acids Res. 13:4011-4027.
53. Petersen, J. G. L., and Holmberg, S. (1986). The *ILV5* gene of *Saccharomyces cerevisiae* is highly expressed. Nucleic Acids Res. 14:9631-9651.
54. Velasco, J. A., Cansado, J., Pena, M. C., Kawatami, T., Laborda, J., and Notario, V. (1993). Cloning of the dihydroxiacid dehydratase-encoding gene (*ILV3*) from *Saccharomyces cerevisiae*. Gene 137:179-185.
55. Casey, G. P. (1986). Cloning and analysis of two alleles of the *ILV3* gene from *Saccharomyces carlsbergensis*. Carlsberg Res. Commun. 51:327-341.
56. Kielland-Brandt; M. C., Gjermansen, C., Tullin, S., Nilsson-Tillgren, T., Sigsgaard, P., and Holmberg, S. (1990). Genetic analysis and breeding of brewer's yeast. In: Heslot, H., Davies, J., Florent, J., Bobichon, L., Durand, G., and Penasse, L. (Eds.), Proc. 6th International Symposium on Genetics of Industrial microorganisms, Vol. II, 1990, Société Française de Microbiology, Strasbourg, pp. 877-885.
57. Villanueva, K. D., Goossens, E., and Masschelein, C. A. (1990). Subthreshold vicinal diketone levels in lager brewing yeast fermentations by means of *ILV5* gene amplification. J. Am. Soc. Brew. Chem. 48:111-114.
58. Goossens, E., Debourg, A., Villanueba, K. D., and Masschelein, C. A. (1993). Decreased dyacetyl production in lager brewing yeast by integration of the *ILV5* gene. Eur. Brew. Conv. Congr., 251-258.
59. Mithieux, S. M., and Weiss, A. S. (1995). Tandem integration of multiple *ILV5* copies and elevated transcription in polyploid yeast. Yeast 11:311-316.
60. Gjermansen C., Nilsson-Tillgren, T., Petersen, J. G. L., Kielland-Brandt, M. C., Sigsgaard, P., and Holmberg, S. (1998). Towards diacetyl-less brewer's yeast. Influence of *ilv2* and *ilv5* mutations. J. Basic Microbiol. 28:175-183.
61. Xiao, W., and Rank, G. H. (1988). Generation of an ilv bradytrophic phenocopy in yeast by antisense RNA. Curr. Genet. 13:283-289.
62. Arndt, G. M., Xiao, W., and Rank, G. H. (1994). Antisense RNA regulation of the *ILV2* gene in yeast: a correction. Curr. Genet. 25:289.

63. Godtfredsen, S. E., Lorck, H., and Sigsgaard, P. (1983). On the occurrence of □-acetolactate decarboxylase among microorganisms. Carlsberg Res. Commun. 48:239-247.

64. Godtfredsen, S. E., and Ottesen, M. (1982). Maturation of beer with alpha-acetolactate decarboxilase. Carlsberg Res. Commun. 47:93-102.

65. Godtfredsen, S. E., Rasmussen, A. M., Ottesen, M., Mathiasen, T., and Ahrenst-Larsen, B. (1984). Application of the acetolactate decarboxylase from Lactobacillus casei for accelerated maturation of beer. Carlsberg Res. Commun. 49:69-74.

66. Sone, H., Fujii, T., Kondo, K., Shimizu, F., Tanaka, J., and Inoue, T. (1988). Nucleotide sequence and expression of the *Enterobacter aerogenes* □-acetolactate decarboxylase gene in brewer's yeast. Appl. Environ. Microbiol. 54:38-42.

67. Fujii, T., Kondo, K., Shimizu, F., Sone, H., Tanaka, J-I., and Inoue, T. (1990). Application of a Ribosomal DNA integration vector in the construction of a brewer's yeast having □-acetolactate decarboxylase activity. Appl. Environ. Microbiol. 56:997-1003.

68. Blomqvist, K., Suihko, M.-L., Knowles, J., and Penttilä, M. (1991). Chromosome integration and expression of two bacterial □-acetolactate decarboxylase genes in brewer's yeast. Appl. Environ. Microbiol. 57:2796-2803.

69. Andrews, J. and Gilliland, R. B. (1952). Super-attenuation of beer: a study of three organisms capable of causing abnormal attenuation. J. Inst. Brew. 58-189-196.

70. Tamaki, H. (1978). Genetic studies of ability to ferment starch in Saccharomyces: gene polymorfism. Mol. Gen. Genet. 164:205-209.

71. Polaina, J., and Wiggs, M. Y. (1983). *STA10*: A gene involved in the control of starch utilization by *Saccharomyces*. Curr. Genet. 7:109-112.

72. Gagiano, M. Van Dyk, D., Bauer, F. F., Lambrechst, M. G., and Pretorius, I. S. (1999). Divergent regulation of the evolutionary closely related promoters of the *Saccharomyces cerevisiae STA2* and *MUC1* genes. J. Bacteriol. 181:6497-6508.

73. Yamashita, I., Suzuki, K., and Fukui, S. (1985). Nucleotide sequence of the extracellular glucoamylase gene *STA1* in the yeast *Saccharomyces diastaticus*. J. Bacteriol. 161:567-573.

74. Innis, M. A., Holland, M. J., McCabe, P. C., Cole, G. E., Wittman, V. P., Talk, R., Watt, K. W. K., Gelfand, D. H., Holland, J. P., and Meade, J. H. (1985). Expression, glycosylation, and secretion of an *Aspergillus* glucoamylase by *Saccharomyces cerevisiae*. Science 228:21-26.

75. Scott, R. W. (1972). The viscosity of worts in relation to their content of □-glucan. J. Inst. Brew. 78:179-186.

76. Cantwell, B. A., and McConell, D. J. (1983). Molecular cloning and expression of a *Bacillus subtilis* □-glucanase gene in *Escherichia coli*. Gene 23:211-219.

77. Murphy, N., McConnell, D. J., and Cantwell, B. A. (1984). The DNA sequence of the gene and gene control sites for the excreted *B. subtilis* enzyme □-glucanase. Nucleic Acids Res. 12:5355-5367.

78. Penttilä, M., Lehtovaara, P., Nevalainen, H., Bhikhabhai, R., and Knowles, J. (1986). Homology between cellulase genes of *Trichoderma reesei*: complete nucleotide sequence of the endoglucanase I gene. Gene 45:253-263.

79. Arsdell, J. N., Kwok, S., Schweickart, V. L., Ladner, M. B., Gelfand, D. H., and Innis, M. A. (1987). Cloning characterization and expression in *Saccharomyces cerevisiae* of endoglucanase I from *Trichoderma reesei*. Bio/Technology 5:60-64.

80. Penttilä, M. E., Suihko, M. –L. Lehtinen, U., Nikkola, M., and Knowles, J. K. C. (1987). Construction of brewer's yeast secreting fungal endo-□-glucanase. Curr. Genet. 12:413.420.

81. Fincher, G. B., Lock, P. A., Morgan, M. M., Lingelbach, K., Wettenhall, R. E. H., Mercer, J. F. B., Brandt, A., and Thomsen, K. K. (1986). Primary structure of the (1□3,1□4)-□-D-glucan 4-glucanohydrolase from barley aleurone. Proc. Natl. Acad. Sci. 83:2081-2085.

82. Jackson, E. A., Balance, G. M., and Thomsen, K. K. (1986). Construction of a yeast vector directing the synthesis and release of barley (1□3,1□4)-□-glucanase. Carlsberg Res. Commun. 51:445-458.

83. Olsen, O., and Thomsen, K. K. (1989). Processing and secretion of barley (1-3,1-4)-beta-glucanase in yeast. Carlsberg Res. Commun. 54:29-39.

84. Jensen, L. G., Olsen, O., Kops, O., Wolf, N., Thomsen, K. K., and von Wettstein, D. (1996). Transgenic barley expressing a protein-engineered, thermostable (1,3-1,4)-beta-glucanase during germination. Proc. Natl. Acad. Sci. 93:3487-3491.

85. Horvath, H., Huang, J., Wong, O., Kohl, E., Okita, T., Kannangara, C. G., and von Wettstein, D. (2000). The production of recombinant proteins in transgenic barley grains. Proc. Natl. Acad. Sci. 97:1914-1919.

86. Hansen, J., and Kielland-Brandt, M. C. (1996a). Inactivation of *MET10* in brewer's yeast specifically increases SO_2 formation during beer production. Nature Biotechnol. 14:1587-1591.

87. Cherest, H., and Surdin-Kerjan, Y. (1992). Genetic analysis of a new mutation conferring cysteine auxotrophy in *Saccharomyces cerevisiae*: updating of the sulfur metabolism pathway. Genetics 130:51-58.

88. Korch, C., Mountain, H. A., and Byström, A. S. (1991). Cloning, nucleotide sequence and regulation of *MET14*, the gene encoding the AP quinase of *Saccharomyces cerevisiae*. Mol. Gen. Genet. 229:96-108.

89. Thomas, D., Barbey, R., and Surdin-Keryan, Y. (1990). Gene-enzyme relationship in the sulphate assimilation pathway of *Saccharomyces cerevisiae*. Study of the 3'-phosphoadenylylsulfate reductase structural gene. J. Biol. Chem. 265:15518-15524.

90. Cherest, H., Thao, N. N., and Surdin-Kerjan, Y. (1985). Transcriptional regulation of the *MET3* gene of *Saccharomyces cerevisiae*. Gene 34:269-281.

91. Thomas, D., Rothstein, R., Rosenberg, N., and Surdin-Kerjan, Y. (1988). *SAM2* encodes the second methionine *S*-adenosyl transferase in *Saccharomyces cerevisiae*: physiology and regulation of both enzymes. Mol. Cell. Biol. 8:5132-5139.

92. Thomas, D., Cherest, H., and Surdin-Kerjan, Y. (1989). Elements involved in *S*-adenosyl methionine-mediated regulation of the *Saccharomyces cerevisiae MET25* gene. Mol. Cell. Biol. 9:3292-3298.

93. Hansen, J., and Johannesen, P. F. (2000). Cysteine is essential for transcriptional regulation of the sulfur assimilation genes in *Saccharomyces cerevisiae*. Mol. Gen. Genet. 263:535-542.

94. Hansen, J., Muldbjerg, M., Cherest, H., and Surdin-Kerjan, Y. (1997). Siroheme biosynthesis in Saccharomyces cerevisiae requires the products of both the MET1 and MET8 genes. FEBS Lett. 401:20-24.

95. Hansen, J., and Kielland-Brandt, M. C. (1996b). Inactivation of *MET2* in brewer's yeast increases the level of sulfite in beer. J. Biotechnol. 50:75-87.

96. Baroni, M., Livian, S., Martegani, E., and Alberghina, L. (1986). Molecular cloning and regulation of the expresión of the *MET2* gene of *Saccharomyces cerevisiae*. Gene 46:71-78.

97. Miki, B. L. A., Poon, N., James, A. P., and Seligy, V. L. (1982). Possible mechanism for flocculation interactions governed by the *FLO1* gene in *Saccharomyces cerevisiae*. J. Bacteriol. 150:878-889.

98. Miki, B. L. A., Poon, N., and Seligy, V. L. (1982). Repression and induction of flocculation interactions in *Saccharomyces cerevisiae*. J. Bacteriol. 150:890-899.

99. Javadekar, V. S., Sivaraman, H., Sainkar, S. R., and Khan, M. I. (2000). A mannose-binding protein from the cell surface of flocculent *Saccharomyces cerevisiae* (NCIM 3528): its role in flocculation. Yeast 16:991.

100. Stratford, M., and Assinder, S. (1991). Yeast flocculation: Flo1 and NewFlo phenotypes and receptor structure. Yeast 7:559-574.

101. Teunissen, A. W. R. H., and Steensma; H. Y. (1995). Review: the dominant flocculation genes of *Saccharomyces cerevisiae* constitute a new subtelomeric gene family. Yeast 11:1001-1013.

102. Caro, L. H. P., Tettelin, H., Vossen, J. H., Ram, A. F. J., van den Ende H., and Klis, F. M. (1997). *In silico* identification of glycosyl-phosphatidylinositol-anchored plasma-membrane and cell wall proteins of *Saccharomyces cerevisiae*. Yeast 13:14771489.

103. Teunissen, A. W. R. H., van den Berg, J. A., and Steensma, H. Y. (1993). Physical localization of the flocculation gene *FLO1* on chromosome I of *Saccharomyces cerevisiae*. Yeast 9:1-10

104. Teunissen, A. W. R. H., Holub, E., van der Hucht, J., van den Berg, J. A., and Steensma, H. Y. (1993). Sequence of the open reading frame of the *FLO1* gene from *Saccharomyces cerevisiae*. Yeast 9:423-427

105. Watari, J., Takata, Y., Ogawa, M., Sahara, H., Koshino, S., Onnela, M. L., Airaksinen, u., Jaatinen, R., Penttilä, M., Keranen, S. (1994). Molecular cloning and análisis of the yeast flocculation gene *FLO1*. Yeast 10:211-225.

106. Bidard, F., Bony, M., Blondin, B., Dequin, S., and Barre, P. (1995) The *Saccharomyces cerevisiae FLO1* flocculation gene encodes for a cell surface protein. Yeast 11:809-822.

107. Bony, M., Thines-Sempoux, D., Barre, P., and Blondin, B. (1997). Localization and cell surface anchoring of the *Saccharomyces cerevisiae* flocculation protein Flo1p. J. Bacterio. 179:4929-4936.

108. Kobayashi, O., Hayashi, N., Kuroki, R., and Sone, H. (1998). The region of the *FLO1* proteins responsible for sugar recognition. J. Bacteriol. 180:6503-6510.

109. Kobayashi, O, Suda, H., Ohtani, T., and Sone, H. (1996). Molecular cloning and analysis of the dominant flocculation gene *FLO8* from *Saccharomyces cerevisiae*. Mol. Gen. Genet. 251:707-715.

110. Kobayashi, O., Yoshimoto, H., and Sone, H. (1999). Analysis of the genes activated by the *FLO8* gene in *Saccharomyces cerevisiae*. Curr. Genet. 36:256-261.

111. Rupp, S., Summers, E., Lo, H. J., Madhani, H., and Fink, G. R. (1999). MAP kinase and cAMP filamentation signalling pathways converge on the unusually large promoter of the yeast *FLO11* gene. EMBO J. 18:1257-1269.

112. Pan, X., and Heitman, J. (1999). Cyclic AMP-dependent protein kinase regulates pseudohyphal differentiation in *Saccharomyces cerevisiae*. Mol. Cell. Biol. 19:4874-4887.

113. Guo, B., Styles, C. A., Feng, Q., and Fink, G. R. (2000). A *Saccharomyces cerevisiae* gene family involved in invasive growth, cell-cell adhesion, and mating. Proc. Natl. Acad. Sci. 97:12158-12163.

114. Schoeman, H., Vivier, M. A., Du Toit, M., Dicks, L. M., and Pretorius, I. S. (1999). The development of bactericidal yeast strains by expressing the *Pediococcus acidilactici pediocin* gene (*pedA*) in *Saccharomyces cerevisiae*. Yeast 15:647-656.

115. Young, T. W. (1981). The genetic manipulation of killer character into brewing yeast. J. Inst. Brew. 87:292-295.

116. Bussey, H., Vernet, T., and Sdicu, . M. (1988). Mutual antagonism among killer yeast: competition between K1 and K2 killers and a novel cDNA-based K1-K2 killer strain of *Saccharomyces cerevisiae*. Can. J. Microbiol. 34:38-44.

117. Dickinson, J. R., and Dawes, I. W. (1992). The catabolism of branched-chain amino acids occurs via 2-oxoacid dehydrogenase in *Saccharomyces cerevisiae*. J. Gen. Microbiol. 138:2029-2033.

118. Fujii, T., Nagasawa, N., Iwamatsu, A., Bogaki, T., Tamai, Y., and Hamachi, M. (1994). Molecular cloning, sequence analysis, and expression of the yeast alcohol acetyltransferase gene. Appl. Environ. Microbiol. 60:2786-2792.

119. Santayanarayama, T., Umbarger, H. E., and Lindengren, G. (1968). Biosynthesis of branched-chain amino acids in yeast: regulation of leucine biosynthesis in prototrophic and leucine auxotrophic strains. J. Bacteriol. 96:2018-2024.

120. Ulm, E. H., Böhme, R., and Kohlhaw, G. (1972). ø-isopropyl-malate synthase from yeast: purification, kinetic studies, and effect of ligands on stability. J. Bacteriol. 110:118-1126.

121. Lee, S., Villa, K., and Patino, H. (1995). Yeast strain development for enhanced production of desirable alcohol/esters in beer. J. Am. Soc. Brew. Chem. 53:153-156.

Applied Mycology and Biotechnology
Volume 2. Agriculture and Food Production

Genetic Diversity of Yeasts in Wine Production

Tahía Benítez and Antonio C. Codón
Department of Genetics, Faculty of Biology, University of Seville, Apartado 1095, E-41080 Seville, Spain (e-mail: tahia@cica.es).

Wine elaboration is a complex multipopulational process in which several microbial species are successively involved. At early stages of fermentation, a high number of non-*Saccharomyces* yeast species predominate in the musts. These yeasts can actually be precisely identified by several molecular techniques among which polymorphism of rDNA and internal transcribed spacers (ITS) at rDNA regions have proved to be the most useful. This identification has also made it possible to analyse the contribution of these yeasts to the final organoleptic characteristics of the wine produced. As the concentration of ethanol increases in the must, different strains of *Saccharomyces cerevisiae* predominate. At the end of fermentation they represent almost 100% of the microbial population. The different strains are highly polymorphic with regards to their DNA content, chromosomal size, and DNA sequence of their mtDNA. This polymorphism seems to result from chromosome reorganizations (duplications, deletions, translocations), homologous recombination and gene conversion, occurring both at mitosis and meiosis and, in some cases, mediated by the presence of DNA repeats such as Y' or X subtelomeric regions or Ty transposable elements. Reorganizations and changes in DNA sequences might be favoured by DNA breaks caused by ethanol and DNA repair via recombination. The lack of proof-reading ability of mtDNA polymerase could explain preferential alteration of mtDNA caused by ethanol. In some cases, specific genomic organizations or phenotypic features seem to reflect adaptation to specific conditions, i.e., specific chromosome or gene amplification, or capabilities such as tolerance to CO_2 or acetaldehyde during wine ageing.

1. INTRODUCTION

Wine fermentation is perhaps one of the earliest use of a biotechnological process. The history of winemaking goes back to Mesopotamia as early as the 7th century BC [cited in 1]; however, since last 150 years process of winemaking have been gradually established and improved. Pasteur stated that alcoholic fermentation was a process which correlated with the life and organization of yeast cells [cited in 2], and with the knowledge that yeasts were responsible for the biotransformation of grape sugars into ethanol and carbon dioxide, winemakers initiated the control of wine elaboration and introduced selected wine strains to direct fermentation. Wine results from microbial fermentation of grape juice that is undergone by different species of yeasts, which are naturally present on grape skin, in soil and air, transported by insects and also present in the winery environment [3]. From these different species, the task of the winemaker is to ensure that only the desired fermentative yeasts predominate in the juice and carry out the fermentation. Different yeast strains are also responsible for specific fermentations, which give rise to bread, wine, beer, sake, cider, distilled drinks and other fermented products; they are also suppliers of enzymes, flavours, essences, proteins and vitamins, and also are the vectors of heterologous protein production. The world production of wine yeasts is only about 1 to 5 thousand tons, whereas production of bakers' yeasts has reached almost 5 million tons per year. The main reason for this difference is that selected wine yeasts have only been produced for the last 15 years, whereas bakers' yeasts have been produced for more than 100 years. However, wine yeast production is continuously increasing [4].

2. YEAST ASSOCIATED WITH GRAPES AND WINE

Wine fermentation differs from other industrial fermentation processes as the juice contains a varying number of microorganisms [1]. In the course of fermentation and also in some cases of wine maturation, the changing composition of the medium selects for various

Table 1. Studies about wine yeast identification using molecular biology techniques. Modified from [5].

Methodology	Genus
δ elements	*Saccharomyces*
Intron splice site	*Saccharomyces*
Karyotype	*Saccharomyces*
	Hanseniaspora
	Zygosaccharomyces
Microsatellite	*Saccharomyces*
Nested PCR	*Brettanomyces*
Plasmids	*Saccharomyces*
	Zygosaccharomyces
RAPDs	*Saccharomyces*
	Metschnikowia
	Rhodotorula
	Zygosaccharomyces
	Candida
	Pichia
	Torulaspora
	Hansenula
RFLP-karyotype	*Candida*
	Kloeckera
	Schizosaccharomyces
RFLP-mtDNA	*Saccharomyces*
	Kluyveromyces
	Zygosaccharomyces
	Brettanomyces
RFLP-ITS/5.8S rRNA gene	*Candida*
	Hanseniaspora
	Saccharomyces
	25 different genera

types of yeasts (Table 1), bacteria and other microorganisms of which, only some are able to grow and compete with the highly specialized fermentative yeasts which quickly dominate the alcoholic fermentation. Most of these desirable populations of yeast belong to the species ***Saccharomyces cerevisiae.*** There are controversies of whether in many cases, uncontrolled growth of yeasts other than *S. cerevisiae* and bacteria negatively affects the final wine quality [4] or whether the presence of non-*Saccharomyces* species could be important. The reason for this importance is the production of secondary metabolites which contribute to the final taste and flavour [5]. Therefore, the wealth of yeast biodiversity with unknown enological properties may still be largely hidden [1].

2.1. Wine Yeast Diversity

Yeasts are defined as unicellular fungi which usually grow and divide asexually by budding. Taxonomically they are grouped under different genera of ascomycetes, basidiomycetes and deuteromycetes [2]. This diversity indicates that yeasts have been favourably selected in nature which has kept appearing throughout evolution [6]. Yeast nomenclature has changed considerably in the last few years. Originally, yeasts were classified according to morphological and metabolic criteria e.g. capacity to ferment and/or assimilate certain nitrogen and carbon sources. By using these criteria it was possible to identify different genera and species, although it was not possible to differentiate among strains [2]. However, due to their strong special selection, strain differences among yeasts are often more important than species differences [3]. Recent application of molecular techniques have represented an alternative to the traditional methods of yeast identification. Those new procedures allow a specific desired strain to be identified unequivocally, even in the absence of morphological or biochemical indicators.

Electrophoretic karyotypes (Fig. 1), which allow the separation of individual chromosomes, are particularly useful because some industrial strains have their own characteristic pattern [7]. This technique, in combination with other analytical techniques, also allow to determine the number of copies of a specific chromosome present in a strain, and variations in the size of homologous chromosomes. These techniques used to determine genomic constitution are complemented with analysis of the restriction fragment lengh polymorphism (RFLP) pattern using mtDNA [8, 9, 10], polymorphism after hybridization with specific probes, RAPD, ITS, polymorphism of rDNA, etc.

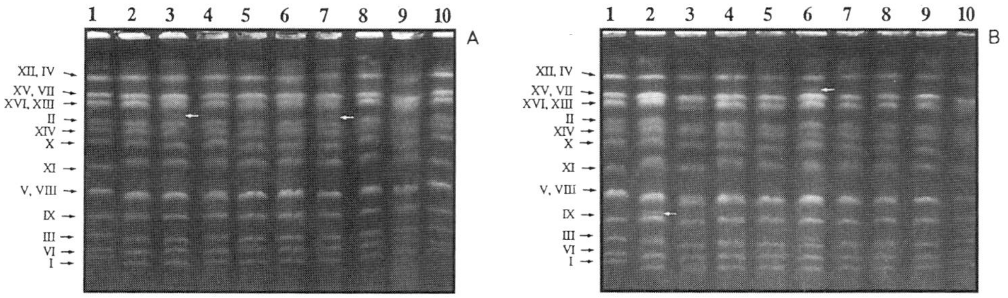

Fig. 1. (A) Chromosomal patterns of the laboratory yeast DS81 (lanes 1 and 10) and the flor strains: *S. cerevisiae var. beticus* (lanes 2 and 4); *montuliensis* (lanes 3, 5, 6, 7 and 9) and *cheresiensis* (lane 8). (B) Chromosomal patterns of the laboratory yeast DS81 (lane 1) and the flor strains: *S. cerevisiae var. beticus* (lanes 2, 3, 4, 6, 8 and 10); *rouxii* (lane 5); *montuliensis* (lane 9) and *cheresiensis* (lane 7). Reprinted with permission from ref. [12].

The RFLP of mtDNA has been used successfully as the best technique of individual characterization in multipopulational processes such as wine elaboration [8, 9]. Such identification and control have allowed researchers to establish the role that a selected strain plays in the aroma, flavour and organoleptic characteristics of wine as compared to the natural microbiota during the vinification process, which will be described below.

2.2. Interspecific Diversity (Non-*Saccharomyces* yeasts)

2.2.1. Presence and effects of non-*Saccharomyces* yeasts on musts and wines

Fermentation of grape must and production of wine, conducted by traditional methods, include the interaction not only of yeasts but with other microorganisms such as fungi, lactic acid bacteria, acetic acid bacteria, mycovirus, bacteriophages etc. [1]. Must fermentation gives rise to a final product which results from the combined action of all these microbes, which grow more or less in succession throughout the fermentation. Several studies have described the isolation and identification of yeast species from the grape surface, soil and wineries. The frequency of species and their proportion depends largely on the isolation process [5]. In addition, it varies with the country, wine variety and methods of wine elaboration. Adverse climatic conditions at time of harvest also affect composition of the yeast population, or the population shows atypical evolution through the fermentation process [11]. Freshly crushed grapes yield a must with a yeast population of 10^3 to 10^5 CFU/ml. It is generally admitted that the genera *Kloeckera* and *Hanseniospora* are predominant on the surface of the grapes (50 to 70% of the isolates), whereas fermentative species of *Saccharomyces* occur at extremely low frequency (less than 50 CFU/ml) [1]. By contrast, *S. cerevisiae* is abundant on winery equipment [12]. Grape must endures the growth of only a limited number of microbial species because the low pH and high sugar concentration [13] exert a strong selective pressure on the microorganisms resulting proliferation of only a few yeast and bacterial species. Out of over 700 species of yeasts, at least 15 species belonging to 7 genera are associated with wine making [14]. Fermentations are generally initiated by the growth of various species of *Candida, Debaryomyces, Hanseniospora, Hansenula, Kloeckera, Metschnikowia, Cryptococcus, Pichia, Schizosaccharomyces, Torulaspora* and *Zygosaccharomyces* [5]. Sulphur dioxide added as antimicrobial preservative and the increasing levels of ethanol impose additional selection against oxidative and ethanol sensitive microbial species. Growth of these species is limited after 2-3 days of fermentation, and only the most strongly fermenting and ethanol tolerant species, usually *S. cerevisiae*, takes over the fermentation, reaching final populations of about 10^8 CFU/ml. For this reason, *S. cerevisiae* is preferred for initiating must fermentation.

The use of pure yeast cultures of *S. cerevisiae* offers undeniable advantages with regard to the ease of control and homogeneity of fermentations. However, wine fermented by an indigenous population of yeast may have a genuine quality different from that of wine fermented with a pure culture of *S. cerevisiae* [3], as suggested above. In addition, at the beginning of fermentation, low fermentative yeasts produce some important reactions in must which could improve the final flavour of wine [5]. For this reason, it may be preferable to use a mixture of yeast species as starter to produce wines of good quality. It has already been described that, in addition to grape variety, storage and fermentation conditions, the physico-chemical processes involved in fermentation mainly determine the aroma of wine. Particularly, those occurring in the must or wine at any stage during fermentation play the most significant role in the production of volatile compounds [3]. In some cases the importance of yeasts with low fermentation power such as *Kloecckera apiculata* in the production of significant overall amounts of volatile substances in wine obtained from musts of Monastrell grapes has been solidly demonstrated [15]. Sensorial and microbiological analyses of Majorca wines showed significant differences between inoculated wines and those fermented spontaneously by wild yeasts [16]. Fermentation of Verdejo, Palomino or Viura grapes carried out with selected local yeast strains, belonging to the species *K. apiculata, Torulaspora rosei* and *Saccharomyces ellipsoideus* also gave rise to wines with the expected specific organoleptic features [17]. Appreciable quantities of ethyl acetate have been associated with the presence of *Candida glabrata* and *Debaryomyces hansenii* in Tenerife wines [18]. There are also controversial results on the effect of different yeast species involved in spontaneous fermentation: despite the occurrence of indigenous flora, final inoculated yeasts succeeded in taking over the fermentation of Pedro Ximenez grapes and yielding wines similar to those obtained from sterile musts [19] whereas other authors found a high similarity between fermentation of Pedro Ximenez grapes, carried out with mixed cultures with added *S. cerevisiae,* and those carried out with the indigenous yeasts [20].

2.2.2. Identification of non-*Saccharomyces* yeasts

The correct identification of non-*Saccharomyces* yeasts is important to understand and predict the enzymatic reactions, which will occur during the early stages of fermentation. Classification on the bases of phenotypic features, mainly biochemical properties, needs an excessive number of tests and takes at least one to two weeks [2], whereas molecular biology techniques such as RFLP of mtDNA, chromosomal DNA electrophoresis, rDNA restriction analysis or RAPDs allow a quick and highly reliable yeast identification (Table 1) [7, 5].

Some hemiascomycetous yeasts possess rRNA genes located in a single genomic region, composed by clusters of two transcriptional units, one unit encoding 18S, 5.8S and 25S rRNA with two internal transcribed spacers (ITS1 and ITS2), and the second unit encoding 5S RNA [5]. ITS and 5.8S rDNAs are useful to determine close taxonomical relationships because they exhibit greater interspecific differences than the 18S and 25S transcriptional unit. Intraspecific variations have also proved useful for identification of strains within species. Restriction analysis of this region has allowed the identification of 33 wine yeast species and 129 food yeast species belonging to 25 different genera [5]. The recent applications of molecular techniques to the ecological study of wine yeasts have thus demonstrated that strain variation can also occur within the non-*Saccharomyces* species. Different strains of the unique non-*Saccharomyces* species contribute to the fermentation, and different phases of the fermentation are sometimes dominated by different strains of the same species.

Once wine yeasts species are identified unequivocally, it is important to correlate them with specific properties. For instance, there is little information on the production of enzymes such as proteases, lipases, glicosidases, pectinases and other hydrolases, necessary for the formation of organoleptic compounds in the wine, by specific non-*Saccharomyces* yeast species. Extracellular proteases (necessary for must clarification and yeast autolysis) have been described in strains of *K. apiculata.* Strains of this species and of *Metschnikowia pulcherrima* also produced extracellular proteases involved in the breakdown of juice proteins; glucosidases (which liberate terpenes and anthocyanins from their immobilized glycosidic form) were detected in strains of *Candida, Pichia* and *Hanseniaspora:* pectinases (to improve must extraction) were also detected in various species of *Candida, Cryptococcus, Kluyveromyces* and *Rhodotorula*, although none of the latter strains were wine yeasts; the production of ß-1,3-glucanase, necessary for the hydrolysis of different types of glucans, have been described in several wine yeasts including *S. cerevisiae;* finally, esterases and lipases are being searched in wine yeasts. They are important for yeast autolysis, but these enzymes (mainly lipases) can potentially affect wine quality [5].

2.3. Intraspecific Diversity. *Saccharomyces* Yeast

Saccharomyces cerevisiae strains become increasingly abundant on the grape juice, and is by far the most dominant yeast species colonising any type of wine [3]. Due to the fact that *S. cerevisiae* is practically absent from grapes, but predominates in wine, its presence in musts and wines is directly associated with artificial man-made environments such as fermentation plants.

Strains of *Saccharomyces* are detected in the must when there is already a certain ethanol concentration, and become predominant as the ethanol concentration increases. However, one basic contribution of molecular biology to population genetics was the observation that, even under strong selective conditions, as they are wines with a high alcohol content, there is a high degree of genetic variation among the *Saccharomyces* strains present in these wines (Table 2) [21]. Under these extreme conditions, natural selection was expected to favour

Table 2. Ploidy (*n*) of different *S. cerevisiae* industrial strains

	Strain		Ploidy (*n*)
A	YNN295	G	1
	DS81	G	2
	DADI	B	1.5
	VS	B	2.5
	CT	B	2.7
	SB2	B	2.6
	SB11	B	1.3
	V1	B	2.7
	V2	B	2.7
	ACA21	W	1.9
	ACA22	W	1.9
	ALKO1245	D	1.6
	ALKO1523	D	1.3
	ALKO743	B	3
	ALKO554	B	1.3
	ALKO1611	B	3
	TS146	B	1.6
	ATCC9080	Br	1.6
	NCYC396	Br	2.2
B	*S. cerevisiae (beticus)*	V	1.3 - 2
	S. cerevisiae (cheresiensis)	V	1.3 - 2
	S. cerevisiae (montuliensis)	V	1.7 - 2
	S. cerevisiae (rouxii)	V	2

A : Modified from [29]; B: Modified from [12]. G genetic line; B bakers'; W = wine, Br = brewers'; D= distillers'; V = velum.

the best strain and, by removing all other strains from the mixed population, selection acting in this way would eliminate genetic variation. A possible explanation is that selection favours different genotypes under different conditions. Alternatively, several genotypes may work equally well under such extreme conditions. In fact, *Saccharomyces* strains with specific characteristics appear in different types of wine, so that the strains became classified into several different races or varieties [2]. In most cases, a taxonomic linkage exists between races and strain properties: highly ethanol tolerance (var. *ellipsoideus*), tolerance to CO_2 (var. *bayanus*), intensive flavour (var. *capensis*), low volatile acidity (var. *rosei*) [22], film-forming strains with strong oxidative capabilities (var. *beticus* and *cheresiensis*), high concentrations of acetaldehyde production and tolerance (var. *montuliensis* and *rouxii*) and others [12, 23, 24, 25, 26], all varieties belonging to the same species, *S. cerevisiae*.

In conclusion, assignment of traditional wine yeast strains to the single species *S. cerevisiae* does not imply that they are similar, either phenotypically or genetically. The strains differ significantly not only in their metabolic features such as fermentative capacities, or production of aromatic compounds [11, 26] but also in their genetic configuration such as DNA content, chromosomal pattern or mtDNA sequence [12, 23, 27], as will be discussed below. These differences are probably the result of adaptation to specific conditions, and to different industrial environments; for instance, wines with different sugar, alcohol or tannin contents, and even different at subsequent stages of the same type of wine [12].

2.3.1. Identification and diversity of *Saccharomyces* wine yeasts

Saccharomyces cerevisiae wine strains differ in their chromosomal patterns, DNA content or RFLP of their mtDNA. With regards to their DNA content, some authors have reported great variations of different industrial yeasts, and suggested that these variations may respond to specific industrial environments [28]. However, when comparing simultaneously different industrial yeast groups such as bakers', distillers', brewers' and wine strains, similar variations in DNA content have been found among yeast from the same group, i.e., bakers', and those from different groups, i.e., bakers' vs. brewers' (Table 2) [29]. This may indicate that DNA content does not reflect adaptation to specific environments. Exceptionally, flor yeasts isolated from sherry wine are aneuploids, which have a DNA content lower than other industrial yeasts [12]. The reason for this can be that ethanol, present at very high concentration (over 15% v/v) and oxidative conditions [12] favour chromosome loss. In fact, most flor yeasts are aneuploids whose DNA content is generally of less than 2n (Table 2) [12]. Other authors have analyzed flor yeasts by genetic marker segregation after crossing them with laboratory-marked haploid strains [30]. Results indicate the presence of several copies of most chromosomes analyzed in the industrial strains examined (Table 3), but preferential elimination of laboratory strain chromosomes cannot be ruled out. With regards to the chromosomal pattern, the polymorphism found in yeast strains belonging either to different or the same industrial groups is amazingly high [29]. Whereas a standard haploid laboratory strain displays 15 bands corresponding to 16 chromosomes, great variations have been found in all sorts of industrial yeasts, both in the number and size of the chromosomes, resulting in variations both in the number and position of the chromosomal bands [27, 31-34]. Differences in the number of chromosomal bands are mostly the result of homologous chromosomes of different sizes [27]. In some cases polymorphism is so high that nearly each strain can be identified unequivocally by its specific chromosomal pattern [35]. The fact that differences in size and number of chromosomal bands are displayed by yeasts from either the same or different industrial groups might also indicate that global changes both in size and number of chromosomes do not reflect selection to specific environmental conditions.

2.3.2. Genetic constitution of *Saccharomyces cerevisiae* strains and adaptation to specific industrial conditions

Chromosomal polymorphism has been described to be higher in industrial yeasts than in laboratory strain [35]. Among industrial yeasts, the literature concerning chromosomal polymorphism of brewers', bakers' or distillers' yeasts is very scarce, whereas that of wine yeasts is more abundant. Among wine strains, high chromosomal polymorphism has been reported [31-34]. However, among them, film-forming strains do show scarce variability (Fig. 1) [12]. It may be that the selective conditions which are so severe (lack of fermentable carbon sources and over 15% ethanol) have favoured an almost unique chromosomal pattern, similar in all flor yeasts and different from those of other industrial or laboratory yeasts [12, 35]. This lack of polymorphism may also be related to the scarce presence of Ty1 elements (although Ty2 are abundant in these strains) (Fig. 2) [36], as will be discussed below, since, in the absence of recombination, it has been suggested that a population became monomorphic [37]. Some authors support this suggestion by describing sexual isolation of wine yeasts which stop them from mixing their features during wine fermentation and maturation [38]. In order to

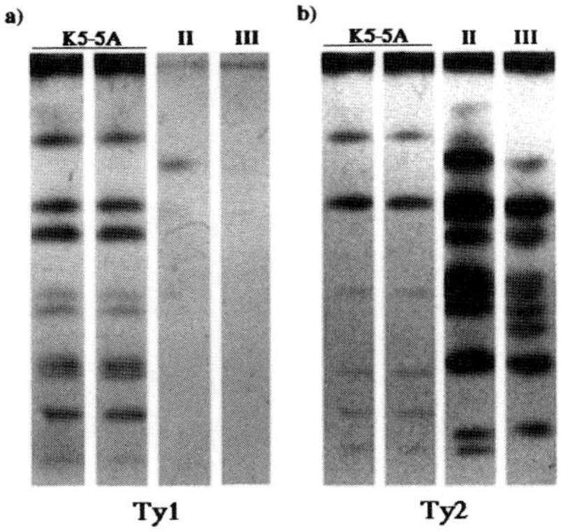

Fig. 2. Hybridization patterns obtained with probes from Ty1 (a) or Ty2 (b) sequences on chromosomes of flor yeasts and of the laboratory strain K5-5A. Flor yeast chromosomes contain only Ty*2* elements while Ty1 is the main element present in the laboratory strain. Reprinted with permission from ref. [36].

identify each chromosome specifically, they are hybridized with probes corresponding to genes which, in laboratory strains, are known to be present once in the genome [27]. Chromosomes of industrial yeasts, including wine yeasts, hybridize in all cases with these probes, indicating that DNA similarity with these genes is very high. Hybridization confirms the existence of several homologous chromosomes of different sizes [35, 36]; sometimes interchromosomal (translocations) and intrachromosomal (deletions, duplications, inversions) reorganizations have also been found.

If polymorphism of chromosomes reflects selection under specific conditions, extreme selective conditions such as the presence of high ethanol concentration or oxidative stress should favour any chromosomal reorganization, which results in an increase of viability or of growth rate under such conditions. In all flor strains examined by Guijo et al., [30], polysomy of chromosome XIII was observed (Table 3). This chromosome contains the *ADH2* and *ADH3* loci which encodes for the ADHII and ADHIII isoenzmes of alcohol dehydrogenase, which are involved in ethanol oxidative utilization during biological ageing of wines. Similar results have been found in other industrial yeast groups, precisely in bakers' yeasts [29, 39]. The procedure for biomass production of these yeasts would favour any chromosomal reorganization, which resulted in an increase in the growth rate [40]. Furthermore, if there is a limited addition of the substrate (molasses) whose main carbon source in sucrose, amplification of the *SUC* gene which encodes invertase would result in a more efficient utilization of the sucrose. In bakers' strains, *SUC* gene has been amplified and translocated to several chromosomes as judged by the presence of several bands which hybridize with the probe [29, 39], and a similar phenomenon seems to have occurred in distillers' yeasts [29]. The wine yeasts analyzed, which in their natural environments ferment glucose and fructose but not sucrose, only possess a single band [29]. Furthermore, the accumulation of *SUC* genes was only observed in populations derived from sources containing sucrose and seemed to be absent in other strains from sources promoting the *MEL* gene [41]. *RTM1* gene, whose expression confers resistance to the toxicity of molasses is also present in bakers' strains in multiple copies which are physically associated with *SUC* telomeric loci (Fig. 3) [42]. *RTM* sequences are not detected in laboratory or wine strains. In dough without addition of sugar, the principal fermentable sugar for bakers' yeasts is maltose liberated from the starch of the flour by amylases [32, 33]. *MAL* loci are necessary to ferment maltose. Each *MAL* locus is a complex of three genes encoding maltose permease (*MALT*), maltase (*MALS*) and a transcription activator (*MALR*) [32, 33, 43]. Families of *MAL* loci, mapping on different chromosomes in several copies, have also been identified in bakers' yeasts, selected for flour fermentation, but not in wine yeasts [43]. In addition, the quality of brewing strains is largely determined by their flocculation properties, and several dominant, semi-dominant and recessive flocculations (*FLO*) genes have been recognized in these brewing strains [44].

Both electrophoretic karyotyping and mtDNA restriction analysis have also revealed a considerable degree of polymorphism in natural yeast populations of *S. cerevisiae* isolated from fermenting musts in El Penedés, Spain [45]. Genetic analysis indicated a strong correlation between selected phenotypes with high tolerance to ethanol and temperature and mtDNA polymorphism. Furthermore, the karyotypes also revealed a correlation between distinct genetic traits and specific microenvironments, so that molecular analysis allowed to study geographical distribution of natural yeast populations as well as the identification of strains with specific properties.

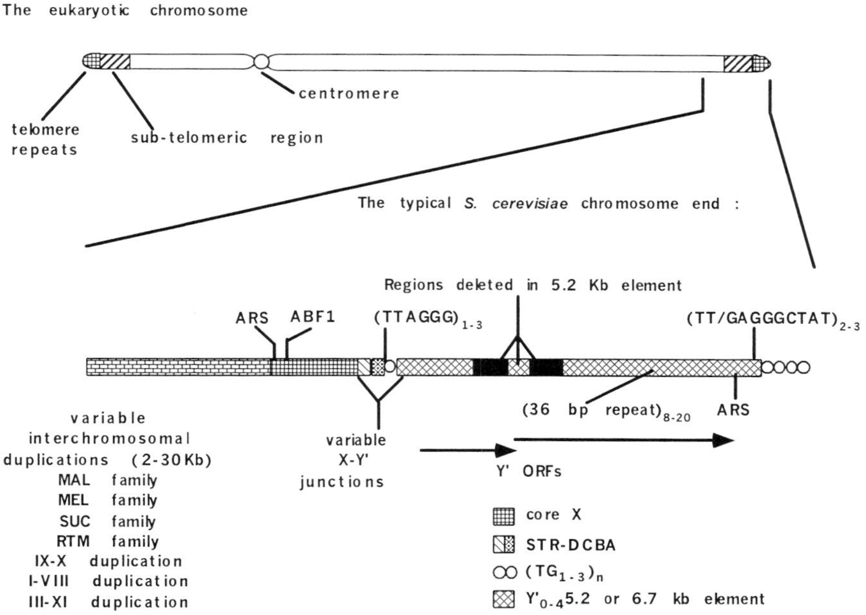

Fig. 3. The general chromosome end of eukaryotes. The *S. cerevisiae* subtelomeric region is expanded in the lower part of the figure (see text for details). Modified from ref. [58].

3. SOURCES OF GENETIC DIVERSITY

3.1. Chromosomal Reorganizations as a Source of Variability

Chromosomal reorganizations are natural phenomena in the life cycle of many organisms, and in some cases, it is part of their development pattern. As an example we may consider the *de novo* telomere formation in *S. cerevisiae*, the mating type switch in fungi or the chromosome reorganization of the immunoglobuline genes in higher eukaryotes [46]. There are other nonprogrammed genome reorganizations which include deletions, translocations etc. These reorganizations give rise in the resultant products to chromosome length polymorphism, to dramatic changes in the viability and to changes in chromosome number. The latter is a consequence of the lack of pairing of homeologous chromosomes and non-disjunction at meiosis. Incorrect disjunctions result in an unequal distribution of the chromosomes in the progeny and low viability and aneuploidies in the surviving products [46]. *Chromosome length polymorphism* has been observed in both asexual and sexual fungi, indicating that it results from both mitotic and meiotic processes [47]. Natural isolates of *S. cerevisiae* also display highly variable chromosomal patterns, as indicated above, whereas industrial strains show chromosome structures, which are highly polymorphic and very different from those of laboratory strains [48]. The exact mechanisms accounting for these variations and rearrangements are unknown, although the chromosomal structures observed are probably the results of multiple events (insertions, delections, translocations) favoured by their specific environment, which have accumulated over a long period of time [29]. Recently, whole genome duplication was proposed as an explanation for the many duplicated chromosomal regions in *S. cerevisiae* [49]. Computer simulation indicated that 8% of the original genes retained in duplicate after genome duplication and 70 reciprocal translocation between chromosomes, produced arrangements very similar to the map of real duplications in yeasts, although the original order of chromosomal rearrangements could not be established [49].

Table 3. Copy number of each chromosome in the *S. cerevisiae* strains isolated from velum (F,G) or during the last phase of an industrial fermentation (E)

	Saccharomyces cerevisiae strains					
Chromosome	F-12	F-21	F-23	F-25	G-1	E-1
I	2	2	2	**3**	3	4
II	2	2	2	**4**	3	2
III	2	2	2	2	2	2
IV	2	2	2	**3**	3	2
V	2	**2 or 3**	2	**3**	3	2
VI	2	2	2	**4**	3	4
VII	**2 or 3**	2 or 4	2 or 3	4	2	2
VIII	2	2	2	**3**	3	2
IX	2	2	2	**3**	3	2
X	2	2	2	**3**	2	2
XI	2	2	2	**3 or 4**	3	2
XII	2	2	2	**4**	3	2
XIII	**4**	3	3	4	4	4 or more
XIV	2	2	2	**4**	3	2
XV	2	2	**3**	4	3	2
XVI	2	2	2	**4**	2	2

Modified from ref. [30].

Gross chromosome rearrangements such as inversions, translocations and deletions, may be stimulated by events involving transposable elements (Fig. 4) [50]. Furthermore, genetic analysis of mutants which increased up to 5000-fold the rate of gross chromosome rearrangement formation suggests that at least three distinct pathways can suppress these arrangements: two of which suppress microhomology-mediated rearrangements, whereas one suppresses nonhomology-mediated rearrangements [51]. However, in *S. cerevisiae* genomic reorganization takes place mostly *via* recombination, and this phenomenon occurs between homologous sequences (Fig. 5). Therefore, the main substrate for recombination leading to genome reorganization are repeat sequences which, in yeasts, are very scarce:

telomeric and subtelomeric Y' and X regions, Ty transposable elements and, to a lesser extent, rRNA and tRNA genes [46]. In laboratory *S. cerevisiae* the recombination frequency between Ty elements present in the same (allelic or asymmetrical) or different (ectopic) chromosomes in diploid cells has been estimated to be 1% in meiosis and 10^{-7} in mitosis [54], whereas the transposition frequency was of 10^{-4} to 10^{-3} per generation [52-54]. Ty elements of the same family are highly homologous, and the reason seems to be due to gene conversion mechanisms, which takes place in addition to transposition and recombination [55, 56]. With regards to the subtelomeric regions X and Y', they appear repeated 4 to 5 times at the end of the chromosomes in laboratory strains, although some strains have 26 to 30 repeated sequences of these subtelomeric Y' families, concentrated in the higher molecular weight chromosomes [57]. Precisely, the only element which appears to be shared by all chromosome ends is part of the previously defined X element which contains an autonomous replication sequence (ARS) consensus (Fig. 3) [58]. Among other possible functions, subtelomeric repeats are suitable regions for adaptive amplification of genes. Also, these tandem repetitions originate genome reorganizations at meiosis as a consequence of telomeric interactions and of mitotic recombinations between Y' regions giving rise to duplications, deletions and insertions (Fig. 4). The sequences of the Y' regions are highly conserved, which may be due to gene conversion [52]. Recombination between homologous chromosomes possessing different copy number of Y' regions

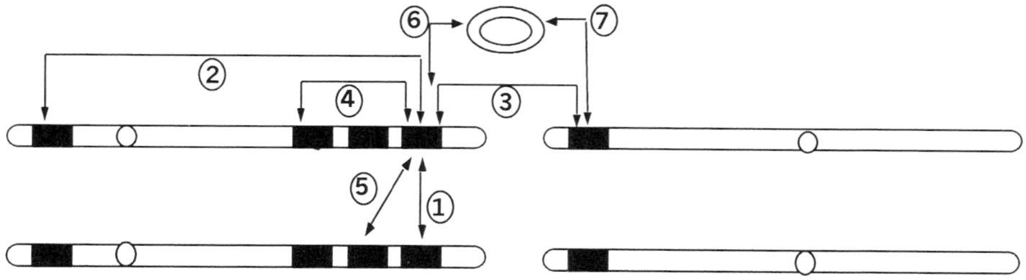

Fig. 4. Chromosomal reorganization caused by Ty and Y' repeat elements. 1: Allelic recombination, 2: intrachromosomal recombination, 3: ectopic recombination, 4: recombination between tandem repeats, 5: asymmetrical recombination, 6: excision, 7: insertion. Modified from ref. [7].

results in unequal distribution of the Y' elements and in consequence, in changes in the size of the chromosomes involved in the event. The frequency of recombination events for these elements is of 2 x 10^{-6} per generation in mitosis and 2% in meiosis [52]. Some gene families such as *SUC, MAL, MEL* and others, located in the telomeric regions, appear to be the result of these genome reorganizations [59]. Experiments to localize the flocculation *FLO* genes, which play an important role in the quality of brewing strains, have revealed that they also belong to a family whose members are located near the end of chromosomes, like the *SUC, MEL* and *MAL* genes, which are also important for good quality baking or brewing strains [44]. Genome rearrangements may also result from recombination between repeated sequences dispersed through the genome, rather than located at telomeric regions specifically, such as Ty or LTR (δ) sequences. Analysis of a high number of different *S. cerevisiae* strains have demonstrated that these strains possessed different duplications and translocations of subtelomeric regions of chromosomes XI and III [60]. *Reciprocal translocations* involving 80 kb of the left arm of chromosome III and 45 kb of the right arm of chromosome I has also been demonstrated to be the cause for observed chromosomal polymorphism in some *S. cerevisiae* laboratory strains [48]. In this case, analysis of the translocation breakpoints indicated a transposition hot-spot on chromosomes I and III and the translocation mechanism involved homologous recombination between Ty2 transposable elements [48]. In other *S. cerevisiae* homothallic strains, DNA length polymorphisms of chromosome III were generated during meiosis, sporulation or the spore germination process and were produced by the loss or addition of a specific DNA unit of ~100 kb, although it has not been elucidated whether the product of the *HO* gene affects the generation of this polymorphism or not [61].

Analyses of subtelomeric *MEL* gene sequences in wine and other industrial and laboratory strains have also shown that this gene family has diverged into two groups [62, 63]: (i) in group one all the *MEL* genes and their flanking regions have nearly identical DNA sequence and restriction pattern, suggesting very rapid distribution of the gene to separate chromosome, probably by transposition [64, 65], and (ii) in the second group sharp change to sequence heterogeneity and restriction pattern occurs, mostly due to the flanking regions, suggesting distribution of the *MEL* genes to new chromosomal locations, probably mediated by δ sequences [59]. Unstable mutations at specific loci of *S. cerevisiae* have been demonstrated which may be due to the insertion of Ty transposable elements [36], whose transposition generated duplications and whose expression was controlled by regulatory genes similar to those controlling elements in maize [37]. Other workers reported the decrease in viability and aneuploidies among the survivors. These phenomena were attributed to different numbers of Ty elements and Y' repeats as well as other subtelomeric elements, which led to impaired chiasma formation [66]. Further studies carried out with meiotic products of a bakers' strain indicate the disappearance of chromosomal bands present in the parental strain [39]. This was partly attributed to the presence of multiple Ty transposable elements, which seemed to undergo inter-chromosomal translocation together with amplification, giving rise to differences in chromosomal size. Mitotic recombination, preferentially between sister chromatides as a mechanism of DNA repair, is an important source of genomic

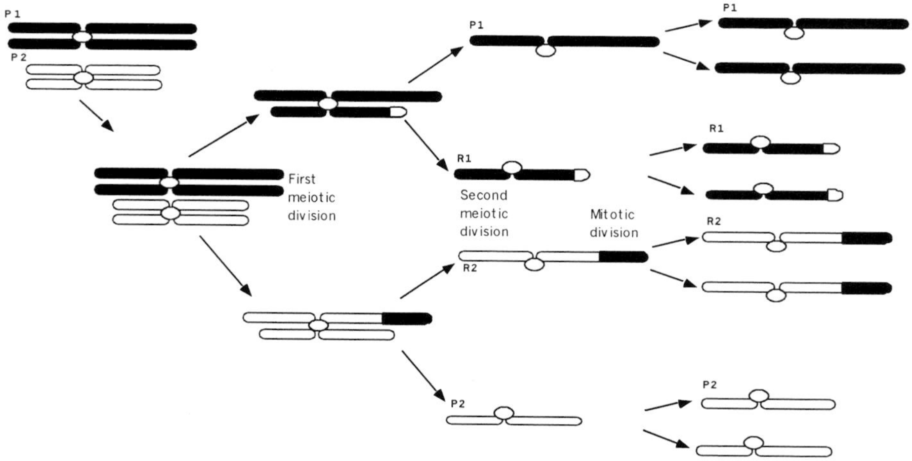

Fig. 5. Recombination occurring between two homologous chromosomes of different size (P1 and P2) results in four meiotic products, two of the same size as the two parental (P1 and P2) and two whose sizes differ from those of the parental (R1 and R2). Modified from ref. [7].

instability in *S. cerevisiae* [55]. Even so, whereas these chromosome reorganizations in bakers' yeasts occur with high frequency during meiosis, genomic stability during long term mitotic growth was very high [43]. It appears that on one hand, the main substrate for recombination leading to genome reorganization is repeat sequences; however, on the other hand, gene conversion has been demonstrated to play a major role in controlling chromosome stability of large tandem chromosome repeats in *S. cerevisiae* [67]. At least for rDNA, gene conversion (rather than unequal sister chromatid exchange) is the predominant recombination mechanism regulating expansion and contraction of the DNA. In addition, phenomena such as mutational homozygosis may also be relevant to genomic stability of yeasts [68]. Mutations at specific loci, for instance *cdc28*, also decreased the mitotic stability of chromosomes probably because of its role on changes in chromatin organization, which depends on the start gene *CDC28* [69].

3.1.1. Chromosomal rearrangements in wine yeasts

Chromosomal patterns for enological *S. cerevisiae* strains evidenced a marked polymorphism for these strains and, when compared with laboratory haploid strains, their patterns were interpreted as some bands being homologous chromosomes of different sizes. This suggestion has been further confirmed by karyotype analysis of meiotic segregates of some strains [34]. In addition, analysis of a number of independent crosses between haploid testers and unselected populations of spores of different wine strains allowed to distinguish between disomic, trisomic or tetrasomic chromosomal complements in the parents, and showed that linkage relationships among the chromosomes of the wine strains were not identical to those of the laboratory testers [31]. Translocations seemed to have placed genes found on specific chromosomes of laboratory strains on different chromosomes present in different copy numbers in wine strains and different copies of a specific chromosome were not identical in wine strains (Fig. 6); these analyses also indicated that aneuploidies are widespread among wine strains, and suggested that *S. cerevisiae* wine yeasts are exceptionally tolerant to spontaneous loss or gain of chromosomes [31], but did not provide an explanation either for such widespread or for mechanisms accounting for it. Aneuploidy of wine strains is in part due to the tolerance that *S. cerevisiae* wine strains have for this condition; as long as meiosis is not a regular part of their life cycle, the maintenance of a stable, although unbalanced, chromosome complement may be advantageous. Other workers [36] have described the genomic complexity and stability at mitosis and meiosis after obtaining yeast hybrids derived from flor yeasts isolated from sherry wine and laboratory strains. Genetic analysis indicated that flor yeasts were aneuploids, and that recombination occurred not only among homologous chromosomes of similar length but also among polymorphic partners with different sizes. Furthermore, polymorphism in chromosome length seemed to be a major source of karyotypic variation, as judged by the new chromosomal variants which were frequently observed in the meiotic products (Fig. 6), recombinational events occurring both mitotically and meiotically.

Flor yeasts isolated from sherry wine had a distinctive molecular karyotype (Fig. 1) [12], and showed a similar distribution of Ty2 sequences, lacking Ty1 sequences (Fig. 2) [36]. These strains do not often sporulate or, when sporulating, produced nonviable spores [12], indicating that flor yeast populations are genetically isolated. Sancho et al. [38] reported that the distribution of characteristics of sugar fermentation and metal tolerance was different in strains isolated during sherry wine fermentation, film formation and ageing, suggesting the existence of sexual isolation in the yeast populations and indicating that this sexual isolation prevented the random distribution of taxonomic characters. Similarities found in flor yeasts in their molecular karyotype (Fig. 1) [12] and Ty2 distribution (Fig. 2) [36] also suggest that the different strains of the isolated flor yeasts are very closely related, in spite of their phenotypic differences [26]. Ty2 sequences were found on many chromosomes in all flor yeasts but on very few chromosomes in laboratory strains and,

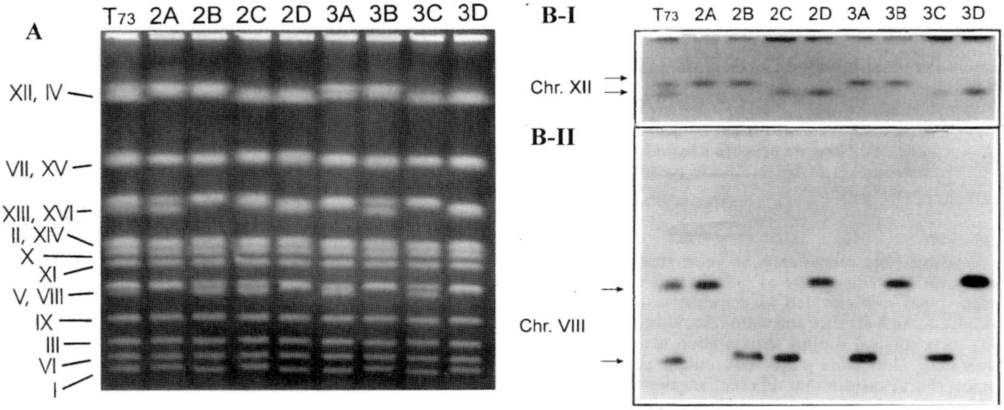

Fig. 6. (A) Electrophoretic karyotypes of a wine yeast and two complete meiotic derivatives. (B) Hybridization with probes from chromosomes XII (rDNA; I) and VIII (CUP1; II). Reprinted with permission from ref. [73].

conversely, Ty1 sequences were only found on chromosomes of laboratory strains [36]. The low polymorphism found in these wine strains suggested that ectopic recombination between Ty2 elements rarely takes place. Although allelic recombination between Ty elements has been reported to occur at similar frequencies as those observed for other genes [53, 54], Ty2 sequences undergo transposition, gene conversion and recombination at very low frequency. Thus, the lack of Ty1 genetic elements and the low mobility and lack of recombination of Ty2 elements could be an explanation for the distinctive karyotype of wine yeasts isolated from sherry. Other authors have found that selected *S. cerevisiae* strains able to carry out the second fermentation of sparkling wine in El Penedés differed enormously in their karyotype and showed high rates of karyotype instability during vegetative growth. The most frequent alteration in the karyotype pattern in wine yeasts was the variation in size of chromosome XII, which reflected changes in size of the rDNA clusters [45, 70]. Furthermore, data indicated a very different organization of rDNA clusters in wine strains from that reported for other yeasts. Whereas rDNA repeats seemed to play an important role in the changes in size observed at least in chromosome XII [70], *SUC* genes or Ty elements did not show amplification or transposition. Processes that could be related to rearrangements seemed to originate from ectopic recombination between repeated sequences interspersed in the genome, and their frequency suggests that they may play an important role in the continuous introduction and maintenance of genetic diversity in wine populations. Analysis of 450 karyotypes of meiotic and mitotic derivatives of one of these strains indicated that derivatives with low karyotype variability at mitosis also showed low rates of chromosomal rearrangements at meiosis, suggesting that both phenotypes were linked. Segregation analysis also indicated that the stable phenotype was determined by only one or two genes. Furthermore, some of the rearrangements acquired selective advantages over the parental strain in rich medium, indicating that chromosome rearrangements might provide selective advantages even under nonselective conditions [70]. Mortimer *et al.* [71] described a new mechanism accounting for genetic variability: they found that several wine strains analyzed were diploid, homothallic and genetically unique, and that they yielded asci that came from genetically different cells. These differences were interpreted to have arisen from heterozygous, homothallic strains, which had sporulated, yielding homozygous spore clones. Those authors suggested that natural heterozygous wine yeast strains undergo meiosis at certain frequency, so that a multiple heterozygote gives rise to homozygous descendants some of which may replace the original heterozygous parental. The process was called "genome renewal", and changes of heterozygotes to homozygous diploids were detected directly in several of the wine strains. However, sexual isolation in yeast population during wine production [12, 36, 38], the high level of heterozygosity [72], and the low sporulation rates of wine yeasts [36] do not favour this hypothesis. More recently, Puig et al. [73] studied the stability of wine yeasts by analyzing electrophoretic karyotype and the *URA3* locus in a *URA3/ura3* wine yeast in several grape must fermentations. These authors found mitotic rearrangements for chromosomes VIII and XII after 30 mitotic divisions, and the presence of *ura3/ura3* homozygous at fairly high frequency, sporulation and meiosis not being involved in these processes. They suggested a process of gradual adaptation to vinification conditions, as chromosomal rearrangements and aneuploidies, acquired following numerous mitotic divisions, are maintained vegetatively, and mitotic recombination instead of sporulation, would eliminate deleterious mutations and introduce genetic diversity. This suggestion does not mean that wine strains never sporulate, but that sporulation is not significant with regards to their genome evolution. The presence of lethal recessive mutations present in heterozygosis in wine yeasts [72] support this view since the frequency of sporulation is lower than that of mutation. Finally, it is worth considering that wine yeasts are in the presence of high concentrations of ethanol and sometimes, of acetaldehyde [26]. Ristow et al. [74] demonstrated that ethanol induced DNA single-strand breaks in *S. cerevisiae,* and that acetaldehyde also had a deleterious effect on chromosomal DNA in cells as well as on isolated DNA. Their results also supported that ethanol is mutagenic *via* its metabolite, acetaldehyde, although both acetaldehyde and ethanol cause DNA strain breaks in living *S. cerevisiae* cells. Furthermore, ethanol is metabolized via acetaldehyde, which was active on chromosomal DNA but the effect of exogenous acetaldehyde was much stronger than that of metabolized ethanol because this is rapidly metabolized to acetic acid [74]. It therefore seems that in such mutagenic environment as, for instance, that of sherry wine, with 16% ethanol and up to 1.2 g/l of acetaldehyde, wine yeasts might undergo continuous changes due to errors induced during the mechanism of DNA repair *via* recombination. This suggestion has proved to be true, at least with regards to the effect of ethanol on the m DNA of wine yeasts (Fig.7), as will be discussed below. If, as suggested, both ethanol and acetaldehyde induce continuous breaks in the DNA, such mechanism of DNA repair *via* recombination would explain the high frequency of chromosome reorganizations (Fig. 6) even in the absence of meiosis and sporulation, and the divergence between laboratory chromosomal organization and that of wine yeasts [73].

3.1.2. The Negative Effects of Chromosomal Reorganizations

3.1.2.1. Limitation of large reorganizations

Though strains of wine yeast have undergone many chromosomal reorganizations [73], most housekeeping genes are usually localized at specific positions in the chromosomes [29, 39]. The frequency of gene conversion due to recombination between Ty elements located at different chromosomes is only 2×10^{-6}, suggesting that there are mechanisms which repress ectopic exchanges. In fact, gene *TOP3* represses meiotic recombination events in chromosomal regions, which are near a δ element [53, 54], whereas allelic recombination between Ty elements is not repressed at meiosis or mitosis. The fairly mitotic stability, shown by the repetitive pattern of the electrophoretic karyotypes of

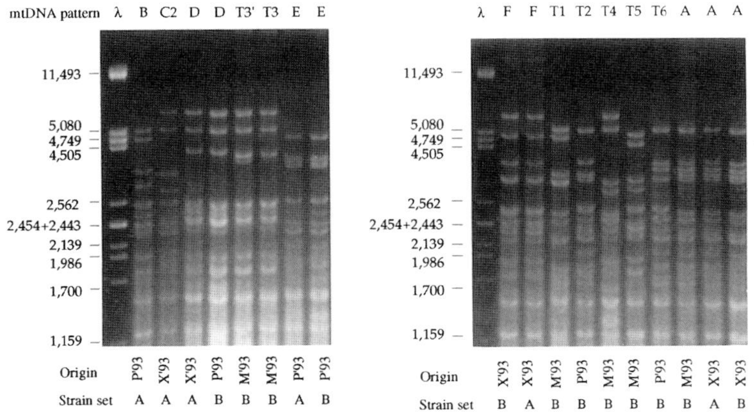

Fig. 7. *HinfI* restriction patterns of different strains from El Penedés. Numbers indicate the size in bp. The two gels show most mtDNA patterns found in different strains. Reprinted with permission from ref. [45].

wine yeasts could be explained by a similar process, so that ectopic recombination between repeat sequences is mostly repressed, and meiosis and sporulation are very scarce [36]. When meiosis takes place, chromosomal reorganization seems to be frequent, as if mechanisms repressing ectopic recombination were not active [66].

3.1.2.2. Reorganization of non essential sequences

Polymorphism in size and number of wine yeast chromosomes within populations belonging to the species *S. cerevisiae* is the rule rather than the exception. Hence, the concept of "genome plasticity" which refers to spontaneous reorganizations at a high frequency, but this phenomenon does not affect strain fitness dramatically [28]. In addition, a variable number of repeat sequences are present in subtelomeric regions, with no apparent disadvantage. Rather, in *est1* mutants the senescence caused by telomeric shortening is avoided by proliferation of Y' subtelomeric elements, which also protect adjacent unique sequences from heterochromatinization [52].

3.1.2.3. Favourable selection of reorganised chromosomal regions

Since most wine yeasts are polyploid or aneuploid, and most chromosomal reorganizations are non reciprocal, these phenomena result in strains heterozygous for the reorganization. The presence of more than one allele in the reorganized region may produce a better fitness to specific environments, or to colonize new environments not available to the wild type.

3.1.2.4. Apomixis

Apomictic strains are those, which have undergone mutations that, prevent them from completing either the first or the second meiotic division [75]. The resulting asci form diads instead of tetrads, with the two meiotic products possessing the same chromosomal karyotype and DNA content as the parental [76]. Most wine yeasts are aneuploid; the well known tolerance of *S. cerevisiae* strains to this condition of spontaneous loss or gain of chromosomes, so long as meiosis is not a regular part of their life cycle has been discussed above. In apomictic wine strains, if sporulation does occur, this mechanism reduces the loss of viability of meiotic products from aneuploid wine yeasts whose DNA content is below 2n [12]. It is noteworthy that the frequency of apomixis in wine yeasts is fairly high, about 10 to

15% [72], also that most of these strains possess more than one apomicitic recessive mutation [77, 78], so that reversion of one of them does not result in complete meiosis and therefore in lack of viability of the meiotic products in strains with a DNA content below 2n. Genes involved in apomixis are located in chromosome VIII, as if recombinations were restricted and these mutated genes have to be inherited as a single block [78]. Finally, the fact that mitochondria are very important for ethanol tolerance in wine yeasts [79], and that meiotic and apomictic sporulation displayed different dependency on mitochondrial function, meiotic sporulation being more strictly dependent, is worth mentioning [80].

3.1.2.5. Homothallism

Many of the *S. cerevisiae* wine yeasts are homothallic [72]. Meiotic products from homothallic strains undergo a mating type switch when it has been divided once. This change allows the cell to mate with cells from the same population (so that polymorphism of homologous chromosome size disappears), in order to double its DNA content (so that aneuploidies of uneven chromosomes also disappear); in addition, they become homozygous for all the genes, except the mating locus. Hence, homothallism of aneuploid wine strains make it possible for sporulation to occur, and so the meiotic products will have an even number of extra chromosomes, decreasing dramatically the frequency of non chromosomal disjunction and of spore mortality. In aneuploid wine yeasts and as a consequence of their being homothallic, the increase in viability for some strains went from 60 to 100% in successive generations of sporulation and analysis of the meiotic products [72]. In some cases, this viability increased from 9 to 90% in only three generations (Table 4) [30].

3.2. Variability of the mitochondrial genome

Variability in RFLP generated from the mtDNA has allowed the unequivocal identification of specific strains from some industrial yeast groups, mainly wine yeasts (Fig. 7) [81]. Bakers' strains cannot be identified by analyzing RFLP of their mtDNA because they displayed the same pattern [35]. One reason for this discrepancy could be the fact that wine elaboration is carried out by multiple yeast species and races. However, bakers' strains displaying high variability of their karyotype still had the same RFLP pattern of their mtDNA [35]. Furthermore, restriction maps of the 2 μm plasmid, of common *S. cerevisiae* strains, described as a prototypical selfish DNA molecule [81], also were produced. All strains were uniquely different when evaluated by their chromosomal patterns, whereas only a few 2 μm variants were defined [81]. An extensive study of wine yeast mitochondrial genomes indicated a high variability

Table 4. Genetic analysis of enological yeasts

Strains	Percentage of sporulation (20 days)	Spore viability (%) 1st Gener[a]	2nd Gener[b]	3rd Gener[b]	Thallism[c]
E-1	15.9	47.7 (11)	68.7	89.6	HOM
F-12	30.6	10.0 (20)	70.8	85.4	HOM
F-21	10.4	9.0 (45)	83.9	91.6	HOM/HET
F-23	16.6	23.2 (28)	31.2	41.6	HOM/HET
F-25	6.2	2.3 (22)	4.2	4.2	HOM?
G-1	8.3	0.0 (12)	<1.0	ND	HOM?

[a]The number of asci dissected is shown in parenthesis. [b]Obtained either by selection of clones with good growth and sporulation, and posterior dissection of asci (E-1, F-12 and F-21 strains), or by obtaining random spores through treatment with diethyl ether (F-23, F-25 and G-1 strains). [c]HOM: homothallic. HET: heterothallic. HOM/HET: heterozygous for the thallism. Modified from ref. [30].

observed in gene order, connected with the presence of long intergenic regions containing recombination sites. Gene order of *Saccharomyces uvarum* suggested that the mitochondrial genome of the *S. cerevisiae*-like strains may had evolved from an ancestral molecule, similar to that of *S. uvarum* through specific genome rearrangements [82]. Furthermore, differences in gene order were due to translocations, confirming the relevant role of gene rearrangements in the evolution of yeast mtDNA genomes. In several of these wine strains, hybridization, as well as the compatibility nucleus-mitochondria have been confirmed by strains possessing the same RFLP of their mtDNA but different chromosomal patterns, and the other way round. Successful restoration of respiration in *S. cerevisiae* petite mutants was achieved by transplacement of mitochondria isolated from *S. bayanus, S. capensis, Saccharomyces delbrueckii, Saccharomyces exiguus, Saccharomyces italicus* and *Saccharomyces oviformis* [83]. In addition, wine yeasts display high tolerance to ethanol and temperature, as well as to the mutagenic effect that ethanol exerts on mtDNA [84, 85], as compared to other yeast industrial groups [72, 79, 86]. It appears that, in some species, most of the genetic information originally located in mitochondria was translocated to the nuclear compartment during evolution. A given functional mitochondrial gene can be either mitochondrially or nuclearly encoded, but in general nuclear information can change the structural organization of mitochondrial genome [87], and mutants affected in nuclear genes which regulate mitochondrial functions, such as those which release glucose repression and control repression of mitochondrial biogenesis, have frequently been described [88]. In spite of such nuclear control, when mitochondria are transferred from wine to laboratory yeast strains they confer to the laboratory recipient strain a considerable increase in tolerance to ethanol, temperature and mitochondrial loss induced by ethanol, indicating that the mitochondrial genome is responsible for this tolerance [79, 86].

Results indicated that mitochondria from wine yeasts have undergone a selective process of adaptation to ethanol, and therefore, are highly tolerant to this compound. However, these mitochondria are permanently exposed to the mutagenic effect of ethanol and probably other compounds such as acetaldehyde [26]. Ethanol acts on mtDNA rather than on the nuclear genome [79, 86]. Furthermore, when yeast strains are incubated in the presence of increasing ethanol or acetaldehyde concentrations, there is a proportional increase in mutants defective in mitochondrial functions (petite mutants) (Fig. 8) [36, 72]. These petites displayed changes in the pattern of the restriction fragment length polymorphism (RFLP) of their mtDNA, followed by a total loss of mtDNA when incubations in ethanol are maintained for long periods (Fig. 8) [Castrejón, unpublished results]. It therefore seems that ethanol and acetaldehyde introduce changes in the mtDNA, and those changes, which give rise to non functional mitochondria lead to complete mtDNA loss. Changes in DNA sequence of the mtDNA could be an indirect effect of ethanol, which introduces DNA breaks as explained above. When this mtDNA is repaired, DNA polymerase introduces mistakes accounting for the observed changes in the RFLP. Due to the fact that mitochondrial DNA polymerase lacks proof-reading capability whereas nuclear DNA polymerase does possess this capacity, the preferential effect of ethanol on mitochondrial rather than on nuclear genome could be the result of a better system to repair breaks existing in the nucleus of the wine yeast cells [46]. Finally, differences of the industrial conditions could account for the differences found when analyzing mtDNA of baker or brewing yeasts (continuously selected under strong aerobic condition during growth in molasses, and invariable, because their mtDNA is hardly ever in the presence of ethanol) [35], and the mtDNA of wine yeasts (also strongly selected in the continuous presence of ethanol and acetaldehyde, but always in the presence of a highly mutagenic environment produced by these compounds) [12, 89]. In consequence, the mutagenic effect of ethanol and acetaldehyde is making the selective process, which would probably give rise to a unique RFLP pattern of the mtDNA.

4. DYNAMICS OF WINE YEAST POPULATIONS DURING MUST FERMENTATION AND WINE MATURATION

4.1. Wine Yeast Populations During Must Fermentation

Out of the multiple populations of microorganisms involved in must fermentation some workers have described the presence of bacteria until the ethanol concentration reached 10-

Fig. 8. Induction of petite mutants (%) of *S. cerevisiae* grown in YPD medium supplemented with increasing concentrations of either acetaldehyde (YPDAc) or ethanol (YPDE) (%). [Castrejón *et al*, unpublished]

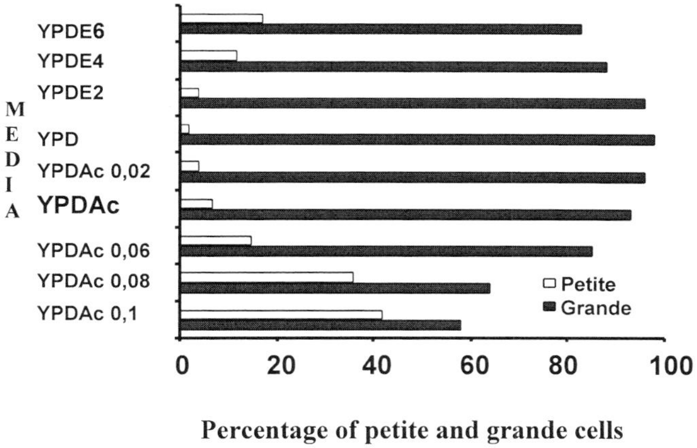

12%, the presence of filamentous fungi and yeasts of different species [90, 91], which are believed to exert a major influence on wine bouquet, being very high at 0 to 2% ethanol but decreasing dramatically above 4% ethanol. As for the yeast population, the osmophylic *Candida stellata* frequently isolated from grapes decreased at ethanol contents above 6-8%, *K. apiculata* present in half-ripe and ripe grapes diminish its concentration at concentrations of ethanol above 10%, *Rhodotorula rubra, Rhodotorula glutinis, Candida guillermondii,* and *Pichia membranaefaciens* at the early stages of fermentation, up to an ethanol content of 8%. Yeasts of the genera *Kloeckera, Candida, Pichia* and others described above eventually died out, leaving *S. cerevisiae* as the dominant species to complete the fermentation [91]. Further inoculation of musts with strains of *S. cerevisiae* selected among indigenous yeasts occurring naturally in the wine making area contributed to a better control of alcoholic fermentation of wines [91]. Furthermore, inoculation of *Torulaspora delbrueckii* and *S. cerevisiae* sequentially, the former with low acetogenic power and the latter with high alcoholgenic power, was beneficial for the production of wines with low volatile acidity [91]. By using molecular analysis, above all that of RFLP of the mtDNA with endonucleases, the dynamics of the natural or inoculated *S. cerevisiae* strains present in spontaneous or inoculated wine fermentation can be perfectly studied [91]. Population dynamics of natural and inoculated industrial wine fermentation are at present routinely studied by using this simple molecular biology technique based on mDNA restriction analysis profile [92].

In spontaneous fermentations, *Zygosaccharomyces bailii* was only isolated in the late stages of fermentation, particularly in wine with the highest alcohol contents (over 14%). This indicated its high tolerance to ethanol that has been related to singularities of its cytoplasmic membrane [90]. Ethanol mainly acts upon membrane structural integrity and membrane permeability. The most important targets of ethanol action have been described to be the yeast cell plasma membrane, mitochondrial membranes, nuclear membranes, vacuolar membranes, endoplasmic reticulum and cytosolic proteins [1]. Wine *S. cerevisiae* yeasts are more resistant than laboratory or other industrial *S. cerevisiae* strains. The physiological responses of wine yeasts to ethanol include alteration in membrane composition as the first adaptive mechanism. It has been proposed that wine *S. cerevisiae* yeasts adapt their membranes to ethanol by changing their composition, but this adaptation might not result in an enhancement of parameters such as growth or fermentation, because the overall kinetics of the inhibition of these processes by ethanol depends upon a number of additional underlying mechanisms different from membrane bound-processes [72]. Acidification curves, which proved to be a reliable measurement of the degree of ethanol interference with cell membranes, allowed to test the ability of the cell membrane to resist ethanol in high and low ethanol tolerance yeasts. Results indicated that highly ethanol tolerant strains were able, after growth in the presence of ethanol to effectively improve the ethanol tolerance of its membrane, and that protein components of the cell membranes played an important role in this ethanol tolerance [72]. This phenomenon also occurs during adaptation and flor formation of wine *S. cerevisiae* strains to the high ethanol concentrations of sherry (over 15%) [12-26].

Differences in membrane fluidity have also been related to mitochondrial activity. Furthermore, high temperatures and inhibitory ethanol concentrations have a lethal effect on yeast cells and it has been proposed that these factors target on the mitochondria [79, 85], so that maintenance of functional mitochondria in wine strains give rise to increased viability under extreme conditions of ethanol and temperature. Thus, wine *S. cerevisiae* strains manifest simultaneously high tolerance to ethanol, thermotolerance and resistance to the mutagenic effects of ethanol on the mitochondrial genome, and this genome influences the above parameters. Enhanced mitochondrial superoxide dismutase activity in wine strains has been proposed to ethanol induced free radical synthesis [1]. Because this enzyme is encoded by a nuclear gene, additional mechanisms encoded in the mitochondrial DNA have to be involved in ethanol tolerance as well.

4.2. Wine Yeast Populations During Post-fermentation and Wine Maturation

After fermentation, some wines are maturated and/or subjected to a second fermentation, with the involvement in most cases of different *S. cerevisiae* strains. For instance, during the elaboration of champagne style wines, the champagne base wine is supplemented with different concentrations of sugar, inoculated with selected yeast strains and bottled, so that a second fermentation takes place. This system in used in the production of cava, a sparkling wine produced according to the rules of the cava appellation in Catalonia and other regions of Spain, and which involves growth, fermentation and death of yeasts [93]. In this case, strains are selected according to their ethanol tolerance, ability to flocculate and produce specific aromatic characteristics [93] and above all, tolerance to the high CO_2 pression which will originated during the second fermentation (5 to 6 atmospheres) [94]. In *S. cerevisiae*, an extreme pressure of 1500 atmospheres causes severe damage to membrane integrity resulting in cell death [95]. Hydrostatic pressure promotes the acidification of vacuoles, caused by the production of carbon dioxide, and trehalose plays a role in cell survival under these conditions [95]. Other desirable features include ability to ferment at low temperatures, capacity for foam formation, and autolysis [94]. Selected strains have been characterized as

S. cerevisiae (*oviformis*). Surprisingly, these selected strains displayed one or two very similar RFLP of their mtDNA, whereas their karyotypes differed [70]. Whereas chromosomal polymorphism was attributed to ectopic recombination occurring between repeat sequences during vegetative growth, the only explanation for the lack of polymorphism of mtDNA has to be a strong selection of these mitochondria and a minor mutagenic effect produced by ethanol.

With regards to yeast strains involved in biological ageing of wine, Martínez et al. [12] described *S. cerevisiae* strains assigned to four different races, *beticus, montuliensis, cheresiensis* and *rouxii*, after using molecular and physiological techniques to classified those yeast strains which form velum on the surface of sherry wine. Furthermore, different isolated populations belonging to the same yeast race displayed a great variability in the RFLP of their mtDNA (more than 85% of the isolates present a different pattern), although the karyotypes were clearly different from that of laboratory strains but with scarce diversity within different isolates (about 15% of the isolates displayed differences in the karyotype) (Fig. 1). Other authors [36], applying molecular and physiological analyses, identified six different flor strains, some of which differed in their molecular karyotype but yielded identical mtDNA restriction patterns, whereas others showed distinguishable mitochondrial genomes but exhibited identical karyotype. This was interpreted as the different strains identified being very closely related. Surprisingly, colonies isolated from individual barrels corresponded in most cases to a single type of strain, different in different barrels, which was the same over a period of at least two years, indicating that the dominant strain was quite stable [36, 89]. The rate of film formation was the main feature in the determination of fitness on sherry wine, influencing the ability to dominate in flor yeast populations. These strains which form velum on the surface of sherry wine grow and survive in the presence of high concentrations of ethanol as an adaptive mechanism that consists in changes in cell hydrophobicity which allow the population to float, the process depending on the synthesis of hydrophobic proteins rather than on lipids [23-25, 96]. However, during wine ageing, yeasts consume some compounds (ethanol, proline) and produce others like acetaldehyde [26]. When studying yeast populations, variations in the frequency of the four races described above were found. Thus, *S. cerevisiae var. beticus,* which was faster at forming velum predominated in younger wines, whereas *S. cerevisiae var. montuliensis*, which produced and resisted higher acetaldehyde concentrations appeared at later stages [23-25]. Initially, *S. cerevisiae var. beticus* was favourably selected against other races in the velum. The dynamic procedure carried out in the cellars (the inoculation of velum from other butts into those where the velum is deteriorated) may had facilitated the new introduction of *var. montuliensis* in the population.

A different source of diversity is being introduced by mankind by the isolation of recombinant wine strains, which displayed heterologous gene expression, sometimes under the control of specific promoters able to work during or at the end of wine fermentation [97]. These new strains produce enzymes which favour the liberation of aromatic compounds [97-99], higher yields of glycerol [100-102], less volatile acidity [103] or better antimicrobial activity [104].

5. CONCLUSIONS

Although *S. cerevisiae* wine yeasts displayed polymorphism of their chromosomes, under extreme conditions (sherry wines with over 15% ethanol) strains displayed an almost unique chromosomal pattern whereas RFLP of their mtDNA was highly variable. It seems that an optimal karyotype has been selected in these strains. The high concentrations of ethanol act preferentially on mtDNA, introducing variations in their mtDNA which do not allow a unique pattern in this case. Although variability with regards to chromosome size and number occurs both at mitosis and meiosis, it seems to occur with higher frequency at meiosis, and to be mediated by Ty, above all Ty1 and Y' repeat sequences. This variability has produced amplification of specific genes or chromosomes, as occurs in wine yeasts witch chromosome XIII. Together with these mechanisms, mutagenic effects of ethanol and chromosomal reorganizations mediated by Ty or Y' elements, recombinant DNA technology has allowed the introduction of wine strains with new genotypes. Molecular analysis of the population dynamics will reveal whether or not these strains are able to predominate over the indigenous flora, and how they will evolve.

Acknowledgements. This study was supported by CICYT project numbers AGL2000-0524, FD97-0820, PTR940022 and PTR95-0198, and Junta de Andalucía PAI CVI-107

6. REFERENCES

1. Pretorius, I. S. (2000). Tailoring wine yeast for the new millennium: novel approaches to the ancient art of winemaking. Yeast 16: 675-729.
2. Barnett, J. A. (1992). The Taxonomy of the genus *Saccharomyces.* Meyen ex Reess: a short review for non-taxonomists. Yeast 8: 1-23.
3. Henick-Kling, T. (1988). Yeast and bacterial control in winemaking. In: Wine Analysis. Modern Methods of Plant Analysis. Linskens, H.F., Jackson, J.F., Eds. New Series Vol.6. Springer-Verlag, Berlin, pp. 279-316.
4. Kamel, B. S. and Stauffer, C. E. (1993). In: Advances in Baking Technology. Blackie, Academical & Professional, London.
5. Esteve- Zarzoso, B., Manzanares, P., Ramón, D. and Querol, A. (1998). The role on non-*Saccharomyces* yeasts in industrial winemaking. Internatl. Microbiol. 1: 143-148.
6. Deak, T. (1991). Foodborne yeasts. Adv. Appl. Microbiol. 36: 179-278.
7. Benítez, T., Castrejón, F., Gasent-Ramírez, J. M. and Codón, A. C. (1996). Development of new strains for the food industry. Biotechnol. Prog. 12: 149-163.
8. Querol, A., Barrio, E., Huerta, T. and Ramón, D. (1992). Molecular monitoring of wine fermentation conducted by active dry yeast strains. Appl. Environ. Microbiol. 58: 2948-2953.
9. Querol, A., Barrio, E. and Ramón, D. (1992). A comparative study of different methods of yeast strain characterization. System. Appl. Microbiol. 15: 439-446.
10. Querol, A., Huerta, T., Barrio, E. and Ramón, D. (1992). Dry yeast strain for use in fermentation of Alicante wines: selection and DNA patterns. J. of Food Sci. 57: 183-186.
11. Querol, A., Jiménez, M. and Huerta, T. (1990). Microbiological and enological parameters during fermentation of musts from poor and normal grape-harvests in the region of Alicante (Spain). J. of Food Sci. 55: 1603-1606.
12. Martínez, P., Codón, A. C., Pérez, L. and Benítez, T. (1995). Physiological and molecular characterization of flor yeasts: polymorphism of flor yeast populations. Yeast 11: 1399-411.
13. Fernández-López, J. A., Almela, L. and Carreño, J. (1997). Maduración de la uva. Investigación y Ciencia junio: 40-42.
14. Cansado, J., Longo, E., Agrelo, D. and Villa, T. G. (1989). Yeasts associated with spontaneous fermentation processes in wines from Ribeiro. Analysis of homo/heterothallism and the killer system of *S. cerevisiae* strains. Microbiologia 5: 79-88.
15. Mateo, J. J., Jiménez, M., Huerta, T. and Pastor, A. (1991). Contribution of different yeasts isolated from musts of monastrell grapes to the aroma of wine. Int. J. Food Microbiol. 14: 153-160.
16. Mora, J., Barbas, J. I. and Mulet, A. (1990). Evolution of the yeast microflora during the first days of fermentation in inoculated Majorcan musts. Microbiologia 6: 65-70.
17. Iñigo, B. and Bravo, F. (1993). Vinificación ecológica industrial como proceso fermentativo de biología múltiple. Alimentación, equipos y tecnología marzo: 85-86.
18. Salvadores, M. P., Diaz, M. E. and Cardell, E. (1993). Autochthonous yeasts isolated in Tenerife wines and their influence on ethyl acetate and higher alcohol concentrations analyzed by gas chromatography. Microbiologia 9: 107-112.
19. Martínez, J., Toledano, F., Millán, C. and Ortega, J. M. (1990). Development of alcoholic fermentation in non-sterile musts from "Pedro Ximenez" grapes inoculated with pure cultures of selected yeasts. Food Microbiol. 7: 217-225.
20. Moreno, J. J., Millán, C., Ortega, J. M. and Medina, M. (1991). Analytical differentiation of wine fermentation using pure and mixed yeast cultures. J. Ind. Microbiol. 7: 181-190.
21. Zeyl, C. (2000). Budding yeast as a model organism for population genetics. Yeast 16: 773-784.
22. Guijo, S., Millán, C. and Ortega, J. M. (1986). Fermentative features of vinification and maturation yeasts

isolated in the Montilla-Moriles region of Southern Spain. Food Microbiol. 3: 133-142.
23. Martínez, P., Pérez, L. and Benítez, T. (1997). Velum formation by flor yeasts isolated from sherry wine. Am. J. Enol. Vitic. 48: 55-62.
24. Martínez, P., Peréz-Rodríguez, L. and Benítez, T. (1997). Factor with affect velum formation by flor yeast isolated from sherry wine. System. Appl. Microbiol. 20: 154-157.
25. Martínez, P., Pérez-Rodríguez, L. and Benítez, T. (1997). Evolution of flor yeast population during the biological ageing of fino sherry wine. Am. J. Enol. Vitic. 48: 160-168.
26. Martínez, P., Valcarcel, M. J., Pérez, L. and Benítez, T. (1998). Metabolism of *Saccharomyces cerevisiae* flor yeasts during fermentation and biological ageing of fino sherry: by-products and aroma compounds. Am. J. Enol. Vitic. 49: 240-250.
27. Naumov, G. I., Naumova, E. S., Lantto, R. A., Louis, E. J. and Korhola, M. (1992). Genetic homology between *Saccharomyces cerevisiae* and its sibling species *S. paradoxus* and *S. bayanus*: electrophoretic karyotypes. Yeast 8: 599-612.
28. Adams, J., Puskas-Rozsa, S., Simlar, J. and Wilke, C.M. (1992). Adaptation and major chromosomal changes in populations of *Saccharomyces cerevisiae*. Curr. Genet. 22: 13-19.
29. Codón, A. C., Benítez, T. and Korhola, M. (1998). Chromosomal polymorphism and adaptation to specific industrial environments of *Saccharomyces* strains. Appl. Microbiol. Biotechnol. 49: 154-163.
30. Guijo, S., Mauricio, J. C., Salmon, J. M. and Ortega, J. M. (1997). Determination of the relative ploidy in different *Saccharomyces cerevisiae* strains used for fermentation and 'flor' film ageing of dry sherry-type wines. Yeast 13: 101-117.
31. Bakalinsky, A. T. and Snow, R. (1990). The chromosomal constitution of wine strains of *Saccharomyces cerevisiae*. Yeast 6: 367-382.
32. Naumov, G. I., Naumova, E. S. and Michels, C. A. (1994). Genetic variation of the repeated MAL loci in natural populations of *Saccharomyces cerevisiae* and *Saccharomyces paradoxus*. Genetics 136: 803-812.
33. Naumov, G. I., Naumova, E. S. and Sancho, E. D. (1994). Sibling species of the *Saccharomyces* sensu stricto complex in Spain. Microbiologia 10: 403-412.
34. Venzihet, F., Blondin, B. and Hallet, J. N. (1990). Chromosomal DNA patterns and mitochondrial DNA polymorphism as tools for identification of enological strains of *Saccharomyces cerevisiae*. Appl. Microbiol. Biotechnol. 32: 568-571.
35. Codón, A. C. and Benítez, T. (1995). Variability of the physiological features and of the nuclear and mitochondrial genomes of bakers' yeasts. System. Appl. Microbiol. 18: 343-352.
36. Ibeas, J. I. and Jimenez, J. (1996). Genomic complexity and chromosomal rearrangements in wine-laboratory yeast hybrids. Curr. Genet. 30: 410-416.
37. Roeder, G. S., Farabaugh, P. J., Chaleff, D. T. and Fink, G. R. (1980). The origins of gene instability in yeast. Science 209: 1375-1380.
38. Sancho, E. D., Hernández, E. and Rodríquez-Navarro, A. (1986). Presumed sexual isolation in yeast populations during production of sherrylike wine. Appl. Environ. Microbiol. 51: 395-397.
39. Codón, A. C., Benítez, T. and Korhola, M. (1997). Chromosomal reorganization during meiosis of *Saccharomyces cerevisiae* baker's yeasts. Curr. Genet. 32: 247-259.
40. Praekelt, U. M. and Meacock, P. A. (1992). MOL1, a *Saccharomyces cerevisiae* gene that is highly expressed in early stationary phase during growth on molasses. Yeast 8: 699-710.
41. Naumov, G. I., Naumova, E. S., Sancho, E. D. and Korhola, M. (1996). Polymeric SUC genes in natural populations of *Saccharomyces cerevisiae*. FEMS Microbiol. Lett. 135: 31-35.
42. Ness, F. and Aigle, M. (1995). RTM1: a member of a new family of telomeric repeated genes in yeast. Genetics 140: 945-956.
43. Gasent-Ramírez, J. M., Castrejón, F., Querol, A., Ramón, D. and Benítez, T. (1999). Genomic stability of *Saccharomyces cerevisiae* baker's yeasts. Syst. Appl. Microbiol. 22: 329-340.
44. Teunissen, A. W. and Steensma, H. Y. (1995). Review: the dominant flocculation genes of *Saccharomyces cerevisiae* constitute a new subtelomeric gene family. Yeast 11: 1001-1013.
45. Nadal, D., Colomer, B. and Pina, B. (1996). Molecular polymorphism distribution in phenotypically distinct populations of wine yeast strains. Appl. Environ. Microbiol. 62: 1944-1950.
46. Louis, E. J. (1995). The chromosome ends of *Saccharomyces cerevisiae*. Yeast 11: 1553-1573.
47. Aguilera, A. and Klein, H. L. (1993). Chromosome aberrations in simpler eukaryotes. In: The Causes and Consequences of Chromosomal Aberrations. Kirsch, I. R. (Ed.): CRC Press, Boca Raton, pp. 51-90.
48. Zolan, M. E. (1995). Chromosome-length polymorphism in fungi. Microbiol. Rev. 59, 686-698.
49. Casaregola, S., Nguyen, H. V., Lepingle, A., Brignon, P., Gendre, F. and Gaillardin, C. (1998). A family of laboratory strains of *Saccharomyces cerevisiae* carry rearrangements involving chromosomes I and III. Yeast 14: 551-564.
50. Seoighe, C. and Wolfe, K. H. (1998). Extent of genomic rearrangement after genome duplication in yeast. Proc. Natl. Acad. Sci. U S A 95: 4447-4452.

51. Finnegan, D. J. (1989). Eukaryotic transposable elements and genome evolution. Trends. Genet. 5: 103-107.
52. Chen, C. and Kolodner, R. D. (1999). Gross chromosomal rearrangements in *Saccharomyces cerevisiae* replication and recombination defective mutants. Nat. Genet. 23: 81-85.
53. Louis, E. J., Naumova, E. S., Lee, A., Naumov, G. and Haber, J. E. (1994). The chromosome end in yeast: its mosaic nature and influence on recombinational dynamics. Genetics 136: 789-802.
54. Kupiec, M. and Petes, T. D. (1988). Allelic and ectopic recombination between Ty elements in yeast. Genetics 119: 549-559.
55. Kupiec, M. and Petes, T. D. (1988). Meiotic recombination between repeated transposable elements in *Saccharomyces cerevisiae*. Mol. Cell. Biol. 8: 2942-2954.
56. Aguilera, A., Chavez, S. and Malagon, F. (2000). Mitotic recombination in yeast: elements controlling its incidence. Yeast 16: 731-754.
57. Salyers, A. A., Shoemaker, N. B., Stevens, A. M. and Li, L. Y. (1995). Conjugative transposons: an unusual and diverse set of integrated gene transfer elements. Microbiol. Rev. 59: 579-590.
58. Louis, E. J. and Haber, J. E. (1990). The subtelomeric Y' repeat family in *Saccharomyces cerevisiae*: an experimental system for repeated sequence evolution. Genetics 124: 533-545.
59. Turakainen, H., Kristo, P. and Korhola, M. (1994). Consideration of the evolution of *the Saccharomyces cerevisiae* MEL gene family on the basis of the nucleotide sequences of the genes and their flanking regions. Yeast 10: 1559-1568.
60. Gromadka, R., Gora, M., Zielenkiewicz, U., Slonimski, P. P. and Rytka, J. (1996). Subtelomeric duplications in *Saccharomyces cerevisiae* chromosomes III and XI: topology, arrangements, corrections of sequence and strain-specific polymorphism. Yeast 12: 583-591.
61. Yoda, A., Kontani, T., Ohshiro, S., Yagishita, N., Hisatomi, T. and Tsuboi, M. (1993). DNA-length polymorphisms of chromosome III in the yeast *Saccharomyces cerevisiae*. J. Fermen. Bioeng. 75: 395-398.
62. Turakainen, H., Aho, S. and Korhola, M. (1993). MEL gene polymorphism in the genus *Saccharomyces*. Appl. Environ. Microbiol. 59: 2622-2630.
63. Naumov, G., Naumova, E., Turakainen, H., Suominen, P. and Korhola, M. (1991). Polymeric genes MEL8, MEL9 and MEL10--new members of alpha-galactosidase gene family in *Saccharomyces cerevisiae*. Curr. Genet. 20: 269-276.
64. Klar, A. J. (1993). Lineage-dependent mating-type transposition in fission and budding yeast. Curr. Opin. Genet. Dev. 3: 745-751.
65. Naumov, G., Turakainen, H., Naumova, E., Aho, S. and Korhola, M. (1990). A new family of polymorphic genes in *Saccharomyces cerevisiae*: alpha-galactosidase genes MEL1-MEL7. Mol. Gen. Genet. 224: 119-128.
66. Jäger, D., Gysler, C. and Niederberger, P. (1992). Aneuploidy events in meiotic progeny of *S. cerevisiae*. In 16th Conference on Yeast Genetics and Molecular Biology, Vienna, pp. S640.
67. Gangloff, S., Zou, H. and Rothstein, R. (1996). Gene conversion plays the major role in controlling the stability of large tandem repeats in yeast. Embo. J. 15: 1715-1725.
68. Esposito, M. S. and Bruschi, C. V. (1993). Diploid yeast cells yield homozygous spontaneous mutations. Curr. Genet. 23: 430-434.
69. Devin, A. B., Prosvirova, T., Peshekhonov, V. T., Chepurnaya, O. V., Smirnova, M. E., Koltovaya, N. A., Troitskaya, E. N. and Arman, I. P. (1990). The start gene CDC28 and the genetic stability of yeast. Yeast 6: 231-243.
70. Nadal, D., Carro, D., Fernández-Larrea, J. and Piña, B. (1999). Analysis and dynamics of the chromosomal complements of wild sparkling-wine yeast strains. Appl. Environ. Microbiol. 65: 1688-1695.
71. Mortimer, R. K., Romano, P., Suzzi, G. and Polsinelli, M. (1994). Genome renewal: a new phenomenon revealed from a genetic study of 43 strains of *Saccharomyces cerevisiae* derived from natural fermentation of grape musts. Yeast 10: 1543-1552.
72. Jiménez, J. and Benítez, T. (1987). Adaptation of yeast cell membrane to ethanol. Applied. Environ. Microbiol. 53: 1196-1198.
73. Puig, S., Querol, A., Barrio, E. and Perez-Ortin, J. E. (2000). Mitotic recombination and genetic changes in *Saccharomyces cerevisiae* during wine fermentation. Appl. Environ. Microbiol. 66: 2057-2061.
74. Ristow, H., Seyfarth, A. and Lochmann, E. R. (1995). Chromosomal damages by ethanol and acetaldehyde in *Saccharomyces cerevisiae* as studied by pulsed field gel electrophoresis. Mutat. Res. 326: 165-170.
75. Mitchell, A. P. (1994). Control of meiotic gene expression in *Saccharomyces cerevisiae*. Microbiol. Rev. 58: 56-70.
76. Hugerat, Y. and Simchen, G. (1993). Mixed segregation and recombination of chromosomes and YACs during single-division meiosis in spo13 strains of *Saccharomyces cerevisiae*. Genetics 135: 297-308.
77. Atcheson, C. L. and Esposito, R. E. (1993). Meiotic recombination in yeast. Curr. Opin. Genet. Dev. 3: 736-744.
78. Klapholz, S. and Esposito, R. E. (1980). Isolation of SPO12-1 and SPO13-1 from a natural variant of yeast

that undergoes a single meiotic division. Genetics 96: 567-588.
79. Jiménez, J. and Benítez, T. (1988). Yeast cell viability under conditions of high temperature and ethanol concentrations depends on the mitochondrial genome. Curr. Genet. 13: 461-469.
80. Marmiroli, N. and Bilinski, C. A. (1985). Partial restoration of meiosis in an apomictic strain of *Saccharomyces cerevisiae*: a model system for investigation of nucleomitochondrial interactions during sporulation. Yeast 1: 39-47.
81. Rank, G. H., Casey, G. P., Xiao, W. and Pringle, A. T. (1991). Polymorphism within the nuclear and 2 micron genomes of *Saccharomyces cerevisiae*. Curr. Genet. 20: 189-194.
82. Cardazzo, B., Minuzzo, S., Sartori, G., Grapputo, A. and Carignani, G. (1998). Evolution of mitochondrial DNA in yeast: gene order and structural organization of the mitochondrial genome of *Saccharomyces uvarum*. Curr. Genet. 33: 52-59.
83. Osusky, M., Kissova, J. and Kovac, L. (1997). Interspecies transplacement of mitochondria in yeasts. Curr. Genet. 32: 24-26.
84. Aguilera, A. and Benítez, T. (1985). Role of mitochondria in ethanol tolerance of *Saccharomyces cerevisiae*. Arch. Microbiol. 142: 389-392.
85. Aguilera, A. and Benítez, T. (1989). Synergistic effects of ethanol and temperature on yeast. Curr. Microbiol. 18: 179-188.
86. Jiménez, J., Longo, E., and Benítez, T. (1988). Induction of petite yeast mutants by membrane-active agents. Appl. Environ. Microbiol. 54: 3126-3132.
87. Hartmann, C., Henry, Y., Tregear, J. and Rode, A. (2000). Nuclear control of mitochondrial genome reorganization characterized using cultured cells of ditelosomic and nullisomic-tetrasomic wheat lines. Curr. Genet. 38: 156-162.
88. Brown, T. A., Evangelista, C. and Trumpower, B. L. (1995). Regulation of nuclear genes encoding mitochondrial proteins in *Saccharomyces cerevisiae*. J. Bacteriol. 177: 6836-6843.
89. Ibeas, J. I., Fernández-Pérez, M., Jiménez, J., Lozano-Romero, J. I. and Perdigones-Fernández, F. (1995). Evolución de la población de levadura durante la fermentación espontánea e inoculada de mostos del marco de Jerez. Alimentación, equipos y tecnología marzo: 45-49.
90. Martínez, J., Millán, C. and Ortega, J. M. (1989). Growth of natural flora during the fermentation of inoculated musts from "Pedro Ximenez" grapes. S. Afr. J. Enol. Vitic. 10: 31-35.
91. Querol, A., Barrio, E. and Ramón, D. (1994). Population dynamics of natural *Saccharomyces* strains during wine fermentation. Int. J. Food Microbiol. 21: 315-323.
92. Toledano, F., Martínez, J., Moreno, J., Ortega, J. M. and Medina, M. (1991). Utilisation sequentielle de deux levures pour la fermentation de mouts destines a la production de vins du type sherry. Cerevisiae and Biotechnol. 1: 35-41.
93. Ramón, D., González-Candelas, L., Pérez-González, J. A., González, R., Ventura, L., Sánchez-Torres, P., Valles, S., Pinaga, F., Gallego, M. V. and Fernández-Espinar, M. T. (1995). Applications of molecular biology in the wine industry. Microbiologia 11: 67-74.
94. Bartra, E. (1995). Microbiological aspects of sparkling wine processing. Microbiologia 11: 43-50.
95. Barcenilla, J. (1993). Microorganismos en la elaboración de espumosos. Alimentación equipos y tecnología marzo: 73-75.
96. Abe, F., Kato, C. and Horikoshi, K. (1999). Pressure-regulated metabolism in microorganisms. Trends Microbiol. 7: 447-453.
97. Martínez, P., Valcarcel, M. J., González, P., Benítez, T. and Pérez, L. (1993). Consumo de etanol, glicerina y amionoácidos totales en vinos finos durante la crianza biológica bajo "velo de flor". Alimentación equipos y tecnología marzo: 61-65.
98. Puig, S., Ramón, D. and Pérez-Ortín, J. E. (1998). Optimized method to obtain stable food-safe recombinant wine yeast strains. J. Agric. Food Chem. 46: 1689-1693.
99. King, A. and Richard Dickinson, J. (2000). Biotransformation of monoterpene alcohols by *Saccharomyces cerevisiae*, *Torulaspora delbrueckii* and *Kluyveromyces lactis*. Yeast 16: 499-506.
100. Villanueva, A., Ramon, D., Valles, S., Lluch, M. A. and MacCabe, A. P. (2000). Heterologous expression in *Aspergillus nidulans* of a *Trichoderma longibrachiatum* endoglucanase of enological relevance. J. Agric. Food Chem. 48: 951-957.
101. Eustace, R. and Thornton, R. J. (1986). Selective hybridization of wine yeasts for higher yields of glycerol. Can. J. Microbiol. 33: 112-117.
102. Michnick, S., Roustan, J. L., Remize, F., Barre, P. and Dequin, S. (1997). Modulation of glycerol and ethanol yields during alcoholic fermentation in *Saccharomyces cerevisiae* strains overexpressed or disrupted for GPD1 encoding glycerol 3-phosphate dehydrogenase. Yeast 13: 783-793.
103. Remize, F., Roustan, J. L., Sablayrolles, J. M., Barre, P. and Dequin, S. (1999). Glycerol overproduction by engineered *Saccharomyces cerevisiae* wine yeast strains leads to substantial changes in by-product formation and to a stimulation of fermentation rate in stationary phase. Appl. Environ. Microbiol. 65: 143-

149.
104. Willians, S. A., Hodges, R. A., Strike, T. L., Snow, R. and Kunkee, R. E. (1984). Cloning the gene for the malolactic fermentation of wine from *Lactobacillus delbruekii* in *Escherichia coli* and yeasts. Appl. Environ. Microbiol. 47: 288-293.
105. Ahmed, A., Sesti, F., Ilan, N., Shih, T. M., Sturley, S. L. and Goldstein, S. A. (1999). A molecular target for viral killer toxin: TOK1 potassium channels. Cell 99: 283-291.

Applied Mycology and Biotechnology
Volume 2. Agriculture and Food Production

Fungal Carotenoids

Carlos Echavarri-Erasun[a] and Eric A. Johnson[b]
[a]Department of Food Microbiology and Toxicology, and [b]Department of Food Microbiology & Toxicology and Bacteriology, Food Research Institute, University of Wisconsin, 1925 Willow Dr., Madison, WI 53706, USA (E-mail: eajohnso@facstaff.wisc.edu).

Carotenoids are naturally occurring pigments formed through the isoprenoid pathway that impart attractive color to animals and plants and also have essential biological functions as antioxidants, membrane stabilizers, and as precursors to essential metabolites such as vitamin A. They are biosynthesized as hydrocarbons (carotenes) and as oxygenated derivatives of carotenes (xanthophylls). Various fungi produce carotenoids of biotechnological interest including β-carotene, lycopene, and astaxanthin (see appendix for systematic names). Compared to other groups of organisms that have been evaluated including animals, terrestrial plants, and prokaryotes, several fungal groups have the ability to produce and accumulate intracelullarly remarkably high levels of carotenoids, a property that they share with certain microalgae. Thus, fungi and microalgae appear most promising for industrial production of carotenoids. *Escherichia coli*, other eubacteria, archaebacteria, and plants have also been investigated as heterologous hosts for carotenoid production, and they have been shown to express a variety of carotenogenic genes, but these organisms accumulate relatively low levels of carotenoids compared to microalgal and fungal systems. Structural genes responsible for carotene synthesis have been cloned and characterized from a number of carotenogenic organisms, but the enzymes carrying out biosynthesis of carotenoids as well as the regulatory genes and regulatory circuits controlling synthesis have been poorly characterized. The role of carotenoids in preventing or alleviating human and animal disease, in enhancing immune responses, and the possible role of carotenoids in slowing the aging process in humans and animals has stimulated interest in the biology of carotenoids. Research in the production of carotenoids through fermentation biotechnology is accelerating with consumers' desire for naturally produced foods and because of scarcity and variability of agricultural sources of certain carotenoids.

1. INTRODUCTION

Several filamentous fungi and yeasts are prolific producers of a distinctive variety of carotenoid pigments [1-3]. Several of the carotenoids produced by fungi such as β-carotene, lycopene, canthaxanthin, cryptoxanthin and astaxanthin are important industrially as components of animal feeds for coloration and as precursors of vitamin A and other functional carotenoid metabolites in plants and animals including humans. Some industrially important carotenoids such as bixin and related carotenoids (obtained commercially by extraction of annatto seeds, *Bixa orelana*), lutein (mainly from marigold, *Tagetes erecta*) flowers and seeds and zeaxanthin from corn (*Zea mays*) and other plant and bacterial sources are not known to be produced by fungi, but when the genes and enzymes are elucidated, fungi could serve as heterologous hosts for production of these industrially-important carotenoids. Heterologous production of carotenoids in bacteria, fungi, and plants has attracted considerable interest in recent years as the genetic basis of carotenogenesis is gradually elucidated [4-10]. In general, fungi produce higher levels of carotenoids than are produced in other tested heterologous hosts including *Escherichia coli,* other eubacteria, archaebacteria, and plants [11-13], although plants have the desirable feature over fungi in that they could be cultivated by relatively untrained personnel in countries where vitamin A deficiency is a leading cause of related morbidity and mortality including blindness and premature deaths [12-15].

Scientific interest in carotenoids has also increased in recent years because of their antioxidant activity of certain carotenoids *in vitro* and *in vivo* [16-23] and the potential of carotenoids to prevent chronic diseases associated with reactive oxygen species (ROS) including certain cancers, macular degeneration and cataract formation, cardiovascular disease, certain infections and other maladies associated with aging [18, 23-30]. Carotenoids have also been postulated to enhance immune responses in animals and humans, to alter gene regulation in humans and animals, to induce apoptosis in human cancer cell lines, and to have other biological effects in humans and animals [16, 31-33]. Although epidemiological evidence supports the involvement of mixtures of carotenoids and other phytochemicals in having beneficial roles in human health and disease, studies using individual carotenoids have given confounding results [24, 27, 29], and much more research is needed to evaluate their biological functions and clinical utility.

Although most carotenoids used industrially are manufactured using chemical synthesis or prepared as agricultural extracts, there is much interest in biotechnological production due to consumer demand for high quality and "natural" food additives, environmental concerns associated with chemical manufacture and solvent extraction of agricultural residues, and the political situation in certain areas of the world that could restrict access to agricultural sources [34]. The production of carotenoids by fungal biotechnology, however, is limited by lack of knowledge of the genetics and biosynthesis of carotenoids in biological systems. Generally, biological production of carotenoids is less efficient and more costly processes than routes using chemical synthesis or extraction of agricultural sources. However, biotechnological production has become more feasible in the past 5 years as the molecular biology of isoprenoid biosynthesis is elucidated and the genetic engineering of microbial production systems are developed.

Fungi could serve as excellent model systems for the manufacture of carotenoids as well as the elucidation of their functions in eukaryotes. Phylogenetic studies have indicated that some groups of animals and fungi share common biological features and animals appear to be more closely related to fungi than they are to plants and a wide variety of protists [35, 36]. Fungi possess eukaryotic features found in humans such as the presence of organelles, related regulatory circuitry, and secretory and trafficking systems. The biochemistry, physiology, and genetics of some fungi have become highly advanced and offer a simpler system for evaluation of carotenoid function than animal models or mammalian cell culture systems. In light of the availability of its entire genome sequence [37, 38], the noncarotenogenic yeast *Saccharomyces cerevisiae* could provide an ideal heterologous system for investigation of carotenoid synthesis and function. Genetic systems have also been elucidated in carotenogenic fungi and yeasts [39-42] and these could also become excellent systems in which to study carotenogenesis.

2. BIOLOGICAL DIVERSITY AND CAROTENOIDS

Biological diversity of microorganisms is reflected in molecular diversity, including synthesis of secondary metabolites [43]. Different phyla of organisms differ markedly in synthesis of various classes of secondary metabolites including carotenoids [1-3, 44] (see Table 1). Carotenoids occur universally in photosynthetic organisms but sporadically in nonphotosynthetic bacteria and eukaryotes. Some prokaryotes synthesize carotenoids of diverse structures not found in algae, plants, and fungi, including C_{30}, C_{45} and C_{50} carotenoids as well as carotenoids containing aromatic, glycosidic, and ester groups [2, 45-49]. Certain carotenogenic bacteria synthesize carotenoids of current industrial importance including β-carotene, lycopene, astaxanthin, canthaxanthin, zeaxanthin, and others.

E. coli can produce carotenoids by introduction of genes and heterologous expression, although they generally do not produce the carotenoids in sufficient quantities for industrial manufacture. Algae (Protista) are a diverse assemblage of photosynthetic organisms and correspondingly produce carotenoids with a wide array of structures [2, 50]. The Chlorophyceae is the class of algae of primary industrial interest, and certain genera including *Dunaliella*, *Chlorococcum*, and *Haematococcus* accumulate high levels of β-carotene, canthaxanthin, and astaxanthin [2, 51, 52]. Terrestrial plants have a limited and uniform of naturally occurring carotenoids consisting mostly of α- and β-carotenes and the corresponding xanthophylls with the hydroxyl groups in the 3 and 3' positions (zeaxanthin and lutein), but plants generally do not produce valuable xanthophylls in sufficiently high yields to compete industrially unless downstream processes are used to concentrate the pigments. These xanthophylls are important industrially as colorants [53], and it would be valuable to find efficient biotechnological production systems for their manufacture [54-57].

Filamentous fungi, yeasts, and microalgae have been principally used industrially for the production of natural pigments by fermentation or algal pond culture [2, 10, 34, 52, 58, 59], which is probably mainly due to their natural ability to efficiently and rapidly produce isoprenoids compared to prokaryotes and plants. On the other hand, prokaryotes have been valuable in the cloning of carotenoid genes and expression of enzymes for carotenoid biosynthesis [10, 60-63]. Despite the wide variety of structures, carotenoids are generally produced in relatively low quantities in prokaryotes such as *Escherichia coli* and these heterologous hosts don't appear suitable for industrial production without considerable strain and fermentation developments. For example, β-carotene was produced at levels of up to 1.5 mg g^{1}dry weight in *E. coli*, but formation of β-carotene above these levels was toxic [10, 62]. In contrast, lycopene, β-carotene, and astaxanthin have been produced in fungi and microalgae at levels ranging from 5-50 $mg.l^{-1}$ and 5-43 $mg.g^{-1}$ dry cell weight [2, 7, 8, 52, 64].

Various fungi appear to have the ability to produce high levels of isoprenoids and terpenoids including ergosterol and carotenoids [2, 59, 65] (Table 1). A variety of isoprenoids are synthesized by plants, including more complex isoprenoids than have been detected in fungi [56], but with some exceptions [66] plants are currently not used for biotechnological production in bulk scale because of the slow growth rates, low quantities of the metabolites, and seasonal variation in expression of the compounds of interest. Production of isoprenoids and other secondary metabolites by plant genetic engineering and plant tissue culture has been considered but is currently not conducted to our knowledge on industrial scale, with rare exceptions such as

Table 1. Representative carotenoids reported in slime molds and fungi (modified from [2]).

Phylogenetic group	Representative Carotenoids
Myxomycota (prokaryotes)	
Myxomycetes (slime molds)	3,4-didehydrolycopene, neurosporaxanthin, γ-carotene, β-carotene,
Eumycota	
Mastigomycotina	
Chytridiomycetes	γ-carotene, lycopene, β–carotene
Oomycetes	
Zygomycotina	
Zygomycetes	β–carotene
Ascomycotina	
HemiAscomycetes	None known
Plectomycetes	γ-carotene, lycopene, β-carotene
Pyrenomycetes	γ-carotene, lycopene, β-carotene, torulene, lycoxanthin, neurosporaxanthin
Discomycetes	β-carotene, γ-carotene, 3,4-didehydrolycopene, torulene, phillipsiaxanthin, aleuriaxanthin, plectaniaxanthin
Loculoascomycetes	None known
Basidiomycotina	
Hymenomycetes	γ-carotene, β-carotene, cryptoxanthin, canthaxanthin, ataxanthin (?)
Gasteromycetes	γ-carotene, β-carotene
Teliomycetes	β-zeacarotene, γ-carotene, β-carotene, torulene, cryoptoxanthin
Basidiomycetous yeasts	β-carotene, HDCO, echinenone, astaxanthin
Deuteromycotina	γ-carotene, β-carotene, torulene

the dye and antimicrobial shikonin, produced by plant tissue culture [67]. Plants do have considerable potential for synthesis of carotenoids in countries where microbial fermentation industry is not highly developed and plants can be grown, harvested, and carotenoids extracted at low costs.

3. CAROTENE PROPERTIES AND GENERAL CONCERNS

The term carotenoids represent a class of structurally related pigments, comprised mainly of C_{40} isoprenoids containing a characteristic polyene chain of conjugated double bonds [68-70]. The two major groups are the hydrocarbons (carotenes) and their oxygenated derivatives (xanthophylls). More than 600 carotenoids have been isolated [68. 70, 71]. The primordial carotene molecule phytoene is formed by head-to-head condensation of two geranylgeranyl precursors and has an acyclic $C_{40}H_{56}$ structure analogous to ζ-carotene (Fig. 1). All C_{40} carotenoids are derived from the acyclic $C_{40}H_{56}$ structure by hydrogenation, dehydrogenation, cyclization, oxidation, esterification, halogenation, and other reactions [45, 72]. For most of the carotenoids, trivial names are usually used [73], although rules for the nomenclature have been published by the IUPAC (International Union of Pure and Applied Chemistry) [74], also see Appendix.

3.1. Physicochemical Properties of Carotenoids

3.1.1. The chromophore and color

Color is a defining attribute of most carotenoids and is primarily due to a conjugated polyene system or chromophore. Typical chromophores from different carotenoids are portrayed in Figure 1. The absorption maxima (λ_{max}) of carotenoids depends on the number of conjugated double bonds present in the chromophore [69, 75]. Most carotenoids absorb light maximally in the 400-600 nm range (7-15 conjugated double bonds). The addition of additional conjugated carbon-carbon double bonds without other structural changes results in a chromophore in which there is a displacement of the absorption maxima towards higher wavelengths (see Table 2). For example, colorless phytoene (maximum absorption in hexane = 286 nm) is transformed by four dehydrogenation steps to red-colored lycopene (maximum absorption in hexane = 473 nm). The spectrum shift of 186 nm is due to the addition of eight carbon-carbon conjugated double bonds resulting in a chromophore absorbing light at higher wavelengths. The carotene absorption maximum can be estimated from the equation λ_{max}(nm)=300.5+65.5xN, in which N is the number of conjugated double bounds present.

Non-conjugated double bonds have little or no effect on the position of the absorption maxima [75], but a bathochromic shift of 2-3 nm results from the introduction of olefinic bonds some three positions away from, and on either side of the chromophore [75]. When the lycopene acyclic termini (ψ-end group) undergo ring closure to form end groups, there is a displacement of the absorption maxima to shorter wavelengths, with a concomitant loss of fine structure in the absorption spectrum. This results from a configurational change that occurs on transformation of ψ- to β-groups, decreasing the degree of coplanarity that is characteristic of acyclic chromophores and affecting light absorption. Such representative changes are shown for the common carotenes lycopene, γ-carotene and β-carotene (Table 2). It is observed that although the number of conjugated double bonds remains constant, β cyclizations decrease the absorption maxima. The red-colored lycopene chromophore with ψ-ends changes into orange and to yellow-orange colors on _-end cyclizations. Similar shifts to lower wavelengths occurs when ψ-ends are cyclized to ε-, γ-, φ-, χ-, or κ-end groups.

Normally, the polyene chain of acyclic and cyclic carotenes has a characteristic absorbance with three peaks, referred to as having high persistence. The introduction of carbonyl groups results in loss of persistence and fine structure. This modification yields a rounded symmetrical peak of absorption (for all-*trans* isomers). Other chemical modifications in the basic structure, with the exception of introduction of epoxide groups, has little effect on absorption characteristics, since they do not sterically interfere with the polyene chain. The introduction of epoxide groups decreases the absorption maxima by 10-20 nm (a process termed a hypsochromic shift). In summary, carotenoid molecules have strong light absorption and solutions may be visible to the human eye with only microgram quantities of pigment present [74]. Absorption at the maximum wavelength and characteristic molar absorptivities for individual carotenoids forms the basis for quantitative determination of carotenoid quantities. The intense light absorption and low detection limits are useful to monitor purification steps and chemical reactions. Loss or change of color provides a warning of decomposition or structural modifications.

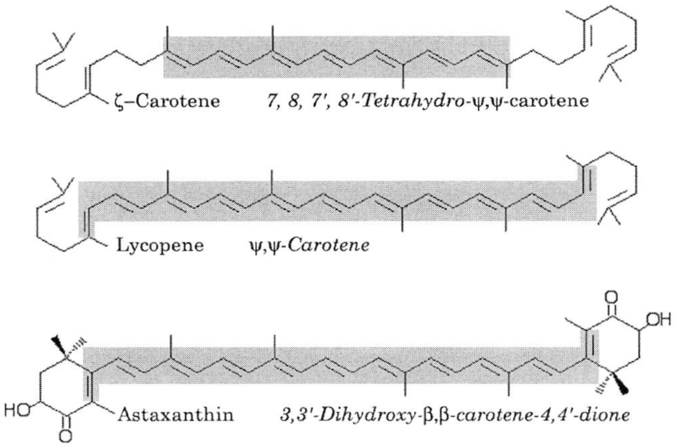

Fig. 1. Chromophore localization in three representative carotenoids. z-Carotene shows its seven conjugated double bond chromophore and four additional double bonds while lycopene shows its eleven conjugated double bond with additional two double bonds. The astaxanthin chromophore is identical to lycopene, even with the b-ring cyclized ends.

3.1.2. *Cis/Trans* (*E/Z*) Geometric Isomers and Photoisomerization

Because of the multiple conjugated double bonds and the variety of end groups (both affecting free rotation about certain C-C bonds) as well as the presence of asymmetric (chiral) C-atoms, carotenoids constitute a variety of geometric and configurational isomers with different chemical and physical properties. The most common isomeric forms among the carotenoids are geometric (*cis/trans* or *E/Z*) and optical (*R/S*) isomers [71, 74]. A double bond will link the two residual parts of the molecule either in a *Z*-configuration (*cis*), with both carbons on the same side of the plane, or in *E*-configuration (*trans*), in which they occur on opposite sides of the plane. The common designations are *cis*- or *trans* and are used in this review, but *E/Z* are the preferred

Table 2. Changes in absorption maxima as influenced by the number of conjugated double bonds and end-group cyclizations.

Carotene	Molecule	No. Double Bonds	Conjugated Double Bonds	No. ß-rings	Absorption maxima (nm)	Color
Phytoene	$C_{40}H_{64}$	9	3	0	276-286-297	colorless
ζ-Carotene	$C_{40}H_{60}$	11	7	0	380-400-425	pale yellow
γ-Carotene	$C_{40}H_{56}$	12	11	1	437-462-492	orange-pink
Lycopene	$C_{40}H_{56}$	13	11	0	448-473-504	red
β-Carotene	$C_{40}H_{56}$	11	11	2	(425)-450-477	yellow
Astaxanthin	$C_{40}H_{52}O_4$	11	11	2	468	red-pink

*The data was obtained using hexane as the solvent and carotenoid concentrations of about 20 μg/ml [75].

systematic nomenclature from the chemical perspective. In general, most natural carotenoids (except for 15-*cis*-phytoene) from nonphotosynthetic organisms are all-*trans*, and the presence of *cis*-isomers may be indicative of chemical alteration or degradation. In photosynthetic organisms, the all-*trans* configuration occurs by light-harvesting complexes, whereas the 15-*cis* configuration occurs in the low energy state of the reaction complexes [76]. The all-*trans* carotenoids generally are the isomers with the lowest solubility and highest melting point [69]. The geometrical configuration of carotenoids have a marked influence on their adsorption affinity to various matrices. In practice, it is possible to separate complex mixtures of isomers by chromatographic techniques [68]. In recent years, HPLC methods have been successfully applied to rapidly and quantitatively separate *cis/trans* isomeric xanthophylls such as astaxanthin [68, 77, 78].

Exposure to light, especially direct sunlight or to ultraviolet light induces *cis/trans* photoisomerization leading to carotenoid destruction, and as a consequence light exposure must be avoided. Low intensity diffuse daylight or artificial light is acceptable for most carotenoid manipulations but should be minimized. It is recommended that all glass containers, chromatography columns be covered with an opaque material such as aluminum foil. Also, exposure to oxygen or reactive oxygen species (ROS) and heat also promotes degradation of carotenoids.

3.1.3. Optical/Configurational Isomers

Carotenoids that contain an asymmetric carbon, *i. e.* a carbon with four different groups attached, can exist in two distinct configurational forms [74]. Both configurations have a mirror-image relationship to one another; such isomers are referred as optical isomers or enantiomers. Optical isomers of carotenoids are commonly referred to as *R* and *S*. Normally; natural carotenoids only exist in one of the possible optical isomers probably due to selectivity of the enzymes during biosynthesis. Andrewes and Starr [79] refuted the generalization that carotenoids exist as one optical isomer in nature regardless of the source, when they isolated (3*R*,3'*R*)-astaxanthin from the yeast *Phaffia rhodozyma,* in which astaxanthin had the opposite optical configuration from (3S,3'S)-astaxanthin isolated from *Haematococcus pluvialis* and from the lobster, *Homarus gammarus*. The finding of (3*R*,3'*R*)-astaxanthin in *P. rhodozyma* led Andrewes and colleagues [80] to hypothesize that the sequential reactions and responsible enzymes leading to hydroxylation are different in the organisms that form different optical isomers [79, 80]. A rare and still unresolved instance of the presence of two optical isomers from a single biological source was noted when zeaxanthin localized in the human macula was present as both optical *R* and *S* isomers, while only one enantiomeric form was present in human plasma [81]. Interest has increased in recent years regarding the bioavailability of isomeric forms of carotenoids [82-84]. The available evidence indicates that geometrical and optical isomers are deposited selectively depending on the organism and tissue in which deposition occurs, but more research is needed to understand the basis of selective bioavailability.

3.2. Solubility of Carotenoids

Carotenoids are lipophilic compounds that are nearly insoluble in water [85]. In aqueous systems they tend to form aggregates or adhere to non-specific surfaces. Carotenes are soluble in polar organic solvents such as acetone and methanol acetone/methanol mixtures that are frequently used for extraction of pigment from biological materials including fungi, which have high water content. Crystalline carotenes can be difficult to dissolve in such solvents; but they can be dissolved in non-polar organic solvents such as hexane, toluene, benzene or halogenated hydrocarbons such as dichloromethane [78, 86]. Dimethylsulfoxide has successfully been used for extraction of carotenoids from the yeast *P. rhodozyma* [87] and could be useful for extracting carotenoids from other tissues.

Hoffmann-La Roche Limited (Basel, Switzerland) has succeeded in manufacturing stable, water-dispersible carotenoid- and apo-carotenoid-rich products with high bioavailability and excellent mixing properties. Free-flowing granulated carotenoid powders are commonly used for animal nutrition. The powders consist of small spherical particles or "beadlets" in which the carotenoid or apo-carotenoid is finely distributed in a starch-encapsulated gelatin and carbohydrate matrix. The industrial manufacturing process is termed *spray-dried coating* and it relies on antioxidant chemicals such as ethoxyquin and ascorbyl palmitate to avoid carotene degradation [88].

3.3. Antioxidant Properties of Carotenoids

Interest in the chemical reactions of carotenoids with radicals and particularly reactive oxygen species (ROS) derived partly from their integral roles in photosynthesis in plant systems [89, 90]. However, much interest and effort has also been devoted to their actions as *in vitro* and as *in vivo* antioxidants and related photoprotectants in humans, animals, and microorganisms [17, 21, 91-97], and their potential to prevent or alleviate human and animal diseases associated with ROS associated *in vivo* responses, signaling, damage and repair [16, 18, 20, 98-100], (also, see cited refs. in section 1). Traditionally, the main functions of carotenoids were believed to be restricted to light absorption and energy transfer, protection against photooxidation in photosynthetic systems, and formation of vitamin A and quenching of singlet oxygen ($^1O_2$) in nonphotosynthetic systems [101]. However, the functions of carotenoids appear to be greatly expanding as their metabolism is studied [16, 20, 24, 27, 33, see additional citations in above sections].

Recent studies have supported that carotenoid structure and their cellular location affects their antioxidant activities [23]. The system of conjugated double bonds in carotenoids constitutes an electron-rich system, which strongly reacts with electrophilic compounds in the proper conditions and environment [102]. In the presence of oxygen or ROS, carotenoids can autooxidize, a process called *bleaching* [103]. Carotenoid antioxidant behavior is dependent on its structure and the nature of the oxidizing specie. The quenching constant for singlet oxygen ($^1O_2$) for a variety of carotenoids in solution has been determined, indicating relationships between carotenoid structure, ROS reactivity, and energy transfer [104]. The conjugated double bond chromophores seem to constitute the major structural entity involved in energy transfer in photosynthetic systems and the chromophore also has a major impact in the quenching of $^1O_2$. It has been suggested that other factors may have important roles when ROS are exposed to carotenoids, including $^1O_2$, superoxide ($O_2^{\bullet-}$), hydrogen peroxide (H_2O_2), and hydroxyl radical ($OH^{\bullet}$) [23]. Carotenoid reactivity with ROS appears to vary with the energy transfer profile, the oxidizing specie, carotene structure, and the electron density profile [105].

Carotenoids can react with ROS in three basic ways, namely electron transfer (eq. [1]), hydrogen abstraction (eq. [2]), and radical species addition (eq. [3]) [23].

$$ROO^{\bullet} + CAR \rightarrow ROO^{-} + CAR^{\bullet+} \qquad [1]$$

$$ROO^{\bullet} + CAR \rightarrow ROOH + CAR^{\bullet} \qquad [2]$$

$$ROO^{\bullet} + CAR \rightarrow (ROO\text{-}CAR)^{\bullet} \qquad [3]$$

Several ROS including $^1O_2$, H_2O_2, $O_2^{\bullet -}$, and $OH^{\bullet}$ are produced in cell systems [18, 98, 106], and they differ in their reactivity with carotenoids. The rate of scavenging of ROS in cells by carotenoids depends on the nature of both the ROS and the carotenoid [107]. For example, β-carotene reacts rapidly and efficiently with peroxyl radicals but less efficiently with $OH^{\bullet}$ and $O_2^{\bullet -}$ [17]. Studies in solution using varying partial oxygen pressures (pO_2) [107] indicated that at low pO_2, β-carotene acted as a chain-breaking antioxidant (eq. [1]), while at higher pO_2, β-carotene acted as a prooxidant. The products of Eqs. [2] and [3] are thought to react with molecular oxygen to yield peroxyl radicals that may lead to lipid peroxidation [17]. Also special conditions, particularly the *in vivo* carotenoid concentration and the presence of other antioxidants and the biological environment, may lead to a shift from antioxidant to prooxidant activity [17, 23, 108].

Most literature regarding carotenoid reactivity with ROS is based on *in vitro* experiments. *In vivo*, the situation is obviously much more complex. Some of the complexities confounding the *in vivo* experiments are the lower carotenoid concentrations, heterogeneous carotenoid composition[23, 68] and possible carotenoid interactions or with other antioxidants such as vitamin E [23]. Furthermore, different classes of organisms appear to differ in their ability to accumulate isoprenoids such as carotenoids. For example, β-carotene was toxic to *E. coli* above 1.5 $mg\,l^{-1}$ [62], whereas certain fungi and microalgae can accumulate quantities of β-carotene and other carotenoids such as lycopene and astaxanthin at levels from 5 to 50 mg^{-1} depending on the species and the strain. Certain plants can also accumulate high levels of carotenoids in selected tissues [12, 66]. The mechanisms by which different organism accumulate different levels of carotenoids is not known, but the differences could involve available cellular space, different biosynthetic capacities, and ability to defend against prooxidant activities of carotenoids present in high cellular quantities. Nonetheless, these observations support that fungi or algae are the current organisms of choice for industrial carotenoid production.

Most *in vivo* experiments on carotenoid metabolism, function and evaluation of pharmaceutical activity have been performed in animal models or in mammalian cell culture [20, 24, 25, 109-111], probably since these models most closely mimic the human system. However, fungal systems should be considered for certain studies, particularly since certain groups of fungi, particularly the basidiomycetes, appear to be closely related to animals [35-36]. The basidomycetous yeast, *Phaffia rhodozyma* (teleomorph *Xanthophyllomyces dendrorhous*) [45] produces astaxanthin and could serve as an excellent model system for studying the biosynthesis, genetics, regulation and function of astaxanthin and other carotenoids in eukaryotic systems [112-114]. Furthermore, yeasts may serve as excellent models for studying aging and oxidative stress responses in eukaryotes [115-117]. The complete genome of *Saccharomyces cerevisiae* has been sequenced [37] (information available in the world wide web http://www.mips.biochem.mpg.de/proj/yeast/). This yeast (although naturally noncarotenogenic) could provide an excellent eukaryotic system to study regulation of oxidative stress and other phenomena associated with carotenoid metabolism, and as a heterologous host for carotenoid production provided that codon usage and regulatory circuits are compatible.

4. ISOLATION, ANALYSIS AND IDENTIFICATION OF FUNGAL CAROTENOIDS

The isolation and analysis of carotenoids has been reviewed [68, 69, 75, 77, 86, 108]. Due to the inherent instability of carotenoids, precautionary measures must be used during isolation and manipulation. In particular, carotenoids are susceptible to light, oxygen, acid, and high temperatures. Precautionary procedures for working with carotenoids have been reviewed [68,

69, 75, 86, 119, 120], and only procedures particularly applicable to isolation and analysis from fungal sources is reviewed here. Furthermore, we will primarily focus this methodological discussion to the yeast *Phaffia rhodozyma* (teleomorph *Xanthophyllomyces dendrorhous*), since this is the organism studied in our laboratory.

Fresh and undamaged fungal cells are required for accurate quantitative work since carotenoids are susceptible to oxidation and enzymic decomposition. Carotenoids are stable in their cell environment for a short period of time, but for long-term storage of cells, they should be lyophilized and kept in a headspace of nitrogen or frozen at low temperatures (e.g. –70°C). Since carotenogenic yeasts and fungi are obligated aerobes and produce ROS as byproducts of metabolism, samples should be kept cold, in an inert gas environment (nitrogen or argon), and protected from light during handling.

4.1. Disruption and Homogenization of Fungal and Yeast Specimens

Quantitative analysis of fungal carotenoids requires well-planned experimental design and proper extraction methodologies. Quantitative extraction of carotenoids from yeast and fungi can be difficult and laborious due to their thick and rigid cell walls that may be surrounded by capsules and slime. There are three main strategies for extracting carotenoids from fungi and yeasts samples (a) direct chemical extraction in organic solvents, (b) mechanical disruption of the cell wall, and (c) enzymatic digestion of the cell wall. Acid hydrolysis of the cell wall followed by extraction [121] is no longer recommended since many carotenoids are labile in acid. In some cases, it may be advantageous to combine one or more of the three methods. In our experience with the encapsulated yeast *P. rhodozyma*, which is relatively recalcitrant to direct chemical extraction and enzyme digestion [122], mechanical disruption using glass beads in a "bead beater" or Braun MSK homogenizer, or by passage through a French press is suitable for extraction of low quantities and small numbers of samples [79, 80, 122-124]. Larger samples (several liters of cell suspension) can be disrupted in larger mechanical disruptors such as a Manton-Gaulin press [125] or in other commercial instruments capable of disrupting fungal cells. Early work showed that *P. rhodozyma* was very recalcitrant to the more gentle method of enzymatic disruption by snail-gut enzymes, Zymolyase, or lytic enzymes from *Bacillus circulans* [122]. Achieving enzymatic digestion was highly desired since it is also required for formation of spheroplasts and for extraction of DNA, intact organelles, and for many procedures used in molecular biology [126, 127]. Although the enzyme mixture from *B. circulans* grown on whole cells or cell walls of *P. rhodozyma* was capable of digesting the yeast over extended incubations, presently most investigators use lytic enzymes from *Trichoderma* spp., which are commercially available and rapidly (~60 min) digest young yeast cells. Many investigators are also extracting carotenoids from *P. rhodozyma* using the hot DMSO method [87]. A combination of chemical and mechanical treatments was used to treat *P. rhodozyma* cells for feeding to rainbow trout [99].

Fungal mycelium cells are usually broken and homogenized directly in the organic solvent with a suitable electric blender [75] or extracted directly with a suitable solvent mixture. An additional mechanical method that has been used for fungal spores, mycelium, yeast cells, and microalgae involves freezing the cells in liquid nitrogen, and grinding with a mortar, followed by organic solvent extraction [128]. The accumulation of astaxanthin in cells of *P. rhodozyma* and particularly in cysts of the microalga *H. pluvialis* restricts bioavailability in feeds, and mechanical or enzymatic disruption may be necessary to liberate the pigments for uptake by salmonids or other animals. Autolysis has been used in *P. rhodozyma* to increase astaxanthin availability in feeds.

4.2. Extraction of Carotenoids from Biological Sources

Methods for extraction of carotenoids from various tissues and preparation of the extracts for pigment purification have been previously reviewed [68, 75, 86, 120]. The choice of a solvent system depends on the biological material, its pre-treatment, and carotenoid composition. Carotenoids are usually extracted from biological samples with water-miscible organic solvents, commonly acetone [77]. Because biological sources usually contain a mixture of carotenoid pathway intermediates and usually a major carotenoid in the mixture, the method must be able to extract a variety of carotenoids of diverse polarity. Extraction should be repeated until the residue and the filtrate are colorless or no more pigment is released, and generally three pigment extractions are sufficient. Pooled samples in organic solvent solution are then concentrated and dried by rotoevaporation or for small samples in a nitrogen gas stream. Carotenoids extracted in this manner are contaminated with various nonpolar substances including colorless lipids. Such compounds may interfere in later methodologies such as mass spectrometry and saponification can be performed to remove contaminants [74]. Saponification is not recommended when it can alter the structure of the carotenoid, e.g. hydrolysis of carotenoid esters or conversion of astaxanthin to astacene by oxidation under the alkaline conditions of saponification [75, 86]. In such cases, an anaerobic saponification procedure is recommended [86].

4.3. Strategies for Isolation, Purification and Identification of Carotenoids

When contaminating lipids have been removed from the sample, chromatographic techniques are the preferred strategy for purifying individual carotenoids from mixtures. The identification of carotenoids requires a combination of techniques fulfilling the following minimum criteria [68]:

- The visible/UV absorption spectrum must agree with the suspected chromophore in at least in two different solvent systems.
- Chromatographic identity must be confirmed in two different TLC or HPLC systems. Ideally, standards should match the chromatographic profile of the carotenoid of interest by TLC (R_f) or by HPLC (retention time).
- The molecular mass and fragmentation pattern must be confirmed by mass spectroscopy.

Among chromatographic methods, high performance liquid chromatography (HPLC), thin layer chromatography (TLC), and column chromatography (CC) are the most popular techniques for separating and purifying carotenoids [129]. HPLC is generally used for separation of low concentrations of carotenoids, since it provides excellent separation, and is highly sensitive and rapid. A typical reverse-phase chromatogram using silica-based matrix column as stationary phase for separating carotenoids extracted from *P. rhodozyma* is illustrated in Figure 2. As expected in this HPLC reverse-phase analysis, the polar carotenoids such as astaxanthin elute more rapidly (astaxanthin elution time 6.72 min) than the non-polar carotenoids such as lycopene (elution time 11.81 min) or β-carotene (elution time 15.61 min). HPLC conditions are described as follows; an Alltech Altima C18 column was used and the mobile phase consisted of a gradient of solvent A (95% acetonitrile, 5% methanol) and solvent B (dichloromethane). The mobile system profile during the 22 min carotene bulk sample run was as follows: (a) elution with 100% solvent A, 8 min; (b) gradient to 65% solvent B, 1 min; (c) isocratic elution with 65% solvent B,

7 min; (d) gradient to 100% solvent A, 1 min; and (e) re-equilibration with 100% solvent A, 4 min. The injection volume was 20 μl and detection was set at 473 nm. Usually, collected peaks are studied by UV/visible scanning for finding carotene identity (see Fig. 3).

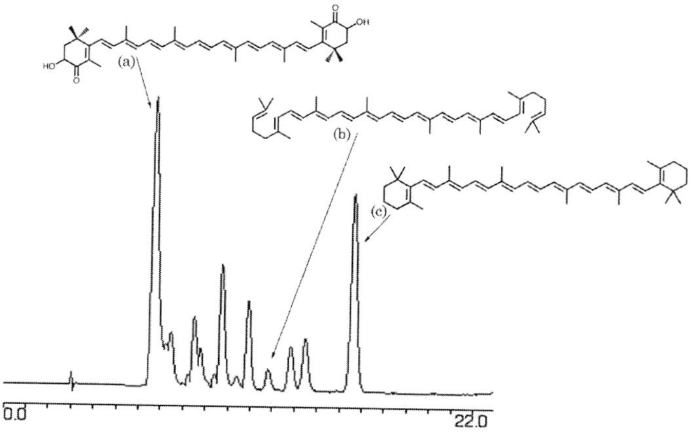

Fig. 2. Reversed-phase chromatographic separation of a crude carotenoid extract from *P. rhodozyma* (natural isolate UCD 67-385). The most representative carotenoids including astaxanthin (a), lycopene (b) and β-carotene (c) are designated by arrows.

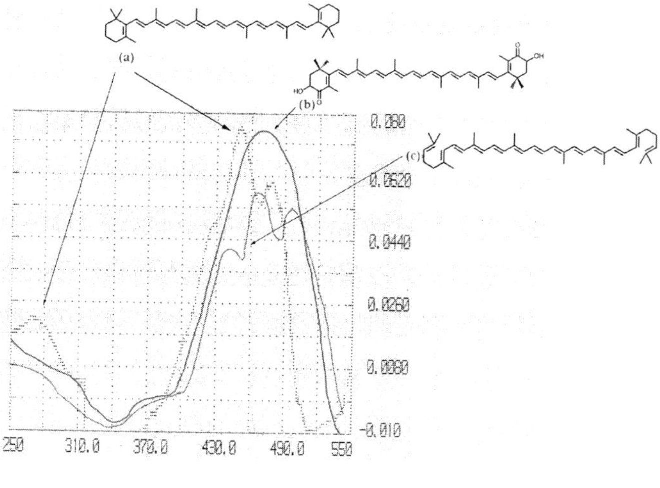

Fig. 3. Absorption spectra of (a) β-carotene, (b) astaxanthin, and (c) lycopene in hexane over the wavelength range of 250-550 nm. A Beckman DU-7 spectrophotometer was used for carotene scans. It is highly recommended to use quartz cuvette and HPLC grade solvents in order to achieve high accurate resolution. Carotene required concentration was low (approximately 10 μg/ml) and hexane was used as solvent.

Subsequent identification of purified carotenoids is performed by a sequence of analyses including UV/visible spectroscopy, which provides valuable information regarding the nature of the chromophore. Concentrations of carotenoids of only a few micrograms per ml are sufficient for UV/visible spectral analysis (see Fig. 3). The shape (e.g. fine structure) of the spectrum and the λ_{max} (maximum wavelength) provide information of the structure of the carotenoid. Several tabulations are available for comparison [69, 75, 130]. UV/visible spectroscopy also provides evidence for the presence of *cis*- and *trans*- geometric isomers. Usually when a *cis*-isomer is present, a peak around 140 nm lower than the λ_{max} for the all *trans*-isomer (in hexane) is present (see spectrum of β-carotene in Fig. 3). When information provided by the UV/visible spectra (see Figure 3) is combined with information already available in carotene databases such Table 1 or already published [75, 131], we may get clues about HPLC or TLC purified elutants or bands identities.

Mass spectroscopy is extremely useful and indispensable in carotenoid identification. It provides the molecular mass of purified carotenoids and can provide much additional information, *i. e.* the fragmentation pattern of the carotenoid of interest, which is valuable for evaluation of purity and structure. A typical mass identification for β-carotene obtained in our laboratory is portrayed in Fig. 4. As emphasized by Britton and colleagues [68, 69], mass confirmation must be determined in order to confirm carotenoid purity and identity.

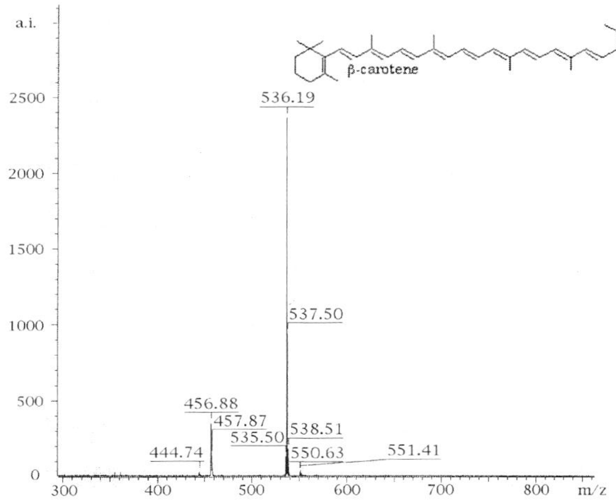

Fig. 4. Example of mass determination of β-carotene HPLC purified from *P. rhodozyma*. The fraction eluting at 15.61 min from reverse-phase HPLC was collected and dried under N_2. This sample was resuspended in acetone to an approximate concentration of 10 μM. Freeze desecation provided good quality crystals that were analyzed. Mass spectrometry was performed on a Brüker Biflex III Maldi-Tof instrument using a 2, 5-dihydroxy benzoic acid as matrix.

Identification of new carotenoids as well as the stereochemistry can be obtained by additional methodologies such as chemical derivatization, infrared and resonance Raman spectroscopy, NMR spectroscopy, circular dichroism, and X-ray crystallography [68, 69, 131, 132]. Discussion of these methods and their applications in structural studies is beyond the scope of this chapter. In practice, most fungi have a relatively standard and consistent array of pigments and UV/visible spectroscopy, TLC or HPLC, and mass spectroscopy are sufficient for identification and quantitation of the complement of carotenoids.

5. BIOSYNTHESIS OF CAROTENOIDS IN FUNGI

Carotenoids belong to the large class of compounds produced by all organisms and referred to as isoprenoids, isopentenoids, or terpenoids. Isoprenoids are the largest class of metabolites known in nature and more than 22,000 distinct compounds have been isolated [133, 134]. A vast array of structurally and functionally distinct groups of metabolites are grouped in the isoprenoids including such diverse substances as essential oils, phytohormones, steroids, cardiac glycosides, vitamins A, D, E, and K, rubber latex, lignans, and others. All isoprenoids share the C_5 hydrocarbon, isoprene, as the common precursor in their biosynthesis. Many of them are produced as secondary metabolites, and have no known function in primary metabolism but certainly have important roles ecologically. As previously mentioned, different groups of organisms (e.g. prokaryotes, fungi, microalgae, plants, animals) differ remarkably in the array of isoprenoids that they produce as well as in their biosynthetic capacity to accumulate various isoprenoids.

5.1. Secondary Metabolism: Carotene Production in Fungi

Fungal species have frequently been reported as sources of secondary metabolism products [135], as well as organisms used in production of human foods and primary metabolites by fermentation processes [135-137]. Many fungi have a high propensity to form isoprenes and to assemble them into various products derived by the isoprenoid pathway. Various fungi have long been known to produce sesquiterpenes [138], sterols [139, 140], and carotenoids [1] by the mevalonate pathway typical of fungi [141] (Fig. 5).

Many plants also produce commercially valuable isoprenoids including essential oils, drugs, agrochemicals, and allelochemicals [134, 142-145]. In eukaryotes including fungi, isoprenoids also have many essential functions in membrane structure, primary metabolism, and response to stress [143, 145-148]. Among fungal sterols, ergosterol is the key component, and its biosynthesis is well established in *S. cerevisiae* [149] and certain other fungi, in which it plays an essential role as a cell membrane component. Ergosterol has been proposed as a fungal marker for detecting fungal contamination in foods and feeds [139]. The biosynthesis of isoprenoids in cells is extremely complex as the pathways to the numerous endproducts must be coordinated to assure optimal cell growth and survival [143, 148] (Fig. 5). For example, sterol and carotenoid biosynthesis need coordination in fungi, and evidence indicates that sterol and carotenoid synthesis are independently regulated in the fungus *Phycomyces blakesleeanus* [140]. Furthermore, carotene and ergosterol were found to accumulate in different compartments in *P. blakesleeanus*, a mechanism probably that contributes in the coordination and separate regulation of the respective pathways after their divergence [142].

5.2. Carotenoid Precursor Pathway

The pathway to isoprenoids has been extensively studied in bacteria, plants and fungi and it has long been known that mevalonate (MVA) is a key intermediate [72]. It was commonly

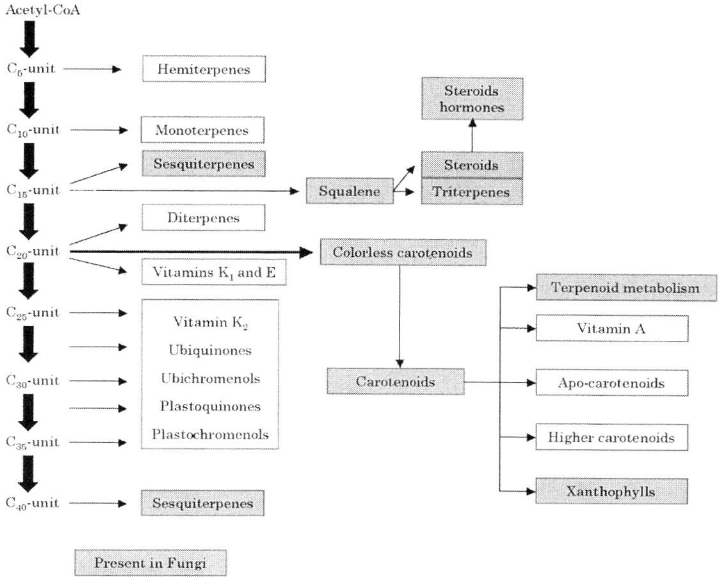

Fig. 5. Schematic of the biosynthesis of carotenoids and other isoprenoids (adapted from Goodwin, 1971 [72]). The figure emphasizes the need for coordination metabolite flow and regulation of the various pathways.

accepted to be formed through condensation acetyl-CoA units (Fig. 5). However, it was later revealed that MVA and acetate were not incorporated at high efficiency into carotenoids and certain other terpenoids in some organisms including plants some bacteria while they were efficiently incorporated into isoprenoids in other organisms including other bacteria, fungi, and mammalian cells. It was discovered using ^{13}C NMR biosynthesis studies that certain bacteria used glyceraldehyde-3-phosphate and pyruvate as precursors to MVA [150], and this group later reported that this alternate pathway existed in several groups of bacteria, algae, and higher plants [151]. The alternate pathway proceeds through 1-deoxy-d-xyulose-5-phosphate (DXP) and 2-methyl-d-erythritol-4-phosphate (MEP). Although many bacteria and plants use this pathway, most yeasts and fungi appear to derive mevalonate by the traditional pathway with acetyl-CoA as the precursor [141, 152].

Considerable physiological studies of regulation of astaxanthin biosynthesis have been performed in the yeast *P. rhodozyma* [153], with the teleomorphic stage named *Xanthophyllomyces dendrorhous*. Although *P. rhodozyma/X. dendrorhous* has often been considered to encompass a small group of yeasts with limited phylogenetic diversity, the studies of Fell and colleagues have demonstrated that this yeast actually comprises a large assemblage, probably consisting of at least three distinct groups [154-156]. The physiological control of astaxanthin biosynthesis and isolation of mutants has been previously been presented in various publications from ours' and others' laboratories [95, 96, 112-114, 123, 124, 157-162] and only certain aspects are considered here. Feeding of 0.1% mevalonate increased astaxanthin yields by 4-fold in *P. rhodozyma* [160], suggesting that inexpensive sources of mevalonate could be valuable in industrial fermentations to increase yields of astaxanthin. Our laboratory demonstrated that (RS)-[2-^{14}C]-mevalonolactone was incorporated at 3-fold greater rate into astaxanthin and β-carotene and that total incorporation into yeast cells increased from 46,850 to 295,220 dpm in the presence of t-butylhydroperoxide, which may have destroyed astaxanthin and perhaps other susceptible isoprenoids. We hypothesized that destruction of key intermediates or the end-product, astaxanthinn, relieved end-product inhibition [113]. This was the first demonstration of potential feedback inhibition in the yeast astaxanthin pathway. The synthesis of MVA must be tightly regulated to ensure a constant flow of MVA into essential isoprenoids and to avoid synthesis of toxic levels of isoprenoid products. It is not known at which step in *P. rhodozyma* and most other fungal carotenogenic systems feedback regulation takes place.

In mammalian cells, the enzymes 3-hydroxy-3-methylglutaryl coenzyme A synthase (HMG-CoA synthase) and 3-hydroxy-3-methylglutaryl coenzyme A reductase (HMG-CoA reductase) are known to be tightly regulated in synthesis of cholesterol and other isoprenoids in mammalian cells [148, 163, 164], but stringent regulation of this enzyme has not been shown in fungi to our knowledge. The isomeration of IPP (isopentenyl pyrophosphate) to DMAPP (dimethylallyl pyrophosphate) by the *idi* gene is a key step in the synthesis of GGPP (geranylgeranyl diphosphate) in *P. rhodozyma* and *Capsicum* (see Figure 6). The presence of the *P. rhodozyma idi* gene on a multi-copy plasmid in an *E. coli* transformed strain, resulted in accumulation of lycopene and other carotenoids [165]. These results indicated that overexpression of *idi* increased formation of DMAPP/IPP and geranylgeranyl pyrophosphate (GGPP). GGPP is usually only present in trace amounts in *E. coli*, but on its overexpression it is apparently partly fed into the carotenoid pathway. It is not clear whether formation of DMAPP and/or IPP were increased by overexpression of *idi* [165]. GGPP is the direct precursor to carotenoids. The enzymes catalyzing the formation of GGPP have not been isolated and characterized in *P. rhodozyma* and most other fungi and the pathway to GGPP is based on analogy to their characterization in animal, plant, and certain bacterial systems.

While mammals contain only 1 isoform of HMG-CoA reductase, 2 forms have been identified in plants as well as in the yeast *S. cerevisiae* [166]. Interestingly, over expression of HMG-CoA reductase leads to significant morphological changes in yeast cells, such as membrane proliferation and formation of membrane layers termed karmellae. *S. cerevisiae* has two isoenzymes of HMG-CoA reductase (Hmg1p and Hmg2p) and both display feedback regulation as well as cross-regulation by oxygen. The catalytic domains of the two isoenzymes are extremely similar (93 % identical). However, examination of the mechanisms of their feedback regulation has revealed striking differences between the enzymes. The regulation of Hmg1p appears to occur at the level of mRNA translation [167], and the signals that control the translation rate are derived from early mevalonate pathway products [168]. In contrast, feedback control of Hmg2p has not been completely elucidated, but it appears that Hmg1p is modulated by molecules synthesized late in the sterol pathway [168].

The expression of both enzymes is regulated by oxygen and the expression pattern has been described as contra-regulation [169, 170], in which two isoenzymes are regulated in opposite fashion by the same stimulus. A model was proposed [168] whereby oxygen was the key controlling extracellular signal. When oxygen tension was high as in aerobic growth, the proportion of Hmg1p in the cells was high whereas Hmg2p was low. When oxygen tension was low as in semi-anaerobic conditions, Hmg1p was low and Hmg2p was high. Thus, both isoenzymes of HMG-CoA reductase work in concert as determined by oxygen availability [168].

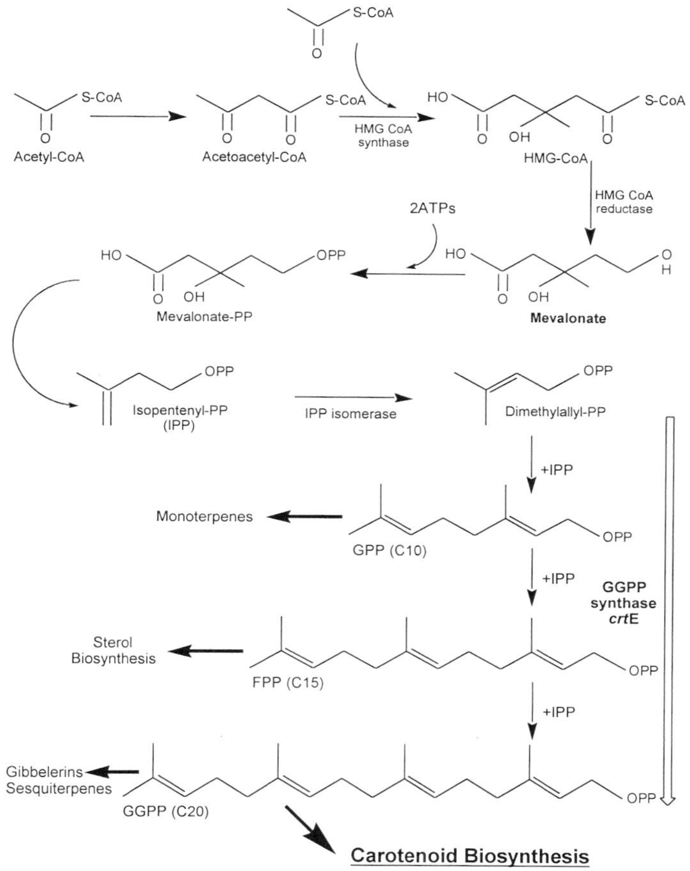

Fig. 6. The probable mevalonate pathway to geranygeranyl diphosphate (GGPP) in the yeast *Phaffia rhodozyma*.

In consideration of this mechanism of regulation of isoprenoid biosynthesis, it is interesting to note that molecular oxygen appears to be mainly required in late steps of the isoprenoid pathway such as in oxygenation of carotenoids, oxidative transformation of squalene to ergosterol, and in other reactions. The direct involvement of oxygen in early steps appears to be minimal. For example, *P. rhodozyma* may grow by fermentation in conditions of low oxygen tension, but colored carotenoids are not produced and the yeast appears beige-colored [124, Echavarri-Erasun and Johnson, unpublished results]. Similarly, under anaerobic conditions *S. cerevisiae* does not epoxidate squalene [168]. Thus, oxygen appears to differentially regulate early and late steps in the isoprenoid pathway and may be a key determinant of the flow of precursors into the pathway.

5.3. Carotenoid Pathway

GGPP (geranylgeranyl diphosphate) is an ubiquitous C_{20} metabolite that is produced in all organisms containing isoprenoids, including non-carotenogenic cells. It serves as a substrate for the synthesis of a variety of essential compounds such as gibberellins, sesquiterpenes, ubiquinones, and dolichols [72, 143, 171]. As suggested by Goodwin in 1971 [72], genetic and enzymic data supported that GGPP formed by GGPP synthase (*crtE*) is the immediate precursor to carotenoid biosynthesis in bacteria and probably other carotenogenic organisms [172]. This conclusion was based on analysis of carotenogenic gene clusters in various bacteria investigated. Overexpression of GGPP synthase in *Capsicum annum* led to a massive increase in carotenoid yield [173]. On the basis of these results, it was suggested that GGPP formation may be a key limiting step in carotenoid production [173]

5.3.1. GGPP Synthase

The formation of GGPP occurs by successive head to tail condensations of C_5 intermediates (Fig. 6). The genes encoding GGPP synthase have been cloned from a variety of fungal species including *Rhizomucor circinelloides* (Velayos *et al.*, unpublished results), *Penicillium paxilli* [174], *Saccharomyces cerevisiae* [175], *Kluyveromyces lactis* [176], *Saccharomyces bayanus* [177], *Zygosaccharomyces rouxii* [178], *Gibberella fujikuroi* [179], and *Neurospora crassa* [180]. Sequence homology analyses of GGPP synthases show a high degree of sequence similarity, especially in four conserved regions, indeed the biochemical mechanism of this-head-to tail joining is basically the conserved in all organisms. The *al-3* gene product of *N. crassa* appears to be regulated by blue light and possibly other signals and codes a GGPP synthase of 428 amino acids in length having 3 domains. The GGPP synthase is highly homologous to other prenyl transferases in organisms ranging from bacteria to humans [180].

5.3.2. Phytoene synthase

Biosynthesis of the primordial carotenoid in the family, 15-*cis*-phytoene, occurs by a tail to tail condensation of 2 GGPP molecules (Fig. 7) and is catalyzed by phytoene synthase. This enzyme catalyzes the condensation of 2 GGPPs to prephytoene and subsequent conversion to 15-*cis*-phytoene. The enzyme has been extensively studied in plants such as pepper [54, 181] and tomato [182].

Amino acid sequence comparisons of phytoene synthases showed a high degree of homology among bacterial and fungal enzymes including *Saccharomyces cerevisiae* [183], and remarkably the enzyme even had considerable homology to human squalene synthases involved in cholesterol biosynthesis [184]. The most detailed characterization of a phytoene synthase was undertaken in tomato [102]. It was demonstrated that Mn^{2+} and ATP are essential for catalytic activity, and that the enzyme structurally interacted with other carotenogenic enzymes [185]. Purification of phytoene synthase from green tomato revealed the existence of close physical associations among phytoene synthase and other enzymes in the isoprenoid pathway including IPP isomerase, GGPP synthase and probably other isoprenoid biosynthesis enzymes. These proteins are anchored to membranes through interaction of hydrophobic domains [40]. The trafficking mechanism of these enzymes to the membrane is unknown, but in eukaryotic cells may involve the 20S SNARE mechanism [186] or the Ypt and Sec4 system that is analogous to the Rabs system in mammalian cells [187], of which both systems of protein trafficking have been demonstrated in *S. cerevisiae*. Interestingly, the Ypt and Sec4 system were shown to bind to membranes because they were modified by a type II geranylgeranyltransferase [187]. It is thought

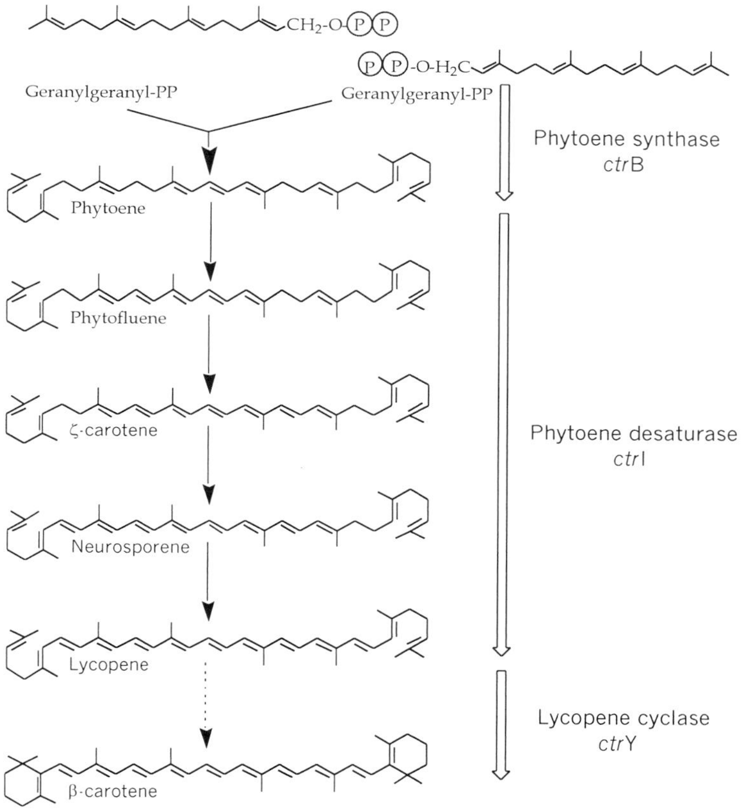

Fig. 7. Postulated biosynthetic pathway of β-carotene from geranylgeranyl diphosphate (GGPP) in the yeast *P. rhodozyma*.

that close association of sequential enzymes in the isoprenoid pathway may have *in vivo* relevance by enabling the channeling of biosynthetic precursors [10, 185].

The genes encoding several fungal phytoene synthases (*psy* genes; closely analogous to bacterial *crtB* genes [10] have been cloned including those from *Neurospora crassa* [188], *Xanthophyllomyces dendrorhous* [189], *Mucor circinelloides* [190], and *Phycomyces blakesleeanus* [40]. Based on the phenotypes of carotene mutants in *Phycomyces*, it was hypothesized that phytoene synthase is coded by a bifunctional gene [191]. It was later confirmed that several fungi including *M. circinelloides* has a bifunctional enzyme having lycopene cyclase and phytoene synthase activities, which are encoded by the *carRP* gene [128]. The *CarRP* protein of *Mucor* is homologous to the *al-2* protein of *N. crassa* in which the C-terminal domain possesses the phytoene synthase activity and the lycopene cyclase activity is probably located in the N-terminal transmembrane domain. Complementation studies in *E. coli* showed that the *carRP* product does indeed have lycopene cyclase activity [128]. It is possible that the N-terminal transmembrane region in *N. crassa* also possesses lycopene cyclase activity, but this has not been confirmed [128, 192]. Additional studies of *crtYB* in *X. dendrorhous* [192] and *crtRA* in *P. blakesleeanus* [40] corroborated the bifunctional nature of the enzyme and its conserved bifunctional structure in fungi. A tabulation of isoprenoid and carotenoid structural genes is presented in Table 3.

As previously discussed, phytoene synthase genes are conserved among bacteria, fungi and plants. However, in bacteria separate genes encode phytoene synthase and lycopene cyclases, whereas in fungi both activities are encoded by a single gene [40]. The fungal gene encoding both activities may have evolved by fusion of separate genes. Fusion of genes involved in the same pathway has previously been observed in fungi, such as for tryptophan biosynthesis in *N. crassa* and *Aspergillus nidulans* [193, 194], and 5-*enol*pyruvylshikimate 3-phosphate production in *A. nidulans* [195]. The protein encoded by the gene *carRP* in *M. circinelloides* may be cleaved into two products from a contiguous transcript in the bifunctional gene model [128, 191]. A putative protease cleavage site was identified in the *crtRA* gene from *Phycomyces* [40]. Future research will be required to determine if the splicing model in *Mucor circinelloides* (*crtRP*) and other homologous genes is also present in other fungal species.

5.3.3. Phytoene desaturase

Phytoene is the first committed carotenoid in the pathway, but it lacks the most appealing characteristic of carotenoids, their color. Phytoene desaturation or dehydrogenation converts the colorless carotenoid phytoene to the first colored carotenoid, ζ–carotene, by two sequential desaturations (see Figure 7). In cyanobacteria [203] and tomato [132], further desaturations to lycopene take place through the intermediate neurosporene, and are catalyzed by a separate enzyme, ζ -carotene desaturase. In contrast, desaturation of phytoene to lycopene in *Rhodobacter capsulatus* [204], *Erwinia uredovora* [205], *Neurospora crassa* [196], *Cercospora nicotianae* [197], *Phycomyces blakesleeanus* [198], *Xanthophyllomyces dendrorhous* [189], and *Mucor circinelloides* [190] are catalyzed by a single phytoene dehydrogenase enzyme. The presence of an enzyme complex carrying out multiple successive dehydrogenations was also supported indirectly in *P. rhodozyma*, by the isolation of mutants that accumulate phytoene, lycopene, or β-carotene but not other carotenoids in the early carotene pathway (206, Echavarri-Erasun and Johnson, unpublished data]. Furthermore, treatment of cultures with nicotine, which is known to inhibit cyclases, resulted in the accumulation of lycopene as the mayor carotenoid. These results suggest that phytoene, lycopene, and β-carotene are the final products of enzymatic conversions, implying that in *P. rhodozyma* only two enzymes (phytoene desaturase and a lycopene cyclase) are required for the synthesis of β-carotene from phytoene.

Our analysis indicates that similarity of published amino acid sequences for phytoene desaturases is greater than 80% among the various prokaryotic and eukaryotic organisms. Interestingly, the degree of homology of the similarity between plant and cyanobacterial phytoene dehydrogenases appears unrelated to their photosynthetic nature, since phytoene desaturase of *Rhodobacter capsulatus*, is more closely related to that of *N. crassa* than are those from

Table 3. Tabulation of isoprenoid and carotenoid encoding structural genes from fungal sources available in the GeneBank.

Current GeneBank Genes Coding for Fungal Carotenogenic Enzymes				
Carotene backbone assembly genes				
Enzyme	Gene	Organisms	Accession N°	References
IPP isomerase	*idi*	*P. rhodozyma*	Y15811	[165]
HMG-CoA reductase	*Hmg1p*	*S. cerevisiae*	M22255	[166]
GGPP synthase	*carB*	*R. circinelloides*	AJ238028	[128]
GGPP synthase	*carB*	*P. paxilli*	AF279807	[174]
GGPP synthase	*BTS1*	*S. cerevisiae*	U31632	[175]
GGPP synthase	*BTS1**	*K. lactis*	AL427778	[176]
GGPP synthase	*BTS1**	*S. bayanus*	AL398163	[177]
GGPP synthase	*BTS1**	*Z. rouxii*	AL395117	[178]
GGPP synthase	*GGPPS*	*G. fujikuroi*	X96943	[179]
GGPP synthetase	*al-3*	*N. crassa*	U20940	[180]
Phytoene synthase	*al-2*	*N. crassa*	L27652	[188]
Phytoene synthase	*crtB*	*X. dendrorhous*	AJ133646	[192]
Phytoene synthase	*carP*	*M. circinelloides*	AJ250827	[128]
Phytoene synthase	*carA*	*P. blakesleeanus*	AJ278287	[40]
Carotenogenic fungal genes				
Phytoene desaturase	*al-1*	*N. crassa*	M57465	[196]
Phytoene desaturase	*PDH1*	*C. nicotianae*	U03903	[197]
Phytoene desaturase	*carB*	*M. circinelloides*	AJ238028	[190]
Phytoene desaturase	*crtI*	*X. dendrorhous*	Y15007	[189]
Phytoene desaturase	*carB*	*P. blakesleeanus*	X78434	[198]
Lycopene cyclase	*carR*	*R. circinelloides*	AJ238028	[128]
Lycopene cyclase	*crtY*	*X. dendrorhous*	Y15007	[192]
Lycopene cyclase	*carR*	*P. blakesleeanus*	X78434	[40]
Oxygenation enzymes (Xanthophyll formation), but non-fungal sources				
β-carotene oxygenase	*crtO*	*H. pluvialis*	X86782	[199]
β-carotene ketolase	*bkt*	*H. pluvialis*	D45881	[200]
β-carotene hydroxylase	*crtZ*	*A. aurantiacum*	D58420	[201]
β-carotene oxygenase	*crtW*	*Alcaligenes PC1*	D58422	[202]

*Similar to *S. cerevisiae* BTS1 (geranylgeranyl diphosphate synthase), but not confirmed by complementation studies.

plants, algae or cyanobacteria [197]. Analysis of phytoene desaturase from fungal specimens indicated that two regions of homology appeared to be conserved in phytoene desaturases [197]. The amino-terminal end of phytoene desaturase is similar to known NAD(H)-NADP(H) and FAD(H) binding motifs [207]. The second domain encodes a dehydrogenase domain [204].

5.3.4. Lycopene cyclase

Certain fungi may produce acyclic lycopene alone, or the organisms may cyclize precursors yielding neurosporene,α-carotene or β-zeacarotene. Fungi producing these carotenoids include *Rhodotorula* and *Phycomyces* [208, 209]. However, cyclization of lycopene appears to predominate in most organisms [45]. Early studies in *P. rhodozyma* suggested that cyclization of acyclic precursors to lycopene may only occur under stress such as in microaerophilic conditions or at low pH [124].

Several fungal lycopene cyclases have been characterized including those from *X. dendrorhous* [192], *Mucor circinelloides* [128], and *P. blakesleeanus* [40]. A gene cyclizing lycopene to _-rings was carefully characterized from *X. dendrorhous*, which was termed *crtYB* [192]. The 5' region of *crtYB* codes for the lycopene cyclization activity. The lycopene cyclase converts lycopene into the monocyclic intermediate γ-carotene and further to β-carotene. A model was proposed in which separate domains of phytoene synthase and lycopene cyclase genetically fused yielding *crtYB* of *X. dendrorhous* [192]. It was proposed that an ancestor to basidiomycetous and ascomycetous fungi possessed the active site of a lycopene cyclase fused, and the gene encoding this catalytic domain fused with the gene encoding phytoene synthase. In the resulting enzyme, the structure of phytoene synthase was conserved, including its membrane-spanning regions that allow membrane integration and favorable conformation for catalysis within both domains.

Although ours' and other laboratories had attempted to isolate mutants of *P. rhodozyma* that accumulate lycopene, these mutants have been very difficult to obtain [112, 206]. At the suggestion of Professor Javier Avalos (Universidad de Sevilla), we successfully isolated lycopene-accumulating mutants from β-carotene mutant parents, based on easily screening red-colored lycopene-accumulating colonies from yellow β-carotene colonies, whereas distinguishing lycopene mutants from astaxanthin red-orange colonies is more difficult. However, there are other possible explanations including the possibility that obtaining β–carotene mutants directly from astaxanthin producers would require more than one mutation.

5.3.5. Formation of Astaxanthin from β-carotene

Although genes and enzymes that convert carotenes to xanthophylls (including astaxanthin, capsanthin, β-cryptoxanthin, neoxanthin, and zeaxanthin) have been isolated and characterized from bacteria, microalgae and plants, [10, 54, 201, 202, 210, 211], very little is known about the genes and enzymes in fungal species that mediate the biosynthesis of xanthophylls from carotene precursors [172, 96,]. β–Carotene probably serves as substrate for oxygenation and conversion to xanthophylls in all the organisms, but the intermediates and enzymes appear to diverge after β-carotene in the limited number of organisms that have been studied (see Figure 8). The genes involved in astaxanthin formation have been studied in the marine bacterium *Agrobacterium aurantiacum* [201, 202, 212], in *Alcaligenes* strain *PC-1* [202], and in the green microalga *Haematococcus pluvialis* [200, 213], each of which appears to convert β-carotene to astaxanthin by different routes. The biological reasons for divergence in xanthophyll biosynthesis with regard to gene and enzyme differences is not understood at this time to our knowledge.

In most studies attempting to reveal xanthophyll biosynthesis genes, *E. coli* strains that were genetically engineered to produce β-carotene or zeaxanthin were transformed with putative xanthophyll encoding genes. By this approach, it was revealed that two genes, *crtW* and *crtZ*, were required for the formation of astaxanthin in *Agrobacterium aurantiacum* [202]. It was shown that *crtW*, the *bkt* gene in *H. pluvialis* [200], and *crtZ* were responsible for the conversion of methylene to keto and hydroxyl-groups respectively in the β--ionone rings of β-carotene. *In*

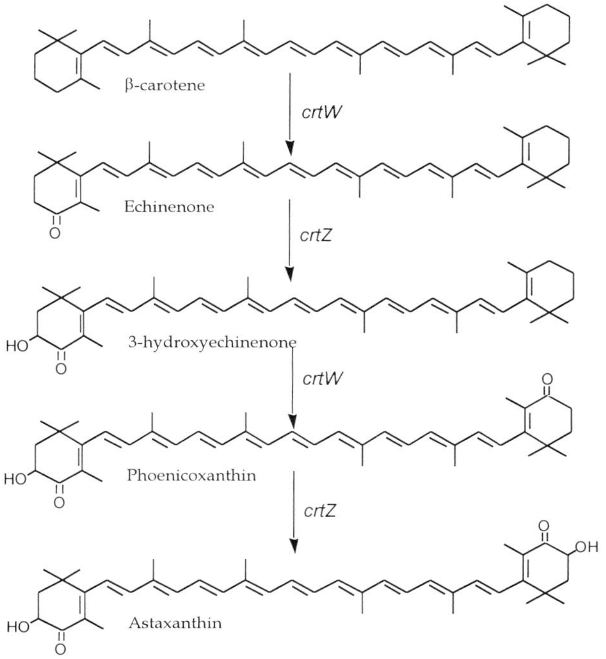

Fig. 8. Hypothetical scheme of astaxanthin biosynthesis from β-carotene in the yeast *Phaffia rhodozyma* [80].

vitro evidence suggested that the responsible enzymes, CrtW and CrtZ, had low substrate specificity in that CrtW oxygenated both β--ionone and 3-hydroxy β-ionone, whereas CrtZ converted β--ionone and 4-keto-β--ionone. The enzyme low specificity would promote formation of numerous intermediates *H. pluvialis*. Also, this seems to be the case in the astaxanthin producing bacterium *A. aurantiacum*, which accumulated 9 different oxygenated compounds from β-carotene [212].

The yeast *P. rhodozyma/X. dendrorhous* is unique among fungal species in its ability to produce high levels of astaxanthin (up to 20 mg g^{-1} cell DCW. In 1976, Andrewes and co-workers proposed a biosynthetic route to astaxanthin based on the isolation of xanthophyll intermediates (Fig. 6). In particular, three compounds, echinenone, 3-hydroxyechinenone, and phoenicoxanthin, were proposed to serve as intermediates to astaxanthin from β-carotene [80]. To our knowledge, these two genes, *crtW* and *crtZ*, and their corresponding enzyme products have not been isolated. Furthermore, *P. rhodozyma/X. dendrorhous* accumulates certain carotenes and xanthophylls under certain conditions [80, 114, 123, 124, 157], in particular the carotenoids torulene, 3-hydroxy-3',4'-didehydro-β-_ψ-carotene-4-one (HDCO) and 3,3'-

dihydroxy-_β,ψ-carotene-4,4'-dione (DCD). It is not clear if the accumulation of these compounds, which would not appear to be intermediates to astaxanthin, is due to nonspecificity of the putative enzymes CrtW and CrtZ or to the presence of other enzymes acting on carotene and xanthophyll substrates. It should be emphasized that the evidence for the biosynthesis of astaxanthin from β-carotene is indirect, and it is puzzling and surprising that despite the industrial interest in *P. rhodozyma/X. dendrorhous* as a biotechnological source of astaxanthin, the enzymes responsible for xanthophyll formation have not been definitively identified since discovery of the yeast more than 25 years ago. Ours' and others' laboratories are attempting to isolate the responsible genes and enzymes and it is hoped that this information will become available within the next few years. Lastly, a further enigma is the biological basis for difference in optical isomers (*R* versus *S*) in different astaxanthin-producing organisms, further supporting differences in the genes and corresponding enzymes responsible for astaxanthin biosynthesis from different sources may be possible.

6. PROPERTIES OF CAROTENOGENIC ENZYMES

Although many carotenogenic genes in fungi have been cloned and sequenced, very little is known about the biochemical properties of the corresponding enzyme products. Although phytoene synthase amino acid sequences have been derived from the corresponding genes cloned from several fungal species and associated codon usage, the only purified phytoene synthase have been obtained from tomato (*Lycopersicon esculentum*) [185] and daffodil [214]. The phytoene synthase appeared to have an interesting requirement for a galactolipid and was induced during flowering.

All phytoene dehydrogenases in nature are thought to be membrane bound proteins, except for those in the extreme halobacteria, which are cytosolic [215]. The biochemical properties of phytoene dehydrogenase from *Phycomyces blakesleeanus* were extensively studied after its purification and solubilization [216]. For purification, the membrane-bound enzyme was solubilized from cell extracts with Tween 60 [217]. The extract was fractionated by ion-exchange chromatography and aliquots were tested for phytoene dehydrogenase actyivity by measuring the conversion of ^{14}C-labelled phytoene into phytofluene. The enzyme was further purified by gel-filtration, and its purity checked by SDS-polyacrylamide gel electrophoresis. The apparent molecular mass of the enzyme was only 14 kDa [216]. This was surprising since amino acid sequence from the nucleotide sequence predicted a protein in the range of 53-66 kDa. Furthermore, the purified phytoene dehydrogenases from *Erwinia* and *Synechococcus* had molecular masses of 56 and 53 kDa, respectively. It was suggested that the 14 kDa protein may aggregate *in vivo* [216]. This hypothesis was supported by quantitative genetic complementation analysis of *Phycomyces* mutants, which suggested that four copies of the *carB* gene product are required for a dehydrogenase complex to convert phytoene to lycopene [216]. The mechanisms of transcription regulation, translation, assembly, and insertion into the membrane are difficult experiments that will require considerably more research.

Biochemical studies suggested that phytoene dehydrogenase from *Phycomyces* was unstable, retaining only 60% activity after 7 days at -70°C. The pH optimum was 7.5, and the enzyme appeared to have a requirement for $NADP^+$. Paradoxically, the highest conversion of phytoene to phytofluene was obtained when $NADP^+$ was combined with FAD [216]. These results are similar to those obtained in crude extracts of *Halobacterium cutirubrum* [218], spinach [219], and tomato [220], but differ from results found with *Anacystis* cell extracts [9].

Current evidence suggests that two classes of phytoene desaturases appear to have evolved from enzymes resembling disulfide oxidoreductases [207]. Recently, it was proposed have that phytoene desaturases belong to the FAD superfamily, in which a consensus motif of approximately 50 residues was observed after sequence alignment [U_2(V/I)UGAGU(G/A)GU(A/S)XAX_2LX_3GX_4(L/V)XEX_5GG (R/K)$X_{7\text{-}9}$GX_3(D/E)XG], where X stands for any residue and U stands for a hydrophobic residue. Such findings suggest that members of this FAD superfamily arose by divergent evolution from a common ancestor and did not acquire the dinucleotide motif from a gene fusion event [221].

As previously mentioned, other fungal carotenogenic have not been biochemically characterized in detail. It is apparent that the isolation and characterization of pure enzymes involved in carotenogenesis is difficult, but the use of modern molecular biology techniques including cloning and overexpression of the proteins heterologous hosts will lead to enzyme characterization.

Interesting and promising genetic *in vivo* and *in vitro* experiments involving xanthophyll biosynthesis enzymes including those for astaxanthin have been achieved [201, 202, 210, 211, 213, 222, 223]. The purpose of these experiments was to compare and elucidate the genes and enzymes involved in xanthophyll formation. The approach was to clone the genes in *Escherichia coli* transformants carrying carotenoid biosynthesis genes, particularly *crtE*, *crtB*, *crtI* and *crtY* responsible for zeaxanthin biosynthesis in *Erwinia* species. It was found that genes cloned from *Agrobacterium aurantiacum*, *Alcaligenes PC-1*, and *Haematococcus pluvialis* resulted in the formation of a variety of xanthophylls including astaxanthin. These experiments have provided a foundation for elucidation of the astaxanthin pathway from a variety of organisms (but not *P. rhodozyma*) at the gene level. Although the organisms studied were distinct phylogenetically, their xanthophyll biosynthetic enzymes (*crtZ*, *crtW*, and *bkt,* see Table 3) had conserved gene sequences. In particular, analysis of *crtZ*, revealed a number of conserved histidine residues that were reported as Fe^{2+} binding domains ($H^{25}X_4H$, $H^{38}X_2HH$, and $H^{123}X_2HH$) [222]. Similar domains have been detected in genes encoding membrane bound fatty acid desaturases [222]. In the case of *bkt* and *crtW* enzymes, such motifs occur at $H^{130}X_3H$ and $H^{294}X_2HH$.

Although the investigators did not obtain purified enzymes catalyzing xanthophyll biosynthesis, some interesting results were obtained *in vitro* in cell extracts. It was confirmed that Fe^{2+} is a cofactor necessary in the biosynthesis of xanthophylls. While Fe^{2+} was essential for xanthophyll formation, other cofactors such as NAD^+, $NADP^+$, FAD, FMN as well as divalent ions Mg^{2+} and Mn^{2+} had no stimulatory effect (in fact, in some cases they revealed as inhibitors). A number of xanthophylls were detected in certain cell extracts such as adonixanthin, phoenicoxanthin (adonirubin), cathaxanthin, 3'-hydroxyechinenone, and 3-hydroxyechinenone, suggesting that the enzymes had low substrate affinities and oxygenated at several carbon sites. Overall, the quantities of xanthophylls produced appeared to be low and enzymic kinetic constants such as K_{cat}, V_{max} and K_m were not reported for xanthophyll synthesis enzymes from the various organisms. The specific enzyme activities reported, e.g. pmol h^{-1} mg $protein^{-1}$ were vanishlingly low [222] indicating that relatively poor substrate conversion and enzymic activity was being obtained. The conditions to study these enzyme reactions will need to be optimized. Perhaps, it will be necessary to perform these reactions in micelles or under other conditions that would mimic their membrane-bound nature. Nonetheless, this work has provided a starting point for further purification and elucidation of the enzymes important in xanthophyll synthesis in from various biological sources.

7. SUBCELLULAR LOCALIZATION OF CAROTENOIDS IN FUNGI

In biological systems carotenoids do not appear localized as free molecules in the cell cytoplasm but instead are present in lipid globules or in membranes. Carotenoids are also associated with proteins or lipoproteins [23]. The cellular location would be expected to depend on their structure, hydrophobicity, and modifications such as hydroxylation or esterification (see Figure 9). Modifications would also determine their orientation and localization in membranes [224, 225] (see Fig. 9). Carotenoid association with proteins or membranes imparts a degree of stabilization [68]. The cellular biology of carotenoid synthesis is poorly understood including cellular location of synthesis, trafficicking to membranes or proteins, insertion into membranes, and biochemical modifications such as hydroxylation and esterification. Elucidation of these processes would be valuable from a fundamental as well as a practical perspective, as carotenoids in most organisms are intracellular and the regulation of their synthesis and maximum yield may be related to available cellular space.

The orientation of carotenoids in a membrane would depends on its structure and on the membrane composition [225]. Based on nuclear magnetic resonance and linear dichroism studies, it was possible to demonstrate that β-carotene and nonpolar carotenoids such as lycopene tend to be located parallel with the membrane surface, but present deep within the lipid hydrophobic core [226, 227]. In contrast, carotenoids possessing dihydroxy and dicarbonyl groups on rings such as astaxanthin are thought to contact the polar sites in the membrane (phosphoryl groups), while the chromophore is located within the hydrophobic core. However, certain other xanthophylls such as lutein locate in an analogous manner as β-carotene and are immersed within the bilayer membrane. This is believed to occur with lutein, since it can freely rotate its ε-ring around C-6 adopting a position parallel with the membrane surface, effectively anchoring the molecule along one side of the membrane [225]. The factors determining the mechanisms of insertion, resulting location, and carotenoids orientation within the membrane are not fully understood, although evidences indicate that incorporation of zeaxanthin instead of violaxanthin in plant thylakoids results in lower susceptibility to lipid peroxidation [228]. The presence of carotenoids in the cytoplasmic membrane of *Staphylococcus aureus 18Z* was demonstrated to decrease membrane fluidity and to increase resistance to killing by oleic acid [229]. Continued research should lead to an enhanced understanding of the structural roles and functions of carotenoids in membranes.

The location of carotenoids in the red yeast *P. rhodozyma/X. dendrorhous* was examined by cell fractionation and by microscopy [2, 158]. Examination of yeast strains containing different carotenoid levels and cultured for various times by laser confocal fluorescence microscopy (LCFM) and transmission electron microscopy (TEM) enabled our laboratory to evaluate the subcellular localization of carotenoids [112]. The presence of fluorescent carotenoids in lipid globules (LGs) and in membranes was confirmed by microscopy and cell fractionation [112, 158]. The degree of fluorescence depended on the strain examined. For example, a mutant rich in β-carotene (emitting at 520 nm) and poor in astaxanthin (emitting at 575 nm) showed strong fluorescence at 520 nm but not at 575 nm when illuminated with an argon laser emitting at fitted with various filters to control the emission wavelength. White or albino mutants showed very

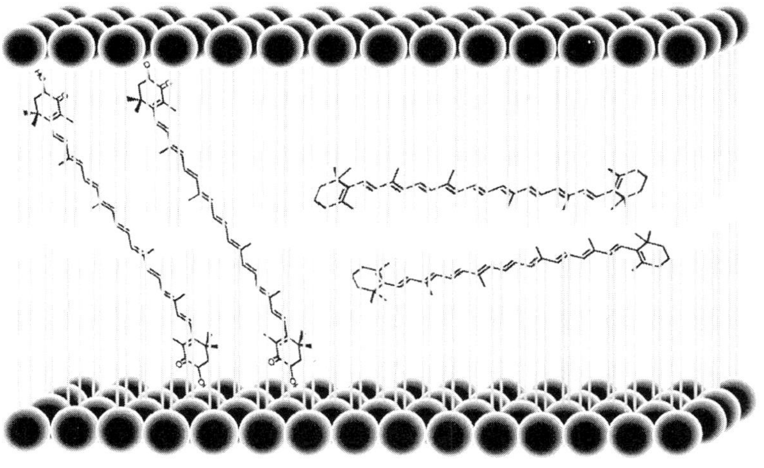

Astaxanthin *β-carotene*

Fig. 9. A hypothetical schematic portraying carotenoids inserted into phospholipid membranes. The orientation would depend on the structure and biophysical properties of the carotenoids, such as polarity and dielectric properties. Astaxanthin is portrayed on the left, where the polar nature of the β-rings with carbonyl and hydroxyl groups is thought to anchor the molecule across the depth of the membrane. In contrast, the comparatively nonpolar β-carotene would be inserted deep within the hydrophobic core [224].

weak fluorescence, as did wild-type yeast grown with piperonyl butoxide, an inhibitor of carotenogenesis in *P. rhodozyma*. The carotenoid content of various yeast strains correlated with the intensities of fluorescence emitted, and prolonged incubation (up to 12 days) containing enhanced carotenoid levels resulted in increased fluorescence [158]. Since the carotenoid content in individual yeast cells correlated with the intensities of fluorescence emitted, it was possible to obtain an estimate of the maximum level in single cells in a population. It was estimated that the maximum yield was approximately 15 mg carotenoid g cell^{-1}. However, there was considerable variation in the number of lipid globules and fluorescence in individual cells, indicating considerable variation within the population. The heterogeneity of fluorescence and apparent carotenoid content is not understood. Interestingly, microscopic images revealed a high concentration of fluorescence in developing bud, suggesting that carotenoids may have an essential role in bud development and growth or survival processes.

The formation and migration of fluorescent LGs and fluorescence in membranes was also investigated. Fluorescent LGs in an astaxanthin-hyperproducing mutant appeared to fragment and migrate to the membrane at an earlier stage of growth than the wild-type. This suggested that the physical fragmentation of LGs and possible surface area of LGs and membranes for deposition was associated with carotenogenesis. LGs may be produced in the endoplasmatic reticulum in mammalian cells and they contain isoprenoids, precursors to carotenoids [230]. Understanding the location of synthesis, trafficking, and deposition of carotenoids could be valuable for development of industrial strains. It is known that HMG-CoA reductase overproducing mutants of *S. cerevisiae* possess large quantities of stacked membranes in endoplasmic reticulum surrounding the nucleus. It is likely that certain enzymes involved in sterol and carotenoid synthesis, particularly in the early stages of precursor and carotene synthesis, share catalytic activities and occur in similar places, such as the endoplasmatic reticulum. Astaxanthin accumulates in the cytoplasm in the laboratory strains of the sweetwater microalga *Haematococcus pluvialis* [231], but massive accumulation of astaxanthin occurs in cysts [232]. Interestingly, astaxanthin appears to initially accumulate in the vicinity of the nucleus [233], possibly in the endoplasmatic reticulum.

8. METABOLIC ENGINEERING FOR CAROTENE PRODUCTION IN FUNGI

Metabolic engineering of isoprenoid pathway in *Candida utilis* and *Saccharomyces cerevisiae* has resulted in increased carotenoid production in common food and feed yeasts with GRAS status [7, 8, 64, 234, 235]. Misawa and colleagues were successful in increasing lycopene yields, primarily by manipulation of HMG-CoA reductase and squalene synthase. In transformed *C. utilis*, the ergosterol content was decreased 65% to the parent strain and accumulated ~6 mg (g cell mass)$^{-1}$, whereas lycopene was increased 7-fold to about 758 μg (g cell mass)$^{-1}$, indicating the need for a selective approach to undesired decrease isoprenoids (ergosterol) and increase the levels of desired products (lycopene). Achieving metabolic engineering for carotenoid production will require a systematic understanding of the coordinate synthesis of various isoprenoids and selective regulation. Furthermore, the results in *C. utilis* and *S. cerevisiae* suggest that it may be preferable to use carotenoid-producing fungi as heterologous host for production since these organisms have evolved the complex cellular biology that is integral to synthesis of carotenoids.

Certain researchers have suggested that plants could serve as ideal systems for industrial production of carotenoids due to their inherent abilities [11-13]. Using metabolic engineering, the genes were introduced for astaxanthin production in tobacco (*Nicotiana tabacum*) [11]. However, the yields of astaxanthin were quite low [205 – 357 μg .(g cell mass)$^{-1}$] compared to production in *P. rhodozyma* and *H. pluvialis* (~5000 to 50,000 μg (g cell mass)$^{-1}$ depending on the strain and culture conditions [2, 59, 78, 157, 236, 237]. Nonetheless, the study of carotenoids in plants could help to understand regulatory mechanisms governing carotenoid biosynthesis and to identify structural genes for production of carotenoids in more efficient heterologous systems (e.g. those encoding lutein). Cultivation of plants containing enhanced levels of carotenoids could also provide nutrients, particularly vitamin A, in economically deprived countries where vitamin A is urgently needed for human health [13].

9. FUNGAL AND ALGAL SPECIES FOR INDUSTRIAL PRODUCTION OF CAROTENOIDS

As pointed out above, certain fungi and microalgae are capable of producing high levels of various carotenoids [1, 2, 59, 65, 238]. Among the lower fungi, the zygomycete *Blakeslea trispora* was used in a process using the stimulators β-ionone and improniazid giving a maximum yield of about 3.2 g l^{-1} of β-carotene [239]. Strain improvement in this fungus was difficult since it is multinucleate in all life stages [240]. Mucorales including *Phycomyces blakesleeanus* and *Choanephora curbibitarium* have been investigated for production of β-carotene, and the basidiomycetous yeasts *Rhodotorula* and *P. rhodozyma*/*X. dendrorhous* have been evaluated for production of β-carotene, torularhodin and astaxanthin [2, 65, 238]. As described above, the genes coding for zeaxanthin from *Erwinia* spp. were used to prepare recombinant strains of *Candida utilis* and *Saccharomyces cerevisiae* that produced lycopene, β-carotene, and zeaxanthin [235, 241].

Microalgae have been used industrially for production of β-carotene and astaxanthin. Natural β-carotene is mainly produced by the halotolerant microalga *Dunaliella salina* [34]. High production depends upon environmental stresses including elevated salt, minerals, temperature, and high light intensities [242]. Under stressful conditions, *Dunaliella* species accumulate β-carotene at up to 10% of their dry weight [34, 243]. The extremely high salt level used in production (~4 M) excludes predatory protozoa and competitive algae and yields carotene-rich dried algal meal or β-carotene-enriched oils. Astaxanthin has been produced by the alga *Haematococcus pluvialis* in pond culture, but production has been problematic due to the freshwater nature of the organism and release of astaxanthin from the cells [59]. The microalga accumulates more than 4% of its dry weight as astaxanthin, in conditions of nitrogen deficiency, oxidative stress, and cyst formation [52]

10. BIOTRANSFORMATIONS USING FUNGAL ENZYMES

Since carotenoids are generally produced as intracellular products, the maximum production is limited by partitioning and available space within the cell. Extractive fermentations or biotransformations using isolated enzymes could lead to a semi-synthetic process for enzymatic synthesis of valuable carotenoids. Synthesis methods using enzymes has been extremely successful in various areas of biotechnology including production of antibiotics, food modifications, and synthesis of industrial chemicals [244, 246]. Generally, the enzyme transformation systems approach or reach industrial feasibility when the substrates and reactions are complex and involve chiral substrates. These considerations indicate that the biotransformation of carotene substrates to more complex and chiral xanthophyll products (e.g. astaxanthin, lutein, zeaxanthin) could be accomplished by biotransformations. Since the substrates are lipophilic and the enzymes are probably membrane-bound in most cases, the syntheses could be performed most efficiently in organic solvent systems [247]. Opportunities for biotransformation of monoterpenes and for bioengineering of isoprenoid biosynthesis have been recently reviewed [134, 248].

11. CONCLUSIONS

Many fungi are prolific producers of a number of isoprenoids including carotenoids, which are a family of terpenoids with economical, aesthetic, and medicinal value. Compared to many other groups of organisms such as bacteria, plants, and animals, several fungal phyla including species within Zygomycetes, Ascomycetes, Basidiomycetes and the asexual Deuteromycetes produce high levels of carotenoids obtaining yields of 5 to 30 mg g^{-1} of biomass. To our knowledge, the only other biological group of organisms that accumulates such high concentrations of carotenoid per unit biomass are microalgae such as *Dunaliella* and *Haematococcus* species and related green microalgae. In contrast to microalgae, fungi can be intensively cultured heterotrophically in fermentors to achieve high biomass and product yields. As the molecular biology of carotenoid and other secondary metabolite biosynthesis as well as fungal regulatory circuits controlling their synthesis become elucidated [39] and heterologous systems are developed [249], well-established fungal systems such *Saccharomyces* could be used for heterologous production of carotenoids and other important industrially important primary and secondary metabolites. However, as discussed above, it may be desirable to use fungi that naturally produce carotenoids as producers of natural or heterologously introduced carotenoids as these fungi have evolved the cellular and molecular biological attributes associated with carotenoid production. Much of the current research on natural isoprenoid biosynthesis is still at the level of gene discovery and characterization. The development of fungal and yeast isoprenoid and carotenoid production systems based on biotechnology will depend on increased research in understanding the genetics of these carotenogenic organisms and heterologous hosts, on the understanding of the remarkably complex and intricate design and regulation of the isoprenoid pathway. Given its diversity and the huge number of compounds that are produced in the isoprenoid pathway, it appears to hold enormous potential for the discovery, production, and availability of numerous compounds for use in animal feeds and for improvement of human health. We are experiencing a revolution in biology with the advent of genomics and proteomics [250] and many tools are becoming available such as monitoring of gene and protein expression *en masse*, and these tools will be exploited for production of isoprenoids through biotechnology. Although most attention of these developing disciplines has been devoted to human disease and human pathogens, the fields are beginning to spread to agricultural and food uses as reflected in the achievement of the genome sequence in *Saccharomyces cerevisiae* [37] and the ongoing sequencing of genomes from plants such as *Arabidopsis* (The Arabidopsis Genome Initiative) [251] and *Lycopersicon*. This information will be valuable in identifying genes in isoprenoid synthesis and the mechanisms governing their expression. Genes encoding secondary metabolites in fungi including carotenoids have been identified using differential display and systematic analysis of gene expression has been studied in fungi including the *S. cerevisiae* [252, 253]. Advances in the research fields of fungal genetics, physiology, and phylogeny [156], the applications of genomics and proteomics to elucidation of isoprenoid synthesis and regulation [2, 5, 10, 118], and the utility of metabolic engineering for carotenoid production [4, 10, 55, 56, 58, 62, 134, 254] should lead to economically important commercial processes for carotenoid production. Lastly, considering the close phylogenetic relation of certain fungi to humans [35], fungi could serve as excellent model systems to study the roles of carotenoids in health and aging.

Appendix

Trivial name	Systematic name
Aleuriaxanthin	(2R)-1',16'-Didehydro-1',2'-didehyro-β,ψ-caroten-2'-ol
Astaxanthin	3,3'-Dihydroxy-β,β-carotene-4,4'-dione
Bixin	Methyl hydrogen 9'-cis-6,6'-diapocarotene-6,6'-dioate
Canthaxanthin	β,β–Carotene-4,4'-dione
Capsanthin	(*3R, 3'S,5'R*)-3,3'-Dihydroxy-β,κ-carotene-6'-one
α-Carotene	(*6'R*)-β,ε-Carotene
β-Carotene	β,β-Carotene
γ-Carotene	β,ψ-Carotene
ζ-Carotene	7,8,7',8'-Tetrahydro-ψ,ψ-carotene
Cryptoxanthin	(3R)-__.__-Caroten-3-ol
3,4-Dehydrolycopene	3,4-Didehydro-_,_-carotene
Echinenone	β,β-Carotene-4-one
HDCO	3-Hydroxy-3',4'-didehydro-β,ψ-carotene-4-one
3-Hydroxyechinenone	3'-Hydroxy-β,β-carotene-4-one
Lutein	(*3R,3R',6'R*)-β,ε-Carotene-3,3'-diol
Lycopene	ψ,ψ-Carotene
Lycoxanthin	ψ,ψ-Carotene-16-ol
Neoxanthin	(3S,5R,6R,3'S,5'R,6'S)-5',6'-Epoxy-6,7-didehydro-5,6,5',6'-tetrahydro-β,β-carotene-3,5,3'-triol
Neurosporaxanthin	4'-Apo-β-carotene-4'-oic acid
Neurosporene	7,8-Dihydro-ψ,ψ-carotene
Phillipsiaxanthin	1,1'Dihydroxy-3,4,3',4'-tetradehydro-1,2,1',2'-tetrahydro-ψ,ψ-carotene-2,2'-dione
Phoenicoxanthin	3-Hydroxy-β,β-carotene-4,4'-dione
Phytoene	15-*cis*-7,8,11,12,7',8',11',12'-Octahydro-ψ,ψ-carotene
Phytofluene	15-*cis*-7,8,11,12,7',8'-Hexahydro-ψ,ψ-carotene
Plectaniaxanthin	(2'R)-3',4'-Didehydro-1',2'-dihydro-β,ψ-carotene-1',2'-diol
Torularhodin	3',4'-Didehydro-β,ψ-carotene-16'-oic acid
Torulene	3',4'-Didehydro-β,ψ-carotene
β-Zeacarotene	7',8'-Dihydro-β,ψ-carotene
Zeaxanthin	(3R,3'R)-β,β-Carotene-3,3'-diol

12. REFERENCES

1. Goodwin, T. W. (1972). Carotenoids in fungi and nonphotosynthetic bacteria, p. 29-88. *In*: D. J. D Hockenhull, (Ed). Progress in Industrial Microbiology. Churchhill -Livingstone, Edinburgh and London.
2. Johnson E. A. and W. Schroeder (1995). Microbial carotenoids, In: Adv. Biochem. Eng. Biotechnol. 53: 119-178.
3. Weedon, B. C. L. (1971). In: O. Isler, (Ed), Carotenoids. Birkhaüser Verlag, Basel und Stuttgart.
4. Barkovich, R. and J. C. Liao (2001). Metabolic engineering of isoprenoids. Metabol. Eng. 3: 27-39.
5. Harker, M. and J. Hirschberg. (1998). Molecular biology of carotenoid biosynthesis in photosynthetic organisms. Meth. Enzymol. 297: 244-263.
6. Lagarde, D., L. Beuf and W. Vermaas (2000). Increased production of zeaxanthin and other pigments by application of genetic engineering techniques to *Synechocystis sp.* Strain PCC 6803. Appl. Environ. Microbiol. 66: 64-72.
7. Miura, Y., K. Kondo, T. Saito, H. Shimada, P. D. Fraserand and N. Misawa (1998a). Production of the carotenoids lycopene, β-caroteneand astaxanthin in the food yeast *Candida utilis*. Appl. Environ. Microbiol. 64: 1226-1229.
8. Miura,, Y., K. Kondo, T. Saito, K. Nakamuraand and N. Misawa. (1998b). Production of lycopene by the food yeast, *Candida utilis*, that does not naturally synthesize carotenoid. Biotechnol. Bioeng. 58: 306-308.
9. Sandmann, G.and S. Kowalczyk (1989). *In vitro* carotenogenesis and characterization of the phytoene desaturase reaction in *Anacystis*. Biochem. Biophys. Res. Commun. 163 (2): 916-921.
10. Sandmann, G. (2001). Carotenoid biosynthesis and biotechnological application. Arch. Biochem. Biophys. 385: 4-12.
11. Mann, V., M. Harker, I. Pecker and J. Hirschberg (2000). Metabolic engineering of astaxanthin production in tobacco flowers. Nature Biotechnol. 18: 888-892.
12. Römer, S., P. D. Fraser, J. W. Kiano, C. A. Shipton, N., Misawa, W. Schuch and P. M. Bramley (2000). Elevation of provitamin A content of transgenic tomato plants. Nature Biotechnol. 18: 666-669.
13. Ye, X., S. Al-Babili, A. Klöti, J. Zhang, P. Lucca, P. Beyer and I. Portrykyus (2000). Engineering the provitamin A (β-carotene) biosynthetic pathway into (carotenoid-free) rice endosperm. Science 287: 303-305.
14. Humphrey, J. H., K. P. West, Jr. and A. Sommer (1992). Vitamin A deficiency and attributable mortality under 5 year olds. WHO Bull. 70: 225-232.
15. Sommer, A. (1997). Vitamin A deficiency, child health and survival. Nutrition 13: 484-485.
16. Bendich, A. and J. A. Olson (1989). Biological actions of carotenoids. FASEB J. 3: 1927-1932.
17. Burton, G. W. and S. K. Ingold (1984). β-Carotene: An unusual type of lipid antioxidant. Science 224: 569-573.
18. Halliwell, B. (1997. Antioxidants and human disease: A general introduction. Nutr. Rev. 55: 544-552.
19. Krinsky, N. I. (1998. The antioxidant and biological properties of the carotenoids. Ann. N. Y. Acad. Sci. 854: 443-447.
20. Krinsky, N. I. and S. T. Many (1999). Current views on carotenoids: biology, epidemiology and trials, p. 46-56. *In*: M. A. Livrea (ed.), Vitamin A and Retinoids: an Update of Biological Aspects and Clinical Applications. Birkhäuser Verlag, Basel.
21. Naguib, Y. M. A. (2000). Antioxidant activities of astaxanthin and related carotenoids. J. Agric. Food Chem. 48: 1150-1154.
22. Palozza, P. and N. I. Krinsky (1992). Antioxidant effects of carotenoids *in vivo* and *in vitro*: an overview. Meth. Enzymol. 213: 403-420.
23. Young, A. J. and G. M. Lowe (2001). Antioxidant and prooxidant properties of carotenoids. Arch. Biochem. Biophys. 385: 20-27.
24. Bertram, J. S. (1999). Carotenoids and gene regulation. Nutr. Rev. 57: 182-191.
25. Chew, B. P., J. S. Park, M. W. Wong and T. S. Wong (1999). A comparison of the anticancer activities of dietary β-carotene, canthaxanthin and astaxanthin in mice *in vivo*. Anticancer Res. 19: 1849-1854.
26. Hinds, T. S., W. L. West and and E. M. Knight (1997). Carotenoids and retinoids: a review of research, clinical and public health applications. J. Clin. Pharmacol. 37: 551-558.
27. Mayne, S. T. (1996). Beta-carotene, carotenoids and disease prevention in humans. FASEB J. 10: 690-701.
28. Miki, W. (1991). Biological functions and activities of animal carotenoids. Pure Appl. Chem. 63: 141-146.

29. von den Bergh. (1999). Carotenoid interactions. Nutr. Rev. 57: 1-10.
30. Wang, X, R., Willén and T. Wadström (2000). Astaxanthin-rich meal and vitamin C inhibit *Helicobacter pylori* infection in BALB/cA mice. Antimicrob. Ag. Chemother. 44: 2452-2457.
31. Chew, B. P. (1993). Role of carotenoids in the immune response. J. Dairy Sci. 76: 2804-2811.
32. Palozza, P., N. Maggiano, G. Calviello, P. Lanza, E. Piccioni, F. O. Ranaletti and G. M. Bartoli (1998). Canthaxanthin induces apoptosis in human cancer cell lines. Carcinogenesis 19: 373-376.
33. Vershinin, A. (1999. Biological functions of carotenoids – diversity and evolution. BioFactors 10: 99-104.
34. O'Callaghan, M. C. (1996). Biotechnology of food colours: The role of bioprocessing. *In*: G. A. F. Hendry and J. D. Houghton, (Eds.), Natural Food Colorants, 2nd Edition, Blackie Academic & Professional, Chapman & Hall, London.
35. Baldauf, S. L. and J. D. Palmer (1993). Animals and fungi are each other's closest relatives: Congruent evidence from multiple proteins. Proc. Natl. Acad. Sci. USA 90: 11558-11562.
36. Zhang, S., A. Varma and P. R. Willianson (1999). The yeast *Cryptococcus neoformans* uses 'mammalian'enhancer sites in the regulation of the virulence gene, *CNLAC1*. Gene 227: 231-240.
37. Goffeau. A., B. G. Barrell, H. Bussey, R. W. Davis, B. Dujon, H. Feldmann, F. Galibert, J. D. Hoheisel, C. Jacq, M. Johnston, E. J. Louis, H. W. Mewes, Y. Murakami, P. Philippsen, H. Tettelin and S. G. Oliver. (1996). Life with 6000 Genes. Science 274: 546-567.
38. Dujon, B. (1996). The yeast genome project: what did we learn? Trends Genet. 12: 263-270.
39. Tudzynski, P.and B. Tudzynski (1997). Fungal genetics: novel techniques and regulatory circuits, p. 229-249. *In*: Anke, T. (Ed.), Fungal Biotechnology. Chapman & Hall, Weinheim.
40. Arrach, N., R. Fernández-Martín, E. Cerdá-Olmedo and J. Avalos (2001). A single gene for lycopene cyclase, phytoene synthaseand regulation of carotene biosynthesis in *Phycomyces*. Proc. Natl. Acad. Sci. USA 98: 1687-1692.
41. Golubev, W. I. (1995). Perfect state of *Rhodomyces dendrorhous* (*Phaffia rhodozyma*). Yeast 11: 101-110.
42. Kurtzman, C. P.and J. W. Fell, (Eds.). The Yeasts, A Taxonomic Study. 4th edition, Elsevier, Amsterdam (1998).
43. Caporale, L. H. (1995). Chemical ecology: a view from the pharmaceutical industry. Proc. Natl. Acad. Sci. USA 92: 75-82.
44. Liaaen-Jensen, S. and A. G. Andrewes (1972). Microbial carotenoids. Annu. Rev. Microbiol. 26: 225-248.
45. Goodwin, T. W. (1980). The biochemistry of the carotenoids. Vol 1, Chapman and Hall, London, p. 377.
46. Krubasik, P. and G. Sandmann (2000). A carotenogenic gene cluster from *Brevibacterium linens* with novel lycopene cyclase genes involved in the synthesis of aromatic carotenoids. Mol. Gen. Genet. 263: 423-432.
47. Krügel, H., P. Krubasik, K. Weber, H. P. Saluz and G. Sandmann (1999). Functional analysis of genes from *Streptomyces griseus* involved in the synthesis of isorenieratene, a carotenoid with aromatic end groups, revealed a novel type of carotenoid desaturase. Biochim. Biophys. Acta 1439: 57-64.
48. Raisig, A. and G. Sandmann (1999). 4,4'-Diapophytoene desaturase: catalytic properties of an enzyme from the C_{30} carotenoid pathway of *Staphylococcus aureus*. J. Bacteriol. 181: 6184-6187.
49. Viveiros, M., P. Krubasik, G. Sandmann and M. Houisaini-Iraqui (2000). Structural and functional analysis of the gene cluster encoding carotenoid biosynthesis in *Mycobacterium aurum* A+. FEMS Microbiol. Lett. 187: 95-101.
50. Rowan, K. S. (1989). Photosynthetic pigments of algae. Cambridge University Press, Cambridge.
51. Liu, B.-H. and Y.-K. Lee (1999). Composition and biosynthetic pathways of carotenoids in the astaxanthin-producing green alga *Chlorococcum* sp. Biotechnol. Lett. 21: 1007-1010.
52. Margalith, P. Z. (1999). Production of ketocarotenoids by microalgae. Appl. Microbiol. Biotechnol. 51: 431-438.
53. Bauernfeind, J. C., G. B. Brubacher, H. M. Kläui and W. L. Marusisch (1971) Use of carotenoids, p. 743-770. *In*: O. Isler, (Ed). Carotenoids. Birkhäuser Verlag, Basel und Stuttgart.
54. Cunningham, F. X, Jr. and E. Gantt (1998). Genes and enzymes of carotenoid biosynthesis in plants. Annu. Rev. Plant Physiol. Mol. Biol. 49: 557-583.
55. Dixon, R. A. (1999). Plant natural products: the molecular genetic basis of biosynthetic diversity. Curr. Opin. Biotechnol. 10: 192-197.
56. Hirschberg, J. (1999). Production of high-value compounds: carotenoids and vitamin E. Curr. Opin. Biotechnol. 10: 186-191.

57. Cerdá-Olmedo, E. (1989). Production of carotenoids with fungi, p. 27-42. In: E. J. Vandamme, (Ed.), Biotechnology of Vitamins, Pigmentsand Growth Factors, Elsevier Applied Science, London .
58. Misawa, N. and H. Shimada (1998). Metabolic engineering for the production of carotenoids in non-carotenogenic bacteria and yeasts. J. Biotechnol. 59: 169-181.
59. Johnson, E. A. and G. H An (1991). Astaxanthin from microbial sources. Crit. Rev. Biotechnol. 11(4): 297-326.
60. Bohlmann, J., G. Meyer-Gauen and R. Croteau (1998). Plant terpenoid synthases: Molecular biology and phylogenetic analysis. Proc. Natl. Acad. Sci. USA 95: 4126-4133.
61. Sandmann, G. (1997). High level expression of carotenogenic genes for enzyme purification and biochemical characterization. Pure Appl. Chem. 69: 2163-2168.
62. Sandmann, G., M. Albrecht, G. Schnurr, O. Knörzer and P. Böger (1999). The biotechnological potential and design of novel carotenoids by gene combination in *Escherichia coli*. Trends Biotechnol. 17: 233-237.
63. Schmidt-Dannert, C., D. Umeno and F. Arnold (2000). Molecular breeding of carotenoid biosynthetic pathways. Nature Biotechnol. 18: 750-753.
64. Shimada, H., K. Kondo, P. D. Fraser, M. Yutaka, T. Saito and N. Misawa (1998). Increased carotenoid production by the food yeast *Candida utilis* through metabolic engineering of the isoprenoid pathway. Appl. Environ. Microbiol. 64 (7): 2676-2680.
65. Stahmann, K.P. (1997). Vitamins, p. 81-90. *In*: T Anke (Ed.), Fungal Biotechnology, Chapman & Hall, London.
66. Ong, A. S. H. and E. S. Tee (1992). Natural sources of carotenoids from plants and oils. Meth. Enzymol. 213: 142-167.
67. Stafford, A. (1986). Plant cell biotechnology: perspectives. Enzyme Microb. Technol. 8: 578.
68. Britton, G., S. Liaaen-Jensen and H. Pfander, (Eds.) (1995a). Carotenoids. Vol 1A: Isolation and Analysis, Birkhäuser Verlag, Basel.
69. Britton, G., S. Liaaen-Jensen and H. Pfander, (Eds.) (1995b). Carotenoids. Vol 1B: Spectroscopy, Birkhäuser Verlag, Basel.
70. Pfander, H. (1992). Carotenoids: An overview. Meth. Enzymol. 213: 3-13.
71. Eugster, C. H. (1995). History: 175 years of carotenoid chemistry, p. 1-12. *In*: Britton, G., S. Liaaen-Jensen and H. Pfander (Eds.), Carotenoids. Vol. 1A. Isolation and Analysis. Birkhäuser Verlag, Basel
72. Goodwin, T. W. (1971). Biosynthesis, p. 577-636. *In*: O. Isler, (Ed), Carotenoids. Birkhäuser Verlag, Basel and Stuttgart.
73. Pfander, H. (1987). Key to carotenoids, 2nd Edition, Appendix (Vol. 1A, Chapter 8). Birkhäuser, Basle (1987).
74. Weedon, B. C. L. and G. P. Moss (1995). Structure and nomenclature, p. 27-69. *In:* Britton, G., Liaaen-Jensen, S. and Pfander H (Eds), Carotenoids, Vol Ia, Isolation and analysis, Birkhäuser, Basle.
75. Davies, B. H. (1976). Carotenoids, p. 38-165. *In*: Goodwin, T. W (Ed.), Chemistry and biochemistry of plant pigments, Academic Press, London, England.
76. Koyama, Y. and R. Fujii (1999). *Cis-trans* carotenoids in photosynthesis: configurations, excited-state properties and physiological functions, p. 161-168. *In*: H. A. Frank, A. J. Young, G. Britton and R. J. Cogdell (Eds.), The Photochemistry of Carotenoids. Kluwer Academic Publishers, The Netherlands.
77. Matsuno, T. (1992). Structure and characterization of carotenoids from various habitats and natural sources. Meth. Enzymol. 213: 22-31.
78. Yuan, J. –P. and F. Cheng (1999). Isomerization of *trans*-astaxanthin to *cis*-isomers in organic solvents. J. Agr. Food Chem. 47: 3656-3660.
79. Andrewes, A. G. and M. P. Starr (1976). (3*R*,3'*R*)-Astaxanthin from the yeast *Phaffia rhodozyma.* Phytochem. 15: 1009-1011.
80. Andrewes, A. G., H. J. Phaff and M. P. Starr (1976). Carotenoids of *Phaffia rhodozyma*, a red pigmented fermenting yeast. Phytochem. 15: 1003-1007.
81. Stahl, W., W. Schwarz, A. R. Sundquist and H. Sies (1992). *Cis-trans* isomers of lycopene and β-carotene in human serum and tissues. Arch. Biochem. Biophys. 294: 173-177.
82. Krinsky, N. I., M. D. Russett, G. J. Handelman and D. M. Snodderly (1990). Structural and geometrical isomers of carotenoids in human plasma. J. Nutr. 120: 1654-1662.
83. Østerlie, M., B. Bjerkeng and S. Liaaen-Jensen (1999). Accumulation of astaxanthin all-*E*, 9*Z* and 13*Z* geometrical isomers and 3 and 3' *RS* optical isomers in rainbow trout (*Oncorhynchus mykiss*) is selective. J. Nutr. 129: 391-398.

84. Parker, R. S. (1996). Absorption, metabolism and transport of carotenoids. FASEB J. 10: 542-551.
85. Craft, N. E. and J. H. Soares. 1992. Relative solubility, stabilityand absorptivity of lutein and β-carotene in organic solvents. J. Agric. Food Chem. 40: 431-434.
86. Schiedt, K. and S. Liaaen-Jensen (1995). Isolation and analysis, p. 81-108. *In*. Britton, G., S. Liaaen-Jensen and H. Pfander, (Eds.), Carotenoids. Vol. 1A. Isolation and Analysis. Birkhäuser Verlag, Basel.
87. Sedmak, J. J., D. K. Weerasinghe and S. O. Jolly (1990). Extraction and quantification of astaxanthin from *Phaffia rhodozyma*. Biotechnol. Techniq. 4: 107-112.
88. Latscha, T. (1990). Carotenoids in animal nutrition. F. Hoffmann-La Roche Ltd Print, 2175, Basel, Switzerland.
89. Foyer, C. H. and J. Harbinson (1999). Relationships between antioxidant metabolism and carotenoids in the regulation of photosynthesis, p. 327-339. *In*: Frank, H. A., A. J. Young, G. Brittonand R. J. Cogdell (Eds.), The Photochemistry of Carotenoids. Adv. Photosyn. vol. 8. Kluwer Academic Publishers, Dordrecht (1999).
90. Telfer, A., J. De las Rivas and J. Barber (1991). Carotene within the isolated phostosystem II reaction center: Photooxidation and irreversible bleaching of this chromophore by oxidised P680. Biochim. Biophys. Acta 1060: 106-114.
91. Burton, G. W. (1989). Antioxidant action of carotenoids. J. Nutr. 119: 109-111.
92. Foote, C. S. and R. W. Denny (1968). Chemistry of singlet oxygen. VII. Quenching by beta-carotene. J. Am. Chem. Soc. 90: 6233-6235.
93. Krinsky, N. I. (1989). Antioxidant functions of carotenoids. Free Radical Biol. Med. 7: 617-635.
94. Mortensen, A., L. H. Skibsted and T. G. Truscott (2001). The interaction of dietary carotenoids with radical species. Arch. Biochem. Biophys. 385: 13-19.
95. Schroeder, W. A. and E. A. Johnson (1993). Antioxidant role of carotenoids in *Phaffia rhodozyma*. J. Gen. Microbiol. 139: 907-912.
96. Schroeder, W. A. and E. A. Johnson (1995a). Carotenoids protect *Phaffia rhodozyma* against singlet oxygen damage. J. Indust. Microbiol. 14: 502-507.
97. Tuveson, R. W., R. A. Larson and J. Kagan (1988). Role of cloned genes expressed in *Escherichia coli* in protecting against inactivation by near-UV light and specific phototoxic molecules. J. Bacteriol. 170: 4675-4680.
98. Ames, B. N., M. K. Shigenaga and T. M. Hagen (1993). Oxidants, Antioxidantsand the Degenerative Diseases of Aging. Proc. Natl. Acad. Sci. USA 90: 7915-7922.
99. Nakano, T., T. Kanmuri, M. Sato and M. Takeuchi (1999). Effect of astaxanthin rich yeast (*Phaffia rhodozyma*) on oxidative stress in rainbow trout. Biochim. Biophys. Acta 1426: 119-125.
100. Peto, R., R. Doll, J. D. Buckley and M. B. Sporn (1981). Can dietary β-carotene materially reduce cancer rates? Nature (London) 290: 201-208.
101. Horton, P., A. V. Ruban and A. J. Young (1999). The electronic states of carotenoids, p. 137-157. *In*: Frank, H. A., A. J. Young, G. Britton and R. J. Cogdell, R, (Eds), The photochemistry of carotenoids, Kluwer Academic, Dordreacht, The Netherlands.
102. van den Berg, H., R. Faulks, F. H. Granado, J. Hirschberg, B. Olmedilla, G. Sandman, S. Southon and W. Stahl (2000). The potential for the improvement of carotenoid levels in food and the likely system effects. J. Sci. Food. Agr. 80: 880-912.
103. Rice-Evans, C. A., J. Sampson, P. M. Bramley and D. E. Holloway (1997). Why do we expect carotenoids to be antioxidants *in vivo*? Free Radical Res. 26: 381-398.
104. Baltschun, D., S. Beutner, K. Briviba, H. D. Martin, J. Paust, M. Peters, S. Röver, H. Sies, W. Stahl and F. Stenhorst (1997). *Singlet Oxygen Quenching Abilities of Carotenoids.* Liebigs Ann-Recueil. 9: 1887-1893.
105. Woodall, A. A., S. W. Lee, R. J. Weesie, M. J. Jackson and G. Britton (1997). Oxidation of carotenoids by free radicals: relationship between structure and reactivity. Biochim Biophys Acta. 1336: 33-42.
106. Halliwell, B. and J. M. C. Gutteridge (1999). Free radicals in biology and medicine, 3rd ed., Oxford Univ. Press, Oxford.
107. Mortensen, A., L. H. Skibsted, J. Sampson, C. Rice-Evans and S. A. Everett (1997). Comparative mechanisms and rates of free radical scavenging by carotenoid antioxidants. FEBS Lett. 418:91-97.
108. Palozza, P. (1998). Proxidant actions of carotenoids in biologic systems. Nutr. Rev. 56: 257-265.
109. Jyonouchi, H., S. Sun, K. Iijima and M. D. Gross (2000). Antitumor activity of astaxanthin and its mode of action. Nutr. Cancer 36: 59-65.
110. Nishino, H. (1998). Cancer prevention by carotenoids. Mutat. Res. 402: 159-163.

111. Wolz, E., H. Liechti, B. Notter, G. Oesterhelt and A. Kistler (1999). Characterization of metabolites of astaxanthin in primary cultures of rat hepatocytes. Drug Metab. Dispos. 27: 456-462.

112. An, G. -H., J. Bielich, R. Auerbach and E. A. Johnson (1991). Isolation and characterization of carotenoid hyperproducer mutants yeast by flow cytometry and cell shorting. Bio. Technol. 9: 70.

113. Schroeder, W. A. and E. A. Johnson (1995b). Singlet oxygen and peroxyl radicals regulate carotenoid biosynthesis in *Phaffia rhodozyma.* J. Biol. Chem. 270: 18374-18379.

114. Schroeder, W. A., P. Calo, M. L. DeClerq and E. A. Johnson (1996). Selection for carotenogenesis in the yeast *Phaffia rhodozyma* by dark-generated singlet oxygen. Microbiol. 142: 2923-2929.

115. Gershon, H. and D. Gershon (2000). The budding yeast, *Saccharomyces cerevisiae*, as a model for aging research: a critical review. Mech. Ageing Develop. 120: 1-22.

116. Guarente, L. and C. Kenyon (2000). enetic pathways that regulate ageing in model organisms. Nature 408: 255-262.

117. Jamieson, D. J. (1998). Oxidative stress responses of the yeast *Saccharomyces cerevisiae.* Yeast 14: 1511-1527.

118. Packer, L. (Ed.). (1992). Carotenoids. Part A. Chemistry, Separation, Quantitation and Autoxidation. Meth. Enzymol. vol. 213. Academic Press, San Diego.

119. Britton, G. (1995). Structure and properties of carotenoids in relation to function. FASEB J. 9: 1551-1558.

120. Schmitz, H. H., C. L. Poor, E. T. Gugger and J. W. Erdman, Jr. (1993). Analysis of carotenoids in human and animal tissues, p. 102-116. *In*: Packer, L. (Ed.), Meth. Enzymol., vol. 214. Academic Press, San Diego.

121. Simpson, K. L., C. O. Chichester and H. J. Phaff (1971). Carotenoid pigments of yeasts, p. 493-515. In: Rose, A. H. and J. S. Harrison, (Eds.), The Yeasts, Vol. 2. Academic Press, New York.

122. Johnson, E. A., T. G. Villa, M. J. Lewis and H. J. Phaff (1978). Simple method for the isolation of astaxanthin from the basidiomycetous yeast *Phaffia rhodozyma.* Appl. Environ. Microbiol. 35: 1155-1159.

123. An, G. H., D. B. Schuman and E. A. Johnson. (1989). Isolation of *Phaffia rhodozyma* mutants with increased astaxanthin content. Appl. Environ. Microbiol. 55: 116-124.

124. Johnson, E. A. and M. J. Lewis (1979). Astaxanthin formation by the yeast *Phaffia rhodozyma.* J. Gen. Microbiol. 115: 173-183.

125. Johnson, E. A., M. J. Lewis and C. R. Grau (1980). Pigmentation of egg yolks with astaxanthin from the yeast *Phaffia rhodozyma.* Poult. Sci. 59: 1777-1782.

126. Pera, L. M., L. Rubinstein, M. D. Baigorí, L. I. C. de Figueroa and D. A. Callieri (1999). Influence of manganese on cell morphology, protoplast formation and D-glucosidase activity in *Phaffia rhodozyma.* FEMS Lett. 171: 155-160.

127. Wery, J., M. J. M. Dalderup, J. Ter Linde, and A. J. J. van Ooyen (1996). Structural and phylogenetic analysis of the actin gene from the yeast *Phaffia rhodozyma.* Yeast 12: 641-651.

128. Velayos, A., A. P. Eslava and E. A. Iturriaga (2000b). A bifunctional enzyme with lycopene cyclase and phytoene synthase activities is encoded by the *carRP* gene of *Mucor circinelloides.* Eur. J. Biochem. 267: 5509-5519.

129. Pfander, H. Chromatography: Part I. (1995). General Aspects. *In*: Britton, G., S. Liaaen-Jensen and H. Pfander, (Eds), Carotenoids. Vol 1A: Isolation and analysis, Birkhäuser, Basle, pp 109-116.

130. Straub, O. (1987). Key to Carotenoids, 2nd edition. H. Pfander (Ed.). Birkhäuser Verlag, Basel.

131. Isler, O. (Ed.) (1971). The Carotenoids. Bitkhäuser Verlag, Basel.

132. Pecker, I., D. Chamovitz, H. Linden, G. Sandmann and J. Hoirschberg (1992). A single polypeptide catalyzing the conversion of phytoene to □-carotene is transcriptionally regulated during tomato fruit ripening. Proc. Natl. Acad. Sci. USA 89: 4962-4966.

133. Connolly, J. D. and R. A. (1991). Hill. Dictionary of terpenoids. Chapman and Hall, New York (1991).

134. McCaskill, D. and R. Croteau (1997). Prospects for the bioengineering of isoprenoid biosynthesis. Adv. Biochem. Eng. 55: 107-146.

135. Anke, T. (Ed.) (1997). Fungal Biotechnology. Chapman & Hall, London.

136. Vandamme, E. J, (Ed.) (1989). Biotechnology of Vitamins, Pigments and Growth Factors. Elsevier Applied Science, London.

137. Vandamme, E. J. (1992). Production of vitamins, coenzymes and related biochemicals by biotechnological processes. J. Chem. Tech. Biotechnol. 53: 313-327.

138. Cai, P., D. Smith, B. Cunningham, S. Brown-Shimer, B. Katz, C. Pearce, D. Venables and D. Houck. (1998). Epolones: novel sesquiterpene-tropolones from fungus OS-F69284 that induce erythropoietin in human cells. J. Natural Products 61: 791-795.

139. Bailly, J. D., P. Le Bars, A. Pietri, G. Benard and J. Le Bars (1999). Evaluation of a fluorodensitometric method for analysis of ergosterol as a fungal marker in compound feeds. J. of Food Protect. 62: 686-690.
140. Bejarano, E. R. and E. Cerdá-Olmedo (1992). Independence of the carotene and sterol pathways of *Phycomyces*. FEBS Letters 306: 209-212.
141. Hirai, N., R. Yoshida, Y. Todoroki and H. Ohigashi (2000). Biosynthesis of abscisic acid by the non-mevalonate pathway in plants and by the mevalonate pathway in fungi. Biosci. Biotechnol. Biochem. 64: 1448-1458.
142. Atta, R. U., A. Nesreen, F. Akhtar, M. S. Sheckhani, J. Clardy, M. Parvez and M. I. Choudhari (1997). Antifungal diterpenoid alkaloids from *Delphinium denudatum*. J. Natural Prod. 60: 472-474.
143. Dewick, P. M. (1997). Medicinal Natural Products. A Biosynthetic Approach, John Wiley & Sons, Chichester .
144. Milanova, R. and M. Moore (1993). The hydroxylation of plant diterpene analogues by the fungus *Syncephalastrum racemosum*. Archives of Biochem. Biophys. 303: 165-171.
145. Nes, W. D., E. J. Parish and J. M. Trzaskos (Eds.) (1992). Regulation of isopentenoid metabolism. ACS Symp. Series 497, Amer. Chem. Soc., Washington, DC.
146. Azevedo, J. E. and A. Videira (1994). Characterization of a membrane fragment of respiratory chain complex I from *Neurospora crassa*. Insights on the topology of the ubiquinone-binding site. Int. J. Biochem. 26: 505-510.
147. Brunt., S. A and J. C. Silver (1991). Molecular cloning and characterization of two distinct *hsp85* sequences from the steroid responsive fungus *Achlya ambisexualis*. Curr Genet. 19: 383-388.
148. Grünler, J., J. Ericcson and G. Dallner (1994). Branch-point relationships in the biosynthesis of cholesterol, dolichol, ubiquinone and prenylated proteins. Biochim. Biophys. Acta 1212: 259-277.
149. Bocking T, K., D. Barrow, A. G. Netting, T. C. Chilcott, H. G. Coster and M. Hofer (2000). Effects of singlet oxygen on membrane sterols in the yeast *Saccharomyces cerevisiae*. Eur. J. Biochem. 267: 1607-1618.
150. Rohmer, M., M. Knami, P. Simonin, B. Sutter and H. Sahm (1993). Isoprenoid biosynthesis in bacteria: a novel pathway for the early steps leading to isopentenyl diphosphate. Biochem. J. 295: 517-524.
151. Rohmer, M. (1999). The discovery of a mevalonate-independent pathway for isoprenoid biosynthesis in bacteria, algae and higher plants. Nat. Prod. Rep. 16: 565-574.
152. Disch, A. and M. Rohmer (1998). On the absence of the glyceraldehyde 3-phosphate/pyruvate pathway for isoprenoid biosynthesis in fungi and yeasts. FEMS Microbiol. Lett. 168: 201-208.
153. Miller, M. W., M. Yoneyama and M. Soneda (1976). *Phaffia*, a new yeast genus in the Deuteromycotina (Blastomycetes). Int. J. Sys. Bacteriol. 26: 286-291.
154. Fell, J. W. and G. M. Blatt (1999). Separation of strains of the yeasts *Xanthophyllomyces dendrorhous* and *Phaffia rhodozyma* based on rDNA IGS and ITS sequence analysis. J. Indust. Microbiol. Biotechnol. 23: 677-681.
155. Fell, J. W., T. Boekhout, A. Fonseca, G. Scorzetti, and A. Statzell-Tallman (2000). Biodiversity and systematics of basidiomycetous yeasts as determined by large-subunit rDNA D1/D2 domain sequence analysis. Int. J. Syst. Evol. Microbiol. 3:1351-1371.
156. McLaughlin, E. G. McLuaghlin and P. A. Lemke (Eds.) (2001). The *Mycota*: a comprehensive treatise on fungi as experimental systems for basid and applied research. VII. Systematics and Evolution. Springer-Verlag, Berlin.
157. An, G. H., M. -H. Cho and E. A. Johnson (1999). Monocyclic carotenoid biosynthetic pathway in the yeast *Phaffia rhodozyma* (*Xanthophyllomyces dendrorhous*). J. Biosci. Bioeng. 88: 189-193.
158. An, G. H., O. S. Suh, H. -C. Kwon, K. Kim and E. A. Johnson (2000). Quantification of carotenoids in cells of *Phaffia rhodozyma* by autoflorescence. Biotech. Lett. 22: 1031-1034.
159. Bon, J. A., T. D. Leathers and R. K. Jayaswal (1997). Isolation of astaxanthin-overproducing mutants of *Phaffia rhodozyma*. Biotechnol. Lett. 19: 109-112.
160. Calo, P., T. De Miguel, B. Jorge and T. G. Villa (1995). Mevalonic acid increases *trans*-astaxanthin and carotenoid biosynthesis in *Phaffia rhodozyma*. Biotechnol. Lett. 17: 575-578.
161. Chan, H. Y. and K. P. Ho (1999). Growth and carotenoid production by pH-stat cultures of *Phaffia rhodozyma*. Biotechnol. Lett. 21: 953-958.
162. Gu, W. –L., G. –H. An, and E. A. Johnson. (1997). Ethanol increases carotenoid production in Phaffia rhodozyma. J. Indust. Microbiol. Biotech. 19: 114-117.
163. Endo, A. and K. Hasumi (1993). HMG-CoA reductase inhibitors. Nat. Prod. Rep. 10: 541-550.
164. Goldstein, J. L. and M. S. Brown (1990). Regulation of the mevalonate pathway. Nature 343: 425-430.

165. Verdoes, J. C. and A. J. J. van Ooyen (1999). Isolation of the isopentenyl diphosphate isomerase encoding gene of *Phafia rhodozyma*; improved carotenoid production in *Escherichia coli*. Acta Bot. Gallica 146: 43-53.
166. Basson, M. E., M. Thorsness and J. Rine (1986). *Saccharomyces cerevisiae* contains two functional genes encoding 3-hydroxy-3-methylglutaryl coenzyme A reductase. Proc. Natl. Acad. Sci. USA 83: 5563-5567.
167. Dimster, D. D., M. K. Thorsness and J. Rine (1994). Feedback regulation of 3-hydroxy-3-methylglutaryl coenzyme A reductase in *Saccharomyces cerevisiae*. Mol. Biol. Cell. 5: 656-665.
168. Hampton, R., D. Dimster-Denk and J. Rine (1996). The biology of HMG-CoA reductase: the pros of contra-regulation. Trends Biochem. Sci. 21: 140-145.
169. Casey, W. M., G. A. Keesler and L. W. Parks (1992). Regulation of partitioned sterol biosynthesis in *Saccharomyces cerevisiae*. J. Bacteriol. 174: 7283-7288.
170. Thorsness, M., W. Schalfer, L. D'ari and J. Rine (1989). Positive and negative transcriptional control by heme of genes encoding 3-hydroxy-3-methylglutaryl coenzyme A reductase in *Saccharomyces cerevisiae*. Mol. Cell. Biol. 9: 5702-5712.
171. Bramley, P. M. and A. Mackenzie (1988). Regulation of carotenoid biosynthesis, Vol. 29, p. 291-343. *In*: B. L. Horockerand E. R. Stadtman (Eds.), Current topics in cellular regulation, Academic Press, San Diego, California.
172. Sandmann, G. (1994). Carotene biosynthesis in microorganisms and plants. Eur. J. Biochem. 223: 7-24.
173. Kuntz, M., S. Romer, C. Suire, P. Hugueny, J. H. Weil, R. Schantz and B. Camara. 1992. Identification of a cDNA for the plastid-located geranylgeranyl pyrophosphate synthase from *Capsicum annum*: correlative increase in enzyme activity and transcript level during fruit ripening. Plant J. 2: 25-34.
174. Young, C., L. McMillan, E. Telfer and B. Scott (2001). Molecular cloning and genetic analysis of an indole-diterpene cluster from *Penicillium paxilli*. Mol. Microbiol. 39: 754-764.
175. Jiang, Y., P. Proteau, D. Poulter and S. Ferro-Novick (1995). BTS1 encodes a geranylgeranyl diphosphate synthase in *Saccharomyces cerevisiae*. J. Biol. Biochem. 270: 21793-21799.
176. Bolotin-Fukuhara, M., C. Toffano-Nieoche, F. Artiguenave, G. Dechateau-Nguyen, M. Lemaire, R. Marmeisse, R. Montrocher, C. Robert, M. Termier, P. Wincker and M. Wesolowski-Louvel (2000). Genomic exploration of the hemiascomycetous yeast: 11. *Kluyveromyces lactis*. FEBS Letts. 487: 66-70.
177. Bon, E., C. Neuveglise, S. Casaregola, F. Artiguenave, P. Wincker, M. Aigle and P. Durrens (2000). Genomic exploration of the hemiascomycetous yeasts: 5. *Saccharomyces bayanus var. uvarum*. FEBS Lett. 487:37-41.
178. de Montigny, J., M. Straub, S. Potier, F. Tekaia, B. Dujon, P. Wincker, F. Artiguenave and J. Souciet (2000). Genomic exploration of the hemiascomycetous yeasts: 8. *Zygosaccharomyces rouxii*. FEBS Lett. 48:52-55.
179. Mende, K., V. Homann and B. Tudzynski (1997). The geranylgeranyl diphosphate synthase gene of *Gibberella fujikuroi*: isolation and expression. Mol. Gen. 255:96-105.
180. Carattoli, A., C. Cogoni, G. Morelli and G. Macino (1994). Molecular characterization of upstream regulatory sequences controlling the photoinduced expression of the albino-3 gene of *Neurospora crassa*. Mol. Microbiol. 13: 787-795.
181. Dogbo, O. and B. Camara (1987). Purification of isopentenyl pyrophosphate isomerase and geranylgeranyl pyrophosphate synthase from *Capsicum* chromoplasts by affinity chromatography. Biochim. Biophys. Acta. 920: 140-148.
182. Misawa, N., M. R. Truesdale, G. Sandmann, P. D. Fraser, C. Bird, W. Schuch and P. M. Bramley (1994). Expression of a tomato cDNA coding for phytoene synthase in *Escherichia coli*, phytoene formation *in vivo* and *in vitro*and functional analysis of the various truncated gene products. J. Biochem. 116: 980-985.
183. Armstrong, G. A., M. Alberti and J. E. Hearst (1990). Conserved enzymes mediate the early reactions of carotenoid biosynthesis in nonphotosynthetic and photosynthetic prokaryotes. Proc. Natl. Acad. Sci. USA 87: 9975-9979.
184. Summers, C., F. Karst and A. D. Charles (1993). Cloning, expression and characterization of the cDNA encoding human hepatic squalene synthaseand its relationship to phytoene synthase. Gene 136: 185-192.
185. Fraser, P. D., W. Schuch and P. M. Bramley (2000). Phytoene synthase from tomato (*Lycopersicon esculentum*) chloroplasts-partial purification and biochemical properties. Planta 211: 361-369.
186. Sutton, R. B., D. Fasshauer, R. Jahn and A. T. Brunger (1998). Crystal-strucuture of a SNARE complex involved in synaptic exocytosis at 2.4 angstrom resolution. Nature 395: 3470-353.
187. Jiang, Y. and S. Ferro-Novick (1994). Identification of yeast component A: reconstitution of the geranylgeranyltransferase that modifies Ypt1p and Sec4p. Proc. Natl. Acad. Sci. USA 91: 4377-4381.

188. Schmidhauser, T. J., F. R. Lauter, M. Schamacher, W. Zhou, V. E. Russo and C. Yanofsky, C. (1994). Characterization of *al*-2, the phytoene synthase gene of *Neurospora crassa*. Cloning sequence analysis and photoregulation. J. Biol. Chem. 269: 12060-12066.
189. Verdoes, J. C., N. Misawa and A. J. J. van Ooyen (1999). Cloning and characterization of the astaxanthin biosynthetic gene encoding phytoene desaturase of *Xanthophyllomyces dendrorhous*. Biotechnol. Bioeng. 63: 750-755.
190. Velayos, A., J. L. Blasco, M. I. Alvarez, E. A. Iturriaga and A. P. Eslava (2000). Blue-light regulation of phytoene dehydrogenase (*carB*) gene expression in *Mucor circinelloides*. Planta 210 (6): 938-946.
191. Torres-Martínez, S., F. J. Murilloand E. Cerdá-Olmedo (1980). Genetics of lycopene cyclization and substrate transfer in β-carotene biosynthesis in *Phycomyces*. Genet. Res. 36: 299-309.
192. Verdoes, J. C., P. Krubasik, G. Sandmann and A. J. J. van Ooyen (1999). Isolation and functional characterization of a novel type of carotenoid biosynthetic gene from *Xanthophyllomyces dendrorhous*. Mol. Gen. Genet. 262: 453-461.
193. Schechtman, M. G., C. and Yanofski (1983). Structure of the tri-functional *trp-1* gene from *Neurospora crassa* and its aberrant expression in *Escherichia coli*. J. Mol. Appl. Genet. 2:83-99.
194. Mullaney, E. J., J. E. Hamer, K. A. Roberti, M. Yelton and W. E. Timberlake (1985). Primary structure of the *trp*C gene from *Aspergillus nidulans*. Mol. Gen. Genet. 199: 37-45.
195. Hawkins, A. R.and M. Smith (1991). Domain structure and interaction within the pentafunctional arom polypeptide. Eur. J. Biochem. 196: 717-724.
196. Schmidhauser, T. J., F. R. Lauter, V. E. Russo and C. Yanofski (1990). Cloning, sequenceand photoregulation of *al*-1, a carotenoid biosynthetic gene of *Neurospora crassa*. Mol. Cell. Biol. 10: 5064-5070.
197. Ehrenshaft, M. and M. E. Daub (1994). Isolation, sequenceand characterization of the *Cercospora nicotianae* phytoene dehydrogenase gene. Appl. Environ. Microbiol. 60: 2766-2771.
198. Ruiz-Hidalgo, M. J., E. P. Benito, G. Sandmann and A. P. Eslava (1997). The phytoene dehydrogenase gene of *Phycomyces*: Regulation of its expression by blue light and vitamin A. Mol. Gen. Genet. 253: 734-744.
199. Harker, M. and J. Hirschberg (1997). Biosynthesis of ketocarotenoids in transgenic cyanobacteria expressing the algal gene for beta-C-4-oxygenase, *crtO*. FEBS Lett. 404: 129-134.
200. Kajiwara, S., T. Kakizono, T. Saito, K. Kondo, T. Ohtani, N. Nishio, S. Nagai and N. Misawa (1995). Isolation and functional identification of a novel cDNA for astaxanthin biosynthesis from *Haematococcus pluvialis*and astaxanthin synthesis in *Escherichia coli*. Plant Mol. Biol. 29: 343-352.
201. Misawa, N., S. Kajiwara, K. Kondo, A. Yokohama, Y. Satomi, T. Saito, W. Miki and T. Ohtani (1995). Canthaxanthin biosynthesis by the conversion of methylene to keto groups in a hydrocarbon β-carotene by a single gene. Biochem. Biophys. Res. Comm. 3: 867-876.
202. Misawa, N., Y. Satomi, K. Kondo, A. Yokohama, S. Kajiwara, T. Saito, T. Ohtani and W. Miki (1995b). Structure and functional analysis of a marine bacterial carotenoid biosynthesis gene cluster and astaxanthin biosynthetic pathway proposed at the gene level. J. Bacteriol. 177: 6575-6584.
203. Chamovitz, D., I. Pecker and J. Hirschberg (1991). The molecular basis of resistance to the herbicide norflurazon. Plant. Mol. Biol. 16: 967-74.
204. Armstrong, G. A., M. Alberti, F. Leach and J. E. Hearst (1989). Nucleotide sequence, organizationand nature of the protein products of the carotenoid biosynthetic gene cluster of *Rhodobacter capsullatus*. Mol. Gen. Genet. 216: 254-268.
205. Misawa, N., M. Nakagawa, K. Kobayashi, S. Yomono, Y. Izawa, K. Nakamura and K. Karashima (1990). Elucidation of the *Erwinia uredovora* carotenoid biosynthetic pathway by functional analysis of gene products expressed in *Escherichia coli*. J. Bacteriol. 172: 6704-6712.
206. Girard, P., B. Falconnier, J. Bricout and B. Vladescu (1994). Carotene producing mutants of *P. rhodozyma*. Appl. Microbiol. Biotechnol. 41: 183-191.
207. Bartley, G. E., T. J. Schmidhauser, C. Yanofsky and P. A. Scolnick (1990). Carotene desaturases from *Rhodobacter capsulatus* and *Neurospora crassa* are structurally and functionally conserved and contain domains homologous to flavoprotein disulfide oxidoreductases. J. Biol. Chem. 265: 16020-16024.
208. Simpson, K. L., T. O. M. Nakayama and C. O. Chischester (1964). The biosynthetic origin of the carboxyl oxygen atoms of the carotenoid pigment, torularhodin. Biochem. J. 92: 508-510.
209. Davies, B. H., J. Villoutreix, R. J. H. Williams and T. W. Goodwin (1963). The possible role of zeacarotene in carotenoid cyclization. Biochem. J. 89: 96.

210. Fraser, P. D., H. Shimada and N. Misawa (1998). Enzymic confirmation of reactions involved in routes to astaxanthin formation, elucidated using a direct substrate *in vitro* assay. Eur. J. Biochem. 252: 229-236.
211. Linden, H. (1999). Carotenoid hydroxylase from *Haematococcus pluvialis*: cDNA sequence, regulationand functional complementation. Biochim. Biophys. Acta 1446: 203-212.
212. Yokoyama, A. and W. Miki (1995). Composition and presumed biosynthetic pathway of carotenoids in the astaxanthin-producing bacterium *Agrobacterium aurantiacum*. FEMS Microbiol. Lett. 128: 139-144.
213. Lotan, T. and J. Hirschberg (1995). Cloning and expression in *Escherichia coli* of the gene encoding β-C-4-oxygenase, that converts β-carotene to the ketocarotenoid canthaxanthin in *Haematococcus pluvialis*. FEBS Letts. 364: 125-128.
214. Schledz, M., S. al-Babili, J. von Lintig, H. Haubruck, S. Rabbani, H. Kleinig and P. Beyer (1996). Phytoene synthase from Narcissus pseudonanarcissus: functional expression, galactolipid requirement, topological distribution in chromoplastsand induction during flowering. Plant J. 10: 781-792.
215. Bramley, P. M. (1985). The *in vitro* biosynthesis of carotenoids. Adv. Lip. Res. 21: 243-279.
216. Fraser P. D and P. M. Bramley (1994). The purification of phytoene dehydrogenase from *Phycomyces blakesleeanus*. Biochim. Biophys. Acta 1212: 59-66.
217. Bramley, P. M. and R. F. Taylor (1985). The solubilization of carotenogenic enzymes of *Phycomyces blakesleeanus*. Biochim. Biophys. Acta 839: 155-160.
218. Kushwaha, S. C, M. Kates and J. W. Porter (1976). Enzymatic synthesis of C40 carotenes by cell-free preparation from *Halobacterium cutirubrum*. Can. J. Biochem. 54:816-23.
219. Subbarayan C., S. C. Kushwaha, G. Suzueand and J. W. Porter (1970). Enzymatic conversion of isopentenyl pyrophosphate-4-14C and phytoene-14C to acyclic carotenes by an ammonium sulfate-precipitated spinach enzyme system. Arch. Biochem. Biophys. 137: 547-57.
220. Qureshi, A. A., A. G. Andrewes, N. Qureshi, and J. W. Porter (1974). The enzymatic conversion of cis-(14C)phytofluene, trans-(14C)phytoflueneand trans-zeta-(^{14}C)carotene to more unsaturated acyclic, monocyclicand dicyclic carotenes by a cell-free preparation of red tomato fruits. Arch. Biochem. Biophys. 162: 93-107.
221. Dailey, T. A. and H. A. Dailey (1998). Identification of an FAD superfamily containing photoporphyrinogen oxidases, monoamine oxidases and phytoene desaturase. J. Biol. Chem. 273: 13658-13662.
222. Fraser, P. D., Y. Miura and N. Misawa (1997). *In vitro* characterization of astaxanthin biosynthetic enzymes. J. Biol. Chem. 272: 6128-6135.
223. Yokoyama, A., Y. Shizuri and N. Misawa (1998). Production of new carotenoids, astaxanthin glucosides, by *Escherichia coli* transformants carrying carotenoid biosynthetic genes. Tetrahedron Lett. 39: 3709-3712.
224. Havaux, M. (1998). Carotenoids as membrane stabilizers in chloroplasts. Trends Plant Sci. 3: 147-151.
225. Gruszecki, W. I. (1999). Carotenoids in membranes, p. 363-379. *In*: Frank, H. A., A. J. Young, G. Brittonand R. J. Cogdell, R, (Eds), The photochemistry of carotenoids, Kluwer Academic, Dordreacht, The Netherlands (1999).
226. van de Ven, M., M. Kattenberg, G. Van Ginkel and Y. K. Levine (1984). Study of the orientational ordering of carotenoids in lipid bilayers by resonance-Raman spectroscopy. Biophys. J. 45: 1203-1209.
227. Johansson, L. B. A., G. Lindblom, A. Wieslander and G. Arvidson (1981). orientation of β–carotene and retinal in lipid bilayers. FEBS Lett. 128: 97-99.
228. Sarry, J. E.and J. L. Montillet, Y. Sauvaire and M. Havaux (1994). The protective function of the xanthophyll cycle in photosynthesis. FEBS Letts. 353(2): 147-150.
229. Chamberlain, N. R., B. G. Mehrtens, Z. Xiong, F. A. Kapral, J. L. Boardman and J. I. Rearick (1991). Correlation of carotenoid production, decreased membrane fluidity and resistance to oleic acid killing in *Staphylococcus aureus* 19Z. Infect. Immun. 59: 4332-4337.
230. Carrol, M (Ed) (1989). Organelles. The Guilford Press, N. Y.
231. Lang, N. J. (1968). Electron microscopic studies of extraplastidic astaxanthin in *Haematococcus*. J. Phycol. 4: 12.
232. Kobayashi, M. and Y. Sakamoto (1999). Singlet oxygen quenching abilitiy of astaxanthin esters from the green alga *Haematococcus pluvialis*. Biotechnol. Lett. 21: 265-269.
233. Santos, M. F. and J. F. Mesquita (1984). Ultrastructural study of *Haematococcus lacustris* (*girod.*) *Rostafinski* (Volvocales). 1. Some aspects of carotenogenesis. Cytologia 49: 215.
234. Boze, H., G. Moulin and P. Galzy (1992). Production of food and fodder yeasts. Crit. Rev. Biotechnol. 12: 65-86.

235. Yamano, S., T. Ishii, M. Nakagawa, H. Ikenaga and N. Misawa (1994). Metabolic engineering for production of β-carotene and lycopene in *Saccharomyces cerevisiae*. Biosci. Biotechnol. Biochem. 58: 1112-1114.

236. Boussiba, S., W. Bing, J. P. Yuan, A. Zarka and F. Chen (1999). Changes in pigment profiles in the green algae *Haematococcus pluvialis* exposed to environmental stresses. Biotechnol. Lett. 21: 601-604.

237. Chumpolkulwong, N., T. Kakizono, S. Nagai and N. Nishio (1997). Increased astaxanthin production by *Phaffia rhodozyma* mutants isolated as resistant to diphenylamine. J. Ferment. Bioeng. 83: 429-434.

238. Cerdá-Olmedo, E. (1989). Production of carotenoids with fungi, p. 27-42. *In*: E. J. Vandamme, (Ed.) Biotechnology of Vitamins, Pigmentsand Growth Factors, Elsevier Applied Science, London.

239. Ninet, L. and J. Rinaut (1979). Carotenoids, p. 529-543. *In*: Peppler, H. J.and D. Perlmann, (Eds.), Microbial Technology, 2nd edition. Academic Press, New York.

240. Mehta, B. J. and E. Cerdá-Olmedo (1995). Mutants of carotene production in *Blakeslea trispora*. Appl. Microbiol. Biotechnol. 42: 836-838.

241. Ausich, R. L., F. Brinkhaus, I. Mukharji, J. Proffitt, J. Yarger and H.CH Chen (1991). Biosythesis of carotenoids in genetically engineered hosts. Patent WO 91/13078.

242. Borowitzka, M. A. (1994). Large-scale algal culture systems: the next generation. Australas. Biotechnol. 4: 212-215.

243. Borowitzka, L. J. and M. A. Borowitzka. β–carotene production with algae, p. 15-26. *In*: E. J. Vandamme, (Ed.) (1989). Biotechnology of Vitamins, Pigmentsand Growth Factors. Elsevier Applied Science, London.

244. Koeller, K. M. and C.-H. Wong (2001). Enzymes for chemical synthesis. Nature 409: 232-240.

245. Walsh, C. (2001). Enabling the chemistry of life. Nature 409: 226-231.

246. Schmid, A., J. S. Dordick, B. Hauer, A. Kiener, M. Wubbolts and B. Witholt (2001). Industrial biocatalysis: today and tomorrow. Nature 249: 258-268.

247. Klivanov, A. M. (2001). Improving enzymes by using them in organic solvents. Nature 409: 241-246.

248. van den Werf, M., J. A. M. de Bont and D. J. Leak. 1997. Opportunities in microbial biotransformation of monoterpenes. Adv. Biochem. Eng./Biotechnol. 55: 147-177.

249. Jari, G. (1997). Heterologous gene expression in filamentous fungi, p. 251-264. *In*: Anke, T. (Ed.), Fungal Biotechnology, Chapman & Hall, Weinheim.

250. Broder, S. and J. C. Venter (2000). Sequencing the entire genomes of free-living organims: The foundation of pharmacology in the new millennium. Annu. Rev. Pharmacol. Toxicol. 40: 97-132.

251. The Arabidopsis Genome Initiative. 2000. Analysis of the genome sequence of the flowering plant *Arabidopsis thaliana*. Nature 408:796-815.

252. Appleyard, V. C. L., S. E. Unkles, M. Legg and J. R. Kinghorn (1995). Secondary metabolite production in filamentous fungi displayed. Mol. Gen. Genet. 247: 338-342.

253. Ferea, T. L., D. Botstein, P. O. Brown and R. F. Rosenzweig (1999). Systematic changes in gene expression patterns following adaptive evolution in yeast. Proc. Natl. Acad. Sci. USA 96: 9721-9726.

254. Schmidt-Dannert, C. (2000). Engineering novel carotenoids in microorganisms. Curr. Opin. Biotechnol. 11: 255-261

Applied Mycology and Biotechnology
Volume 2. Agriculture and Food Production

Edible Fungi: Biotechnological Approaches

Raj D. Rai and O. P. Ahlawat
National Research Centre for Mushroom, Chambaghat, Solan 173 213, India (E-Mail: rdrai@rediffmail.com).

Edible mushrooms, due to their commercial and environmental importance, have lately attracted the attention of the researchers. Modern biotechnological techniques and tools like protoplasting, RAPD, RFLP, electroporation and ballistic guns have been used to varied levels of success in the genetic improvement of edible mushrooms with respect to the yield, quality, disease resistance and also their degrading ability of lignocellulose. In the area of the researches aimed at the understanding of the molecular mechanisms underlying the fruiting and 'flushing' of the mushrooms, low molecular weight carbohydrates like trehalose and mannitol have been found to play the critical role but the genetic control giving the stimulus for triggering the fruiting has not yet been fully deciphered though significant and important leads have been obtained. The substrate preparation (composting) technologies for the common mushroom (*Agaricus bisporus*) have witnessed significant advancements leading to the completely indoor and significantly improved technologies with respect to quality of the end-product as well as environment-friendliness; use of genetically engineered microbes and also the enzymes have been attempted and may further shorten the period of composting and improve the product with almost eliminating the deleterious effect on the environment. Edible fungi including the commercial mushrooms have vast and unique degradative potential and, expectedly, have been used in the bioremediation particularly in the xenobiotics of chlorinated polyphenols, polyaromatic hydrocarbons and other harmful and calcitrant molecules.

1. INTRODUCTION

The cultivation of edible fungi, commonly called mushrooms, is a true microbial technology and represents large scale controlled application of microbial technology for the profitable conversion of lignocellulosic wastes into food and feed and in the economic terms the importance may be next to the yeast only [1]. Global trade in edible mushrooms is over 15 billion US dollars and currently the trade is growing at a phenomenal rate of 5-6% per annum. Mushroom cultivation represents a very large scale solid substrate fermentation (SSF) of the agro and forestry wastes. Biotechnological significance of the process has been decribed by many workers [1,2,3]. Currently, the mushrooms are regarded as the most profitable and environment-friendly mehtod for recycling of the vast lignocellulosic wastes. A Global Network on mushrooms was mooted in the First International Congress for the Characterization, Conservation, Evaluation and Utilization of Mushroom Genetic Resources, under the aegis of the FAO, held in Bordeaux in 1998. Mushroom genetic resources have been considered important for agriculture, human food, cattle feed, human health, and for chemical and pharmaceutical industry [4]. Keeping in view the economic and environmental importance of the mushrooms, many biochemical and biotechnological investigations have, of late, attracted the attention of the researchers. Authors have endeavoured to review and update the knowledge in some of the frontier areas of current interest in mushrooms.

Unlike other food crops representing a single species, many genetically and morphological distinct genera and species of mushrooms, are being cultivated world over. Of about 2000 edible fleshy fungi, about 20 types are being artificially cultivated and five are being traded in sizeable quantities and represent 85.3% of total mushroom production [5,6]. For convenience of the readers the five "leaders' have been depicted in Fig.1. Though there are many aspects of mushrooms having biotechnological significance but the authors have tried to restrict themselves to four most active frontier areas namely molecular techniques in genetic improvement of mushrooms, molecular mechanisms underlying growth and fruiting, recent advances in substrate preparation technologies, and edible fungi as tools for bioremediation.

Table 1. World production of cultivated mushrooms [1994].

Species	Common Name	Production (x 1000 tonnes)	Percent
Agaricus bisporus/ bitorquis	Button mushroom	1946.0	37.6
Lentinula edodes	Shiitake	826.2	16.8
Pleurotus spp.	Oyster mushroom	797.4	16.3
Auricularia spp.	Black ear mushroom	420.1	8.5
Volvariella volvacea	Paddy straw mushroom	298.8	6.1
Others	-	578.8	14.7
Total		4903.3	100.0

Source: Chang [5].

2. BIOCHEMICAL AND MOLECULAR ASPECTS OF FRUITING

Differentiation in multicellular eukaryotes is an interesting and intriguing phenomenon not well understood, primarily due to inability of many organisms to differentiate under defined conditions. Fungi, due to genetically controlled simple morphological eukaryotic organisms suitable for studies on differentiation of multicellular eukaryotes. Understanding the molecular mechanisms underlying fructification of mushrooms, [7,8] besides basic biological interest, has vast commercial importance because it would have the way for enhancing and regulating the yield of fruitbodies and also in 'domesticating' the wild mushrooms, which have hitherto defined the human efforts to fructify them under controlled conditions. Unfortunately, biochemistry of mushroom fructification has been studied in the fungi of lesser commercial importance like *Schizophyllum commune*, *Coprinus* spp. and *Flammuina velutipes* because of the ease in their fruiting and relative difficulties in obtaining reproducible fruiting in commercial mushrooms under defined *in vitro* conditions. Nevertheless, significant advances have been made in understanding the fruiting of the commercial mushrooms at biochemical level and some excellent reviews on the subject have published [9-13].

2.1 Cell Wall Expansion and Elongation

Formation of basidiomycete fruitbody from the vegetative mycelium occurs in two easily recognisable steps: the initiation phase characterised by the aggregation of the mycelium to form pinheads or 'primordia' and the morphogenetic phase where primordia develop and differentiate into fruitbodies with stipe, pileus, volva and gills. In contrast to the growth of the vegetative mycelium by apical extension only, fruitbody hyphae of *Agaricus bisporus* are reported to grow by diffuse extension over whole wall surface especially during rapid expansion and elongation phase. The differences between the growth modes of the vegetative and fruitbody hyphae could be due to variations in chemical composition, physical structure and enzymes in their cell walls. Chemical composition of cell walls of mycelium and stipes are reported to be essentially similar - polymers of glucose (glucan 44.8%) and N-acetyl-glucosamine (chitin 35.6%) are the main constituents [14] . Chitin synthesis is essential for elongation of stipe hyphal walls and N-acetyl glucosamine, the chitin precursor, is incorporated all over the surface in elongating hyphae of expanding fruitbodies [15]. Chitosomes, the vesicular organelles containing chitin synthase activity, have been isolated from *A. bisporus* [16]; higher activities of the enzyme were recorded in upper stipe (rapid growth zone) than in the lower stipe and also during rapid expansion

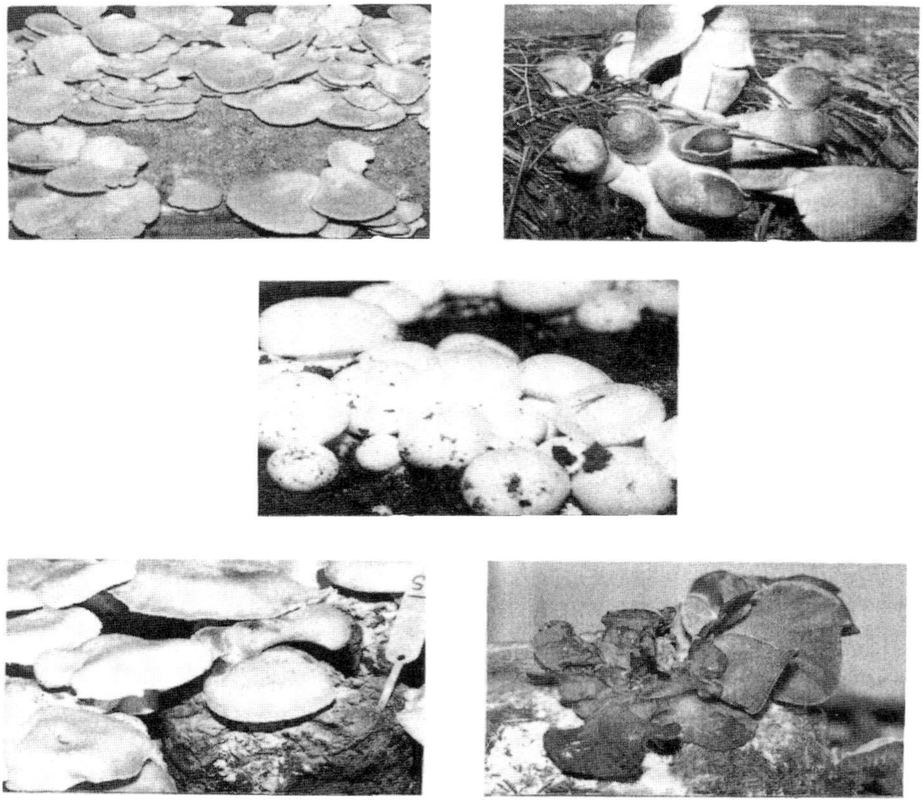

Fig. 1. Important commercial edible mushrooms

than at the end of the elongation. Chitin synthase of *A. bisporus* was similar in properties to those reported from the other fungi and is proteolytically activated [17]. Mol and Wessels [14] reported that the differences in the modes of wall expansion of vegetative mycelium and stipes of *A. bisporus* lied in the susceptibility of the chitin: chitin in elongating stipe wall was found to be more susceptible to chitinase and a positive correlation was demonstrated between the competence of hyphae to expand, susceptibility of the 1-3-β-glucan in the wall to enzymatic degradation and the presence of linked glucose side chains [1-6]. It was suggested that H-bonding among homologous chains in the chitin-glucan complex of the wall of the elongating stipe hyphae may be weaker than that in the non-elongating hyphae and the vegetative hyphae. Besides, chitin chains in the elongating hyphae of stipe were transversely oriented and not organised in distinct microfibrils whereas walls from the vegetative hypahe showed randomly oriented microfibrils in an arnorphous matrix. Based on the above observations, a tentative model for diffuse wall extension in fruitbody stipe of *A .bisporus* was proposed but authors [18] seem to have under- emphasized the role of autolytic enzymes in plasticising the wall in the model.

2.2 Chemical Composition

Changes in chemical composition during fruitbody development of various mushrooms have been extensively studied [9-23]. Many workers have observed increase in moisture content of the fruitbody during its development and it has been opined that the fruitbody development in basidiomycetes is partly due to osmotically driven cell-elongation. Changes in protein content of the fruitbodies have also been studied [19-23]. Intracellular and 'shockable' protein contents decreased at the transition stage in *Lentinus edodes* [24]. Above observations suggest significant turn over and reorganisation of proteins during fruitbody development of mushrooms. Intracellular proteases are highly activated during fruitbody development of *A. bisporus* [25]. New proteins are reported to be formed both in different tissues and at different stages of development of *V. volvacea* [26]. Protein-N may be channelised towards site of active protein and nucleic acid biosynthesis, e.g. in the gills for spore formation. Depletion of N in growth medium is suggested to trigger the transition from mycelium to fruitbody and also results in accumulation of glucosamine which is suggested to be utilized later for chitin synthesis during fruiting of *L.edodes* [24]. Doubling of DNA content suggestive of active meiosis in the gill tissues just before the sporulation was observed in *A. bisporus* fruitbodies growing at the normal time in phase with the flush but the fruitbodies growing out of phase with the flush did not show such effect. Minamide and Hammond [27] suggested that fruitbodies received stimulus related to the progress of flushing cycle from the mycelium for meiosis to proceed.

2.3 Carbohydrate Metabolism

In differentiation of the organisms nucleic acids and proteins normally play initial and the crucial role. Differential genetic activity expresses itself in differential synthesis of structural and functional proteins, the enzymes which may later regulate the levels of other constituents like carbohydrates, lipids etc. But fungal walls are not only made up of modified carbohydrates as discussed earlier, carbohydrates have been found to play crucial role in the life cycle of many fungi. It is, therefore, not surprising that the study of carbohydrate metabolism during growth and fructification of the mushrooms, especially the *A. bisporus*, has received considerable attention [9,10,28]. Structural carbohydrates have been dealt with in the section on cell wall.

Non-structural carbohydrates in relation to fruiting will be discussed here. Sugar alcohol, mannitol, a disaccharide, trehalose and a glycogen-like soluble polysaccharide are the principal non-structural carbohydrates in vegetative mycelium and fruitbodies of *A. bisporus*. Differences observed in the contents of these soluble carbohydrates between vegetative mycelium and fruitbodies provided the first insight, which led to detailed investigations, into the crucial role of carbohydrate metabolism in fructification. Vegetative mycelium contained about 1.5-4.5% mannitol on dry weight basis. Primordia started accumulating mannitol and contained double the amount than their supporting mycelium; accumulation of mannitol continued during further development and mature fruitbodies contained 24.5-30%, even upto 50% mannitol as their dry weight [29]. Logically, such enormous accumulation cannot be without significance. Storage function of mannitol is unlikely in view of its very slow turnover [30]. Another interesting observation regarding mannitol content in the fruitbodies was that though on dry weight basis it increased but on fresh weight basis its concentration remained more or less constant between 150-200 mm [29]. It was, therefore, suggested that mannitol plays the role of an osmoticant to attract water in fruitbody hyphae and create hydrostatic pressure for hyphal extension. The resultant turgor pressure would also physically support the fruitbodies [9,31]. Similar role has been suggested for urea in *C. cinereus* [22,33], but the urea content in *A. bisporus* was found to be very low [31]. Accumulation of huge quantities of mannitol may be related to diffuse extension of fruitbody hyphal wall all along the surface which may require greater hydrostatic pressure as compared to localised pressure required at the tip of vegetative hyphae. Mannitol content of the fruitbodies of *A. bisporus* is also reported to be positively correlated with the fruitbody yield [33]. Though mannitol can be translocated in sporophore but in view of its uneven distribution it is more likely to be synthesized *in situ*. It is synthesized by reduction of fructose by the enzyme mannitol dehydrogenase (MDH) using NADPH as cofactor [34,35]. Under cellular conditions MDH system is favourable to synthesis of mannitol and poised against its oxidation [34,36,37]. It is oxidised and used under severe stress conditions when NADPH is depleted [36-39]. As there was no difference in MDH activity between fruitbody and vegetative hyphae [40], supply of fructose the substrate and cofactor NADPH may be acting as controls for mannitol synthesis [9,10,36,37]. Carbohydrate metabolism observed in fruiting and flushing of mushrooms is depicted in Fig. 2 and 3.

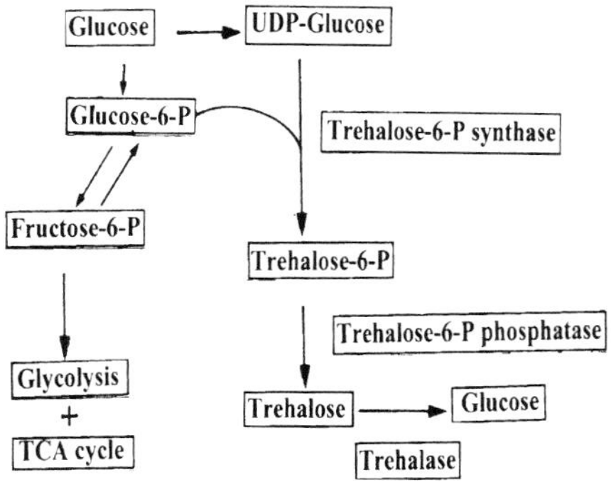

Fig. 2. Carbohydrate metabolism in *Agaricus bisporus* [28].

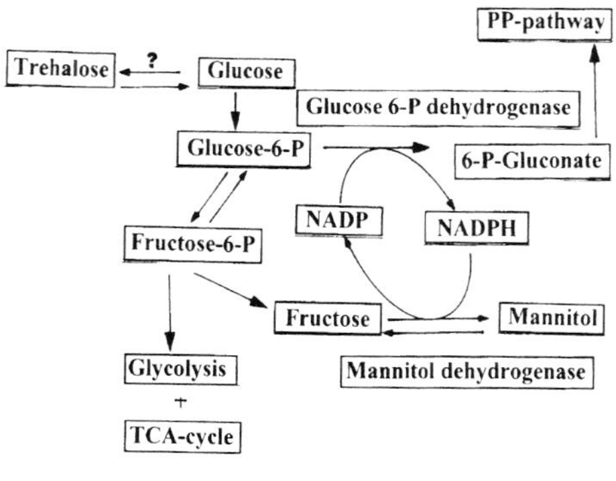

Fig. 3. Trehalose sythesis and degradation in mushrooms [28]

Hexose catabolism occurs via well-known routes like Embden-Meyerhof Pathway, Pentose phosphate pathway (PPP) and Kreb's cycle. Early studies indicated a significant [56-57%] turnover of hexoses via PPP in the fruitbodies of *A. bisporus*. PPP activity, as measured by radiorespirometry and enzymes was found to be greater in fruitbodies than in vegetative mycelium of *A. bisporus* [40,41]. Glucose-6 phosphate dehydrogenase [G6PDH], the first and key enzyme of PPP and a major site of NADPH generation, showed several fold higher activity in fruitbodies than in vegetative mycelium [40]. NADPH generated in PPP was shown to be used for mannitol [42]. Higher availability of NADPH and also of fructose [discussed later] may be playing important roles in enhanced synthesis and accumulation of mannitol in the fruitbodies. NADPH and pentoses generated in activated PPP in fruitbodies may also be utilized for other biosynthetic processes like nucleic acids and lipids [9].

2.4 Metabolism of Flushes

It is common observation that during commercial cropping of *A. bisporus* and also of some other mushrooms, there is heavy and synchronous appearance called 'flushes' of sporophores at about weekly intervals, with very little fruitbody production between the flushes which is termed as 'flush break' Recent work done at the Glasshouse Crops Research Institute [now IHR], Littlehampton, on carbohydrate metabolism during fruiting rhythm of *A. bisporus* has thrown some light on this intriguing phenomenon. In a mathematical model of flushing, a hypothetical substance which is absorbed and accumulated in the mycelium to a threshold level sufficient to trigger flush initiation was proposed [43,44]. Flush growth then depletes the intra-mycelial substrate and prevents new initiation. The substrate accumulates again after harvesting and triggers another flush. Harvesting practice is known to regulate the next flush. It appears plausible that development of primordia into mature fruitbodies depends on stimulus from mycelium as well as feedback from growing fruitbodies.

Analysis of fruitbodies at primordium and button stage daily throughout several flushes during the commercial crop of *A. bisporus* showed that trehalose and glycogen levels in fruitbodies were highest (ca. 20% of dry wt.) at time of flush emergence i.e., when pins of a new flush were just emerging from the casing but the levels declined to 10-25% of the peak level during mid flush i.e., by the time button stage appeared 3-4 days later and reached a minima and rose again after harvesting. Trehalose and glycogen behaved similar to the hypothetical substrate discussed earlier. Trehalose and glycogen content of sporophores at flush emergence correlated well with ultimate yield of the flush. Mannitol did not vary as per flush but fluctuated according to size of fruitbody as discussed in previous section [45].

It has been suggested [9,10] that mycelium absorbs glucose produced in the compost by action of cellulase and other lytic enzymes [46], converts it in trehalose which is then transported in the fruitbodies; excess carbohydrate is stored as glycogen. An interesting observation was that the extracellular endoglucanase production in compost by *A. bisporus* was found to be developmentally regulated and followed the flush pattern i.e. high activity when fruitbodies were growing and peak activity coincided with the peak harvest [47,48]. Besides fruitbody biomass regulation of EG production, Hammond [9] has suggested that high levels of trehalose observed between the flushes could repress the cellulase which may be derepressed due to its consumption during fruitbody growth and respiration. Between the flushes, EG level is though low but high accumulation of trehalose and glycogen may be due to very few and small sporophore initials. G-6-PDH activity in fruitbodies during flushes followed a pattern similar to trehalose and glycogen level; its activity was 20 times higher at flush initiation than at mid flush i.e., peak activity coinciding the peak trehalose and glycogen level. G-6-PDH did not vary in mycelium and MDH did not show any definite pattern [27,45]. High trehalose and glycogen, with high G-6-PDH activity allowing high PPP flux and NADPH production would provide substrates for mannitol synthesis at flush initiation. G-6-PDH activity was found to be linked to trehalose levels under all conditions of fruitbody growth but how this control was exercised remains unclear [9].

Elevated activity of G-6-PDH was due to increased synthesis of enzyme protein and also appearance of a new and different G-6-PDH protein (G-6-PDH2), which appeared only at flush initiation [27]. Trehalase and glycogen phosphorylase showed a pattern opposite to the G-6-PDH i.e., more active during mid-flush and very low at flush initiation [49]. Very high activity of extracellular EG during the mid-flush would meet the supply of high amounts of trehalase and glycogen phosphorylase at this stage to produce substrates for mannitol and PPP [47]. Though a good picture of biochemistry of flushing has emerged, the nature of stimulus which starts off the flush growth remains to be investigated [10]. Nevertheless, interrelations of various processes as described above compel us to believe that the flushing rhythm is genetically programmed which will have to be deciphered with a view to harnessing biotechnology for regulating the flushes if desired. Morphogenesis in organisms including fungi is result of differential genetic activity [50,51]. While studying the changes in trehalose phosphorylase activity and carbohydrate levels during axenic fruiting in the *A. bisporus* strains indicated the enhanced synthesis of trehalose before hyphal aggregates formed and translocation of trehalose from mycelium to the aggregate occurred thereafter [52,53].

2.5 Extracellular Enzymes

Though extracellular enzymes in various basidiomycetes have been vastly studied mainly to understand the fungal-nutrition and lignocellulose degradation, but the treatment of the subject here would be restricted as related to the morphogenesis of mushrooms [54]. Polyphenol oxidases, especially the laccase, have been most extensively studied and implicated in fruitbody development of basidiomycetes. The enzyme is reported to be of importance in *Agaricus* systematics also [55]. Laccase protein constituted 2% of extracellular protein secreted by *A. bisporus* [56]. Extracellular laccase activity in the compost during cropping of *A. bisporus* increased up to the first flush but declined rapidly thereafter [46,57]. The activity did not decline and remained high in non-fruiting cultures and in cultures without fruiting due to non-casing [58]. Decline was observed in only those cultures in which fruitbodies were allowed to grow beyond 1-2 cm. Above observations suggest that extracellular laccase activity in *A. bisporus* is developmentally regulated. Inactivation of the enzyme in fruiting cultures has been reported to be due to degradation [60-70%] of enzyme protein probably to meet the N demand of developing fruitbodies [59,60]. Elevated levels of proteases during this stage of development of *A. bisporus* are also reported [58]. Laccase is reported to be developmentally regulated in *L.edodes* [24,61]. Extracellular polyphenol oxidase in *P. ostreatus* was maximal during exponential growth but declined thereafter [62]. Role of the enzyme in basidiomycetes is though not very clear, laccase and peroxidases often in cooperation with other enzymes are suggested to be involved in lignin degradation by white-rot basidiomycetes [63,64]. Role of laccase in oxidation of phenols to quinones for oxidative polymerisation of cell surface components of fungi has also been suggested [65].

Extracellular cellulase (endoglucanase, EG) is also reported to be developmentally regulated in *A. bisporus* [46,47,48,66]. During fruiting, EG activity in the substrate behaves inversely to laccase i.e. it remains low before fruiting but increases sharply at fruitbody formation. EG production was in direct proportion to the frutibody biomass produced during each flush;

fruitbody harvested paralleled the EG activity in the substrate [47]. Fruitbody-regulated EG production may be acting as a mechanism of cell economy to produce as per the demand. Further investigations [48] on laccase and EG production by *A.bisporus* in deep troughs during cropping cycle has revealed that EG and laccase activity near the fruiting surface followed the typical pattern discussed above: distinct peak of EG with sharp fall of laccase during the flushes and reverse pattern during the interflushes. The intensity of above changes lessened with the compost depth. EG activity declined steadily with the compost depth and was almost undetectable beyond a depth of 0.9m in early stages of mushroom cropping. Laccase activity close to the fruiting surface declined rapidly at the onset of fructification but was four fold higher at 0.9m depth and expected decline was also delayed. Decline in laccase coincided with increase in EG throughout the cropping. EG level in top layer was proportional to the mushroom yield. Findings were consistent with the observations of Smith [67] showing progressive downward utilization of substrate nutrients during successive flushes. In *P.sajor-caju* also, laccase activity appeared, peaked and declined earlier when cellulase activities started rising at the pinhead initiation [68]. Peaking of extracellular cellulase and xylanase at the time of fruiting of *L.edodes* cultures has also been reported [69,70]. Sermanni *et al.* [71] comparing the laccase and EG production by *P.ostreatus* and *A.bisporus* during fruiting found that while *P.ostreatus* followed the pattern of *A. bisporus* in EG production, but not in laccase production which remained low without much change during entire cropping.

Transcriptional regulation of laccase and cellulase genes in *A. bisporus* during fruitbody development in wheat straw compost has been studied [72]. Gene expression for laccase and cellobiohydrolase correlated with the known changes of laccase and cellulase activities during the fruitbody life cycle. Extracellular laminarinase, xylanase, acid and alkaline phosphatases and alkaline protease production in *A. bisporus* did not follow any trend specific to fruiting [58]. However, intracellular proteases were highly activated in *A. bisporus* fruitbody; acid protease was active in the pileus, while neutral and alkaline proteases were maximal in the stipe. Gills showed lowest activity of all the proteases. Mature fruitbodies showed upto 40-fold higher activity than the primordia [58]. It was suggested that intracellular proteins were being degraded to meet the requirements of N and C in the region of active protein and nucleic acid synthesis in the gills for spore formation.. Increases in protease activity of *P. ostreatus* was observed at priomordium formation induced by light [73]. Proteases are developmentally regulated in *S. commune* also [74]. Extracellular acid phosphatase pesaked at the time of fruiting after N depletion in medium in *L. edodes* [24,75]. No definite pattern in activities of extracellular cellulase and xylanase of *P.sajor-caju* was reported [76].

Extracellular proteinases from the mycelium of the *A. bisporus* have been studied by Burton *et al.* [77] with a view to understanding the derivation of N from proteins in the compost. One chymotrypsin like specificity was found to be a 27 Kda serine proteinase and the activity increased with the mycelial growth on the liquid medium containing a humic fraction [from the mushroom compost] as the C source but remained low in the filtrates of the cultures grown on glutamate or casein. Regulation of the activity in compost was suggested by a doubling during the early stages of sporophore growth.

1. MOLECULAR TECHNIQUES IN GENETIC IMPROVEMENT OF EDIBLE MUSHROOMS

Owing to the economic importance of the edible mushrooms and various biotechnological tools available with the researchers, genetic improvement programmes of mushrooms have gained a strong momentum. The pre-requisite step in the genetic improvement programme of fungi is the availability of large stocks of germplasm having variability in different attributes of commercial importance and also the cultures can be raised from the wild germplasm, either by tissue culture or spore culture techniques and can be maintained by any of the routine techniques. Culture multiplication by either of the way did not have any problem in basidiomycetes but, in the nature, the ideal way of natural survival and distribution is through the formation of fruitbodies in the reproductive cycle which comprises three cardinal events-plasmogamy, karyogamy and meiosis.

3.1 Life Cycle

The knowledge of life cycle of a particular species is very important before starting the breeding activity. Though the life cycle in different species of the basidiomycetes varies from one another, still some general pattern can be drawn suiting to maximum of the species [78,79] (Fig. 4).

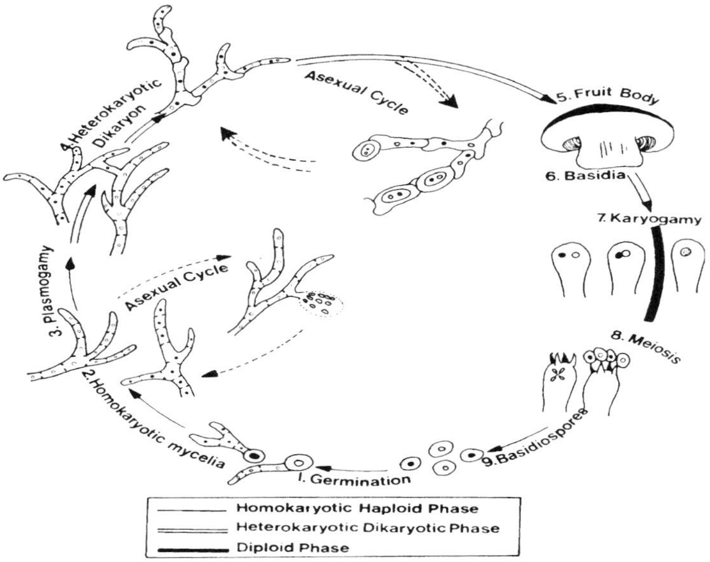

Fig. 4. Life cycle of a typical basidiomycete steps-1 to 9. 1. Basidiospore-germination; 2. Development of homo-karyotic mycelium; 3. Plasmogamy through mating of two compatible homokaryons; 4. Development of hetero-karyotic dikaryon; 5. Initiation of basidiocarp (fruit body) 6. Development of basidia in the fruit bodies; 7. Karyogamy i.e. fusion of the two parental nuclei; 8. Fusion nucleus undergoes meiosis; 9. Formation of basidiospores each receiving one of the daughter nuclei. [78, 79]

In case of basidiomycetes, the uniques feature is the existence of 3 nuclear phases instead of two as is the case in the higher plants. The first which follows meiosis is the haploid, homokaryotic or monokaryotic, the second is the result of plasmogamy between two monokaryons and is called heterokaryotic dikaryon phase and the last is the heterokaryotic diploid phase which happens because of karyogamy between two nuclei. The knowledge about the life cycle of different species of the basidiomycetes has helped in understanding of the following aspects.

3.1.1 Heterothallism

The first report on heterothallism in basidiomycetes came in case of *Coprinus fimetarius* and *S. commune* based upon the mating of monosporous mycelia and the formation of clamp connections but, later molecular studies helped in dividing the heterothallism into two major forms; bifactorial heterothallism and unifactorial heterothallism (Fig. 5). The first form consisted of two unlinked mating type factors with heteroallelic condition at both the loci and is also called as tetrapolar incompatibility or heterothallism with bifactorial control.

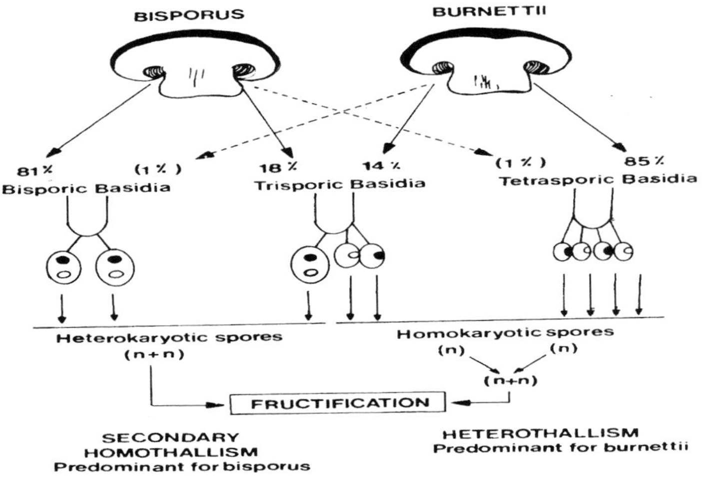

Fig. 5. Two life cycles of the varieties, *bisporus* and burnettii of *Agaricus bisporus* [79,87]

The second form consisted of only single mating type factor with heteroallelic condition at that locus and is called as bipolar incompatibility or heterothallism with unifactorial control.

3.1.2 Homothallism

In homothallism, the single viable basidiospore is having capability to germinate and form fruitbody without the requirement of a mating process involving sharing of nuclei from opposite mating types. So the whole life cycle can be completed with the involvement of a single basidiospore. Based upon the number of basidiospore formation on a basidium and the involvement of incompatibility factor(s), the homothallism can be divided into two forms: primary and secondary. The phenomenon of primary homothallism in *Volvariella volvacea* was ascertained with the help of malachite green-resistant mutants and aspartate requiring auxotrophs [80]. In this mushroom the heterokaryotic mycelium can be raised from a single spore containing a single post-meiotic nucleus. However, the primary homothallism assigned to *Volvariella* spp. is still not final because of variability in single spore isolates .

The phenomenon of secondary homothallism is almost similar to that of unifactorial heterothallic species, but the 2 spored nature of basidia conceals the possibility of heterothallism and forced the researchers to designate this typical type of phenomenon as secondary homothallism. The best example of secondary homothallism is the common mushroom *Agaricus bisporus* (Fig. 5). The homokaryotic phase peculiar for primary homothallism is found only in spores of 3 and 4 spores per basidium and the frequency of occurrence is approximately 5%.

3.2 Genetic Improvement/Development of Newer Strains

The process of mushroom breeding has remained integral part of all research and development activities aimed at improving the cultivation technology of mushrooms; but in the beginning, the process of strainal improvement centered round identification and selection of better performing strains. However, with the release of two hybrids U_1 & U_3 by the conventional techniques of crossing homokaryons and selection of high yielding hybrids [81,82], the mushroom breeding programmes all over the world took a big leap and newer techniques supporting the breeding programme were brought in light [83]. The newer techniques, like protoplasting, have also improved the methodology as well as frequency of homokaryons isolation having utility in hybrid formation. The distinct features of different mushroom species like existence or non-existence of clamp connections in heterokaryons, variations in number of nuclei per cell and also in basidiospore per basidium have created confusion and forced to adopt different breeding strategies for each species.

Though the conventional method of mushroom breeding like introduction and selection has remained the main tool in the mushroom improvement programmes but other techniques like mutations, single spore culture, multispore culture and hybridization have also been tried with different levels of success. The multispore culture technique is said to be useful in rejuvenation of the degenerated cultures while single spore culture and hybridization techniques have generated some of the commercially popular strains all over the world. The best example of hybridization is the development of U_1 and U_3 hybrids of *A.bisporus* [81,82] in Holland. Similarly, some of the single spore isolates from *A.bisporus* [84] and *Volvariella* spp. [85] were also reported to have higher yield potential than their parents.

In traditional hybridization system of mushroom improvement, the homokaryons obtained either from single spore isolates or from protoplasting, are crossed to make hybrids which are then evaluated for the desirable traits. Though the technique seems to be very simple but, in the absence of variability in the available germplasm, and the dearth of genetic markers to differentiate between the homokaryons and heterokaryons, success has been seldom achieved in obtaining the hybrids. But the new developments in the culture preservation techniques and emergence of DNA-based genetic markers have solved many of these problems. In order to tackle some of these problems, *Agaricus* Recovery Programme or ARP was founded by Dr.Rick Kerrigan in 1988 which has been recently renamed as *Agaricus* Resource Programme [86]. Presently, more than 206 *Agaricus* strains including the wild ones are available with the ARP and most of these strains have been collected from major spawn companies and researchinstitutes from all over the world. The strains available with ARP have been identified by using isozyme marker techniques. The strains available with ARP are an excellent source of genetic variability and best example is the four spore per basidia strain obtained from Sonaran desert of California by Callac *et al.* [87]. Moleecular techniques can be helpful in removing the existing species specific barriers in developing the new hybrids of mushrooms.

The mushroom breeding in the past was based upon the morphological features of the homokaryons and heterokaryons, which are bound to vary depending upon the growing conditions. However, there are features which do not vary with the growing conditions because these are based upon the genetic make-up of an organism [88]. With the advent of DNA based markers, the identification of homokaryons as well as heterokaryons has become much easier and the certification in mushroom industry has become possible [89,90,91].

The first gene based technique used in mushrooms was isozyme analysis. Isozymes are different forms of same enzyme having differences in their structure but catalyse the same type of reaction or having same level of activity. These have been extensively studied in Pennsylvania State University in the early 1980's [92-96]. The changes in the amino acid sequence in a protein results in change in the net charge on the enzyme which affects the overall mobility of the enzyme and separation of isozyme under the electric field. Under controlled conditions, isozyme marker is an excellent tool for differentiating the strains because the allozyme loci behave in a Mandelian manner in crosses. More than a dozen isozyme markers have so far been identified in *A. bisporus*. In comparison to other techniques, this technique is quite cheap and data generation per unit effort is very high. The isozyme technique in combination with other techniques like cell wall polysaccharides, nuclear and mitochondrial RFLPs helped in differentiating two Spanish species of *A. bisporus* [97]. Similar line of work has also been carried out on other basidiomycetes like *Lentinula edodes* [98,99] and *Pleurotus* spp. [100,101].

Next to isozymes, other DNA-based molecular marker used in mushroom breeding is the Restriction Fragment Length Polymorphism [102,103]. RFLP comprises of two interlinked techniques: first is the restriction of DNA with site specific restriction endonucleases which make nick at specific restriction nucleotide sequences, and the second technique is the southern hybridization [104], in which the DNA fragments generated by restriction endonucleases get separated on the gel, are made single stranded followed by their transfer on a nitrocellulose based membrane for hybridization. A small segment of radiolabelled DNA homologous usually less than 10,000 bp cloned on bacterial plasmid is used as a probe. The fragments of genomic DNA hybridizing to the radioactive probe are detected by autoradiogram. In addition to karyotyping, RFLP also has utility in construction of genetic linkage maps or in genetic mapping [95,105-107]. The natural population of *P. ostreatus* belonging to distinct geographical regions were found to have genetic distinctness and this type of genetic relationship was established with mDNA RFLP [108,109]. The *P. ostreatus* was further analysed by using other techniques like pulsed-field gel electrophoresis in combination with the use of single copy DNA probes [110]. The differentiation of primary homothallic species *Volvariella* spp. is also possible through RFLP and AP-PCR or combination of the two [111].

In case of *A. bisporus* the RFLP and RAPD have helped in studying the genetic diversity, isolation of homokaryons, inheritance of nuclear and mitochondrial markers and for establishing genetic linkage maps [112]. Similar studies have also been carried out with *L. edodes* [113]. This technique has also helped in identification and characterization of a number of *A. bisporus* genes and has opened the path for obtaining genetic transformants for different phenotypic and production related characters [111]. The phylogenic relationships between different species of *Lentinula* such as *L. edodes, L. lateritia, L. novaezelandiae* and *L. boryana* originally isolated from Asia-Australian regions was established by carrying out RFLP of the internal transcribed spacer [ITS] regions and intergenic regions [IGR] of the ribosomal DNA [rDNA] repeats [114]. In order to increase the efficiency of this technique a miniprep method involving simultaneous detection of dsRNA and RFLP analysis was developed by Barroso *et al.* [115] which helped in testing 12 strains per day and was found suitable for characterizing dikaryotic and homokaryotic strains as well as basidiocarps. But high cost, time-consumption and high labour input are some of the drawbacks of this technique.

3.2.1. RAPD

Another method of wide use is Random Amplified Polymorphic DNA (RAPD) analysis [116]. RAPD is basically based on technique of rapid multiplication of small fragment of DNA by polymerase chain reaction [PCR]. The requirement of small quantity of DNA, fast speed and ability to handle multiple number of samples at a time this technique more user- friendly than the RFLP. The ability of this technique to provide high number of markers with high efficiency of reproducibility makes it a powerful tool in the construction of linkage maps [117,118]. The RAPD related protocols for mushrooms like extraction of DNA, amplification through PCR and manual sequencing was first standardized by Vilgalys and Sun [119]. The role of RAPD in studying the phylogenetic relationship at genus or species level became evident with the work of Moncalvo [120], where he studied the relationship between a newly described taxon *Pleurotus cystidiosus* var. formosensis and other members of the *P.cystidiosus* complex. The RAPD analysis has also been successful in studying the protoplast fusants obtained from *Auricularia auricula* + *A. polytricha* and *P. ostreatus* + *P. cornucopiae* [121]. In case of shiitake this technique was used to differentiate commercial shiitake *L. edodes* isolates [122] and crossing progenies [123] which could have positive implications on mushroom breeding and genetic identification of new isolates [124]. In inheritance study it was found that inheritance of RAPD markers depends upon the source of homokaryons and the germination rate of mushroom spores [117]. The genetic diversity among different isolates of *A. aegerita* has also been studied by using RAPD markers [125-127].

3.2.2 Other advance techniques

A simple but more powerful technique like pulsed field gel electrophoresis in combination with the use of single copy DNA probes was found to have greater potential in proper karyotyping of *P. ostreatus* [110]. The occurrence of genetic recombination in nature and high level of genetic polymorphism in *Lentinus* spp. belonging to remote mountain areas of China was proved by sequencing of ITS regions and rDNA lineage studies [128]. At the same time, by using sequence specific oligonucleotide probes, Lee *et al.* [129] reported that *P. ostreatus* is different from other species like *P. florida* and *P. pulmonarius*. The template mixing method devised by Loftus *et al.* [130] was successfully employed to differentiate between the two alleles inherited by F_1 homokaryons through the presence or absence of the heteroduplex by converting the uninformative SCAR into a co-dominant marker. These markers have made the breeding programme much more easier [131]. In other mushrooms like *Morchella*, *Verpa* and *Disciotis* the phylogenetic relationship was established by restriction enzyme analysis of the 28s ribosomal RNA gene [132]. The two Spanish species of *A. bisporus* were differentiated by using multiple markers like cell wall polysaccharides, isoenzyme patterns, nuclear and mitochondrial RFLP's. The strains were found to differ in xylose content of cell walls, types of glycosidic linkages in the inner polysaccharide fractions, in esterase patterns and mitochondrial and nuclear DNA RFLP [97].

3.2.3 Cloning, sequencing and transformation

The studies carried out on *A.* elucidated that *A. bisporus* contained 13 chromosomes [160]. However, at a later stage Sonnenberg *et al.* [161] mapped 25 genes to individual chromosome and out of which 17 were identified by using pulsed field electrophoresis. But the strainal improvement programmes require information beyond the number of chromosomes and this is possible through isolation, cloning and sequencing of genes responsible for yield, quality, disease-resistance and substrate utilization for their future use in transformation purposes. In case of mushrooms, transformations have been possible through recombinant DNA technology involving incubation with the protoplast of the recipient cell, electroporation [162] and by use of the ballistic gun system [163].

In order to study the expression of genes responsible for morphological changes during growth of button mushroom and expression of cellulase, endoglucanase and laccase at different growth stages different strategies have been adopted by different workers. The expression of genes responsible for the production of various enzymes like cellulase, endoglucanase, laccase in *A. bisporus* [164,165], only laccase in *P. ostreatus* and *L. edodes* [166,167] have been studied and were found to vary in expression at different stages of growth. Different approaches like the use of DNA library for studying growth regulatory genes and substrate utilizing genes in case of *A. bisporus* [164,168], detection of double stranded viral RNA for developing resistant species of mushrooms [169] and use of rescued plasmid vector from *P. ostreatus* for identifying fungal genome sequence and transformation in other strains of *P. ostreatus* [170] have been tried with varied level of success. In the process of cloning and sequencing it was recorded that a DNA fragment of 1635 bp from *S. commune* having fruitbody inducing activity can be utilized other mushrooms of commercial importance. Similarly the sporelessness in case of *Pleurotus* spp. was found to be linked to a limited number of loci but not with loci responsible for unfavourable appearance of fruitbodies [171,172].

Schizophyllum commune an edible mushroom but not of commercial importance has been used as a prototype for genetic studies and manipulation of the other edible fungi [173]. Attempts have been made to develop transformants of *S. commune* [174], *C. cinereus* [175] and *Agrocybe aegerita* [176,177]. Despite several attempts, not much success could be achieved in case of *A.bisporus* [178,179,180]. The failure in case of *A. bisporus* can be attributed to any of the following reasons like inability of the transforming DNA to enter into the cell, to integrate into the chromosome, to replicate or to express [181]. Strategy like positive selection strategy by using $trp3^{iar}$ gene from *Coprinus cinereus* for transformation in *P. ostreatus* and *V. volvacea* has helped in obtaining successful transformations [182] and by using homologous DNA-mediated transformation system higher efficiency of about 50 fold has also been achieved [183].

In ballistic gun method the small fragment of DNA with desired genes are coated on a tungestun chip and then chips are bombarded to any tissue of mushroom with very high speed. This technique is mostly adopted where other techniques like protoplast or *Agrobacterium*-mediated transformation fail. The technique has been successfully used in case of *Coprinus* spp. [184]. The transformations in case of *P. ostreatus* and *A. bisporus* are also possible by using *A. tumefaciens*-mediated transformation system [185]. Still there is a long list of genes which can be cloned in *A. bisporus* by evolving the suitable transformation system (Table 2).

Table 2. Possible useful genes for incorporation into *Agaricus bisporus.*

Gene / Product	Source	Function
Amylase	*Bacillus amyloliquefaciens*	Starch hydrolysis
Cellulases	*Trichoderma* spp.	Cellulolysis
Fungiside resistant	Various	Fungicide resistance
Hemicellulases	*Aspergillus* spp. *Trichoderma* spp.	Hemicellulolysis
Ligninase [s]	*Phanerochaete chrysosporium*	Lignin degradation
Lipases	Various	Lipid degradation
Mating gene	Various	Homokaryotic fruiting
Protease	Various	Protein degradation

Source: Sodhi *et al.* [186]

4. RECENT ADVANCES IN SUBSTRATE PREPARATION

Lignocellulosics account for about 60% of the total plant biomass and are the most abundant natural raw material present on the Earth. These are produced through photosynthetic reactions operating inside the cells of green plants and the net dry biomass production by plants is estimated to be 155 billion tons per year [187]. Lignocellulosics contain mainly cellulose, hemicellulose and lignin. This inexhaustible and renewable natural source can be utilized through microbes for the production of industrial chemicals, liquid fuel, protein rich food (mushroom or single cell protein) and feeds.

Mushrooms, particularly white button mushrooms *A. bisporus* and *A. bitorquis*, draw the nutrients from a specially fermented materials of agricultural origin. All the necessary nutrient components like carbon, nitrogen, phosphorus, iron, sulphur, potassium and vitamins required for mushroom growth are assimilated from this specially fermented substrate called compost and the process of its artificial preparation by adding all the requisite ingredients is called composting. It is a multistep process guided by a consortium of microorganisms including fungi, bacteria and actinomycetes. Role of thermophilic fungi in composting has been well-elucidated and the reports have also proved their role in obtaining desired selectivity in the substrate [188]. Some of these fungi not only help in obtaining the desired selectivity in the mushroom substrate but also contribute themselves to the *Agaricus* nutrition [189]. During the process of composting, also called substrate preparation, the readily available nutrients like carbohydrates and nitrogen are assimilated by the microorganisms present in compost ingredients and organisms multiply with the process of composting and as a result large fraction of nutrients is locked up inside the microbial cells present in the substrate and mushrooms have the ability to utilize these nutrients with the aid of their cell hydrolyzing enzymes [190]. The lignin-humus complex formed after fermentation can only be attacked by mushrooms because of the specific degradative enzymes producing capacity and thus selectivity in the mushroom substrate is achieved.

The fermentation process operating in the mushroom substrate preparation is different from other fermentation processes because of the defined attributes of the mushroom substrate which help in obtaining highest possible yield of mushrooms with little chance of failure because of undesired moulds, insect and pest infestation. It has been reported that the end product obtained in mushroom substrate fermentative process must have a desired bulk density of about 550 to 600 kg m^3 [191], selective only for *A. bisporus* [192], sufficient water holding capacity [193], sufficient air space for the growth of aerobic mycelium [194], specific nitrogen level of about 1.75% [195] and suitable pH ranging from 7.2 to 7.8 [195].

4.1 Factors Affecting Conversion of Lignocellulose into Fungal Biomass

Important factors affecting lignin degradation are temperature, pH, water content [196], carbon dioxide and oxygen levels [197]. Carbon dioxide, in general, is taken as inhibitory and oxygen as stimulatory in such processes [196,198]. The addition of other nutrient sources like nitrogen promotes the bioconversion of non-lignin fractions and inhibit lignin bioconversion [196,199]. But the most important factor which affects compost quality is the microbial population stimulating the fermentative process. Therefore, for the improvement in methods of mushroom substrate preparation, the knowledge of basics underlying this process is very important. The degradative process in the composting passes through phases like mesophilic in the beginning, followed by thermophilic in the middle and again mesophilic in the end. All these phases ultimately lead to the production of CO_2, water, minerals and a stabilized organic matter called compost. The process the composting finally results in 15-35% weight loss due to the release of metabolic by products like CO_2, H_2O and other gaseous materials [200].

4.2 Methods of Composting

With due consideration to the quality of the product expected and parameters involved in composting, following three methods of compost preparation were evolved over the time as a process of improvement.

Long method of composting. This is the oldest method and now exists only in few pockets of the world because of the low mushroom yields, proneness to attack by pathogens, and more time as well as labour consuming process [194]. This method is a completely outdoor activity and takes about 28 days, though production of this compost in lesser duration has also been reported [201]. The biomass loss in this method is also very high [30-35%] and also creates more environmental problems like release of unwanted harmful gases. The microbiological and chemical reactions operating inside the substrate to convert it in a suitable medium for mushroom growth remain unsystematic under uncontrolled environmental conditions of composting. The quality of compost is poor and often it causes heavy crop losses under unhygienic conditions of cropping rooms due to which it has presently limited relevance.

Short method of composting. With many shortcomings associated with the long method of composting and with the realization of importance of temperature and microbes in composting process, improved method of substrate preparation with scope of temperature modulation came in existence near 1950's. The original thinking behind this process was to have a temperature of 50 to 60°C for obtaining a selective substrate with very less chances of infestation by problematic moulds, insects and pests. This was based upon the work of an American Scientist, Lambert [202], who studied the compost from different zones having different temperature and oxygen conditions and found that productive compost came from regions having temperature between 50-60°C and adequate supply of O2. Based upon the findings of Lambert, Sinden and Hauser [203,204] came with a new concept of substrate preparation having pasteurization as the integral part of composting process. This process of composting was later on called as the short method of composting because it took comparatively less time than the long method of composting. The short method of composting mainly consists of two phases: outdoor composting for 10-15 days (Phase-1), followed by pasteurization and conditions inside an insulated well built structure called as the pasteurization tunnel taking around another seven days. In this method, the remixing/turning of substrate ingredients is carried out almost on every alternate day so that alteast 5 to 6 turnings can be achieved in 10 days of outdoor composting (Phase-I). Another important parameter is the maintenance of aerobic condition by blowing oxygen @1 to 15 m^3/ton of wet compost/hour which facilitates high temperature, production of more homogenous substrate in comparatively 30% less time than long method of composting [205]. The desired quantity of air can be inserted either by making compost stacks on the GI pipes with holes and connected to an air blower or by ventilating the stacks by inserting vertical ducts in the stack [206]. During phase-I of composting the temperature can rise up to 80°C into the core of stack resulting into charring of the ingredients giving a dark brown colour of the compost. However, the outer surface having a temperature around 50-60°C facilitates the growth of thermophilic microorganisms. The compost after completion of Phase-I of composting should be of blackish colour having moisture around 72-75%, very heavy smell of ammonia, pH around 8-8.5, nitrogen content between 1.5- 2.0 with ammonia concentration around 800-1000 ppm [207]. Based upon the temperature conditions maintained inside the pasteurization tunnel (phase-II of composting) it can be divided into two sections i.e. pasteurization and conditioning.

Pasteurization. During phase-I of composting, different zones of the stack possessdifferent levels of temperature and different types of microorganisms both qualitatively as well as quantitatively. However, during pasteurization a homogenous environmental conditions is achieved and temperature is maintained between 57 to 60°C for 6-8 h. Temperature beyond 60 °C is not recommended because it leads to elimination of some beneficial thermophilic microorganisms having role later in compost conditioning. The temperature of 60°C can be achieved either by self-heating of the compost or by injecting steam; in either of the cases this temperature is sufficient to kill the problematic insect/pests of the mushroom crop. The time and duration of pasteurization has less bearing on the crop yield and it can be effected immediately after tunnel filling or later on [208].

Conditioning. During conditioning the temperature of the bulk is maintained in the range of 45 to 52°C suited for the growth of desirable thermophilic microorganisms. The excess of ammonia released during this stage is utilized by the growing microorganisms and gets entrapped as microbial protein/or biological nitrogen. The high population of thermophiles achieved at this stage is helpful for whole of the composting process because majority of them perish during the pasteurization process and some of the left outs will help in building up the requisite population during post-pasteurization conditioning process [209]. Besides the temperature, another important factor affecting the outcome of the conditioning process is the sufficient supply of oxygen. Oxygen level of about 10% during this process is essential to build up requisite number of thermophiles and assimilation of ammonia in the form of microbial protein. Both pasteurization and conditioning are essential for achieving the selectivity in the compost at the expense of harmful organisms. There are many advantages of short method of composting: (i) more compost per unit weight of the ingredients is produced, (ii) more mushroom yield per unit area of cropping room or per unit weight of compost, (iii) less chances of infestation from insects and infection from competitor or pathogenic moulds, (iv) shorter duration of substrate preparation and cropping cycle, and (v) less environmental pollution.

Indoor composting. The problems of environmental pollution associated with the long method and to a lesser extent with short method of composting drew the attention of researchers to evolve alternative methods where production of stinking gases like ammonia, methane, hydrogen sulfide and other methylated sulfur compounds can be brought to their minimum level [210-213]. The gases are usually produced under high temperature and anaerobicity inside the compost mass [214]. The production of these gases also depends upon the type and quantity of the basal ingredients used for substrate preparation [215]. In Europe and Australia, the strict environment safety related legislations forced many of the companies to close their composting units. So a 'clean' process was in demand which could produce the high quality mushroom substrate without any adverse effect on the environment [216]. Because first time the work on such composting system started using completely indoor system for compost preparation, it is also called as indoor composting [217], or environmentally controlled composting [218], rapid indoor composting and aerated rapid composting [219]. Though the work initially started with omitting the phase-I completely [220] which resulted in sharp decrease in mushroom yield [221] but, later on, the phase-I of composting was tried inside tunnel with proper arrangement of forced air and satisfactory results were obtained [222,223]. On the basis of temperature conditions maintained inside the tunnel, the indoor composting can be divided into two categories i.e. INRA methods and Anglo-Dutch method.

INRA method is mainly followed in France, Italy, Belgium and Austria and large quantity of compost is being prepared by this method [224]. In this process the phase-I is carried out at a constant temperature of 80°C for 2-3 days under indoor conditions followed by phase-II in which the temperature is kept at 50°C for 5-7 days [225]. As very high temperature attained during phase-I results in killing of all desirable microbes also [226] reinoculation with mature compost or thermophilic fungi becomes necessary in this process [227,228,229]. In situations where proper ventilation in the tunnel during phase-I is maintained or the phase-I is being carried out in the bunker system, reinoculation with thermophiles can be dispensed with because comparatively cooler regions in the stack support the existence of thermophiles [193,230]. Work of evolving a suitable formula for inoculation of *Scytalidium thermophilum,* desirable for this important thermophilic fungus for obtaining compost of desired quality has also been carried out [231].

The another method called as Anglo-Dutch method was the result of continuous research in several European countries and Australia [216,232,233,234] which basically involves the maintenance of an optimum temperature for the growth of thermophilic micro-organisms particularly *Scytalidium thermophilum* and two species of *Humicola* [235]. In this method, a short pasteurization phase of 4-6 hours at 60°C is followed by a week-long conditioning phase at 41°C. This method has been reported to provide excellent selectivity with excellent outdoor control and substantial saving of raw material [191,193]. However, on the contrary some problems like light texture of finished compost which poses problems during compressing [236] and low yield with lower level of selectivity have also been reported [210]. The importance of physical parameters like thorough mixing of ingredients, adequate moisture and good oxygen supply was studied in detail in two Dutch firms, Christiaens and Gieoam and all these factors were found to be important for minimising toxic compounds and malodorous gas production [237]. Some advatages of indoor composting over other methods are: (i) composting takes lesser time than the short and long methods [236], (ii) the mushroom yield per unit weight of compost is very high [30-35kg/100kg of compost [224], (iii) efficient control and management of environmental pollutants conforming to civic laws, (iv) lesser loss of raw materials [10-15%] and thus increased end-product [193], and (v) improved selectivity of mature compost for *A. bisporus* [191]

4.3 Role of Microbes and Their Metabolites in Composting

Compost being a rich source of nutrients harbours a diversified microbial flora which includes bacteria, fungi, actinomycetes, viruses, helminths and protozoa [238,239] and their qualitative as well as quantitative existence depends upon the ingredients and the geographical location [240]. Generally, a microbial progression from mesophilic to thermophilic and again to mesophilic is observed during composting. The factors which determine the dominance of a particular microbial flora during composting depend upon the temperature, moisture, aeration, hydrogen ion concentration (pH) and microbial interactions prevailing inside the compost substrate [241]. Recently molecular techniques like sequencing of rDNA genes have also been employed in monitoring the microbial diversity in mushroom compost [242]. In fact, the compost ingredients harbour varied types of microflora but the fungi have been studied by majority of the researchers. The fungi found growing saprophytically on the cereal straw can be divided into three major groups [207]: (i)rimary saprophytes: *Alternaria alternata, Cladosporium cladosporoides, Doratomyces stemonitis, Stachybotrys atra, Stilbum nanum,* (ii) fungi growing on stored materials: *Aspergillus* spp., *Penicillium* spp., *Absidia ramosa etc.,* and (iii) opportunist fungi: *Mucor hiemalis, Rhizomucor pusillus and Trichoderma viride* etc.

Work [243,244] has also been carried out on mesophilic fungi of chicken manure, an important ingredient of modern day mushroom compost, but majority of these fungi disappear as soon as the temperature of compost pile attains the thermophilic range. Fungi flourish where growth conditions are sub-optimal for other microorganisms [bacteria and actinomycetes] for example lower moisture contents of substrate [245] and lower pH levels. In the initial stage, mesophilic organisms, mainly bacteria, begin to breakdown and utilize the easily available carbohydrates and proteins [246] and the heat generated leads to rise in the stack temperature which is further enhanced by the insulating properties of the compost mass [247]. This rise in temperature results in change in the microflora from mesophilic to thermophilic and less diverse [248,249]. With enhanced metabolic activities, the temperature continues to rise and optimum degradation rates are reported at a temperature range between 45 to 55°C [250]. Excessive temperature is not conducive for proper composting because it inactivates most of the microbial population as well as their activities [211].

Several studies have been conducted to elucidate the role of thermophilic fungi in the composting, which led to the development of the latest method i.e. indoor composting. During the rapid multiplication of thermophilic fungi cellulose and hemicellulose are broken down to expose the available lignin for selectivity to the compost. The thermophilic fungi also utilize available proteins and other nitrogenous substances during their growth which results in production of ammonia which further raises the temperature and softens the straw. The available nitrogen and ammonia during indoor composting is 'locked' inside the microbial cells as this microbial protein is utilized by the mushroom mycelium for its nutrition [190]. The role of thermophilic fungi in composting and as nutrient of *Agaricus bisporus* has been studied by several workers [188, 251-254]. Among different thermophilic fungi, the role of *S. thermophilum* and *Humicola* spp. has been highlighted in most of the cases [231,253-257]. Specific studies like enhanced growth of *A. bisporus* mycelium on *H. insolens* encapsulated grains and subsequent better growth of such spawn in mushroom compost further confirmed the importance of thermophilic fungi during composting [258]. Based upon the findings with thermophiles Nair [259], Nair and Price [233], Savoie and Libmond [260] and Savoie *et al.* [261] used PRITAN, an accelerator, bacterial polysaccharides and Express^{TM} for further improving the composting techniques. In these studies several beneficial effects like enhanced process control, less time, better environment protection and good yields were obtained.

5. EDIBLE MUSHROOMS AND BIOREMEDIATION

Bioremediation here signifies removal or detoxification of xenobiotic compounds present in soil or water ecosystem through normal physiological and biochemical processes catalyzed by the existing or introduced microflora. The solid state fermentation technology employed for substrate preparation for the cultivation of *A. bisporus* and *A. bitorquis* can help in answering many questions related to the environmental problems. Recently, the solid state fermentation has widened its horizons, and culturing and delivery of bio-pesticides for biocontrol of insects and pests and bioremediation of xenobiotic compounds have been attempted. But here we will confine ourselves to the SSF technologies related to mushrooms and their potential use in bioremediation. Two approaches can be considered for the bioremediation purpose [262,263].

1. Adding of compost to the contaminated land for augmenting the bio-remedial action involving indigenous micro-organisms or adding of selected potential biodegrading micro-organisms with compost [264].
2. Composting of excavated contaminated soil and raw materials for mushroom compost production.

Recently, considerable interest in solid phase technologies for bioremediation has emerged [265-267] which has led to emergence of three strategies like i] land farming ii] composting and iii] engineered soil cells [266]. The land farming has been explained by Semple *et al.* [263] and are similar in both cases where the treatment of contaminated soil is done 'on site' through conventional agricultural practices. The aim of both the techniques is to create favourable environmental conditions for the existing microflora to biodegrade the organic contaminants, through their catabolic pathways (Fig. 6).

The process with the involvement of engineered microbes is basically a combination of two techniques namely land farming and composting, where the potential microbes are first screened for their catabolic potential and then improved further for their enhanced catabolic potential through modern biotechnological methods. However, compared with *in situ* techniques, the kinetics of biodegradation of toxic chemicals in *ex situ* bioremediation have been found more favourable and economical.

A white rot fungus, *Phanerochaete chrysosporium*, has been widely studied in the area of bioremediation because of its ability to degrade a wide range of compounds [268,269] namely several number of polycyclic aromatic hydrocarbons (PAHs), polychlorinated biphenyls (PCBs), 2,3,7,8-tetrachlorodibenzo- p-dioxin, DDT, indane, chlordane and explosives like TNT, RDX and HMX (265). The TNT mineralizing potential varied in different wood decaying, litter decaying and micromycetous fungi [270]. The another basidiomycete in light is *Pleurotus* spp., a commercial mushroom, which has the potential to biodegrade several xenobiotic compounds [271]. The catabolic pathways (Fig. 7) existing in these fungi are the basis of their existing potential of bioremediation [272,273] because most of the xenobiotic compounds have chemical structure similar to that of the lignin biodegradation products [268]. Considering the potential of *Phanerochaete* spp. and *Pleurotus* spp. several studies were carried out to see their direct seeding effects on soil contaminants like polychlorinated phenols (PCPs), polychlorinated biphenyls and the pesticides like atrazine and carbofuran [274-276]. The incorporation of spent mushroom

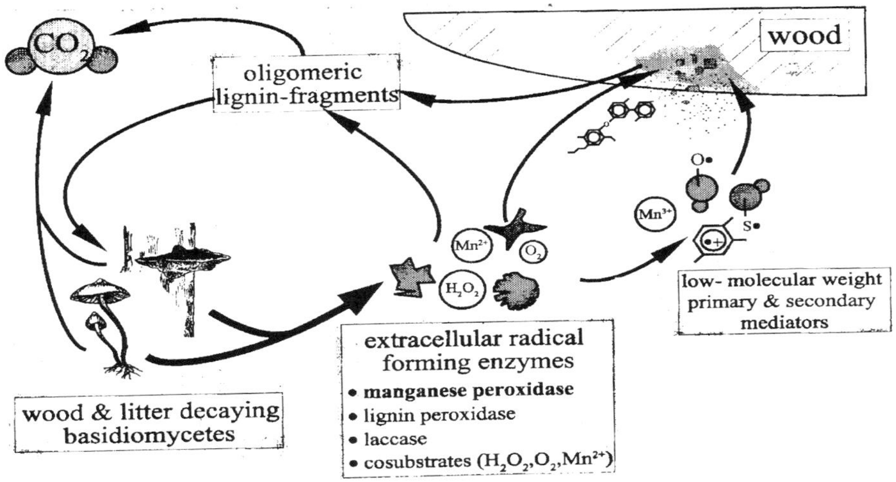

Fig. 6. Breakdown and mineralisation mechanism of macromolecular lignin by wood and litter decaying fungi (Source: Ref. 270).

substrate (SMS) in soil has been reported to enhance the disappearance of alachor and help in obtaining a good crop of garden pea cv. Taichung. The role of SMS in alachlor disappearance was found to be indirect because it stimulated soil microbial flora capable of alachlor degradation [277, 278]. However, contrary to this, the incorporation of SMS has been found to increase the persistence of chloropyrifos, chlorfenvinphos and carbofuran like synthetic chemicals in soil [279]. A simple and less expensive technique to dispose off the pesticide wastes was evaluated by Berry *et al.* [280] by involving a mixture of cotton and wheat straw inoculated with *P. pulmonarius* [281]. The results showed reduced levels of methanol extractable atrazine and only 30% of the total radioactivity of added ^{14}C-ring labeled atrazine was recovered suggesting a potential role of *P. pulmonarius* in mineralization of the pesticide. Another significant contribution in this direction was made by Zafar *et al.* [282] who reported the influence of nutrients on biodegradative potential of *Trametes versicolor.* The detoxification of carbofuran (2-3-dihydro-2, 2-dimethyl-7-benzofuranyl methylcarbamate) and the fate of labeled carbofuran was determined under controlled conditions, it was recorded that the carbofuran-7-phenol was susceptible to oxidative coupling by the fungus in the presence of horse radish peroxidase (HRP) and H_2O_2 [283].

Two more species of white rote fungi, *Dichomitus squalens* and *Pleurotus* sp. were tested for the mineralization potential of radio-labeled pyrene [284]. With both the species the mineralization of pyrene was found higher in presence of other soil microorganisms. However, the mineralization with pure cultures was higher with *Pleurotus* _sp. than with the *D. squalens*. In a separate study the mineralization potential for pyrene of four fungi including three white rot basidiomycetes namely *Phanerochaete chrysosporium, Dichomitus squalens, Pleurotus florida* and one brown rot fungus, *Flammulina velutipes* was compared and maximum mineralization of

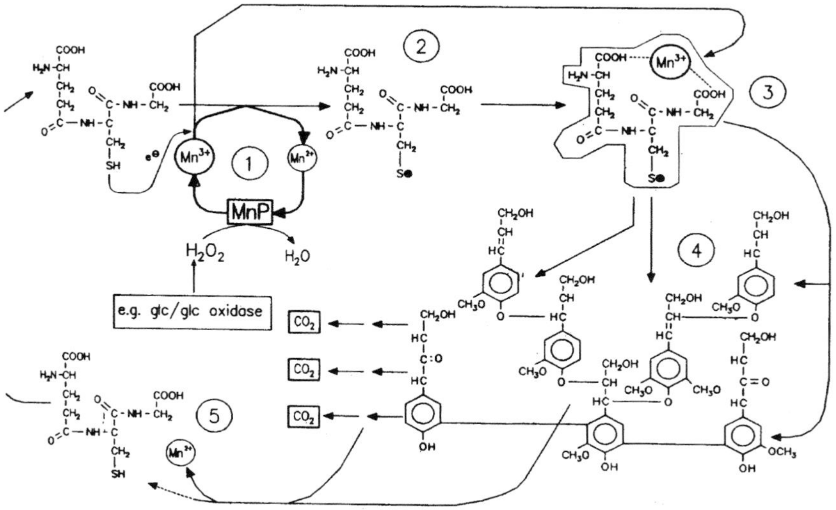

Fig. 7. The enzymatic combustion of lignin by manganese peroxidase-a hypothetical model. [270].

48 to 52% was recorded in *P. florida* and *P. chrysosporium* followed by *D. squalens* and no mineralization with *F. velutipes* [285]. The other basidiomycete, *L. edodes* has also showed dechlorination and mineralization of xenobiotic compounds on mixing in the contaminated soil [286]. In laboratory experiments the mixing of PCP contaminate sterile soil with aliquots of spent sawdust cultures of shiitake mushroom, supplemented with a nutrient solution of glucose, thiamine and mineral salt resulted in disappearance of about 44.4-60.5% of PCP in 21 days of incubation but the disappearance was not correlated with the polyphenol oxidase activity in the inoculum [286,287]. However, during the same period of incubation, Bushwell [288] reported only 50% breakdown of PCP. The potential of SMS in remediation of contaminated land sites has also been highlighted by Bushwell, [264]. Inoculation of *Phanerochaete chrysosporium* grown on pith from sugarcane bagasse had the potential of bioremediation of contaminated soil with aromatics [289]. The bio-accumulation property of basidiomycetes also makes them suitable in removal of heavy metals like lead and copper [290]. The spent mushroom substrate has the ability to chemically adsorb the organic and inorganic pollutants and in addition it harbours diverse category of microbes having potential to detoxify the potential xenobiotic compounds present in soil or water [270,288,291].

An easy and cost effective method of developing bioremediation system is through use of composts and composting technologies and recently Fermor *et al.* [292] have described the use of this system for bioremediation of pentachlorophenol, polycyclic aromatic hydrocarbons and aromatic monomers. Various studies have proved that the SMS has all the desirable attributes of a good soil conditioner and bioremediator of xenobiotic compounds in soil as well as in water . The desirable attributes of SMS are: permeability to air, alkaline to neutral pH, high water-holding capacity, nutritional richness, harbouring diverse category of microflora which help in soil aggregate formation and degradation of xenobiotics. The increasing environmental awareness combined with imposition of strict laws [293] and comparatively low cost of land remediation by composting or compost amendment as compared to that of removal of landfill [upto £ 100/tonne] or incineration (£50-1200/tonne) has prompted the use of SMS for bioremediation activities [263]. The possible use of composting technologies to remediate polluted soil has been tried at various levels like respirometers [10-100 g; 294,295], bench composter (1 kg; 296] environmentally controlled compost bins [20-40 kg; 297] and outdoor tarpaulin windows (tonnes; 298). The three types of compounds namely chlorophenols, polycyclic aromatic hydrocarbons and BTEX compounds were used in the study and it found that the fate of the pollutants in the soil system depends upon the chemical and physical properties of the pollutants and environmental factors [temperature, moisture content] [263]. Study on the role of compost microflora in bioremediation of chlorophenol was carried out by Watts [299], Amner *et al.* [300], Semple & Fermor [301] and concluded that several types of actinomycetes, bacteria survives in SMS and the population of each type is affected by the availability of a particular type of chlorophenol in the system. However, in some cases, the calcitrant compounds formed during the catabolism of the toxic compounds have been found to persist in the environment for a much longer period than the original one. One example is the catabolism of PCP by *P. chrysosporium* and *L. edodes* which resulted in formation of a strong calcitrant like penta chloroanisole [301]. However, the compound could be further detoxified by the existing soil microflora [302]. From various studies it has been concluded that PCP degradative potential of mushroom compost can be increased by pre-exposure to PCP [294]. In a significant study it was found that the spent mushroom substrate of *Pleurotus pulmonarius* could remove higher levels of pentachlorophenol than the pure cultures of other fungi like *Armillaria gallica, A. mellea, Ganoderma lucidum, L. edodes, P. chrysosporium, P. pulmonarius, Polyporus* sp., *C. cinereus* and *V. volvacea and* the removal of PCP was related to the other microflora associated with SMS and its physical properties [303]. The biodegradative potential of phase-II mushroom compost for pure PAH or in PAH contaminated soil has also been investigated [292,304,305,306]. In addition, the residues obtained after mushroom cultivation could also be utilized for large scale production of phenol oxidases which have role in environmental purification [307].

The another potential use of SMS has been found in cleaning of sulphur deposits from coal mines. The limestone, gypsum, organic matter and "bulk" that constitute SMS make it an ideal choice for passive treatment of coal mines drainage. The quality of water coming out of such sites can be improved by construction of a wet land with SMS. The water released from these wet lands meats clean steam specifications [308,309]. Now with the increasing importance of environment safety newer bio-indicators have also been discovered to assess the effect of SMS land applications on adjacent surface water quality [310]. Spent mushroom substrate, if used for bioremediation solves the twin problems of its disposal as well as of environment cleansing which is the reason of very active research in this area of edible mushrooms.

6. CONCLUSIONS

Mushrooms are being very actively researched mainly due to their commercial importance. Transformation of the vegetative mycelium into the fruitbodies represents an interesting phenomenon for the researchers to study the morphogenesis of multi cellular eukaryotes. Though the beginning has been made but the findings on carbohydrate metabolism seen more to be the effect than the cause for differentiation; needless to emphasize that much more needs to be done to decipher the phenomenon at the genetic level expressing itself through the protein route. Newer biotechnological te:hniques and tools are likely to play important role in the genetic improvement of mushrooms in future. Though economically profitable level has been achieved in the substrate preparation technology but understanding at the microlevel may result in a breakthrough breaking the yield barrier which seems to have currently plateaued. Nutritional, medicinal, commercial and environmental importance of edible fungi will continually force the researchers to study these wonderful microbes at the molecular level.

7. REFERENCES

1. Wood, D.A. (1989). Mushroom Biotechnology. International Industrial Biotechnology.9: 1.
2. Chang, S.T. and Miles, P.G. (1989). Edible Mushrooms and their cultivation. CRC Press, Florida.
3. Jong, S.C. and Birmingham, J.M. (1992). Edible mushrooms in biotechnology. Proc. Asian Mycol. Symp. Seoul, Korea: 18.
4. Labarere, J. and Menini, U. (2000). Characterisation, conservation, evaluation and utilization of mushroom genetic resources for food and agriculture. The Global Network on Mushrooms. (FAO) Bordeaux, France: 239.
5. Chang, S.T. (1996). Mushroom research and development – equality and mutual benefit. In: Mushroom Biology and Mushroom Products. ed. Royse, D.J. Pennsylvania State University, Pennsylvania: 1.
6. Rai, R.D. and Verma, R.N. (1998). Mushroom Production in India – Status and Prospects. In: Trends in Microbial Exploitation. eds. Rai, B., Uradhyay, R.S. and Dubey, N.K. BHU, Varanasi. 9: 159.
7. Smith, J.F. and Berry, D.R. (1974). An Introduction to Biochemistry of Fungal Development. Academic Press, London: 326 .
8. Turian, G. (1983). Concepts of fungal differentiation. In: Fungal Differentiation. ed. Smith, J.E. Marcel Dekker, New York, pp. 1.
9. Hammond, J.B.W. (1985). Biochemistry of *Agaricus* fructification. In: Developmental Biology of Higher Fungi. ed. Moore, D., Casselton, L.A., Wood, D.A. and Frankland, J.C. Cambridge University Press, Cambridge: 389.
10. Hammond, J.B.W. (1986). Carbohydrates and mushroom growth. Mushroom J. 165: 316.
11. Hammond, J.B.W. and. Wood, D.A. (1985). Metabolism, biochemistry and physiology. In: Biology and Technology of cultivated Mushroom. ed. Flegg, P.B., Spencer, D.M.and Wood, D.A. Wiley, Chichester: 63.
12. Rai, R.D. and Saxena, S. (1991). Biochemistry of mushroom fructification. In: Advances in Mushroom Biotechnology. ed. Gukulapalan, C. and Lulu Das. Scientific Publsihers, Jodhpur: 31.
13. Groot, P.W.J., Visser, J., Griensven, L.J.L.D. and Schaap, P.J. (1998). Biochemical and molecular aspects of growth and fruiting of the edible mushroom *Agaricus bisporus*. Mycol. Res. 102: 1297.
14. Mol, P.C. and Wessels, J.G.H. (1990). Differences in wall structure between substrate hyphae and hyphae of fruitbody stipes in *Agaricus bisporus*. Mycol. Res. 94: 472.
15. Craig, G.D., Newsam, R.J., Gull, K. and Wood, D.A. (1979). An ultrastructural and autoradiographic study of stipe elongation in *Agaricus bisporus*. Protoplasma. 98: 15.
16. Barrtnicki – Garcia, S., Brakers, C.E., Reyes, E. and Ruiz-Herrera, S. (1978). Isolation of chitosomes from taxonomically diverse fungi and synthesis of microfibrils in vitro. Expl. Mycol. 2: 173.
17. Hanseler, E., Nyhlen, L.E. and. Rast, D.M. (1983). Isolation and properties of chitin synthetase from *Agaricus bisporus* mycelium. Expl. Mycol. 7: 17.
18. Mol, P.C., Vermulen, C.A. and Wessels, J.G.H. (1990). Diffuse extension of hyphae of stipes of *Agaricus bisporus* may be based on a unique wall structure. Mycol. Res. 94: 480.
19. Rai, R.D. and Sexana, S. (1988). Growth and morphogenesis of mushrooms. In: Report. National Centre for Mushroom Research and Training, Solan, India: 42.

20. Rai, R.D. and Sexana, S. (1989). Protein, amino acids and sugars during fruitbody development of *Pleurotus citrinopileatus*. Indian J. Mushrooms: 15.
21. Kaul, T.N. (1987). Fruiting in larger fungi – a review. In: Science and Cultivation Technology of Edible Fungi. ed. Kaul, T.N. and Kapur, B.M. RRL, Jammu Tawi: 230.
22. Moore, D., Elhiti, M.M.Y. and Butler, R.D. (1979). Morphogenesis of carpophore of *Croprinus cinereus*. New Phytologist. 83: 695.
23. Li, G.S.F. and Chang, S.T. (1982). Nutritive value of *Volvariella volvacea*. In: Tropical Mushrooms – biological nature and cultivation methods. ed. Chang, S.T. and Quimio, T.H. Chinese University Press, Hong Kong: 199.
24. Leatham, G.F. (1985). Growth and development of *Lentinus edodes* on a chemically defined medium. In: Developmental Biology of Higher Fungi. ed. Moore, D., Casselton, L.A., Wood, D.A. & Frankland, J.C. Cambridge University Press, Cambridge: 403.
25. Wood, D.A. (1979). Biochemical changes during growth and development of *Agaricus bisporus*. Mushroom Sci. 10(1): 401.
26. Chang, S.T. and Chen, K.Y. (1973). Quantitative and qualitative changes in proteins during morphogenesis of the basidiocarp of *Volvariella volvacea*. Mycologia. 65: 355.
27. Minamide, T. and Hammond, J.B.W. (1985). Evidence for stimulation of G-6-phosphate dehydrogenase synthesis of during initiation of periodic fruitbody growth of *Agaricus bisporus*. Expl. Mycol. 9: 116.
28. Wannet, W.T.B. Drift, C.V., Camp, H.J.M.O. and Griensven, L.J.L.D. (2000). Trehalose and mannitol metabolism in *Agaricus bisporus*. Mushroom Sci. 15(1): 63.
29. Hammond, J.B.W. and Nichols, R. (1976). Carbohydrate metabolism in *Agaricus bisporus* Lange (Sing.): Changes in soluble carbohydrates during growth of mycelium and sporophore. J.Gen. Microbiol. 93: 309.
30. Hammond, J.B.W. and Nichols, R. (1977). Carbohydrate metabolism in *Agaricus bisporus* Lange (Imbach): Metabolism of (C^{14}) labelled sugars by sporophore and mycelium. New Phytologist. 79: 315.
31. Hammond, J.B.W. (1979). Changes in composition of harvested mushrooms (*Agaricus bisporus*). Phytochemistry. 18: 415.
32. Ewaze, J.O., Moore, D. and Stewart, G.R. (1978). Coordinated regulation of enzymes involved in ornithine metabolism and its relation to sporophore morphogenesis in *Coprinus cinereus*. J.Gen. Microbiol. 107: 343.
33. Parrish, G.K. Beleman, R.B. and Kneebone, L.R. (1976). Relationship between yield and mannitol content during the crop cycle of cultivated mushrooms. Hort. Sci. 11: 32.
34. Edmundowicz, J.M. and Wriston, J.C. (1963). Mannitol dehydrogenase from *Agaricus campestris*. J. Biol. Chem. 238: 3539.
35. Ruffner, H.P., Rast, D., Tobler, H. and Karesch, H. (1978). Purification and properties of mannitol dehydrogenase from *Agaricus bisporus* sporocarps. Phytochemistry. 17:865.
36. Morton, N., Dickerson, A.G. and Hammond, J.B.W. (1985). Mannitol metabolism in *Agaricus bisporus*: purification and properties of mannitol dehydrogenase. J.Gen. Microbiol. 131: 2885.
37. Morton, N., Hammond, J.B.W. and Dickerson, A.G. (1985). The *Agaricus bisporus* mannitol pathway during sporophore growth. Trans. Brit. Mycol. Soc. 85: 671.
38. Hammond, J.B.W. (1978). Carbohydrate catabolism in harvested mushrooms. Phytochemistry. 17: 1717.
39. Hammond, J.B.W. and Nichols, R. (1975). Changes in respiration and soluble carbohydrates during postharvest storage of mushrooms (*Agaricus bisporus*). J.Sci.Fd. Agri. 26: 835.
40. Hammond, J.B.W. (1977). Carbohydrate metabolism in *Agaricus bisporus*. Oxidative pathways in mycelium and sporophore. J. Gen. Microbiol. 102: 245.
41. Hou, H.H., Wu, L.C and Chen, C.C. (1975). Respiratory pathways of the cultivated mushroom *Agaricus bisporus*. Mushroom Sci. 9: 37.
42. Dutsch, G.A. and Rast, D. (1972). Biochemische Bezihung Swischen Mannitbuilding und Hexose mono phosphatzyklus in *Agaricus bisporus*. Phytochemistry. 11: 2677.
43. Chanter, D.O. (1979). Harvesting the mushroom crop – a mathematical model. J.Gen. Microbiol. 11: 79.
44. Chanter, D.O. and Thornley, J.M.H. (1978). Mycelial growth initiation and growth of sporophores in the mushroom crop: a mathematical model. J.Gen. Microbiol. 106: 55.
45. Hammond, J.B.W. (1981). Variations in enzyme activity during periodic fruiting of *Agaricus bisprous*. New Phytologist. 89: 419.
46. Wood, D.A. and Goodenough, P. (1977). Fruiting of *Agaricus bisporus*: Changes in extracellular enzyme activities during growth and fruiting. Arch. Microbiol. 114: 161.
47. Claydon, N., Allen, M. and Wood, D.A. (1988). Fruitbody biomass regulated production of extracellular endocellulase during periodic fruiting of *Agaricus bisporus*. Trans. Brit. Mycol. Soc. 90: 85.

48. Smith, J.F., Claydon, N. Love, M.E., Allan, M. and. Wood, D.A. (1989). Effect of substrate depth on extracellular endocellulase and laccase production of *Agaricus bisporus*. Mycol. Res. 93: 292.
49. Wells, T.K., Hammond, J.B.W. and Dickerson, A.G. (1987). Variations in the activities of glycogen phosphorylase and trehalase during the periodic fruiting of the edible mushroom *Agaricus bisporus* (Lange) Imbach. New Phytologist. 105: 273.
50. Griffin, D.W. (1981). Fungal Physiology. Wiley, New York.
51. Wright, B.E. (1978). Concepts of differentiation. In: Filamentous fungi vol. 3. Developmental Mycology. ed. Smith, J.E. and Berry, D.R.. Edward Arnold, London: 1.
52. Wannet, W.J.B., Aben, E.M.J., Drift, C., Griensven, L.J.L.D. and Vogels, G.D. (1999). Trehalose phosphorylase activity and carbohydrate levels during axenic fruiting in three *Agaricus* strains. Current Microbiol. 39: 205.
53. Stoop, J.M.H and Mooibroek, H. (1999). Cloning and characterization of NADP-mannitol dehydrogenase cDNA from the button mushroom *Agaricus bisporus* and its expression in response to NaCl stress. Appl. Env. Microbiol. 64: 4689.
54. Wood, D.A. (1985). Production and role of extracellular enzymes during morphogenesis of basidiomycete fungi. In: Developmental Biology of Higher Fungi, ed. Moore, D. Casselton, L.A., Wood, D.A. and Frnakland, J.C. Cambridge University Press, Cambridge: 375.
55. Kerrigan, R.W. and Ross, I.K. (1988). Extracellular laccases: biochemical markers for *Agaricus* systematics. Mycologia. 80: 689.
56. Wood, D.A. (1980). Production, purification and properties of extracellular laccase of *Agaricus bisporus*. J.Gen.Microbiol. 117: 327.
57. Turner, E.M. (1974). Phenol oxidase activity in relation to substrate and developmental stage in mushroom *Agaricus bisporus*. Trans. Brit. Mycol. Soc. 63: 541.
58. Wood D.A. (1979). Biochemical changes during growth and development of *Agaricus bisporus*. Mushroom Sci. 10(1): 401.
59. Flegg, P.B. and Wood, D.A. (1985). Growth and fruiting. In: Biology and Technology of Cultivated Mushrooms. ed. Flegg, P.B., Spencer, D.M. and Wood, D.A.. Wiley, Chickester: 141.
60. Wood, D.A. (1980). Inactivation of extracellular laccase during fruiting of *Agaricus bisporus*. J. Gen. Microbiol. 117: 339.
61. Leatham, G.F. and Stahmann, M.A. (1981). Studies on laccase of *Lentinus edodes*: specificity, localization and association with the development of fruiting bodies. J.Gen. Microbiol. 125: 147.
62. Kim, K.J., Shin, K.S. and Hong, S.W. (1986). Induction of extracellular polyphenol oxidase from two white rot fungi. Korean J. Mycol. 14: 43.
63. Szklarz, G.D. and Leonowicz, A. (1986). Cooperation between fungal laccase and glucose oxidase in degradation of lignin derivatives. Phytochemistry. 25: 2537.
64. Szklarz, G.D. Antibus, R.K. Sinsbaugh, R.L. and Linkins, A.K. (1989). Production of phenol oxidases and peroxidases by wood rotting fungi. Mycologia. 81: 234.
65. Leatham, G.F., King, V. and Stahmann, M.A. (1980). *In vitro* protein polymerisation by quinones or free radicals generated by plant or fungal oxidative enzymes. Phytopathology. 70: 1134.
66. Turner, E.M., Wright, M., Ward, T., Osborne, D.M. and Self, R. (1975). Production of ethylene and other volatiles and changes in cellulase and laccase activities during the life cycle of cultivated mushroom *Agaricus bisporus*. J.Gen. Microbiol. 91: 167.
67. Smith, J.F. (1984). Nutrient transport in trough system of mushroom culture. Scientia Horticulturae. 27: 257.
68. Rai, R.D. and Saxena, S. (1990). Extracellular enzymes and non-structural components during growth of *Pleurotus sajor-caju* on rice straw. Mushroom J. Tropics. 10: 69.
69. Ishikawa, H., Oki, T. and Senba, Y. (1983). Changes in activities of extracellular enzymes during fruiting of the mushroom *Lentinus edodes* (Berk) Sing. Mokuzai Gakkaishi. 29: 280.
70. Tokimoto, K., Kawai, A. and Komatsu, M. (1977). Nutritional aspects of bed logs of *Lentinus edodes* (Berk) Sing. during fruitbody development. Report. Tottori Mycological Institute (Japan). 15: 65.
71. Sermanni, G.G., Basile, G. and Luna, M. (1979). Biochemical changes occurring in the compost during growth and reproduction of *Pleurotus ostreatus* and *Agaricus bisporus*. Mushroom Sci. 10: 37.
72. Ohga, S., Smith, M., Thurston, C.F. and Wood, D.A. (1999). Transcriptional regulation of laccase and cellulase genes in the mycelium of *Agaricus bisporus* during fruitbody development on a solid substrate. Mycol. Res. 103: 1557.
73. Netzer von, U. (1979). Investigations of primordia formation in the *Pleurotus ostreatus* dikaryon '869x381'. Mushroom Sci. 10: 703.

74. Schwalb, M.N.(1977). Developmentally regulated proteases from the basidiomycete *Schizophyllum commune.* J.Biol. Chem. 252: 8435.
75. Leatham, G.F. and Hasselkus, J.C. (1989). Extracellular acid phosphatases of *Lentinus edodes*: correlation of increased activity with fruitbody development and enzyme localization, substrates, effectors and stability. Mushroom J. Tropics. 9: 55.
76. Madan, M. and Bisaria, R. (1983). Cellulolytic enzymes from an edible mushroom, *Pleurotus sajor-caju.* Biotechnol. Lett. 5: 601.
77. Burton, K.S., Smith, J.F. Wood, D.A. and Thurston, C.F. (1997). Extracellular proteinases from the mycelium of the cultivated mushroom *Agaricus bisporus.* Mycol. Res. 101: 1341.
78. Raper, C.A. (1976) The Biology and Cultivation of Edible Fungi. ed. Chang, S.T. and Hayes, W.A. Academic Press, New York, USA: 83.
79. Verma R.N., Yadav, M.C., Dhar, B.L. and Upadhyay, R.C. (2000). Strategies for genetic improvement of mushrooms – future perspectives. Mushroom Research. 9:11.
80. Chang, S.T., Li, S.X. and Maher, M.J. (1991). Genetical studies on the sexuality pattern of *Volvariella volvacea.* Mushroom Sci. XIII: 119.
81. Fritsche, G. (1984) Breeding *Agaricus bisporus* at the mushroom experimental station, Horst. Mushroom Journal. 122: 49.
82. Fritsche, G. (1991) A personal view on mushroom breeding from 1957-1991. In : Genetics and Breeding of *Agaricus.* Pudoc. Wageningen: 3.
83. Loftus, M. (1995) Breeding new strains of mushrooms. Mushroom News. 43: 6.
84. Bhandal, M.S. and Mehta, K.B. (1989) Evaluation and improvement of strains of *Agaricus bisporus.* Mushroom Sci. 12 (1): 25.
85. Kalra, R. and Phutela, R.P. (1991) Strain selection and development in *Volvariella.* Advances in Mushroom Sci.: 49.
86. Kerrigan, R.W. (1991) What on earth is the *Agaricus* recovery programme? The Mycologist. 5(1): 21.
87. Callac, P., Billette, C., Imbernon, M. and Kerrigan, R.W. (1993) Morphological, genetic and infertility analysis reveal a novel, tetrasporic variety of *Agaricus bisporus* from the Sonaran desert of California. Mycologia. 85: 835.
88. Kush, R.S., Wach, M.P. and Horgan, P.A. (1995) Molecular strategies for *Agaricus* breeding. In : The Mycota II, Genetics and Biotechnology. ed. Kuck, Springer-Verlag, Berlin, Heidelberg: .321.
89. Tomati, U., Rapana, P. and Fiordiponti, P. (1999) Application of molecular techniques to certification in mushroom growing. Informatore Agrario. 55 : 43.
90. Kerrigan, R.W. and Van Griensven, L.J.L.D. (2000). A brief history of marker assisted selection in *Agaricus bisporus.* Mushroom Science. XV : 183.
91. Kerrigan, R.W. (2001) A brief history of marker assisted selection in *Agaricus bisporus.* Mushroom News. 49(7): 4.
92. Royse, D.J. and May, B. (1982) Use of isozyme variation to identify of *Agaricus brunnescens.* Mycologia. 74: 93.
93. Kerrigan, R.W., Ross, I.K. (1989) Allozyme of a wild *Agaricus bisporus* population : new alleles, new genotypes. Mycologia. 81 : 433.
94. Wang, Z.S., Liao, J.H., Wang, H.C. and. Maher, M.J. (1991) Protoplast fusion and fusant variation in *Agaricus bisporus.* Mushroom Science. XIII : 17.
95. Kerrigan, R.W., Carvalho, D.D., Horgen, P.A. and Anderson, J.B. (1993) Mitochondrial DNA polymorphism and divergence in the sub divided global population of *A.bisporus.* In: Ist Int.Conference on Mushroom Biology and Mushroom Products. Hongkong: 128.
96. Kerrigan, R.W., Billette, C., Callac, P. and Velcko, A.J.J. (1996) A Summary of allelic diversity and geographical distribution and six allozyme loci of *A.bisporus* In: Mushroom Biology Mushroom Products. ed. Royse, D.J. Penn State University, USA: 25.
97. Calonje, M., Garcia-Mendoza, C., Novaes-Ledieu, M., Labarere, J. and Elliott, T.J. (1995) Characterization of two commercial *Agaricus bisporus* strains by cell-wall structure, isozymes patterns, nucler and mitochondrial restriction fragment length polymorphism (RFLP). Mushroom Science. XIV: 133.
98. Ohmasa, M. and Furukawa, H. (1986) Analysis of esterase and malate dehydrogenase isozyme of *Lentinus edodes* by isoelectric focusing for identification and descrimination of stocks. Trans.Mycol.Soc.Japan. 27 : 79.
99. Itavaara, M. (1988) Identification of shiitake strains and some other basidiomyces: protein profile, esterase and acid phosphate zymograms as an aid in taxonomy. Trans. Br. Mycol. Soc. 91: 295.

100. Zervakis, G. and Labarere, J. (1992) Taxonomic relationship with in the fungal genus *Pleurotus* as determined by isoelectric focusing analysis of enzyme pattern. J.Gen.Microbiol.138: 635.
101. Iracabal, B., Roux, P., Labarere, J. and Maher, M.J. (1991) Study on enzyme polymorphism in *Agaricus* and *Pleurotus* spp. for characterization and genetic improvement. Mushroom Science. XIII: 37.
102. Castle, A.J., Horgen, P.A. and Anderson, J.B. (1987) Restriction fragment length polymorphism in the mushroom *Agaricus brunnescens* and *A.bitorquis*. Appl. Environ. Microbiol. 53: 816.
103. Loftus, M.G., Moore, D. and Elliott, T.J. (1988) DNA polymorphism in commercial and wild strains of the cultivated mushroom *Agaricus bisporus*. Theoretical and Applied Genetics. 76 : 712.
104. Southern, E.M. (1975) Detection of specific sequences among DNA fragments separated by gel electrophoresis. J. Molecular Biol. 98: 503.
105. Botstein, D., White, R., Skolnick, M and Davis, R.W. (1980) Construction of genetic linkage map in man using RFLP. American Journal of Human Genetics. 82 : 314.
106. Hulbert, S.H., Ilott, T.W., Legg, E.J., Lincoln, S.E., Landev, E.S. and Michelmaore, R.W. (1988) Genetic analysis of the fungus *Bremia Lactucae* using RFLPs. Genetics. 120 : 947.
107. Chang, C., Bowman, J.L., Dejohn, A.W., Lander, E.S., and Meyerowitz, E. (1988) RFLP map linkage map for *Arabidopsis thaliana*. In: proceedings of the National Academy of Science of the United States of America. 85: 6865.
108. Toyomasu, T., Takazawa, H. and Zennyongi, A. (1992) Restriction fragment length polymorphism of mitochondrial DNAs from the basidiomycetes *Pleurotus* spp. Biosci. Biotech. Biochem. 56: 359.
109. Matsumoto, T. and Fukumasa, N.Y. (1995) Mitochondrial DNA restriction fragment length polymorphisms and phenetic relationships in natural populations of the oyster mushroom, *Pleurotus ostreatus*. Mycological Research. 99: 562.
110. Larraya, L.M., Perez, G., Panas, G.G., Baars, J.J.P., Mikosch, T.S.P., Pisabarro, A.G. and Ramirez, L. (1999). Molecular karyotype of the white rot fungus *Pleurotus ostreatus*. Appl. Environ. Microbiol. 65: 3413.
111. Chiu, S.W., Chen, M.J. and Chang, S.T. (1995). Differentiating homothallic *Volvariella* mushrooms by RFLPs and AP-PCR. Myco. Res. 99: 333.
112. Lin, F.X., and Lin, F.C. (1999). RAPD analysis of parental strains and their crossing progenies of *Lentinula edodes*. Mycosystema. 18: 279.
113. Stoop J.M.H. and Mooibroek, H. (1999). Advances in genetic analysis and biotechnology of the cultivated button mushroom, *Agaricus bisporus*. Appl. Microbiol. Biotechnol. 52: 474.
114. Nicholson, M.S., Thon, M.R. and Royse, D.J. (1995). Phylogeny of *Lentinula* spp. based on RFLP analysis of internal transcribed spacers and intergenic regions of ribosomal DNA. Mushroom Science. XIV: 153.
115. Barroso, G., Perennes, D., Labarere, J. and Elliott, T.J. (1995). A 'miniprep' method for RFLP analysis and ds RNAs detection perfected in the cultivated fungus *Agrocybe aegerita*. Mushroom Science. XIV: 1.
116. Kush, R.S., Becker, E. and Wach, M. (1992). DNA amplification polymorphisms of the cultivated mushroom *Agaricus bisporus*. Appl. Environ. Microbiol. 58: 2971.
117. Legg, E.J. (1995). The effect of homokaryon source on genetic map of *Agaricus bisporus* – a comparison. Mushroom Journal. 541: 18-19.
118. Legg, E.J. (1994). Molecular Biology and the improved hybrid mushroom. Mushroom News. 42(7): 30.
119. Vilgalys, R. and Sun, B.L. (1994). Ancient and recent patterns of geographic speciation in the oyster mushroom *Pleurotus* revealed by phylogenetic analysis of ribosomal DNA sequences. Proc. Natl. Acad. Sci. USA, 91: 4599.
120. Moncalvo, J.M. (1995). *Pleurotus cystidiosus*, var. Jormosensis var. nov : an usual *Pleurotus* collection of subgenus *coremiopleurotus* from Taiwan. Mycol. Res. 99: 1479.
121. Sunagawa, M., Tamai, Y., Neda, H., Miyazaki, K. and Miura, K. (1995). Application of random amplified polymorphic DNA (RAPD) markers. I. Analysis of fusants in edible mushrooms. Mukuzai Gakkaishi. 41: 945.
122. Zhang, Y.F., Molina, F.I. and Zhang, Y.F. (1995). Strain typing of *Lentinula edodes* by RAPD assay. FEMS-Microbiology Letters. 131: 17.
123. Lin, F.X., and Lin, F.C. (1999). RAPD analysis of parental strains and their crossing progenies of *Lentinula edodes*. Mycosystema. 18(3): 279.
124. Huang, R.C., Huang, M.L. and Wang, B.C. (1999). Phylogenetic study and grouping of commercial shiitake isolates in Taiwan. Journal of the Chinese Agricultural Chemical Society. 37: 431.
125. Tan, Q., Yan, P.L., Zhan, C.X., Guo, L.G., Wang, N., He, D.M., Ling, X.F., Chen, M.J. and Pan, Y.J. (1999). Genetic diversity anlaysis of different isolates of *Agrocybe aegerita* using RAPD markers. Acta – Agriculturae Shanghai. 15: 18.

126.Neda, H. and Nakai, T. (1995). Phylogenetic analysis of *Pleurotus* based on data from partial sequences of 18S rDNA and ITS-1 regions. Mushroom Sci. 14: 161.
127.Vilgalys, R., Moncalvo, J.M., Liou, S.R. and Volovsek, M. (1996). Recent advances in molecular systematics of the genus *Pleurotus*. In : Mushroom Biology and Mushroom Products. ed. Royse, D.J. Penn State Univ. USA: 91.
128.Chiu, S.W., Chiu, W.T., Lin, F.C. and Moore, D. (2000). Diversity of rDNA sequences indicates that China harbours the greatest germplasm resource of the cultivated mushroom *Lentinula edodes*. In : Science and Cultivation of Edible Fungi. ed. van Griensven, L.J.L.D. Balkema, Rotterdam, The Netherlands: 239.
129.Lee, H.K., Shin, C.S., Min, K.B., Choi, K.S., Kim, B.G., Yoo, Y.B. and Min, K.H. (2000). Molecular systematics of the genus *Pleurotus* using sequence–specific oligonucleotide probes. In : Science and Cultivation of Edible Fungi. ed. Van Griensven, L.J.L.D. Balkema, Rotterdam, The Netherlands: 207.
130.Loftus, M., McCarty, M., Lodder, S., Pitchford, K. and Legg, E. (1996). Template mixing as a method for analysis of sequence characterized amplified regions (SCAR) in the cultivated mushroom *Agaricus bisporus*. Cultivated Mushroom Research CMR Newsletter. 3: 39.
131.Loftus, M., Bouchti, K.L., Robles, C. and Griensven, L.J.L.D. (2000). Use of a SCAR marker for cap colour in *Agaricus bisporus* breeding programmes. Mushroom Science. XV: 201.
132.Bunyard, B.A., Nicholson, M.S. and Royse, D.J. (1995). Phylogenetic resolution of *Morchella*, *Verpa* and *Disciotis* (Pezizales : Morchellaceae) based on restriction enzyme analysis of the 28S ribosomal RNA gene. Experimental Mycology. 19: 223.
133.Castle, A.J., Horgen, A. and Anderson, J.B. (1998). Crosses among homokaryons from commercial and wild collected strains of the mushroom *Agaricus brunnescens*. Appl.Environ. Microbiol. 54: 1643.
134.Mukherjee, M. and Sengupta, S. (1986). Mutagenesis of protoplasts and regeneration of mycelium in the mushroom *Volvariella volvacea*. Appl. Environ. Microbiol. 52: 1412.
135.Toyomasu, T. and Mori, K. (1987). Fruitbody formation of the fusion products obtained on interspecific protoplast fusion between *Pleurotus* spp. Agric. Biol. Chem. 51: 2037.
136.Eguchi, F. and Higaki, M. (1995). Pigment less mutant induced by the regeneration of protoplasts and cell fusion of basidiomycetes. Mukuzai Gakkaishi. 41(5): 505.
137.Homolka, L., Volakova, I. and Nerud, F. (1995). Variability of enzymatic activities in ligninolytic fungi *Pleurotus ostreatus* and *Lentinus tigrinus* after protoplasting and UV-mutagenization. Biotechnol. Tech. 9(3): 157.
138.Sonnenberg, A.S., Wessels, J.G. and van Griensven, L.J.L.D. (1988). An efficient protoplasting/regeneration system for *Agaricus bisporus* and *Agaricus bitorquis*. Curr. Microbiol. 17: 285.
139.Das, N. and Mukherjee, M. (1995). Conditions for isolation of regenerating protoplasts from *Pleurotus sajor-caju*. J. Basic Microbiol. 35: 157.
140.Chen, X.Y. and Hampp, R. (1993). Isolation and regeneration of protoplaste from gills of *Agaricus bisporus*. Curr. Microbiol. 26: 307.
141.Peng, M., Lemke, P.A. and Shaw, J.J. (1993). Improved conditions for protoplast formation and transformation of *Pleurotus ostreatus*. Appl. Microbiol. Biotechnol. 40: 101.
142.Horgen, P.A., Jin, T., Anderson, J.B. and Van Griensven, L.J.L.D. (1991). The use of protoplast production, protoplast regeneration and restriction fragment length polymorphisms in developing a systematic and highly reproducible breeding strategy for *Agaricus bisporus*. In : Genetic and Breeding of *Agaricus*. Horst, Netherlands: 62.
143.Sonnenberg, A.S.M., Hollander, K.D., van de Mukhof, A.P.J. and van Griensven, L.J.L.D. (1991). In : Genetic and Breeding of *Agaricus*. ed. van Griensven, L.J.L.D. Agr. Publishing and Documentation (Pudoc), Wageningen, The Netherlands: 57.
144.Summerbell, R.C., Castles, A.J., Horgen, P.A. and Anderson, J.B. (1989). Inheritance of restriction fragment length polymorphisms in *Agaricus brunnescens*. Genetics. 123: 293.
145.Royer, J.C., Horgen, P.A., and Van Griensven, L.J.L.D. (1991). Towards a transformation system for *Agaricus bisporus*. In : Genetics and Breeding of *Agaricus*. proceedings of the First Int. Seminar on Mushroom Science, Horst, Netherlands: 135.
146.Toyomasu, T. and Mori, K. (1987). Fruitbody formation of the fusion products obtained on interspecific protoplast fusion between *Pleurotus* spp. Agric. Biol. Chem. 51: 2037.
147.Toyomasu, T., Matsumoto, T. and Mori, K. (1986). Interspecific protoplast fusion between *Pleurotus ostreaus* and *Pleurotus salmoneo-stramineus*. Agric. Biol. Chem. 50: 223.
148.Zhao, J. and Chang, S.T. (1997). Interspecific hybridization between *Volvariella volvacea* and *V. bombycina* by PEG-induced protoplast fusion. World J. Microbiol. Biotech. 13: 145.

149.Sunagawa, M., Masatake, O., Shinso, Y., Nobuo, Y. and Toshinaga, I. (1991). Intra- specific protoplast fusion between auxotrophic mutants of *Auricularia polytricha.* Mokuzai Gakkaishi. 37: 1069.
150.Sunagawa, M., Kiyoshi, M., Masatake, O., Shinso, Y., Nobuo, Y. and Toshinaga, I. (1992). Intraspecific heterokaryon formation by protoplast fusion of auxotroph mutants of *Auricularia polytricha.* Mokuzai Gakkaishi. 38: 386.
151.Sunagawa, M. (1992). Interspecific heterokaryon formation between *Auricularia auricula* – Judae and *Auricularia polytricha* by electrical protoplast fusion. Research Bulletins, College Experiment Forests. 49: 219.
152.Zhang, C.K., Chen, Y.L., Qin, H.M. and Liang, Y. (1996). Identification and mushroom production test of the fusants from protoplasts of different *Lentinula edodes* strains. In : Mushroom Biology and Mushroom Products. ed. Royse, D.J. Penn. State Univ. USA: 103.
153.Wang, F.D., Ye, X.F., Ye, G.Z. and Xia, Z.A. (1992). Interspecific protoplast fusion of straw mushroom and behaviour of regenerated strains, Acta–Agriculturae Shanghai, 8,1: 14.
154.Zhang, J.M., Zhang, Y.P. and Chen, M.Y. (1992). Study on intergeneric electrofusion of protoplast between *Lentinus* and *Pleurotus.* Acta Botanica Yunnanica. 14: 283.
155.Zhao, J. and Chang, S.T. (1996). Intergeneric hybridization between *Pleurotus ostreatus* and *Schizophyllum commune* by PEG-induced protoplast fusion. World J. Microbiol. Biotech. 12: 573.
156.Garcia Mendoza, C. (1992). Cell wall structure and protoplast reversion in basidiomycetes. World J. Microbiol. Biotechnol. 8: 36.
157.Luo, X.C., He, P.X. and Shi, Q.L. (1997). Study on the monokaryotization and application of protoplast in *Auricularia auricula.* Fungal Science. 12: 69-74.
158.Zhang, C.K., Liang, Y. and Qin, H.M. (1994). Study on regenerational conditions of monokaryotic mycelium of *Lentinus edodes.* J. Microbiol. 5: 9.
159.Bhattiprolu, G.R. (1995). Protoplast fusion of *Coprinus cinereus* auxotropic mutants. Mushroom Res. 4(2): 69-71.
160.Royer, J.C., Hintz, W.E., Kerrigan, R.W. and Horgen, P.A. (1992). Genome. 35: 694.
161.Sonnenberg, A.S.M., De Groot, P.W.J., Schaap, P.J., Baars, J.J.P., Visser, J. and van Griensven, L.J.L.D. (1996). The isolation of expressed sequence tags of *Agaricus bisporus* and their assignment to chromosomes. Appl. Environ. Microbiol. 62: 4542.
162.Ward, M., Kodama, K.K. and Wilson, L.J. (1989). Experimental Mycology. 13: 289.
163.Li, A. and Horgen, P.A. (1993). Attempts to develop a transformation system in *Agaricus bisporus*, utilizing particle bombardment and several other novel approaches. Cultivated Mushroom Newsletter. 1: 11.
164.Wood, D.A., Claydon, N., Burton, K.S., Matcham, S.E., Allan, M., Perry, C., Thurston, C.F., Raguz, S. and Yague, E. (1991). Molecular analysis of enzymes of *Agaricus bisporus.* Mushroom Science. 13: 43.
165.Ohga, S., Smith, M., Christopher, F.T. and Wood, D.A. (1999). Transcriptional regulation of laccase and cellulase genes in the mycelium of *Agaricus bisporus* during fruitbody development on a solid substrate. Mycol. Res. 103(12): 1557.
166.Giardina, P., Cannio, R., Martirani, L., Martirani, L., Marzullo, L., Palmieri, G. and Sannia, G. (1995). Cloning and sequencing of a laccase gene from the lignin-degrading basidiomycetes *Pleurotus ostreatus.* Appl. Environ. Microbiol. 61: 2408.
167.Zhao, J. and Kwan, H.S. (1999). Characterization, molecular cloning, and differential expression analysis of laccase genes for the edible mushroom *Lentinula edodes.* Appl. Environ. Microbiol. 65(11): 4908.
168.De Groot, P.W.J., Schaap, P.J., Griensven, L.J.L.D. and Visser, J. (1997). Isolation of developmentally regulated genes from the edible mushroom *Agaricus bisprous.* Microbiology. 143: 1993.
169.Sonnenberg, A.S.M., and Van Griensven, L.J.L.D. (1987). Application of molecular and cell biology techniques in the improvement of mushroom cultivation. Bedrijfsontwikkeling. 18(8): 248.
170.Peng, M., Singh, N.K. and Lemke, P.A. (1992). Recovery of recombinant plasmids from *Pleurotus ostreatus* transformants, Curr. Genet., 22: 53.
171.Baars, J.J.P., Sonnenberg, A.S.M., Mikosch, T.S.P., and Van Griensven, L.J.L.D. (2000). Development of a sporeless strain of oyster mushroom *Pleurotus ostreatus.* Mushroom Science. XV: 317-323.
172.Baars, J., Hollander, K., Zandt, J., Kuenen, J., Camps, E., Lavrijssen, B., Litjens, R., Mikosch, T., Sonnenmans, D., Sonnenberg, A. and Van Griensven, L. (2001). Genetics and Breeding of the Button Mushroom. Mushroom World. 12(1): 11.
173.Ullrich, R.C. and Novotny, C.P. (1984). Genetic manipulation in the exploitation and study of basidiomycetes. Microbiology: 133-135.
174.Munoz-Rivas, A., Specht, C.A., Drummond, B.J., Froeliger, E. and Novotony, C.P. (1986). Transformation of the basidiomycetes *Schizophyllum commune.* Mol. Gen. Genet. 250: 103.

175.Binninger, D.M., Skrzynia, C., Pukkila, P.J. and Casselton, L.A. (1987). DNA-mediated transformation of the basidiomycete *Coprinus cinereus.* EMBOJ. 6: 835.

176.Noel, T. and Labarere, J. (1994). Homologous transformation of the edible basidiomycetes *Agrocybe aegerita* with URA 1 gene : characterization of integrative events and of rearranged free plasmids in transformants. Current Genetics. 25: 432.

177.Noel, T., Ho Huynh, T.D. and Labarere, J. (1991). Genetic variability of the wild imcompatibility alleles of the basidiomycete *A.aegerita.* Theoretical and Applied Genetics. 81: 745.

178.Challen, M.P., Rao, B.J. and Elliott, T.J. (1991). Transformation strategies for *Agaricus.* In : Genetics and Breeding of *Agaricus.* ed. van Griensven, L.J.L.D. Proceedings of the First International Seminar on Mushroom Science, Pudoc, Netherlands: 129.

179.Royse, J.C., Hintz, W.E, and Horgen, P.A. (1991). In : Genetics and Breeding of *Agaricus.* ed. Griensvesn, L.J.L.D. Agril. Publishing and Documentation (Pudoc), Wageningen, The Netherlands: 52.

180.Challen, M.P. and Elliott, T.J. (1994). Evaluation of the 5–fluoroindole resistance marker for mushroom transformation. Cultivated Mushroom Newsletter. 2: 13.

181.Royer, J.C. and Horgen, P.A. (1991). Towards a transformation system for *Agaricus bisporus.* In : Genetics and Breeding of *Agaricus.* ed. van Griensven, L.J.L.D. Proceedings of the First International Seminar on Mushroom Science, Pudoc, The Netherlands: 135.

182.Jia, J.H., Buswell, J.A. and Peberdy, J.F. (1998). Transformation of the edible fungi, *Pleurotus ostreatus* and *Volvariella volvacea.* Mycol. Res. 102(7): 876.

183.Honda, Y., Irie, T., Watanabe, T. and Kawahara, M. (2000). Molecular breeding of oyster mushroom using a homologous DNA–mediated transformation system. Mushroom Science. 15: 151.

184.Moore, A.J., Challen, M.P., Warner, P.J. and Elliott, T.J. (1995). Ballistics for the delivery of transforming DNA to mushrooms. Mushroom Science. 14: 63.

185.Mikosch, T.S.P., Lavrijssen, B., Sonnenberg, A.S.M. and Van Griensven, L.J.L.D. (2000). *Agrobacterium tumefaciens* mediated transformation of *Agaricus bisporus.* In : Science and Cultivation of Edible Fungi. ed. Van Griensven, L.J.L.D. Balkema, Netherlands: 173.

186.Sodhi, H.S., Elliott, T.J. and Wood, D.A. (1997). Genetics and genetic manipulation in *Agaricus bisporus.* In: Advances in Mushroom Biology and Production. ed. Rai, R.D., Dhar, B.L. and Verma, R.N. Mushroom Society of India, NRCM, Solan, India: 415.

187.Goldstein, I.S. (1981). Organic chemicals from biomass. CRC Press, Boca Raton.

188.Straatsma, G., Gerrits, J.P.J., Augustijn, M.P.A.M., Op den Camp, H.J.M., Vogets, G.D. and van Griensvan, L.J.L.D. (1989). Population dynamics of Scytalidium thermophilum in mushroom compost and stimulatory effects on growth rate and yield of *Agaricus bisporus.* J.Gen. Microbiol. 135: 751.

189.Batterley, D.A. (1993). Supplements, composting, mushroom nutrition and future direction. Mushroom News. 41: 8-12.

190.Wood, D.A. and Fermor, T.R. (1989). Nutrition of *Agaricus bisporus.* In Biology and Technology of the Cultivated Mushroom. ed. Flegg, P.B., Spencer, D.M. and Wood, D.A. John Wiley and Sons, England: 43.

191.Miller, F.C., Harper, E.R., Macauley, B.J. and Gulliver, A. (1990). Composting based on moderatelly thermophilic aerobic conditions for the production of commercial mushroom growing compost. Australian Journal of Experimental Agriculture. 30: 415.

192.Ross, R.C. and Harris, P.J. (1983). The significance of thermophilic fungi in mushroom compost preparation. Scientia Horticulturae. 20: 61-70.

193.Miller, F.C. (1997). Enclosed phase-1 mushroom composting systems considerations of the underlying technology and methods of implementation. In advances in Mushroom Biology and Production. ed. Rai, R.D., Dhar, B.L. and Verma, R.N. MSI, Solan: 129.

194.Vijay, B. and Gupta, Y. (1995) Production technology of *Agaricus bisporus.* In Advances in Horticulture. Vol.13, ed. Chadha, K.L. and Sharma, S.R. MPH, New Delhi: 63.

195.Vijay, B. (1996). Investigations on compost mycoflora and crop improvement in *Agaricus bisporus* (Lange) Sing. Ph.D thesis. Himachal Pradesh University, Shimla,India.

196.Reid, I.D. (1985). Appl. Environ. Microbiol. 50: 133.

197.Zadrazil, F. (1983). In : Production and feeding of single cell protein. ed. Ferranti, M.P. and Fiechter, A. Applied Sci. Publ. London: 76.

198.Kamra, D.N. and Zadrazil, F. (1985). Biotechnol. Lett. 7: 335.

199.Levonen – Munoz, E. and Bone, D.H. (1985). Biotechnol. Bioeng. 27: 382.

200.Cannel, E. and Moo-Young, M. (1980). Proc. Biochem. 15: 24.

201.Dhar, B.L. and Kapoor, J.N. (1989b). Short duration vigorous composts for *Agaricus bisporus*. Mushroom Science. 12: 471.
202.Lambert, G. (1941). Studies on the preparation of mushroom compost. J. Agric. Res. 62: 415.
203.Sinden, J.W. and Hauser, E. (1950). The short method of mushroom composting. Mushroom Science. 1: 52.
204.Sinden, J.W. and Hauser, E. (1953). The nature of the composting process and its relation to short composting. Mushroom Science. 2: 123.
205.Macauley, B.J. and Perrin, P.S. (1995). Positive aeration of conventional (Phase-I) mushroom compost stacks for odour abatement and process control. Science and Cultivation of Edible Fungi. ed. Elliott, T.J. Balkema, Rotterdam: 223.
206.Zhou, Q.F. and Chen, J.C. (1989). Two energy saving methods of secondary fermentation of mushroom cultivation substrate. Edible Fungi of China. 1: 22.
207. Vijay, B. and Ahlawat, O.P. (2001). Recent advances made in composting for white button mushroom. In: Recent Advances in the Cultivation Technology of Edible Mushrooms. ed. Verma, R.N. and Vijay, B. NRCM, Solan. India (in press).
208 Gerrits, J.P.G. (1988b) .Duration and time of pasteurization in tunnels. Champignon cultuur. 32: 387.
209 Beyer, D.M. (1984). Managing microbial activity during phase-II. Mushroom News. 7: 7.
210 Vestjens, T. (1994). Tunnel composting technology. Mushroom Journal. 531: 14.
211 Derikx, P.J.L., Op den Camp, H.J.M., Van der Drift, C., Van Griensven, L.J.L.D. and Vogels, G.D. (1990). Identification and quantification of odorous compounds emitted during the production of mushroom compost. Appl. Environ. Microbiol.56: 176.
212 Derikx, P.J.L., Simns, F.H.M., Op den Camp, H.J.M., Vander Drift, C., van Griensven, L.J.L.D. and Vogels, G.D. (1991). Evoluation of volatile sulphur compounds during laboratory scale incubation and indoor preparation of compost used as a substrate in mushroom cutlivation. Appl. Environ. Microbiol. 57: 563.
213 Op den Camp, H.J.M. and Derikx, P.J.L. (1991). Stank Produktie bij buiten-en indoor - composting. De Champignoncultuur. 35: 15.
214 Op den Camp, H.J.M., Pol, A., Van Griensvan, L.J.L.D. and Gerrits, J.P.G. (1993). The production of unpleasant odours during the preparation of indoor compost (IVC) and the effects of the use of an air purifier. Mushroom Information. 718: 5-14.
215 Pecchia, J.A., Beyer, D.M. and Wuest, P.J. (2000). The effect of poultry manure based formulations on odor generation in Phase-1 mushroom composting. In : Science and Cultivaaation of Edible Fungi. ed. Van Griensven, L.J.L.D. Balkema, Rotterdam, The Netherlands: 335.
216 Harper, E., Miller, F.C. and Macauley, B.J. (1992). Physical management and interpretation of an environmentally controlled composting eco-system. Australian Journal of Experimental Agriculture. 32: 657.
217 Laborde, J. (1992). What is new in the use of indoor composting on an industrial scale. Mushroom Information. 5: 12.
218 Gulliver, A., Miller, F.C., Harper, E. and Macauley, B.J. (1991). Environmentally controlled composting on a commercial scale in Australia. Mushroom Science. 13: 155.
219 Laborde, J. (1993). Required qualities of indoor composting. Mushroom Information. 9: 5.
220 Gerrits, J.P.G. (1989). Indoor composting op basis van paardemest of stro. Champignon cultuur. 33: 556.
221 Van Griensven, L.J.L.D. (1991). Concurrentieslag. De Champignoncultuur. 35: 337-339.
222 Gerrits, J.P.G., Amsing, J.G.M., Straatsma, G.D. and Van Griensvan, L.J.L.D. (1993). Indoor compost : Phase-1 processen van 3 of 6 dagen in tunnels. De Champignoncultuur. 37: 339.
223 Labance, S.E. and Heinemann, P. (1998). Phase-1 mushroom substrate preparation using controlled environmental conditions to reduce odors and improve productivity. In : Northeast Agricultural and Biological Engineering Conference. Halifax, Nova Scotia, Canada. : 12.
224 Laborde, I.J. (1994b). Controlled composting indoors (indoor composting) an overview of the current technique. Mushroom Information. 9: 5.
225 Laborde, I.J. (1991). Composting : Current and future techniques in France and abroad. Mushroom Information. 3: 4.
226 Evered, C.E., Noble, R. and Atkey, P.T. (1995). Microbial populations and straw degradation in mushroom composts prepared in controlled environments. Mushroom Science. 14: 245.
227 Houdeau, G., Olivier, J.M. and Chabbert, B. (1991). Improvement of indoor short composting. Mushroom Science. 13: 215.
228 Van Griensvan, L.J.L.D. (1992). The production of indoor compost history and prospects. Mushroom Information. 10(2/3): 6.

229 Laborde, J., Lanzi, G., Francescutti, B. and Giordeni, E. (1993). Indoor composting : general principals and large scale developments in Italy. In : Mushroom Biology and Mushroom Products. ed. Chang, S.T., Buswell J.A. and Chiu, S. Chinese Univ. Press, Hong Kong: 93.
230 Laborde, J. (1992). What is new in the use of indoor composting on an industrial scale. Mushroom Information. 5: 12.
231 Straatsma, G., Olijnsma, T.W., Gerrits, J.P.G., Griensvan L.J.L.D. and Op den Camp, H.J.M. (1995a). Inoculation of indoor Phase-II compost with thermophiles. In : Scienceand Cultivation of Edible Fungi. ed. Elliott, T.J. Balkema, Rotterdam: 283.
232 Gerrits, J.P.G. and Van Griensvan, L.J.L.D. (1990). New developments in indoor composting (Tunnel process). Mushroom J. 205: 21.
233 Nair, N.G. and Price, G. (1991). A composting process to minimize odour pollution. In : Science and Cultivation of Edible Fungi. ed. Maher, M.J. Balkema, Rotterdan: 205.
234 Noble, R. and Gaze, R.H. (1994). Controlled environmental composting in the U.K. Mushroom J. 531: 17.
235 Straatsma, G. and Samson, R.A. (1993). Taxonomy of *Scytalidium thermophilum* an important thermophilic fungus in mushroom compost. Mycol. Res. 97: 321.
236 Perrin, P.S. and Gaze, R.H. (1987). Controlled environment composting. Mushroom Journal. 174: 195.
237 Christiaens, P. (1997). Technology of new methods of substrate production. Champignon. 396: 80.
238 Gray, K.R., Sherman, K. and Biddlestone, A.J. (1971). A review of composting-Part 1. Process Biochem. : 32.
239 Pereira-Neto, J.T. (1987). On the treatment of municipal refuse and sewage sludge using aerated static pile composting - a low cost technology approach. Ph.D Thesis,Leed University, UK.
240 Krogstad, O. and Gudding, R. (1975). Acta Agriculturae Scandinavica. 25: 281.
241 Stentiford, E.I. and Dodds. (1996). In : Solid substrate cultivation, ed. de Bertoldi, M., sequi, P., Lemmes, B. and Papi, T. Champman and Hall, London .
242 Ivors, K.L., Beyer, D.M, Wuest, P.L. and Kang, S. (2000). Survey of microbial diversity within mushroom substrate using molecular techniques. In: Science and Cultivation of Edible Fungi. ed. Van Griensvan, L.J.L.D. Balkema, Rotterdam,The Netherlands: 401.
243 Bacon, C.W., Burdick, Q. and Robbins, J.D. (1974). Fungi in poultry feed and houses. Poultry Sci. 53: 1632.
244 Grewal, P.S., Sohi, H.S. and Vijay, B. (1989). Cost effective pretreatment of chicken manure for controlling nematodes an fungal flora in synthetic compost used for cultivation of *Agaricus bisporus* (Lange) Singer. Indian J. Nematol. 18: 22.
245 Finstesin, M.S. and Morris, M.L. (1975). Adv. Appl. Microbiol. 19: 113.
246 Fermor, T.R., Zadrazil, F. and Reiniger, P. (1988). Significance of micro-organisms in the composting process for cultivation of edible fungi. In : Treatment of Lignocellulosics with White Rot Fungi. ed. Zadrazil, F.: 21.
247 Stentiford, E.I., Kelling, S. and Adams, J.L. (1988). Refuse - derived fuel- improved profitability by composting fines residue. Institute of Mechanical Engineering: 31.
248 Storm, P.F. (1985). Appl. Environ. Microbiol. 50: 906.
249 MaeGregor, S.T., Miller, F.C., Psarianos, K.M. and Finstein, M.S. (1981). Appl. Environ. Microbiol. 41: 1321.
250 de Bertoldi, M., Vallini, G. and Pera, A. (1983).Waste Management and Research. 1: 157.
251 Straatsma, G., Gerrits, J.P.G. and Van Griensvan, L.J.L.D. (1991). Growth of *Agaricus bisporus* on mushroom compost. In: Science and Cultivation of Edible Fungi. ed. Maher, M.J. Balkema, Rotterdam: 761.
252 Wiegant, W.M., Wery, J., Buitenhuis, E.T. and Bout, J.A.M. (1992). Growth promoting effect of thermophilic fungi on the mycelium of the edible mushroom *Agaricus bisporus*. Appl. Environ. Microbiol. 58: 2654.
253 Vijay, B., Sharma, S.R. and Lakhanpal, T.N. (1997). Mycoflora of *Agaricus bisporus* compost. In : Advances in Mushroom Biology and Production. ed. Rai, R.D., Dhar, B.L. and Verma, R.N. Mushroom Society of India, Solan, India.: 139.
254 Ahlawat, O.P. and Vijay, B. (2001). Role of thermophilic fungi in compost production for white button mushroom *Agaricus bisporus*. In : Role of Microbes in the Management of the Environmental Pollution. ed. Tewari, R.P., Mukerji, K.G., Gupta J.K.and Gupta, L.K. A.P.H. Publishing Corporation, 5, Ansari Road, Daryaganj, New Delhi: 83.
255 Straatsma, G., Olijnsma, T.W., Van Griensvan, L.J.L.D. and Op den Camp, H.J.M. (1995b). Growth promotion of *Agaricus bisporus* mycelium by *Scytalidium thermophilum* and CO2. In : Science and Cultivation of Edible Fungi. ed. Elliott, T.J. Balkema, Rotterdam.: 289.
256 Straatsma, G., Samson, R.A., Olijnsma, T.W., Gerrits, J.P.G., Op den Camp, H.J.M. and Van Griensvan,L.J.L.D. (1995c). Bio-conversion of cereal straw into mushroom compost. Canadian J. Bot. 73: 1019.

257 Bilay, V.T. (1995). Interaction of thermophilic fungi from mushroom compost in different agar media and temperature. In: Science and Cultivation of Edible Fungi. ed. Elliott, T.J. Balkema, Rotterdam: 251.
258 Bilay, V.T. (2000). Growth of *Agaricus bisporus* on grain pre-colonized by *Humicola insolens* and growth of mushroom mycelium from this spawn on compost. In Science and Cultivation of Edible Fungi. ed. Van Griensvan, L.J.L.D. Balkema, Rotterdam, The Netherlands: 425.
259 Nair, N.G. (1990). Acclerated assisted composting – an abbreviated method of mushroom compost preparation. Mushroom News. 11: 23.
260 Savoie, J.M. and Libmond, S. (1994). Stimulation of environmentally controlled mushroom composting by polysaccharides. World J. Microbiol, Biotech. 10: 313.
261 Savoie, J.M., Libmond, S., Guillion, M. and Wach, M. (1994). The effect of Express-TM on compost produced under both laboratory and commercial conditions. Mushroom News. 7: 23.
262 Ahlawat, O.P. and Rai, R.D. (2001). Recycling of spent mushroom compost. In: Advances in Mushroom Cultivation Technology of Mushrooms. ed. Verma, R.N. and Vijay, B. National Research Centre for Mushroom, Solan, India. (in press)
263 Semple, K.T., Reid, B.J. and Fermor, T.R. (2000). Impact of composting strategies on the treatment of soils contaminated with organic pollutants : a review, Environmental Pollution. (in press)
264 Buswell, J.A. (1994). Potential of spent mushroom substrate for bioremediation purposes. Compost Sci. Utilization. 2(5): 31.
265 Alexander, M. (1999). In : Biodegradation and Bioremediation. 2nd edition, ed. Alexander, M. Academic Press, USA.: 325.
266 Crawford, R.L. (1996). In : Bioremediation Principles and Applications. ed. Crawford, R.L. and Crawford, D.L. Biotechnology Research Series : 6 Cambridge University Press, UK.
267 Lamar, R.T. (1990). Appl. Environ. Microbiol. 56: 3093.
268 Bumpus, J.A., Tein, M., Wright, D.S. and Aust, S.D. (1985). Science. 228: 1434.
269 Donnelly, P.K., Entry, J.A. and Crawford, D.L. (1993). Appl. Environ. Microbiol. 59: 2642.
270 Hofrichter, M., Scheibner, K., Sack, U. and Fritsche, W. (1997). Degradative capacities of white rot and litter decaying fungi for persistent natural and xenobiotic compounds. In : Advances in Mushroom Biology and Production. ed. Rai, R.D., Dhar, B.L. and Verma, R.N. Mushroom Society of India, Solan, India.: 271.
271 Masaphy, S., Levanon, D. and Henis, Y. (1996). Biores. Technol. 56: 207.
272 Bumpus, J.A. and Aust, S.D. (1987). Appl. Environ. Microbiol.53: 2001.
273 Mileski, G.D., Bumpus, J.A., Jurek M.A. and Aust, S.D. (1988). Appl. Environ. Microbiol. 54: 2885.
274 Lamar, R.T., Davis, M.W., Dietrich, D.M. and Glaser, J.A. (1994). Soil Biol. Biochem. 26: 1603.
275 Glaser, J.A. and Lamar, R.T. (1995). In : Bioremediation Science and Applications. ed. Skipper, H.D. and Turco, R.F. Soil Science Society of America. Madison, WI: 117.
276 Kastanek, F., Demnorova, K., Pazlarova, J., Burkhard, J. and Maleterova, Y. (1999). Int. Biodeterior. Biodegrad. 44: 39.
277 Huang, J.W., Hu, C.K., Tzeng, D.D.S. and Ng, K.H. (1995). Effect of soil amended with spent golden mushroom compost on alleviating phytotoxicity of alachlor to seeding of garden pea. Pl. Pathology Bulletin. 4(2): 76.
278 Huang, J.W., Hu, C.K. and Shih, S.D. (1996). The role of soil microorganisms in alleviation of root injury of garden pea seedlings by alachlor with spent golden mushroom compost. Pl. Pathology Bulletin. 5(3): 137.
279 Rouchand, J., Gustin, F., Steene, F.V., Degheele, D., Gillet, L., Benoit, E., Ceustermans, M. and Vanparys, L. (1992). Influences of the organic fertilizers treatments, the CMC and chemical structures of soil organic matter and the insecticides soil persistences an efficiencies in cauliflower crops. Mededelingen Van de Faculteit Land bouwwetenschappen Rijksuniversiteit Gent. 57(3b): 1173.
280 Berry, D.F., Tomkinson, R.A., Hetzel, G.H., Mullins, O.E. and Young, R.W. (1993). J. Environ. Quality. 22: 366.
281 Masaphy, S., Lev Anon, D. and Henis, Y. (1996). Bioresource Technol. 56: 207.
282 Zafar, S.I., Abdullah, N., Iqbal, M. and Sheeraz, Q. (1996). Int. Biodeterior. Biodegrad. 38: 83.
283 Willems, H.P.L., Berry, D.F. and Mullins, D.E. (1996). J. Environ. Quality. 25: 162.
284 Inder Wiesche, C., Martens, R. and Zadrazil, F. (1996). Appl. Microbiol. Biotechnol. 46: 653.
285 Wolter, M., Zadrazil, F. and Martens, R. (2000). Metabolism of pyrene in wheat straw by brown and white rot fungi. Mushroom Science. 15: 827.
286 Okeke, B.C., Paterson, A., Smith, J.E. and Watsoncraik, I.A. (1997). Appl. Microbiol. Biotechnol. 48: 563.
287 Okeke, B.C., Smith, J.E., Paterson, A. and Watsoncraik, I.A. (1993). Aerobic metabolism of pentachlorophenol by spent sawdust culture of 'shiitake' mushroom (*Lentinus edodes*). Soil. Biotech. Let. 15: 1077.

288 Buswell, J.A. (1995). Potential of spent mushroom substrate for bioremediation purposes. Mushroom News. 43(5): 28.
289 Rodriguez-Vazquez, R., Cruz-Corodova, T., Fernandez-Sanchez, J.M., Roldan-Carrillo, T., Mendoza-Cantu, A., Saucedo-Castana, G. and Tomasini-Campocosio, A. (1999). Folia Microbiologica. 44: 213.
290 Campanella, L., Cardarelli, E., Cordatore, M. and Patrolecco, L. (1994). Mushrooms as bioaccumulators of pollutants. Meded. Fac. Landbouwwet. Rijksuniv. Gent. 59: 1883.
291 Semple, K.T. and Fermor, T.R. (1995). The bioremediation of xenobiotic contamination by composts an associated microflora. Mushroom Sci. XIV(II): 917.
292 Fermor, T.R., Watts, N. U., Duncombe, T., Brooks, R., McCarthy, A., Semple, K.T. and Reid, B. (2000). Bioremediation : Use of composts and composting technologies. Mushroom Science. 15: 833.
293 Gerrits, J.P.G. (1994). Composition, use and legislation of spent mushroom substrate in the Netherlands. Compost Sci. Utilization. 2(3): 24.
294 Semple, K.T. and Fermor, T.R. (1997). Enhanced mineralization of (UL-14C) PCP in mushroom composts. Research in Microbiology. 148: 795-798.
295 Semple, K.T., Watts, N.U. and Fermor, T.R. (1998). Factors affecting the mineralization of (U-14C) benzene in spent mushroom substrate. FEMS Microbiology Letters. 164 : 317.
296 Noble, R., Fermor, T.R., Evered, C.E. and Atkey, P.T. (1997). Bench – scale preparation of mushroom substrates in controlled environments. Compost Science and Utilization. 5(3): 32.
297 Brooks, R.C. (1992). Development of a large scale facility for studying the survival of released bacteria. Ph.D Thesis, University of Liverpool, UK.
298 McCarthy, A.J. (1996). The development of composting system for xenobiotic waste treatment and for the bioremediation of contaminated lands. In : Recycling Technologies, Treatment of Waste and Contaminated Sites. ed. Barton, J. Austrian Research Centre, Siebesdorf, Austria.: 291.
299 Watts, N.U. (1996). The microbiology of chlorophenol degradation in mushroom composts. Ph.D Thesis, University of Liverpool, UK.
300 Amner, W., Edwards, C. and McCarthy, A.J. (1989). Improved medium for recovery and enumeration of farmer's lung organism *Saccharomonospora viridis*. Appl. Environ. Microbiol. 55: 2669.
301 Lestan, D. and Lamar, R.T. (1996). Development of fungal inocula for bioaugmentation of contaminated soils. Appl. Environ. Microbiol. 62: 2045.
302 Haggblom, M.M., Apajalahti, J.H.A. and Salkinoja-Salonen, M.S. (1988). O-methylation of chlorinated para hydroquinones by *Rhotococcus* chlorophenolicus. Appl. Environ. Microbiol. 54: 1818.
303 Chiu, S.W., Ching, M.L., Fong, K.L. and Moore, D. (1998). Spent oyster mushroom substrate performs better than many mushroom mycelia in removing the biocide penta chlorophenol. Mycological Research. 102: 1553.
304 Reid, B.J. (2000). Bioavailability and biodegradabililty of persistent organic pollutants in soil : the key role of contact time. Ph.D Thesis, University of Lancaster, UK.
305 Reid, B.J., Jones, K.C., Fermor T.R. and Semple, K.T. (1998). Assessment of soil spiking and extraction techniques – implication for bioavailability. Proceedings of Eight Annual Meeting of SETAC – Europe. Bordeaux, France.: 3.
306 Reid, B.J., Jones, K.C., Semple K.T. and Fermor, T.R. (1999). Bioremediation potential of PAHs in compost. In : Bioremediation Technologies for Polycyclic Aromatic Hydrocarbon Compounds. Battele Press, Columbas, OH.: 25.
307 Steffen, K.L., Dann, M.S., Fager, K., Fleischer, S.J. and Harper, J.K. (1994). Short-term and long-term impact on an initial large scale SMS soil amendment on vegetable crop productivity and resource use efficiency. Compost Sci. Utilization. 2(4): 75.
308 Pannier, W. (1993). Spent mushroom compost – A natural resource that provides solutions to environmental problems. Mushroom News. 41(11): 10.
309 Stark, L.M. and Williams, F.M. (1994). The role of spent mushroom substrate for the mitigation of coal mine drainage. Compost Sci. Utilization. 2(4): 84.
310 Reed, W. and Keil, C. (2000). The effect of spent mushroom substrate land applications on adjacent surface water using aquatic macroinvertebrates as bio-indicators. Mushroom News. 48(11): 4.

Applied Mycology and Biotechnology
Volume 2. Agriculture and Food Production

Single Cell Proteins from Fungi and Yeasts

U. O. Ugalde[a*] and J. I. Castrillo[b]
[a]Department of Applied Chemistry, Faculty of Chemistry, University of the Basque Country, P. O. Box 1072, 20080 San Sebastian, Spain (E-mail: qppugmau@sc.ehu.es); [b] School of Biological Sciences, Biochemistry Division, University of Manchester, 2.205 Stopford Building, Oxford Road, Manchester M13 9PT, UK. (E-mail: Juan.I.Castrillo@man.ac.uk).

Single Cell Protein (SCP) is a term coined in the 1960's to embrace microbial biomass products which were produced by fermentation. SCP production technologies arose as a promising way to solve the problem of worldwide protein shortage. They evolved as bioconversion processes which turned low value by-products, often wastes, into products with added nutritional and market value. Intensive research into fermentation science and technology for biomass production, as well as feeding, has resulted in a profound body of knowledge, the benefits of which now span far beyond the field of SCP production. The widespread application of plant breeding programmes and agricultural crop production techniques resulted in a high availability of plant food sources, such as soya, maize, wheat and rice in the second half of the 20th century. In addition, political and economic developments, which swayed the world order from a system of blocks to globalisation, facilitated the open trade of agricultural products. These agricultural products outmarketed SCP on the grounds of lower price. However, the combination of sophisticated production with food processing technology yielded a new generation of SCP products which may be used as meat substitutes, texture providing agents and flavour enhancers. Future application of heterologous protein expression may further develop the potential of this food line, resulting in precisely tailored products which meet specific dietary requirements, or simulate high added value specialty products.

1. INTRODUCTION

The pioneering research conducted almost a century ago by Max Delbrück and his colleagues at the Institut für Gärungsgewerbe in Berlin, first highlighted the value of surplus brewer's yeast as a feeding supplement for animals [1]. This experience proved more than useful in the ensuing First World War, when Germany managed to replace as much as half of its imported protein sources by yeast. Since brewers yeast from beer production was not produced in sufficient quantity to meet the demands as a protein feed, a very large proportion of yeast biomass was expressly produced by aerobic fermentations in a semidefined medium containing ammonium salts as the nitrogen source [2]. This methodology was more efficient than brewing, but still resulted in some fermentation of the carbohydrate source, and suboptimal yield of biomass obtained per unit of substrate. In 1919, a process was invented by Sak in Denmark and Hayduck in Germany in which sugar solution was fed to an aerated suspension of yeast instead of adding yeast to a diluted sugar solution [3]. This process was known as 'Zulaufverfahren'. An incremental-feeding or fed-batch process was thus born

* Corresponding Author

which is still successfully used in todays fermentations. After the end of World War I, German interest in fodder yeast declined, but was revived around 1936 by the 'Heeresverwaltung', when both brewer's yeast, and a variety of yeast specially mass cultured, were used to supplement human and animal diets. By then the advantages of aerobic production of baker's yeast in a rich wort had been fully recognised as a rapid means of producing food in large scale industrial installations. A radically different concept to that of agricultural production [4]. Around this time, the nutritive value of yeast was also the intense subject of study, with two important books published [5, 6]. By the begining of World War II, yeasts had been incorporated first into army diets, and later into civilian diets. Ambitious plans were laid for production of well over 100,000 tons per year. This figure never surpassed 15,000, probably because of the extensive disruption which typically accompanies wartime economies.

The sustained interest in fodder yeast initiated in Germany in the inter-war years echoed elsewhere in the World. As part of a larger programme for utilizing natural sources, the Forest Products Laboratory of the United States Department of Agriculture undertook mass cultivation of yeast on sulfite waste liquor, the species used being *Candida utilis*. Production of fodder yeast in the mid-western states of the U. S. A. expanded steadily [7, 8]. The post war period was characterized by the recognised need to tackle the problems of humanity on a global scale. A number of international organisations emerged for this task under the leadership of the United Nations. One such organisation was The Food and Agriculture Organisation of the United Nations (FAO) which brought forward the hunger and malnutrition problem of the world population in 1960, introducing the concept of the protein gap (25% of the world population had a deficiency of protein intake in their diet). The population growth predictions, moreover showed that the number of inhabitants would double between 1960 and 2000, from 2.5 billion to 5 billion (the actual figure reaches 6 billion), and the greater part of this increase would take place in those countries suffering from malnutrition. The Malthusian prospect of a limiting food supply was reinforced by fears that agricultural production would fail to meet the increasing food requirements of humanity. The resumption of peace had also procured a new atmosphere geared towards the academic study in civilian matters, and fermentation processes saw a very important period of progress. A greater involvement of private companies in the marketing of fermentation products had already begun. By the early 60's a number of multi-national companies decided to investigate the production of microbial biomass as a source of feed protein. The basic kinetic mechanisms ruling the growth pattern of microbes had been elucidated [9] and were being established for yeasts and filamentous fungi [10, 11]. However, important technical challenges remained to be solved in industrial fermentations, and the field was boosting with activity.

The relatively low market selling price set for this non-conventional protein steered design towards low product cost and thus, large scale production. Abundant substrates with low prices were sought. By-products as wide ranging as cheese whey, molasses, starch, ethanol and methanol, hydrocarbon substrates and spent sulfite liquor were chosen to sustain commercial processes. The novelty of unwanted waste product consumption added a new economic incentive to SCP production, as the idea of zero cost substrates, or even the obtainment of additional revenues through the concept of waste treatment were argued and incorporated favourably to reduce the production cost estimates. The benefits of SCP production were thus extended from the production of food to the preservation of the environment. However, this same reasoning also conditioned the production volumes to match substrate consumption, as we will discuss in section 5.

By the mid 60's, some quarter of a million tons of food yeast were being produced in different parts of the world and the Soviet Union alone planed an annual production of 900,000 tons by 1970 of food and fodder yeast, to compensate agricultural protein production deficits [12]. By 1980, SCP production processes were operating on a large scale in developed countries, and plans to extend SCP production to underdeveloped countries were being made.

But a number of technical and political developments that occurred in the 80's conditioned the expansion of the promising SCP industry. Marked improvements in plant breeding and crop production on a global basis allowed for a continued increase in agricultural output, beyond the expected ceilings (Fig. 1). Local effects such as agricultural reform implemented in China, also resulted in marked agricultural output. Finally, the prospect of the end of the cold war could first be foreseen at this time, with important liberation of agricultural reserve stocks for market trading [13]. This trend was later materialised by the General Agreement on Tariffs and Trade (GATT) signed in Marrakesh, commiting 118 countries to a new open trade world market in 1994 [14]. This treaty effectively de-regulated the world distribution of goods, opening new market areas and connecting countries with surpluses, and countries with deficits. This treaty had an immense effect on agricultural product trade worldwide.

In view of these developments, the price of the majour agricultural crops did not experience the increases expected in previous decades, and the market price of protein of plant origin continuously decreased in constant currency terms. This effectively outmarketed SCP (Fig. 2).

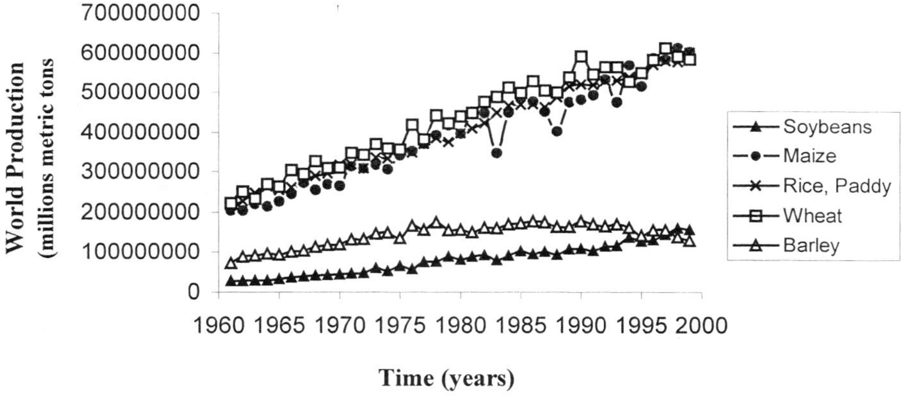

Fig. 1. World production of the main agricultural crops in metric tons. Data obtained from Food and Agriculture Organisation of the United Nations (FAO) [13].

In view of these developments, many industrial SCP processes were discontinued, leaving behind them a wealth of skill and knowledge, which have been successfully benchmarked in other fermentation processes. Specific research in the field also declined in consonance with the market trend (Fig. 3). Some processes evolved towards specialities by applying advanced food processing technologies. One such example is Provesta Corporation (http://www.bpfoods.com), which currently produces a wide variety of food flavours and aromas through the processing of yeast biomass, formerly sold as SCP. But the most notable example of the evolution of SCP processes into new products is perhaps that led by Rank Hovis McDougall (RHM) in cooperation with ICI, founding Marlow Foods (now part of the

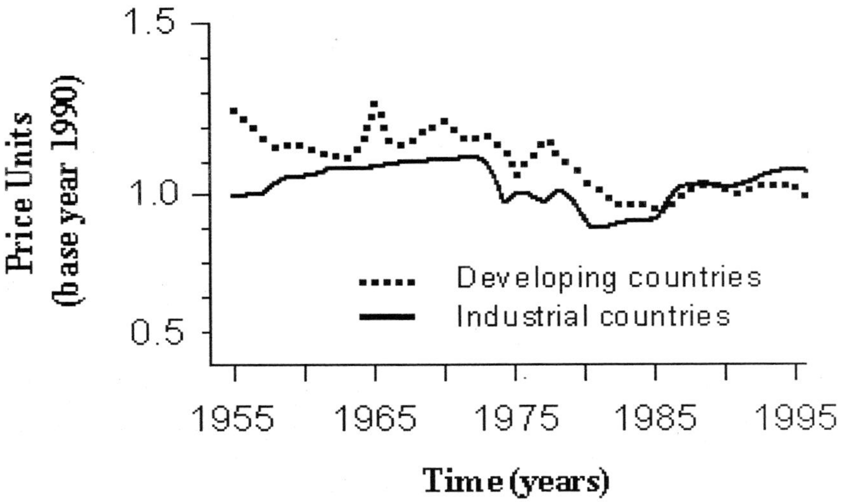

Fig. 2. Real Prices of agricultural exports from industrial and developing countries, (1955-1996). Source: Food and Agriculture Organisation of the United Nations (FAO), based on World Bank data [13].

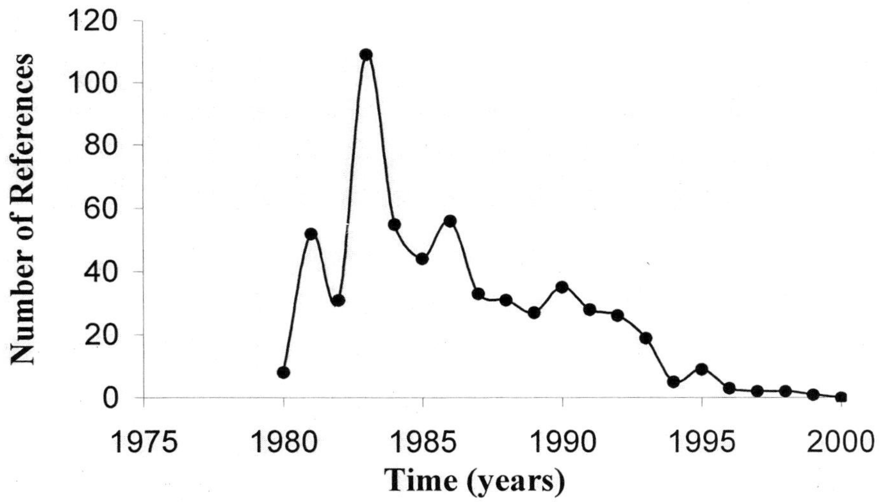

Fig. 3. Number of scientific papers cited including 'SCP' or 'Single Cell Protein' in their title or key words. Data from Cambridge Scientific Abstracts database (CSA) [15].

AstraZeneca group), a company which started producing myco-protein and fungal protein based products under the commercial trademark Quorn™ (http://www.quorn.com). The products were initially derived for human consumption in 1964 from the batch cultivation of *Fusarium venenatum* (formerly *F. graminearum*) strain A3/5 on starch and other waste products. The myco-protein production process experienced an evolution of 20 years and an estimated R+D expenditure of $ 40 million, before unrestricted clearence by the UK Ministry of Agriculture, Fisheries and Foods was granted in 1985. The myco-protein is now produced in continuous culture and the biomass is manipulated to achieve a texture and taste which resemble meat products, covering a market as a meat alternative for vegetarian formulations. Quorn™ products are currently the only SCP-based products exclusively directed at human consumption in the market.

The purpose of this review is to embrace the key aspects of fungal SCP production, consumption and marketing, but at the same time avoiding the exhaustive coverage of technical themes which have already been masterfully dealt with in classic reviews. We intend to present the most up to date developments of this interdisciplinary field, the evolution of which has already been sketched in this introduction. The future prospects of fungal SCP will be considered in the light of our present understanding.

2. SUBSTRATES FOR SCP PRODUCTION

The choice of substrates that are normally abundant and in proximity to the production plant has determined the design and strategy of SCP processes. The most widespread and commonly used substrates for SCP production have been those where the carbon and energy source is derived from carbohydrates. This is due to the fact that their building blocks (mono and disaccharides) are natural microbial substrates, and that carbohydrates are a renewable resource which is widely distributed. Molasses is a by-product of the sugar manufacturing process. The concentrated sugar solution obtained from the milling of sugar cane or sugar beet is cooled allowing the sugar to crystalize. When no more sugar can be crystalized out of solution, the resulting liquid (molasses), containing about 50% sucrose is eliminated. For every 100 Kg of plant, some 3.5 to 4.5 Kg of molasses may be obtained [16]. The fact that molasses may be extracted from at least two sources of plant adapted to tropical and temperate climates, permits the obtainment of molasses in a wide range of geographical locations.

Besides its high sugar content, molasses contains minerals, organic compounds and vitamins which are valuable nutrients in fermentation processes (Table 1) [17]. In fact, about 9% of the dry matter in yeast grown on molasses has been estimated to originate from substances other than sucrose [17]. Nevertheless, biomass production from molasses requires supplementation with a suitable nitrogen source, as well as phosphorus. The traditional nitrogen sources used are ammonia or ammonium salts, and phosphorus can be added in the form of salts.

Table 1. Average values for some constituents in Beet and Cane molasses at 75% (w/w) dry matter [17].

Constituent	Beet molasses	Cane molasses
Total sugars (%)	48-52	48-56
Non sugar organic matter (%)	12-17	9-12
Protein (N x 6.25) (%)	6-10	2-4
Potassium (%)	2-7	1.5-5.0
Calcium (%)	0.1-0.5	0.4-0.8
Magnesium (%)	0.09	0.06
Phosphorus (%)	0.02-0.07	0.6-2.0
Biotin (mg/Kg)	0.02-0.15	1-3
Pantothenic acid (mg/Kg)	50-110	15-55
Inositol (mg/Kg)	5000-8000	2500-6000
Thiamine (mg/Kg)	1.3	1.8

Baker´s yeast was the first microorganism to be produced in aerobic stirred fermentation on molasses as it is still produced today [18, 19]. However, this yeast has seldom been destined as food, but rather for baking purposes. A cheaper, more amenable SCP substrate of carbohydrate origin is starch. This very abundant carbohydrate may be obtained from bulb plants of tropical and temperate regions, or from rice, maize and cereals. In tropical countries, cassava has been proposed as a good source of starch for SCP processes [20]. The Symba process developed in Sweden [16, 21] utilized starchy wastes combining two yeasts in sequential mixed culture: The amylase producing *Endomycopsis fibuligira*, and the fast growing *Candida utilis*. The process consists of three phases: The incoming starch waste from potato tubers is fed through heat exchangers and sterilised. The medium is then fed to a first bioreactor where the starch hydrolysing yeast grows and hydrolyses starch. The hydrolysed solution is then fed to a second reactor where culture conditions favour the proliferation of *C. utilis*. The Quorn™ myco-protein production process is currently supported on glucose, nearly all of which is obtained from maize, but it has been reported earlier to use wheat-starch, a by-product of the production of wheat gluten (the protein fraction) and wheat flour [22, 23]. This means that the process may be applied with various sources of starch as the carbon source [24, 25].

Whey is a residual liquid obtained after the removal of protein and fat from milk. Whey traditionally originates from the curding process in cheese production, but can now be obtained after ultrafiltration procedures for the production of spreading cheeses, where the protein fraction corresponding to lactalbumins and lactoglobulins is incorporated to the casein fraction, and all the proteins are in native form. Approximately 9 Kg of whey may be obtained for every Kg of cheese [26], and the principal component is lactose (4-6%) (w/v), although other nutrients which are also found in significant amounts (Table 2) [27].

Table 2. Whey composition. Data from reference [27].

Component	% (w/v)
Water	92.6-93.5
Lactose	4.5-5.2
Proteins	0.3-1.0
Mineral salts	0.6-0.9
Lactic acid	0.2-0.3

Whey is produced in very large quantities, and can be found in practically every country. Since it is derived from milk, the processing of whey as food technology additive for direct human consumption appears an obvious outlet, but various features hinder this application. The principal sugar, lactose, is in a concentration which is too low to make transport or concentration viable in economic terms. In addition, it presents digestibility difficulties for adults, since the capacity to assimilate lactose in humans diminishes on maturity, especially in African and Asian populations. The disaccharide has a tendency to crystalize in solution at high concentrations, limiting the conditions under which it may be used as a food component. Finally, it has a relatively low sweetening power. Whey has been presented as an extremely suitable substrate for the production of SCP. In 1956 The French dairy company Fromageries Bel pioneered a project to produce yeast from whey, using lactose assimilating *Kluyveromyces marxianus* (formerly *K. fragilis*). In 1983, the company was processing 8000 tons of yeast in continuous culture [28]. Although similar operations were later installed in other western European countries, as well as the US, the Bel process was the largest and longest running operation using whey substrate in the world. The yeast was mostly destined for animal feeding.

Alkanes were considered as an attractive substrate for SCP production, particularly in the former Soviet Union, where the structural deficit in feed protein was compensated by the availability of oil. A large number of microorganisms are able to assimilate n-alkanes and 1-alkenes in liquid culture, and these include yeasts and filamentous fungi (Table 3) [29]. SCP production from hydrocarbons presents practical complications due to the low water solubility of the substrate, as well as the high degree of aeration required for its metabolization (the substrate is essentially oxygen free and therefore presents a high oxidation potential). In consequence the cost of aerating the culture is relatively elevated. In addition, the oxidation processes are highly exothermic and cooling costs are correspondingly higher [30]. This aspect is later dealt with in sections 4.3 and 5. The toxicity of the substrate has raised suspicions on the safety of continuous feeding with SCP containing trace amounts of alkanes. SCP production plants using alkanes are not currently in operation.

Methanol is a by-product of the petrochemical industry and has been used as a substrate for a number of SCP production systems. Methanol tolerance (up to 6 g l^{-1}) and assimilation is a specialised prerequisite which may be found in yeast species such as *Hansenula, Pichia, Candida and Torulopsis*. The advantages of methanol over other petrochemical by-products are many, but the principal one resides in the volatile nature of the substrate, allowing it to be lost in the drying process. A number of industrial examples of SCP production systems using methylotrophic yeasts were implemented in the 70's [31]. However, all have been discontinued.

Table 3. Genera of yeasts and filamentous fungi able to utilize aliphatic hydrocarbons for growth, and relevant for SCP production. Data from reference [29].

	n-alkanes (paraffins)	1-alkenes (olefins)
Yeasts	*Candida* *Hansenula* *Pichia* *Rhodotorula* *Saccharomyces* *Torulopsis* *Trichosporon*	*Candida* *Debaryomyces* *Hansenula* *Rhodotorula*
Filamentous fungi	*Aspergillus* *Cephalosporium* *Fusarium* *Spicaria* *Cunninghamella* *Monilia* *Mucor* *Paecilomyces* *Penicillium* *Rhizopus* *Trichoderma*	*Aspergillus* *Cephalosporium* *Fusarium* *Spicaria*

Another very important feature of methylotrophic yeasts is their enormous capacity to synthesize proteins which participate in methanol assimilation. This feature has been exploited for the heterologous expression of proteins at the pilot and industrial scales [32]. However, the production of heterologous proteins is a field with a separate orientation to SCP production, and remains besides the outlook of this review.

Cellulose from agriculture and forestry sources constitutes the most abundant renewable resource in the planet. The productivity of forests and woodlands, amounts to 40% of the world net productivity, while cultivated land amounts to a mere 6%. Most of this productivity is in the form of cellulose and lignin (approximately 30% of most woody plants is composed of lignin). However, when considering these materials as potential substrates for SCP production, it must be taken into account that both share a structural and protective role, and are therefore designed to withstand hydrolytic attack. SCP production from these substrates, as in any other case, requires their breakdown into assimilable forms, which can be rapidly taken up in solution by the growing organism. The substrate must therefore be subjected to pretreatments in a number of steps which include milling and chemical or enzymatic hydrolysis. The cost of these steps has prevented the generalised production of SCP from cellulose, although some examples have been tested, such as the Louisiana State University process [33]. Despite the difficulties, efforts to understand and develop ways to degrade cellulosic materials have been pursued as a constant trend which carries through to the present day [34, 35]. The manufacture of industrial cellulose for paper and tissue production includes a number of steps designed to remove the lignin and hemicelluloses from wood where they act as cell wall constituents and as cementing agents joining fibers to larger aggregates. Wood is cooked in a medium containing calcium sulfite with excess free sulfur dioxide. Lignin is thus converted to lignosulfonates and hemicellulose is hydrolysed to monosaccharides. These in turn, may be further broken down to furfurols.

The amount of free sugars in the spent liquor is variable with the type of procedure chosen, as various cellulose fibers may be obtained with different degrees of degradation, as well as the differences in the type of wood used. However, up to half of the matter contained in wood may be extracted out of the fibers, and converted to biodegradable material (Table 4) [36]. Spent sulfite liquor has been used as a substrate for fermentations since 1909 in Sweden, and later in many other parts of the world. The first organism to be used was *Saccharomyces cerevisiae,* although this organism is unable to metabolise pentoses which are found in considerable amounts in this waste product. Later, other organisms better suited for the assimilation of all the sugar monomers were chosen. Namely *Candida tropicalis* and *Candida utilis*. Nevertheless, microorganisms are susceptible to sulfite, which is removed

Table 4. The composition of spent sulfite liquors derived from soft and hard wood. Data from reference 36.

Constituent	Spent Sulfite Liquor Source	
	Spruce	Beech
Total solids (%)	12-14	4-16
Reducing substances (RS, %)	2.5-5.0	3.5-5.0
Therefrom monosaccharides (%)	75	70
RS fermented by yeast (%)	60-70	10-12
RS assimilated by yeast (%)	80-90	80-85
Acetic acid (%)	0.2-0.6	1.0-1.8
Furfurol (g/L)	0.2-0.6	1.0
Total SO_2 (g/L)	3.5-8.0	3.0-7.0
Free SO_2 (g/L)	0.5-1.5	0.5-1.0
Total nitrogen (mg/L)	80-180	260-400
P_2O_5 (mg/L)	45	100-150
Methanol (g/L)	0-1.1	

previous to the fermentation process [37]. Yeast produced from sulfite liquor has been used for feeding at war periods, but lost favour in peace time, being destined for fodder in most instances. However, experiences of baker's yeast produced from sulfite liquor exist in Finland [16]. It is in Finland that an innovating process consuming sulfite liquor was put into operation in 1975. The Pekilo process is a continuous fermentation consuming pulp mill effluent using the filamentous fungus *Paecilomyces variotii*. It transforms sulfite liquor containing 32 g l^{-1} assimilable sugars with a yield of 55% (weight in biomass over unit weight of carbon substrate, see section 4). The protein content of the fungus exceeds 55% (w/w), and has been officially approved as a food in Finland. In 1983, the projected biomass production of the process was estimated to be around 7000 tons per year [16]. The process is not currently in operation.

3. SCP CONSUMPTION AND USES

SCP is normally considered as a source of protein. However, like any other biological material, it also contains nucleic acids, carbohydrate cell wall material, lipids, minerals and vitamins. Nevertheless, these contributions are given little importance by nutritionists, who generally value SCP in terms of Kjeldhal nitrogen x 6.25 (standard factor relating amino nitrogen to protein content). However, about 10-15 % of the total nitrogen in fungi and yeasts is in the form of nucleic acids. These are not metabolized in the same way as proteins but follow a different route. Amino N represents approximately 80% of total microbial nitrogen, and is composed of all essential amino acids required for human growth and nutrition (Table 5). With respect to egg albumin, which is considered a well balanced source of essential amino acids for human nutrition, fungal SCP compares well, except that it is defficient in sulfur containing amino acids. However, they are relatively rich in lysine and threonine with respect to other traditional protein sources of agricultural origin, such as wheat (Table 6) [29].

Table 5. Daily requirements (g) of essential aminoacids for the human adult
Data retrieved from FAO (http://www.fao.org).

Essential aminoacids	FAO recommendation	Minimum
Phenylalanine	2.2	1.1
Methionine	2.2	1.1
Leucine	2.2	1.1
Valine	1.6	0.8
Lysine	1.6	0.8
Isoleucine	1.4	0.7
Threonine	1.0	0.5
Tryptophan	0.5	0.25
Total	12.7	6.35

Table 6. Essential aminoacid content of wheat, egg albumin and some fungal SCP sources (g per 16g N). Data from reference [29].

Amino acids	Wheat	Egg white	*S. cerevisiae*	*C. lipolytica*	*P. notatum*
Lysine	2.8	6.5	7.7	7.8	3.9
Threonine	2.9	5.1	4.8	5.4	---
Methionine	1.5	3.2	1.7	1.6	1.0
Cystine	2.5	2.4	---	0.9	---
Tryptophan	1.1	1.6	1.0	1.3	1.25
Isoleucine	3.3	6.7	4.6	5.3	3.2
Leucine	6.7	8.9	7.0	7.8	5.5
Valine	4.4	7.3	5.3	5.8	3.9
Phenylalanine	4.5	5.8	4.1	4.8	2.8

The protein value of SCP has been determined through three basic parameters generally used in feed evaluation: the total quantity of microbial nitrogen ingested (I), the nitrogen of faeces (F) and urine (U). From these parameters, Digestibility, Biological Value and Protein efficiency can be calculated. Digestibility (D) is the percentage of the total nitrogen consumed which is absorbed from the digestive tract.

$$D = 100 \times (I - F / I)$$

Biological Value (BV) is the percentage of the total nitrogen assimilated which is retained by the organism, taking into account the simultaneous loss of endogenous nitrogen through excretion in urine.

$$BV = 100 \times (I - [F + U]) / (I - F)$$

Protein Efficiency (PE) is the proportion of nitrogen retained when the protein under test is fed and compared with that retained when a reference protein, such as egg albumin, is fed.
Yeast and fungal SCP products register high digestibility values, and these can be substantially increased when supplemented with methionine (Table 7). In addition to these studies, direct feeding trials for long periods of time are performed, followed by regular checks on physiological and health parameters. Checks on gastro-intestinal disturbances, the appearance of ulcers or skin rashes are commonly performed. Many such studies have shown that fungal SCP is better tolerated than other SCP sources, with consumption levels near 150 g per day on a sustained basis.

Table 7. Nutritional parameters of yeast foods in rats. Data from reference [29].

Organism	Digestibility (%)	Biological Value (%)	Protein efficiency
S. cerevisiae	81	59	-
C. utilis	85-88	32-48	0.9
C. utilis + 0.5 % DL-methionine	90	90	2.3

In order to obtain best results in the uses of SCP as feed, it is normally necessary to undertake specific pretreatments which improve the digestion and acceptability of the product. Treatment is directed towards killing cells, and liberating the internal contents. Autolysis has been the most popular method in the commercial production of yeast extract. The procedure involves heating the concentrated cell suspension after harvesting, to 45-50 °C for 24h at pH 6.5. Under these conditions, the internal enzymes hydrolyse the cell wall in part, and also attack proteins, resulting in smaller better digestible peptides. However, the process must be carefully monitored, as many of the resulting peptides may confer undesired tastes and smells to the product, thus limiting its application. Another procedure which avoids undesired peptide formation is mechanical disruption. This may be achieved by milling or grinding with glass beads, or with ultrasonic vibration. Shearing forces thus disrupt cell integrity liberating the internal contents. Alternative methods of shearing include freezing the cell conglomerate and forcing it through narrow dies at pressures of up to 4000 Kg/cm^2. Such methods are currently inapplicable at high scale, due to cost and scaling problems.

At the beginning of this section, we mentioned the nucleic acid content of SCP. This feature is particularly relevant with regards to the treatment of the final product. Nucleic acids are a necessary component of all cells, but present relatively high levels in rapidly dividing cells. Thus, the nucleic acid content of yeast (around 10% of dry wheight) is approximately five times greater than in the average mammalian organ. When nucleic acids are ingested, they are first attacked in the stomach by pancreatic nuclease. The resulting nucleotides are then attacked by nucleotidases in the intestine, resulting in nucleosides and phosphate. These in turn are further degraded to purine and pyrimidine bases. The degradation or purine bases in man results in the production of uric acid. Accumulation of uric acid beyond the excretion capacity of the kidney results in the formation of crystalline deposits in the joints and soft tissues, leading to gout-like manifestations and calculi in the urinary tract. Pyrimidines are degraded to orotic acid, the accumulation of which results in liver damage. Given these risks, the administration of SCP for human or animal consumption appears to be primarily limited by the amount of nucleic acid which would result in significant changes in plasma levels of uric acid (2-7 mg/100 ml in males). The administration of 130 g of yeast daily for one week results in uric acid levels ranging between 4.8 and 8.3 mg/100 ml in human volunteers.

Alkaline hydrolysis destroys RNA but also diminishes the nutritive value of the protein component. A compromise method which is effective consists of an incubation at pH 9.5 followed by a heat shock which precipitates the protein. Sodium chloride extraction follows. In some yeast products, thermal shock at 60 °C is applied followed by pancreatic ribonuclease, reducing the nucleic acid content from 9% to 2%. Similar results have been achieved by a series of short heat bursts which activate intracellular ribonucleases in yeast [38]. Toxicological tests taking into account aspects other than RNA content, such as allergenicity and mutagenicity, show that SCP is an acceptable product in the case of yeasts [39]. In conclusion, SCP from fungal origin may be used as a source of protein but its nucleic acid content, as well as its deficiencies in sulfur containing amino acids render it as a food factor which must be formulated along with other compensating sources of protein. In most cases, yeast and fungal SCP has been included in animal feeds with excellent results, and it is normally accepted that a 10% contribution of this product in mixtures with other sources providing carbohydrates, lipids and vitamins, is acceptable. Yeast protein is most commonly included in poultry food formulations. However, the advent of aquaculture has recently seen the emergence of ever more sophisticated feeds for fish cultivation, and SCP of fungal origin has proved to be well digested by fish. In addition, the ingestion of yeast and fungal biomass appears to increase resistence to mycoses which often decimate fish farms, particularly at water temperatures above 10 °C [40, 41].

Since the technical characteristics of Quorn™ myco-protein as a product obtained from a filamentous fungus and exclusively designed for human consumption are distinct from other SCP products, it deserves specific treatment in this review. Since fungal biomass proliferates at slower rate than yeast, the starting nucleic acid content subject to removal is also lower (8-9%). The RNA content reduction of myco-protein is effected by a heat shock raising the temperature to 64 °C for 30 min. This treatment also results in important losses in dry weight [22], but RNA levels are reduced well below the levels which limit consumption to 100 gr per day (2% of dry weight). In addition to this very important point, myco-protein presents very high Protein Efficiency values, reaching 75% with respect to egg albumin. In experimental tests where myco-protein was supplemented with 0.2% methionine, this value rose to 100%. Thus, myco-protein could be used as a total replacement for the human diet, in comparison with a mere 10% replacement considered as safe for yeast protein [22, 23]. Quorn™ products are not supplemented with methionine, but egg protein as explained below. Another favourable feature which differentiates Quorn™ products from other SCP products is the advantage taken from the filamentous nature of the microorganism in product design. *Fusarium venenatum* A3/5 filaments are aligned in parallel by a specially designed mechanical process which renders the product a texture very similar to that of meat fibers once set in a light matrix of egg white protein and heated. The final product has a bland taste and light colour which render it susceptible to the addition of flavouring and colouring agents [24].

Toxicity testing of Quorn™ myco-protein has shown that the product can be consumed as the sole source of protein on a continued basis, without any adverse effects. Given the unconventional nature of this product, the tests undertaken for its approval were especially thorough, lasting ten years, with trials on eleven different types of animal. Human trials involved 2500 people with no adverse effects. The product is approved for consumption in the European Union and FDA approval in the United States is due in 2001 [42].

4. THE BIOTECHNOLOGY OF SCP PRODUCTION

4.1. The Biochemistry and Physiology of Biomass Production

Filamentous fungi and yeasts are heterotrophic organisms. The production of biomass through their culture therefore requires the supply of an organic source of carbon and energy, as well as sources of nitrogen, sulfur, phosphorus and other elements in much smaller amounts. The latter, however, may be generally supplied in inorganic forms which are readily assimilated [43]. Although these organisms display a high degree of metabolic heterogeneity, there are some basic aspects which can be applicable to the physiology of most SCP production processes involving filamentous fungi and yeasts, especially when cultured on carbohydrate substrates. The classic example of baker's yeast will be used below for explanatory purposes, but should not be considered as the all-embracing model. Some extreme variants will be also referred to in order to illustrate the variety of physiologies. In terms of energy metabolism, it takes 2 mol of ATP to produce 100 g of dry baker's yeast biomass in aerobic cultures grown on glucose [16]. When the conditions vary, this relationship can change considerably [44, 45]. The synthesis of ATP in heterotrophic organisms growing on glucose (the most common building block of all SCP substrates) arises from two connected metabolic pathways. Glycolysis, which only partially oxidises glucose to pyruvate (later reduced to ethanol and therefore termed oxido-reductive pathway) with a net ATP yield of 2 mols per mol of glucose consumed. The respiratory pathway, which consumes pyruvate, yields 38 mols of ATP and 6 mols of CO_2 through the total oxidation of glucose with 6 mols of O_2 (thus called oxidative pathway) [45]. Since the yield of ATP obtained through the total oxidation of glucose is considerably higher than partial oxidation, the oxidative metabolic pattern is much more favourable for biomass production [16, 45, 46].

Under the oxidative metabolic pattern, two substrates are required, however: the carbon source and oxygen. The latter is a gas which constitutes approximately 21 % of the total air volume, and must be supplied to the culture through sparging. The diffusion rate through the air/liquid interface is the rate limiting step for oxygen availability to cells in large scale liquid culture [47]. The dissolved oxygen concentration in a reactor must always remain above a critical point below which the rate of oxygen consumption becomes dependent on oxygen concentration. In most yeasts and filamentous fungi, that concentration ranges beween 17 and 20% of the air saturation value [48].

Yeasts normally display a high capacity to assimilate sugar substrates through specific transport mechanisms and oxido-reductive metabolism, thus promoting a fast (though low yielding) rate of growth. The resulting ethanol is generally toxic to competing microbes, but can be withstood by yeasts up to concentrations normally reaching 12% (v/v). Ethanol may subsequently be metabolised aerobically by yeast [45]. Thus, when presented with a high glucose containing medium, yeast will metabolise the sugar substrate through oxido-reductive metabolism at a rate considerably higher than that of the matching oxidative pathway, even when oxygen is available above the critical point. This combination of capacities, in addition to recognised osmotolerance, provide many yeasts a competitive edge under high sugar concentration in their natural habitats, such as flower nectars, fruits and seed hydrolysates.

Filamentous fungi share the fermenting capacities of yeast, and also show an important capability to assimilate sugars quickly and turn accumulating metabolic intermediates into toxic by-products which hamper the growth of competitors. These products are always undesirable in commercial SCP production processes [46]. They generally tend to display a higher tendency towards aerobic growth, however. The objective of SCP process designers is to enforce culture conditions which ensure the installment of oxidative metabolic patterns. Thus, maximum oxidation of the carbon source resulting in high ATP yields and minimal side product synthesis prevail. This objective is attained by ensuring the coupling between glycolytic and subsequent oxidative pathways through the skilled control of the availability of the carbon source and oxygen (Fig. 4) [49 and references within]. The methods to attain this control will be sketched in section 4.3.

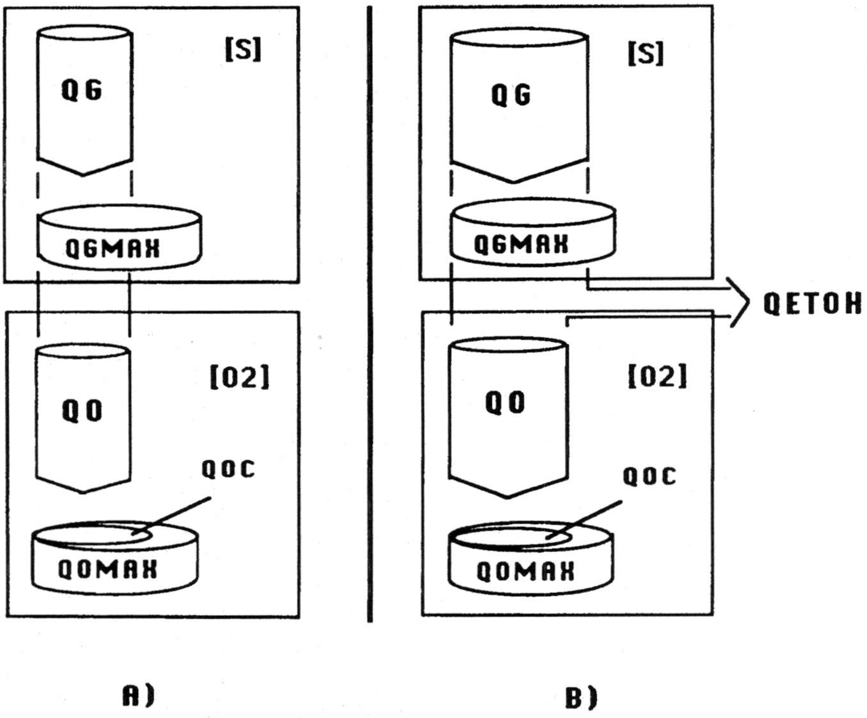

Fig. 4. Schematic description of a model of coupling between glycolytic and oxidative subunits of energy metabolism. A) Oxidative metabolism. B) Oxido-reductive metabolism. ([S], sugar concentration. [O_2], oxygen concentration. Q_i, carbon fluxes, gram atom C g^{-1} h^{-1}). [49 and references within]. Reprinted from *Journal of Biotechnology*, 22, Castrillo, J. I. and Ugalde, U. O., pages 145-152. Copyright (1992), with permission from Elsevier Science.

In a perfectly balanced aerobic culture, up to half of the carbon source supplied is assimilated to build cell material. Thus, the optimum yield is 50% (w/w) (biomass / initial substrate) [16]. The rest is used to fuel the maintenance and growth of cells. Under less efficient metabolic patterns, this element has a greater contribution in terms of carbon substrate invested. Many fungal and yeast strains can only metabolise the substrate oxidatively, for lack of an oxido-reductive metabolic pattern. In other words, they can only process pyruvate oxidatively and lack the reduction branch which turns it to ethanol, lactate or other acids. This is the case of *Trichosporon cutaneum*, which has been proposed as a candidate for SCP production in large scale, precisely because of this physiological feature [50]. For long-term fermentation procesess however, the absence of an additional energy yielding pathway may constitute a drawback, since transitory or local oxygen-limited conditions during the fermentation process will result in high decreases in viability and cell death. Other yeast species are capable of metabolising hexoses oxido-reductively, but not pentoses or specific disaccharides, which are only oxidized (Kluyver effect) [51, 52]. The Kluyver effect has been proposed as an interesting physiological feature which enables the attainment of high biomass concentrations under conditions of oxygen limitation, with the possibility of reducing aeration costs in biomass production [53]. The scope of this effect however, is restricted to certain microorganisms growing on specific sugars. The rather limited physiological information available about this effect in different organisms (mainly in the form of taxonomic tests [54]) has to be confirmed in specific experiments in fermenters with presence of different carbon sources under controlled conditions, in synthetic and complex media [53]. In any case, SCP production involves the use of heterogeneous substrates. This situation usually gives rise to preferred consumption of one, followed by the second. This pattern may be easily recognised kinetically, and it is termed diauxic effect, and it is highly undesirable in biomass production proceses. In some cases, some yeast and fungal strains are capable of assimilating two substrates at the same time. This is the case of lactose utilizing yeast *Kluyveromyces marxianus*, which hydrolyses the carbon source intracellularly. This example highlights the importance of the selection of strains adapted for the complete simultaneous oxidation of the carbon source in a biomass production process. The metabolic diversity encountered among nonconventional yeasts is well documented [55, 56]. Another alternative is to tailor the metabolic capability of strains by genetic engineering, in order to express an oxidative metabolic pattern even under low aeration conditions or to adapt them to new substrates [57 and references within]. This is a promising approach which is yielding results at the laboratory level, but poses difficulties when geared at the production of food and feed protein, given the current restrictions in the use of genetically modified organisms, especially in Europe.

Since alcohols such as ethanol and methanol are products of oxido-reductive metabolism, they may only be catabolized through the oxidative pathway, and growth rate in the case of these substrates has a constant and high yield, but is determined by the tolerance to the substrate by the organism [31, 56].

Inorganic nutrients are assimilated through membrane transport mechanisms and then incorporated to organic molecules [43]. A mention of nitrogen sources is relevant in this review, since the ultimate value of SCP is derived from the conversion of inorganic nitrogen into organic forms through the growth process. Inorganic nitrogen sources are relatively plentiful and available at economic prices in most countries, since they are also used as the nitrogen suplements in soil for agriculture. Inorganic nitrogen may be supplied in the form of ammonia or ammonium salts, urea or nitrates [58, 59]. Organic nitrogen sources that may be found are also assimilated efficiently in most cases [60]. Ammonia and ammonium salts are assimilable by all commonly used yeasts and fungi. Urea assimilation involves either urease degradation extracellularly, leading to ammonia production, or transport and assimilation through the urea amydolyase pathway [61]. In this case, the medium requires supplementation with biotin, which is a cofactor of the enzyme [61, 62]. Nitrate assimilation is reserved for those species possesing a transport system and nitrate reductase complex. In these cases, the adequate molybdenum supplementation is important, for this metal ion is part of the active complex [43, 62]. Aside from aspects related to assimilation of the nitrogen sources, they also influence the pH profile of the culture, an aspect which will later be dealt with in the section on process control. The assimilation of one ammonium ion generates one

Table 8. Optimal concentrations of micronutrients to be supplemented for growth of *K. marxianus* on whey [64]. Nitrogen source, urea (5 g l^{-1}). Phosphorus source, KH_2PO_4 (4 g l^{-1}).

Component	Concentration	g per gram-atom of carbon
$Cl_2Ca.3½H_2O$	0.2 g l^{-1}	0.14 g
$MgSO_4.7H_2O$	0.4 g l^{-1}	0.28 g
NaCl	0.2 g l^{-1}	0.14 g
$FeCl_3.6H_2O$	0.58 mg l^{-1}	0.41 mg
$ZnSO_4.7H_2O$	0.53 mg l^{-1}	0.38 mg
$CuSO_4.5H_2O$	0.12 mg l^{-1}	0.08 mg
$MnSO_4.H_2O$	0.06 mg l^{-1}	0.04 mg
$Na_2MoO_4.2H_2O$	0.01 mg l^{-1}	0.01 mg
Nicotinic acid	5.60 mg l^{-1}	4.0 mg
Calcium pantothenate	1.35 mg l^{-1}	1.0 mg
Biotin	0.04 mg l^{-1}	0.028 mg

of a nitrate ion consumes one proton. Urea assimilation is neutral with respect to the proton balance [63]. The pH of the process is normally in the range pH 4.5-5.5, since yeasts and filamentous fungi are acidophiles. The low pH range is especially indicated to avoid bacterial contaminations in long term cultures.

Besides the nitrogen source, other components are supplied in relatively low amounts. The general requirements of these nutrients vary, although they are usually within the same range. In Table 8 we present the requirements described for *K. marxianus* growing on whey [64]. A low quantitative requirement does not mean that these elements are not important. Limitations in a single one of these constituents often derives in suboptimal yields and productivity values, with consequent economic repercussions [43, 65].

The temperature range at which SCP production processes operate is 25-35 °C. Process temperature is a relevant parameter affecting growth rate, oxygen diffusion and the metabolic pattern of the culture. The metabolic activity associated with substrate oxidation and biomass synthesis yields an exothermic balance which is dependent on the type of substrate used. (Table 9). There is a general stoichiometric relationship between oxygen utilization and heat evolution which approximates 13 kJ g^{-1} cell mass or 17 kJ g^{-1} oxygen utilized [20, 66]. Thus, SCP production processes are generally refrigerated, and this aspect has direct repercussions on the economic viability, as will be discussed below.

4.2. Growth Kinetics

The production of biomass and related products by fermentation has instigated a massive effort to understand the basis of microbial growth in quantitative terms. The extent and sophistication achieved in these studies falls outside the scope of this chapter. However, a number of authoritative reviews are available to the reader [48, 67, 68]. A good introduction to this subject for beginners is that written by S. John Pirt [69]. The main parameters which characterize the performance of a process for SCP production are: Specific growth rate (μ), Growth Yield ($Y_{x/s}$) and Biomass Productivity (P_x).

Table 9. Influence of substrate and cell concentration on oxygen requirement and heat production [29].

Microorganism	Substrate	Cell concentration ($g\ l^{-1}$)	O_2 required ($g\ 100g\ cells^{-1}$)	Heat released ($kJ\ 100g\ cells^{-1}$)
Yeast	carbohydrates	0.5	67	591
Yeast	n-alkanes	1.0	197	3345

Under conditions favourable for the sustainment of microbial growth, a given amount of biomass (X, $g\ h^{-1}$) is expected to experiment an increase (dX) for a given period of time (dt) according to the following expression:

$$dX = \mu X\, dt \qquad [=]\ g\ l^{-1} \text{ or,}$$

$$dX/dt = \mu X \qquad [=]\ g\ l^{-1}\ h^{-1}$$

where μ is the specific growth rate (h^{-1}), characteristic for the organism and the conditions applied. The specific growth rate μ has the dimension of reciprocal time (h^{-1}). It is analogous to the compound interest rate on an investment, thus a specific growth rate of 0.1 h^{-1} is equivalent to a compound interest rate of 10% per hour [69].

For a batch culture, under conditions of exponential growth in which the specific growth rate can be considered constant, $\mu=\mu_{max}$, integration of this expression results in the following equation which describes the amount of biomass produced at any one time during exponential phase.

$$X = X_o\, e^{\mu max\ (t-to)} \qquad [=]\ g\ l^{-1}$$

where X_o is the amount of biomass at initial reference time (t_o), X is the amount of biomass at time t, and μ_{max} the maximum specific growth rate for an experiment under determined conditions. It has an upper limit for every organism. The maximum absolute values obtained for μ experimentally (μ_{max}) may reach values as high as 0.55 h^{-1} in yeast and filamentous fungi. However, in batch culture, growth at exponential phase ($\mu=\mu_{max}$) occurs during a limited period of time. In fact, μ has a variable value, being affected by different growth parameters. Thus, the substrate concentration has a notable influence on it, according to the Monod equation [9]:

$$\mu = \mu_{max}\, S / (S + K_s) \qquad [=]\ h^{-1}$$

where μ_{max} is the maximum specific growth rate of the organism and K_s (mol l^{-1}) the saturation constant, a measure of the organism's affinity for the limiting substrate. It can be considered that K_s is the substrate concentration at which the growth rate is half μ_{max}. Control of substrate concentration has been used to control the value of μ in culture.

Growth Yield ($Y_{x/s}$). The rate at which microbial growth takes place is independent from the efficiency of the bioconversion process from substrate to biomass. Growth Yield is defined by the quotient:

$$Y_{x/s} = \Delta x/\Delta s \qquad [=]\ g \text{ biomass } (g \text{ substrate})^{-1}$$

where Δx is the increase in biomass consequent on utilization of the amount of substrate Δs. The maximum growth yield obtained for a yeast and fungal culture growing on carbohydrates under aerobic conditions ranges between 0.4 and 0.5 g biomass per g of substrate. This is also considered the theoretical maximum in thermodynamic terms [16]. In physiological terms, these yields are achieved under oxidative metabolic patterns, with conditions of no limiting oxygen transfer. Maximum growth yields are normally achieved under conditions of carbon substrate limitation, with μ values around 0.1-0.25 h^{-1}.

Biomass productivity (P_x). The Biomass Productivity can be defined as the amount of biomass produced per volume and per unit of time for a given fermentation process.

$$P_x = \mu X \qquad [=]\ g\ l^{-1}\ h^{-1}$$

This parameter has important repercussions in process design, since it combines two parameters which can be estimated in economic terms: amount of product and process time.

4.3. Process Design and Control

The physiological features of yeast and fungal organisms recommend the control of the carbon source concentrations, as a limiting substrate, as well as an adequate supply of oxygen for the maintenance of balanced growth under an oxidative metabolic pattern.However, since microbial growth is a time dependent process, it exerts continuous modifications on all process parameters which influence physiology, but most dramatically, over substrate concentration. Therefore, an adequate technology which maintains appropriate growth conditions for a prolonged period of time must be implemented specifically for the purpose of obtaining high yield and productivity values.

Batch fermentations are clearly inadequate for the purpose of biomass production, since the conditions in the reaction medium change with time [70]. Fed-batch fermentations are better suited for the purpose of biomass production, since they involve the control of the carbon source supply through feeding rates. However, as the biomass concentration increases, the oxygen demand of the culture reaches a level which cannot be met in engineering or economic terms. Fed-batch culture is still in use for bakers yeast production using well established and proven models [19, 44]. However, they have not been favoured for the production of SCP at a large industrial scale. Prolonging a microbial culture by continuous addition of fresh medium with the simultaneous harvesting of product has been implemented successfully in industrial fermentations destined to biomass production. The most commonly used principle has been the chemostat: a perfectly mixed suspension of biomass into which medium is fed at a constant rate, and the culture is harvested at the same rate so that the culture volume remains constant.

The technical implications of chemostat culture are various and extremely relevant. They will be explained briefly below, although the mathematical demonstrations may be found in classic reviews [71-74]. Since the rates of incoming and outflowing medium to and from the reactor are identical, the volume remains constant. When these conditions are maintained, a steady-state is attained where the specfic growth rate and all parameters remain constant. If the conditions are carefully controlled, the process may be maintained at production settings which are optimal for long periods. Production periods as long as six weeks have been implemented in many fungal and yeast SCP production processes based on carbohydrate carbon sources [22, 26]. This practical time limit has not been surpassed due to the increasing risk of contamination with time, as well as the appearance of undesirable genetic variations in the culture after a critical number of generations, as will be discussed below. Another very important implication is that in chemostat culture the specific growth rate μ may be set by the dilution rate (D, h^{-1}) of the process (D = F/V = flow rate/reactor volume). Other technical measures may be implemented to control substrate and oxygen concentrations under steady-state state conditions for any set dilution rate. Thirdly, the continuous removal of biomass relieves the limitations on oxygen supply that apply for batch and fed-batch culture systems. Finally, a continuously operating installation requires smaller operating volumes than batch operated installations. The reaction volumes also determine the dimensions of all surrounding facilities, with important consequences on capital investment, as will be discussed in section 5.

Under chemostat culture, high yielding (Y = 0.45-0.5) SCP processes have been operated at μ values ranging between 0.2 and 0.3 h^{-1} in yeast cultures [64, 70] and between 0.15 and 0.2 h^{-1} in cultures using filamentous fungi (see below) [46].

The limitations imposed by oxygen transfer through the gas-liquid interface have provoked an important scientific and technological effort directed at optimizing this bottleneck process. Massive aeration is not recommended on economic grounds, but on the technical side also, as the higher the proportion of gas pumped through the solution, the greater is the partial volume occupied by the gas, and therefore the reactor volume. Evaporation and cooling of the medium is another undesired consequence of overaeration. In addition, a common problem of industrial fermentations is the profuse appearance of foam on the head space of the reactor, causing reactor pressurization, spillages and contamination hazard. Among the various designs which have been put to effect, the deep-jet fermenter and the air-lift fermenter have been the most successfully applied [75]. But it is the air-lift in its different variations that has enjoyed the greatest success as the configuration of choice for continuous SCP production. This configuration is presently used in the production of myco-protein which is the basis for Quorn™ products. The control of key process variables is a critical element of SCP production, from oxygen transfer, substrate and product concentration, to the appearance of minimal amounts of toxic compounds through undesired metabolic processes, which may compromise the quality of the final product. Many of these are carried out automatically as a result of a rapid development in all aspects of control, from sensor design to the computer algorithms which modulate the control responses [76-79]. However, since the cost effectiveness of SCP production is under constant scrutiny due to the price competition with plant proteins, simplified control devices are preferred. These often take advantage of existing conventional technology such as oxygen and pH control. Experimental evidence of extracellular medium acidification by cell cultures has pointed towards the existence of a relationship between proton production and cell growth, and the formal relationship between these two parameters has been recently demonstrated [63]. This has enabled for the on-line estimation of biomass and growth-linked product synthesis through pH control analysis using formal relationships which are applicable for a wide range of organisms [80].

Oxygen concentration is a key parameter which must be subject to monitoring and control. Specific electrodes placed at different reactor locations give an account of the levels of dissolved oxygen in the medium. In SCP fermentation processes, the oxygen concentration must never fall below the critical point (see section 4.1). The various control devices used to maintain oxygen levels within the oxidative physiological range, span from increased agitation and aeration, to more sophisticated control of carbon source dosage and oxygen gas injection [81-84].

The biomass from yeast fermentation processes is harvested normally by continuous centrifugation. This process results in biomass concentrations around 30% (w/v). Filamentous fungi are harvested by filtration [46]. The biomass is then treated for RNA reduction and dried in steam drums of spray driers. Drying is expensive, but results in stabilized product with shelf lives of years. This is a key feature in the animal feed and fodder business. For a more detailed revision of the different processes, flow charts and reaction configurations used for commercial SCP production the reader can be referred to the work of Ward [85].

4.4. Practical Example: The Myco-protein Production Process (Quorn™ Products)

The production of *Fusarium venenatum* strain A3/5 takes place in turbidostat culture using air-lift fermenters of 155 m^3 in volume and 50 m tall, weighing over 250 tons each. The Quorn™ fermenters are the largest operating air-lift fermentation facility in the world to date [42]. Each fermenter operates as a loop where culture medium is circulating (Fig. 5). As the liquid flows through the bottom of the loop, air is pumped in. Circulation is induced by a rising column of air bubbles providing good oxygen transfer conditions. This circulation is maintained due to mean density difference between riser and downcomer. Once the top of the fermenter is reached the pressure is reduced. This pressure reduction helps release CO_2

through the loop top. The fermenter broth passes down the downcomer tube which contains additional oxygen supply, with bubbling acting against the current flow, ensuring very high residence times for the bubbles. The bottom of the downcomer hosts the glucose, biotin and mineral salts. The nitrogen supply is done separately in the form of ammonia along with sterile air at the base of the riser. The supply of ammonia to the culture is regulated by a pH monitor set to give a culture of pH of approximately 6.0. The dilution rate of the process ranges between 0.17 and 0.2 h^{-1} and is operated so that glucose is always in excess and the fungus always grows at μ_{max} at a biomass concentration of 10 to 15 g l^{-1} [23]. The culture is kept at a temperature of approximately 30 °C by a heat exchanger set into the riser.

The dilution rate of the fermenter results in an output of 30 tons of liquid per hour. Harvesting by filtration and RNA reduction ensues. The harvested biomass (Fig. 6) containing 8-9% (w/w) RNA is heated to 68 °C for 25 min at a pH of 5-6. This results in the reduction of RNA to ca. 1% (w/w), at the expense of losing up to one third of the total mass, including dissolved salts, RNA, internal water, carbohydrates and protein. Since the startup of the process takes four days and it is relatively unproductive, it was judged most convenient for the production phase to be prolonged as long as possible to minimise the number of startup

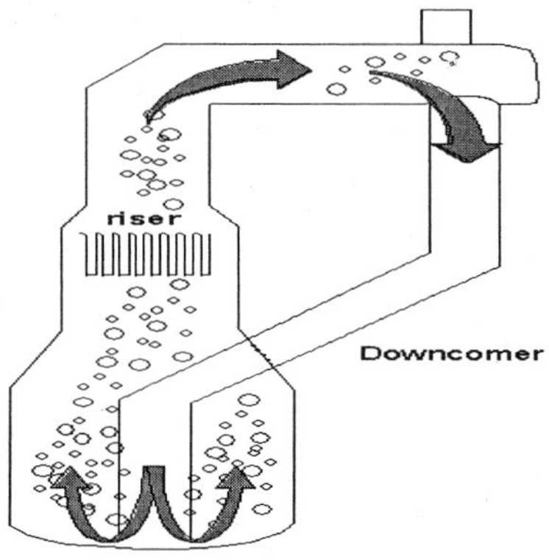

Fig. 5. Diagramatic representation of the Quorn™ air-lift fermenter used by Marlow Foods at Billingham for the production of myco-protein in continuous flow culture. Diagram kindly provided by Marlow Foods.

phases. Studies on the incidence of the production phase duration on product cost and comercial viability had indicated that periods above 200 hours operation can be necessary to result in consistent unit costs of production [86]. In principle, a continuous culture may run indefinitely, as long as contamination is kept under check. However, cultivation beyond 100 generations (about 400 h) of *F. venenatum* may result in the appearance of highly branched colonial mutants which can alter the texture of the final product [87, 88]. The details behind the appearance of mutations in continuous culture, with specific reference to *F. venenatum* strain A3/5 have been studied by external research independent to the production process [22, 23]. From these studies the authors get the conclusion that spontaneous mutant appearance may be managed by careful manipulation of the selective pressure imposed through culture conditions. Thus, a reduction in dilution rate delays the appearance of highly branched mutants. Another strategy is the systematic change in selective pressure by changing the conditions in the culture periodically. In this case programmed changes in the nutrient which becomes limiting constitute an effective strategy.

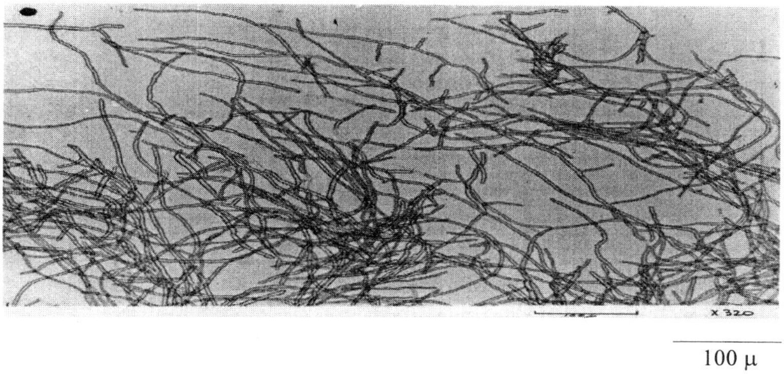

Fig. 6. The appearance of *Fusarium venenatum* A3/5 as collected from the outlet of the Quorn™ fermenter. Photograph kindly provided by Marlow Foods.

The isolation of sparsely branched mutants from the culture medium, which persist in the culture even after prolonged cultivation well after the appearance of highly branched colonial mutants, has also been used with positive results. Thus, the selection of these mutants in prolonged continuous culture, followed by their use as inoculum, resulted in the delay in the appearance of the morphological mutant by up to 53 generations (211 h) [89]. The above described strategies however have been sucefully tested in laboratory conditions. These strategies do not conform to the industrial production process and are not used commercially [42].

5. THE ECONOMICS OF SCP PRODUCTION AND MARKETING

The spectacular development of fermentation technology in the last few decades has not taken place without sophisticated evaluation of the economic viability of processes. This aspect, which stems from chemical engineering, is particularly important taking into account the investment required. In the case of SCP production, the need for accurate cost estimations is very relevant, since in the majority of cases the product is competing against protein sources of plant origin, and the profit margins are predictably low. In other cases, such as that of Quorn™ myco-protein process, fungal protein is competing against meat as a meat substitute, but an added economic effort is required to promote the product against such an established competitor, and the added cost must be compensated for in the production economy. Thus, in all cases, product cost estimation is a central element in the food and feed market industry.

5.1. Parameters Affecting Economic Viability

Several parameters are used as key elements in the estimation of economic viability. They will be briefly outlined, but the reader is referred to specialysed references where full details of this complex area, together with application of these concepts to different SCP processes are presented [90-92]. Total product cost includes all the costs incurred, and it may be divided by the annual production in order to estimate the cost per unit of product. It is normally broken down into Manufacturing cost + General expenses. The former includes all aspects directly related to production, such as Direct operating costs, Labour and supervision and Utilities. General Expenses include such concepts as administration, R and D, and marketing. Sometimes detailed information on these elements is not readily available. However, empirical formulae which relate the unknown values of some parameters to other obtainable ones are used in order to build an approximate picture which enbraces all these elements.

All of the funds required to build start and test the production facility before the product is put to the market are included. This parameter may be further subdivided into Fixed capital, or capital invested in hardware, land and equipment, and working capital, which includes inventory of raw materials, products and supplies, receivable and payable accounts. The easiest way to calculate this parameter is to calculate the return on the investment as a percentage. Despite the elaborate skills with which cost estimation may be carried out, it is still vulnerable to deviations which are sometimes strong, due to the appearance of unaccounted variables. One such variable of technical nature mentioned already can be the appearance of highly branched colony mutants in the myco-protein production process. Other very important variables are more conventional, but they can make or brake a business venture, in the same way as they influence private family economies. Labour costs, fuel prices or interest rates are but a few variables which can unpredictably change as a consequence of local or global developments.

5.2. Practical Aspects of Economic Viability

In this section we will mention the most relevant aspects which have determined the economic viability of SCP from fungal origin. The objective of all fermentation process designers is to make plants operating at the lowest possible *Total Product Cost*. In the case of SCP production, the raw material accounts for 62% of the total product cost, followed by fixed charges attributed to the production process, with 19% [75]. Thus, the main influencing factor has been the cost of the substrate, and this explains the quest for the processing of different substrates explained in section 2. A related question is the matter of scale. It is most logical to build equipment as large as possible because of the economy of scale. However, there is an empirical relationship between cost and size of an item of equipment. According to this relationship, as facility size increases from size 1 to size 2, cost increases thus:

$$Cost1/Cost2 = (Size1/Size2)^n$$

where n is an exponent or scale factor. The scale factor for SCP plants has been estimated to be between 0.7 and 0.8 [93, 94]. Thus, when a process capacity is increased 10 times, the cost of the installation increases approximately 5 times. This margin has an incidence in the total product cost too, since the capital cost influences the rate of payback and interests. Taking into consideration these aspects, continuous culture is by far the most advantageous in economic terms. Besides the physiological and design differences which favour continuous cultures, and which have been alluded to above, the concept of continuous operation leaves minimal time for zero production (the times at which a plant is not producing also involve a cost). It can be concluded therefore, that a continuous culture plant renders much greater profitability for comparable Capital Investment than does a discontinuous plant. This becomes increasingly true, as the degree of process automation increases, thus reducing the comparatively higher labour costs related to continuous operation due to night shift overpay.

The overwhelming majority of SCP production processes which were ever implemented industrially have adjusted to continuous culture designs [85, 91]. Besides the largest elements influencing Total Product Cost, there is a myriad of details which can cut the cost of production. Small though their contribution may seem, the additive effects of all the adequate measures may represent the difference between favourable and unfavourable economic balance. Fine adjustment of the aeration levels to values still higher to the critical oxygen concentration limit, below which the organism no longer supports an oxidative metabolic pattern, reduces aeration consumption, as well as foaming and evaporation of the medium. Fine adjustment of the medium required to sustain growth results in savings in some growth factors, such as vitamins, which are expensive. Cheaper sources of vitamin, where they are found in impure mixtures (yeast extract, soya bean extract, etc) often make all the difference in the cost of the supplement. Many processes sacrifice part of the biomass to make an extract which is fed back as a source of growth factors. Care in the choice of the nitrogen source may be relevant. Some sources of nitrogen (i.e. urea) contain higher amounts of nitrogen per unit weight than ammonium salts. The savings come through transport costs. Hydrated forms of salts are not recommended for the same reason. In addition, since urea consumption does not involve proton extrusion [63], savings can also be made in pH adjustment reagents.

5.3. Advantages and Constraints of SCP as a Market Product

Besides the aspects cited above, the variability in the market price of other products against which SCP is competing, clearly determines the market price and hence the profitability. Most of the SCP products that appeared in the food and feed market from 1960 till 1994 were destined for use as protein factors, in formulations. This means, that they were wholesale products which competed against other sources of protein which could be equally substituted in the formulations, without apparent changes in the final product. One direct competitor for SCP in western countries was brewers yeast. Identical in almost every feature, brewers yeast had a bitter taste which carried through to feed formulations, as the only differing characteristic from SCP yeast. However, this competitor was a by-product, the production of which was independent from the market strategy of the producers. This lead to policies of high turnover, low stockage of the by-product and consequently low market prices. Another competitor was excess bakers yeast. Thus, yeast and fungal SCP had to fall in the by-product market.

One common feature of SCP processes was that they often eliminated waste products, thus covering the function of expensive waste treatment installations. This led to the logic that the substrate may not only be provided at low prices or free, but received with payments by SCP producers. In the case of public wastes, an environmental quota could be payed to SCP producing companies. Such payments would add to those for the final product, with important repercussions on profitability. Though these reasonings made some sense, market reality proved to be very different. Since a profit was expected to materialise from SCP production, the wastes which the process consumed passed on to become substrates, and little interest was payed on their potential environmental hazard once consumed. The use of wastes, in addition brought additional problems in cases where the interest in waste treatment prevailed: The production volumes were not determined by the market demand of the product, but by the need to eliminate the waste. In those instances, waste treatment was the product and SCP was a true by-product, which accumulated until buyers could negotiate bargain sales which liberated stock capacity for the producer. Processes using whey and sulfite liquor were examples vulnerable to these constraints [16, 91]

While SCP protein coexisted with its competitors in the 70's and early 80's, mainly due to the limitations in the availability of brewers yeast, the emergence of cheap protein from soya bean and maize in the late 80's and 90's tilted the balance against SCP processes in most countries. Soya bean protein was available at prices which were 50% lower than SCP, with no restrictions on dosage due to high nucleic acid content. The incidence of the price of competitor protein clearly determined the outmarketing of SCP.

5.3.1. Specificity of the Myco-protein Production Process (Quorn™ Products)

If the myco-protein production process was specific in technical terms, as seen earlier, it is because the business outlook of the product had been initially set with qualitatively different objectives to other SCP processes. This product was directed at human consumption exclusively, and more importantly, at retail sales. The producing company packaged and marketed the product. The product thus embraces each and every aspect, from production to packaging, and including a most important food processing step. The success of the Quorn™ myco-protein process to date is thus principally determined by the product concept and its development. In this respect, it is a modern example of business venturing in biotechnology, where the business idea determines the design and development of process and product.

6. CONCLUSIONS

SCP has a proven record as a source of protein which may be obtained with large productivities in compact installations. The daunting prospects of world famine in the 60's lead to great expectations about the social and economic relevance of microbial proteins as a food source. These expectations were happily not fulfilled due to the opening of trade barriers and marked improvements in the breeding and production of plant protein sources. Nevertheless, there are success stories which paved the way for a new market of specialised foods of microbial origin. The compactness and high degree of control achieved in processes such as Quorn™ myco-protein production process provide a high degree of safety to the consumer, against the uncertainties which regularly surround meat products. In 1995, the first case of the variant Creutzfeldt-Jakob Disease (vCJD), human version of the Bovine Spongiform Encephalopathy (BSE), was diagnosed in the UK [95]. Also, in 1998, high levels of dioxin were found in large lots of chicken meat in Belgium. Other such cases with lower international resonance have constantly arisen, and it is likely that more such cases will continue to surface in the future, requiring a higher degree of control on food quality for animal and human consumption. Substitution of animal products by foods of plant origin is not totally exempt from these dangers, as the use of pesticides or the appearance of carcinogens such as nitrosamines have already been reported in the past. SCP products for animal and human nutrition are safer in this respect, since the components from which they are produced are easily controlled, and their genetic background is well known. They represent a line of non-conventional substitutes which will continue to have a market for these reasons. There is a market for products of microbial origin, aimed at animal and direct human consumption as substitutes for meat or even fish, given the increasing depletion of fish stocks (the EU has recently recommended a reduction of hake catchings in EU waters from 800,000 tons to 300,000 tons for 2001). Aside from this view, the problem of increasing world population and limited food production, may not demand SCP production at this time, but remains as a latent issue. The continued research on the production of microorganisms for animal and human consumption will undoubtedly find application in the future. This research should also incorporate the development of recombinant strains from nonconventional GRAS (Generally Regarded as a Safe) yeasts and fungi [56]. The potential of these techniques to improve the characteristics of foods is considerable and holds a positive prospect [96].

Acknowledgements: We thank Professor A. P. J. Trinci and Professor S. G. Oliver (School of Biological Sciences, University of Manchester, UK), and Dr. P. D. Collins, Technical Director Marlow Foods Ltd (Middlesbrough, North Yorkshire, UK) for their suggestions and comments on the first drafts of this review. All our experience in this field was obtained through the development of research and industrial projects which were funded by Gipuzkoa Foru Aldundia/Diputación Foral de Gipuzkoa, and the Basque Government, to whom we are deeply grateful. The first author is indebted to Miss Eider San Sebastian for the preparation of graphs and tables.

REFERENCES

1. Delbrück, M. (1910). Woschschr. Brau. 27: 375.
2. Hayduck, F. (1913). Z. Spiriusind. 36: 233.
3. Sak, S. (1919). Yeast. Danish Patent 28: 507.
4. Barnell, H. R. (1974). Biology and the Food Industry. Studies in Biology No. 45, Edward Arnold Ltd,

London.
5. Weitzel, W. and Winchel, M. (1932). The Yeast, its Nutritive and Therapeutic Value. Verlag Rothgiese und Diesing, Berlin.
6. Schülein, J. (1937). The Brewer's Yeast as a Medicine and Feeding Stuff. Verlag Steinkopf, Dresden.
7. Peterson, W. H., Snell, E. E. and Frazier, W. C. (1945). Fodder yeasts from wood sugar. Ind. Eng. Chem. 37: 30.
8. Harris, E. E. (1949). Food Yeast Production from Wood Processing By-Products. United States Department of Agriculture Forest Service, No. D 1754.
9. Monod, J. (1942). Recherches sur la croissance des cultures bacteriennes. Hermann, Paris.
10. Pirt, S. J. and Kurowski, W. M. (1970). Extension of the theory of the chemostat with feedback of organisms. Its experimental realization with a yeast culture. J. Gen. Microbiol. 63: 357.
11. Pirt, S. J. (1973). Estimation of substrate affinities (Ks values). of filamentous fungi from colony growth rates. J. Gen. Microbiol. 75: 245.
12. Bunker, H. J. (1966). New Foods. 2nd Int. Congr. Food Sci. and Technol., Warsaw. In Biology and the Food Industry. Studies in Biology No. 45 (Barnell, H. R. Ed., 1974). p. 48. Edward Arnold Ltd. London.
13. The State of Food and Agriculture (2000). This document may be obtained from FAO at http://www.fao.org.
14. The General Agreement on Tariffs and Trade (GATT 1994).. This document may be obtained from the World Trade Organisation at http://www.wto.org.
15. Cambridge Scientific Abstracts database: http://www.csa1.co.uk
16. Oura, E. (1983). Biomass from Carbohydrates. In Biotechnology (H.-J. Rehm and G. Reed Eds.). Vol. 3, p. 3. Verlag Chemie, Weinheim.
17. Olbrich, H. (1973). Biotin activity of molasses. Branntweinwirtschaft 113: 270.
18. White, J. (1954). Yeast Technology, Chapman & Hall, London.
19. Chen, S. L. and Chinger, M. (1985). Production of Baker's Yeast. In Comprehensive Biotechnology (H. W. Blanch, S. Drew and D. I. C. Wang Eds.). Vol. 3, p. 429, Pergamon, Oxford.
20. Forage, A. J. and Righelato, R. C. (1979). Biomass from Carbohydrates. In Microbial Biomass. Economic Microbiology. Vol. 4 (A. H. Rose Ed.). p. 289. Academic Press, London, New York.
21. Jarl, K. (1969). Symba yeast process. Food Technol. 23: 1009.
22. Trinci, A. P. J. (1992). Myco-protein: A twenty-year overnight success story. Mycol. Res. 96 (1). 1.
23. Trinci, A. P. J. (1994). Evolution of the Quorn myco-protein fungus Fusarium graminearum A3/5. Microbiology 140: 2181.
24. Anderson, C., Longton, J., Maddix, C., Scammell, G. W. and Solomons, G. L. (1975). The growth of microfungi on carbohydrates. In Single Cell Protein (Tannenbaum and D. I. C. Wang Eds.). p. 314, MIT Press. Cambridge MA.
25. Steinkraus, K. H. (1986). Microbial biomass protein grown on edible substrates: the indigenous fermented foods. In Microbial Biomass Protein (M. Moo-Young and K. F. Gregory Eds.). Section I, p. 33. Elsevier Applied Science, London.
26. Muller, L. L. (1969). Yeast products from whey. Process Biochem. 4: 21.
27. Moebus, O. and Teuber, M. (1983). General aspects of production of biomass by yeast fermentation from whey and permeate. In Production and Feeding of Single Cell Protein (M. P. Ferranti and A. Fiechter Eds.). Applied Science London.
28. Moulin, G., Malige, B. and Galzy, P. (1983). Balanced flora of an industrial fermenter. Production of yeast from whey. J. Dairy Sci. 66: 21.
29. Rivière, J. (1977). Microbial proteins. In Industrial Applications of Microbiology. Chap. 4, p. 105. Surrey University Press.
30. Einsele, A. (1983). Biomass from Higher Alkanes. In Biotechnology (H.-J. Rehm and G. Reed Eds.). Vol. 3, p. 43, Verlag Chemie, Weinheim.
31. Faust, U. and Präve, P. (1983). Biomass from Methane and Methanol. In Biotechnology (H.-J. Rehm and G. Reed Eds.). Vol. 3, p. 83. Verlag Chemie, Weinheim.
32. Gellissen, G. (2000). Heterologous protein production in methylotrophic yeasts. Appl. Microbiol. Biotechnol. 54: 741.
33. Callihan, C. D. and Clemmer, J. E. (1979). Biomass from Cellulosic Materials. In Microbial Biomass. Economic Microbiology. Vol. 4 (A. H. Rose Ed.)., p. 271. Academic Press, London.
34. Ferranti, M. P. and Fiechter, A. (1983). Production and Feeding of Single Cell Protein. Applied Science Publishers, London and New York.
35. First World Conference on Biomass for Energy and Industry, Sevilla 5-9 June 2000. http://www.etaflorence.it
36. Butshek, G. and Zellstoffablaugen, G. K. (1962). Zellstoffablaugen. In Die Hefen (F. Rieff, R. Kautzmann, H. Lüers and M. Lindemann Eds.). Vol 2, p. 121. Verlag Hans Carl, Nuremberg, Germany.
37. Webb, F. C. (1964). Biochemical Engineering. D. Van Nostrand Ltd. London.

38. Solomons, G. L. (1983). Single Cell Protein. CRC Critical Reviews in Biotechnology. CRC Press. Boca Raton, USA.
39. Schoch U., and Schlatter, Ch. (1983). Toxicological evaluation of SCP produced from whey. In Production and Feeding of Single Cell Protein (M. P. Ferranti and A. Fiechter Eds.).. Applied Science Publishers, London and New York.
40. Nell, J. A.(1985). Comparison of some single cell proteins in the diet of the Sidney rock oyster (Saccostrea commercialis). Prog. Fish. Cult. 47, 110.
41. FAO Aquaculture Development programme, Fish feeds and feeding in developing countries. ADCP/REP/83/18. This document may be obtained at the following site: http://www.fao.org.
42. Marlow Foods Personal Communication.
43. Soumalainen, H. and Oura, E. (1971). Yeast nutrition and solute uptake. In The Yeasts 1st edn. (A. H. Rose and J. S. Harrison Eds.). Vol. 2, p. 3. Academic Press, London.
44. Peppler, H. J. (1979). Production of yeast and yeast products. In Microbial Technology 2nd edn. (D. Perlman and H. J. Peppler Eds.)., Vol. 1, p. 157. Academic Press, New York.
45. Gancedo, C. and Serrano, R. (1989). Energy Yielding Metabolism. In The Yeasts (A. H. Rose and J. S. Harrison Eds.). Vol 3. p. 205. Academic Press, New York.
46. Solomons, G. L. (1985). Production of Biomass by Filamentous Fungi. In Comprehensive Biotechnology (H. W. Blanch, S. Drew and D. I. C. Wang Eds.). Vol. 3, p. 429, Pergamon, Oxford.
47. Sinclair, C. G. and Cantero, D. (1990). Fermentation modelling. In Fermentation. A practical approach (B. McNeil and L. M. Harvey, Eds.)., p. 65. IRL Press at Oxford University Press, Oxford.
48. Bailey, J. E. and Ollis, D. F. (1986). Biochemical Engineering Fundamentals. 2nd Edn. Mac Graw Hill, U.K.
49. Castrillo, J. I. and Ugalde, U. O. (1994). A general model of yeast energy metabolism in aerobic chemostat culture. Yeast 10: 185.
50. Alter, N. and Puhan, Z. (1983). Upgrading of mild UF-permeate by yeast fermentation. Semindustrial trials and economy. In Production and Feeding of Single Cell Protein (M. P. Ferranti and A. Fiechter Eds.). Applied Science Publishers, London and New York.
51. Sims, A. P. and Barnett, J. A. (1978). The requirement of oxygen for the utilization of maltose, cellobiose and D-galactose by certain anaerobically fermenting yeasts (Kluyver effect).. J. Gen. Microbiol. 106: 277.
52. Kaliterna, J., Weusthuis, R. A., Castrillo, J. I., van Dijken, J. P. and Pronk, J. T. (1995). Transient responses of Candida utilis to oxygen limitation: Regulation of the Kluyver effect for maltose. Yeast 11: 317.
53. Castrillo, J. I., Kaliterna, J., Weusthuis, R. A., van Dijken, J. P. and Pronk, J. T. (1996). High-cell-density cultivation of yeasts on disaccharides in oxygen-limited batch cultures. Biotechnol. Bioeng. 49: 621.
54. Barnett, J. A. (1981). The utilization of disaccharides and some other sugars by yeasts. Adv. Carbohydr. Chem. Biochem. 39: 347.
55. Phaff, H. J. (1985). Biology of yeasts other than Saccharomyces. In Biology of Industrial Microorganisms (A. L. Demain and N. A. Solomon Eds.). p. 537. USA Benjamin Cummings Publishing Company.
56. Wolf, K. (1996). Nonconventional Yeasts in Biotechnology. Springer-Verlag, Berlin
57. Rubio-Texeira, M., Castrillo, J. I., Adam, A. C., Ugalde, U. O. and Polaina, J. (1998). Highly efficient assimilation of lactose by a metabolically engineered strain of *Saccharomyces cerevisiae.* Yeast 14: 827.
58. Burn, V. J., Turner, P. R. and Brown, C. M. (1974). Aspects of inorganic nitrogen assimilation in yeasts. Ant. van Leeuwenhoek 40, 1425.
59. Cooper, T. G. (1982). Nitrogen metabolism in Saccharomyces cerevisiae. In The Molecular Biology of the Yeast Saccharomyces. Vol. 2. Metabolism and Gene expression. (J. N. Strathern, E. W. Jones and J. R. Broach Eds.). p. 39. Cold Spring Harbour Laboratory, Cold Spring Harbour, New York.
60. Large, P. J. (1986). Degradation of organic nitrogen compounds by yeasts. Yeast 2: 1.
61. Roon, R. J. and Levenberg, B. (1972). Urea amydolyase. I. Properties of the enzyme from Candida utilis. J. Biol. Chem. 247: 4107.
62. Phaff, H. J., Miller, M. W. and Mrak, E. M. (1978). Nutrition and growth. In, The Life of Yeasts. 2nd edn. p. 163. Harvard University Press.
63. Castrillo, J. I., De Miguel, I. and Ugalde, U. O (1995). Proton production and consumption pathways in yeast metabolism. A chemostat culture analysis. Yeast 1: 1353.
64. Castrillo, J. I. and Ugalde, U. O. (1993). Patterns of energy metabolism and growth kinetics of Kluyveromyces marxianus in whey chemostat culture. Appl. Microbiol. Biotecnol. 40: 386.
65. Hacking, A. J. (1987). Economic Aspects of Biotechnology. Cambridge Studies in Biotechnology 3 (A. J. Hacking Ed.). Cambridge University Press, Cambridge.
66. Cooney, C. L., Wang, D. I. C., and Mateles, R. I. (1969). Measurement of heat evolution and correlation with oxygen consumption during microbial growth. Biotechnol. Bioeng. 11, 269.
67. Herbert, D. (1976). Stoichiometric aspects of microbial growth. In Continuous culture 6: Applications and new fields (A. C. R. Dean, D. C. Ellwood, C. G. T. Evans and J. Melling Eds.). p. 1. Ellis Horwood Limited.

68. Sinclair, C. G. and Kristiansen, B. (1987): Fermentation kinetics and modelling (J. D. Bu' Lock Ed.). Open University Press. Milton. Keynes.
69. Pirt, S. J. (1975). Principles of Microbe and Cell Cultivation. Blackwell, Oxford.
70. Fiechter, A., Käppeli, O. and Meussdoerffer, F. (1987). Batch and continuous culture. In The Yeasts 2nd edn. (A.H. Rose and J. S. Harrison, Eds.). Vol. 2. p. 99. Academic Press, London.
71. Kubitschek, H. E. (1970). Introduction to research with continuous cultures. Prentice Hall. Biological Techniques Series. Englewood, Clifs, New Jersey.
72. Fiechter, A. (1975). Continuous cultivation of yeasts. In Methods in Cell Biology. Vol. XI. (Yeast cells). (D. M. Prescott Ed.). p. 97. Academic Press.
73. Calcott, P. H. (1981). Continuous culture of cells. Vol. I (P. H. Calcott Ed.). CRC Press, Boca Raton.
74. Cooney, C. L. (1986). Continuous culture: A tool for research, development and production. In Perspectives in Biotechnology and Applied Microbiology (D. I. Alanl and M. Moo-Young Eds.). p. 271. Elsevier Applied Science Publishers. London.
75. Stanbury, P. F., Whitaker, A. and Hall, S. J. (2000). Principles of Fermentation Technology. 2nd edn. Butterworth-Heinemann. Oxford.
76. Wang, H. Y. (1986). Bioinstrumentation and Computer Control of Fermentation Processes. In Manual of Industrial Microbiology and Biotechnology (A. L. Demain and N. A. Solomon Eds.). p. 308. American Society for Microbiology. Washington D.C.
77. Ritzka, A., Sosnitza, P., Ulber, R. and Sheper, T. (1997). Fermentation monitoring and process control. Curr. Opin. Biotechnol. 8 (2)., 160.
78. Royce, P. N. (1993). A discussion of recent developments in fermentation monitoring and control from a practical perspective. Crit. Rev. Biotechnol. 13, 117.
79. Sonnleitner, B. (1996). New concepts for quantitative bioprocess research and development. In Advances in Biochemical Engineering and Biotechnology (T. Sheper Ed.). Vol. 54, p. 155. Springer-Verlag, Berlin.
80. Vicente, A., Castrillo, J. I., Teixeira, J. A. and Ugalde, U. (1998). On-line estimation of biomass through pH control analysis in aerobic yeast fermentation systems. Biotechnol. Bioeng. 58: 445.
81. Zhang, X-C., Visala, A., Halme, A., Linko, P. (1994). Functional state modelling and fuzzy control of fed-batch aerobic baker's yeast process. J. Biotechnol. 37: 1.
82. de la Broise, D. and Durand. A. (1989). Osmotic, biomass and oxygen effects on the growth rate *of Fusarium oxysporum* using a dissolved oxygen-controlled turbidostat. Biotechnol. Bioeng. 33: 699.
83. Matsumura, M., Umemoto, K., Shinabe, K. and Kobayashi, J. (1982). Application of pure oxygen in a new gas entraining fermentor. J. Ferment. Technol. 60: 565.
84. von Schalien, R., Fagervik, K., Saxen, B., Ringbom, K., Rydstroem, M. (1995). Adaptative on-line model for aerobic Saccharomyces cerevisiae fermentation. Biotechnol. Bioeng. 48: 631.
85. Ward, O. P. (1992). Biomass production. In Fermentation Biotechnology. Chap. 6, p. 91. John Wiley & Sons, Chichester.
86. Trilli, A. (1977). Prediction of costs in continuous fermentation. J. Appl. Chem. Biotechnol. 27: 251.
87. Wiebe, M. G. and Trinci. A. P. J. (1991). Dilution rate as a determinant of mycelial morphology in continuous culture. Biotechnol. Bioeng. 38: 75.
88. Wiebe, M. G., Robinson, G. D., Cunliffe, B., Trinci, A. P. J. and Oliver, S. G. (1992). Nutrient-dependent selection of morphological mutants of *Fusarium graminearum* A3/5 isolated from long-term continuous flow cultures. Biotechnol. Bioeng. 40 (10)., 1181.
89. Wiebe, M. G., Robson, G. D., Oliver, S. G. and Trinci, A. P. J. (1995). Evolution of *Fusarium graminearum* A3/5 grown in a series of glucose-limited chemostat cultures at a high dilution rate. Mycol. Res. 29: 173.
90. Kalk, J. P. and Langlykke, A. F. (1986). Cost Estimation for Biotechnology Projects. In Manual of Industrial Microbiology and Biotechnology (A. L. Demain and N. A. Solomon Eds.). p. 363. American Society for Microbiology. Washington.
91. Rose, A. H. (1979). Microbial Biomass. Economic Microbiology. Vol. 4 (A.H. Rose Ed.). Academic Press. London.
92. Cooney, C. L., Rha, C. and Tannenbaum, S. R. (1980). Single-Cell Protein: Engineering, economics and utilization in foods. Adv. Food Res. 26, 1.
93. Humpfrey, A. E. (1975). Product outlook and technical feasability of SCP. In Single Cell Protein II (S. R. Tannenbaum and D. I. C. Wang Eds.)., p. 1. MIT Press. Massachusetts.
94. MacLennan, D. G. (1976). Single cell protein from starch. In Continuous Culture 6: Applications and new fields (A. C. R. Dean, D. C. Ellwood, C. G. T. Evans and J. Melling Eds.). p. 69. Ellis Horwood Publisher, Chichester.
95. The BSE Inquiry: The Report. This document may be obtained at the following site: http://www.bseinquiry.gov.uk
96. Uzogara, S. G. (2000). The impact of genetic modification of human foods in the 21st century: A review. Biotechnol. Adv. 18: 179.

Applied Mycology and Biotechnology
Volume 2. Agriculture and Food Production

Cereal Fermentation by Fungi

Cherl-Ho Lee[a] **and Sang Sun Lee**[b]
[a]Graduate School of Biotechnology, CAFST, Korea University, Seoul, 136-701 Korea;
[b]Department of Biology, Korea National University of Education, Chungbuk, 363-791, Korea (E mail: chlee@mail.korea.ac.kr).

The traditional solid-state fermentation starters made by fungi grown on cereals and beans in different Asia-Pacific region are classified, and their processing conditions, microbial composition and biochemical reactions are characterized in this chapter. The types of cereal fermented foods made from the starters include rice wine and beer, alcoholic paste and seasonings. The Korean soysauce, *kanjang*, and soybean paste, *doenjang*, made from meju are compared to those of Japanese *shoyu* and *miso*. Other soybean products like Indonesian *temper* and Chinese *sufu* are also mentioned. Chinese red rice, *anka*, and "enzyme foods" made from different cereals are also discussed.

1. INTRODUCTION

Fungi have long relationship with human food system. They have been used as food ingredient like mushrooms, *tempe* and monascus rice as well as fermentation starter like *nuruk/koji* and *meju*. On the other hand, fungi are considered as food spoilage and toxin producing organisms, which draw great attention in utilization of these organisms in food fermentation. The origin of cereal fermentation by using fungi appears to start from the primitive pottery age (B.C.8000-3000) of Northeast Asia [1]. In areas with high temperature and humidity, mold growth is a natural process in a container storing wet starch materials, for example plant seeds, millet, beans, nuts and tubers. Some molds like *Rhizopus* species produce enzymes, which can hydrolyze raw starch and convert into sugar. When sufficient amount of moisture is provided, the sugar is transformed into alcohol by the yeast existing in nature. In the Asia-pacific region, the consumption of rice as a staple food, and the high population density, which limits animal husbandry practices, have resulted in a typical food processing technology-cereal fermentation with molds. Molds and other microorganisms convert unpalatable carbohydrates of low digestibility and proteins into palatable sugars and amino acids with a high conversion efficiency. The soybean protein conversion ratio into amino acids in a traditional Korean soysauce fermentation for example, is over 75%, which is approximately 15 times higher than the feed protein conversion ratio in beef production, and 6 times higher than in pork production [2].

The type of cereal fermentation in the Asia-Pacific region varies with the raw material available and the climatic conditions. Inhabitants of the tropical Southeastern regions consume primarily rice, while those in subtropical and temperate zone of the Northeastern region including Northern China, Korea and Japan consume wheat, buckwheat, barley, corn, millet and soybeans in addition to rice. Countries of the Mecong delta basin, known for the origin of fish fermentation technology, derive up to 80% of their total caloric intake from rice [3]. On the other hand, countries of the far east, China, Korea and Japan, known as soybean sauce culture zone, consume less rice than countries of Southeastern Asia. According to statistical data, 47% of the total caloric intake in Korea and 43% in Japan were supplied by rice in 1965, which decreased to 23-35% in 1995 due to incorporation of other foods in the diet. China on the other hand, utilizes particularly wheat and shorgum. India has similar cereal consumption pattern with China, but much less meat consumption. These differences in cereal consumption patterns have resulted in variations in cereal fermentation practices in countries of the Asia-Pacific region. The type of fungi grown in fermented food and fermentation starters varies with the geographic climatic conditions. In the northeastern countries, especially Korea and Japan, *Aspergillus*, *Rhizopus* and *Mucor* are often found, whereas in the southeast Asian countries like Thailand, Indonesia and Philippines *Amylomyces*, and *Endomycopsis* are more often encountered [1].

2. FERMENTATION STARTERS

Unlike fruit and milk fermentations, cereal fermentation in most cases requires saccharification process, which is accomplished with some difficulty. One primitive method of cereal saccharification would be chewing raw cereals and spitting them into a vessel in order to allow saccharification to occur through the action of salivary amylase, followed by alcoholic fermentation via natural yeasts. Another method of cereal saccharification is through the malting process. Malting occurs naturally through wet damage of cereals during storage, and is used for beer making in Europe. However, in Asia the malting process is rarely used in traditional fermentation processes. Instead, fermentation starters prepared from the growth of molds on raw or cooked cereals is more commonly used.

2.1. History of Solid State Fermentation Starters in Northeast Asia

The fermentation starter is called *chu* in Chinese, *nuruk* in Korean, *koji* in Japanese, and *ragi* in Southeast Asian countries and *bakhar ranu* or *marchaar* (*murcha*) in India (1, 4). The first documentation of *chu* was found in Shu-Ching written in the Chou dynasty (1121-256 B.C.), in which it is stated that *chu* is essential for making alcoholic beverages. It is speculated that man must have discovered *chu* much earlier than it was documented in the literature [5]. In Chi-Min-Yao-Shu written by Jia-Si-Xie of Late-Wei kingdom in the 6th century, dozens of preparation methods for *chu*, the cereal fermentation starter, were described [6]. Methodology for *chu* preparation is very similar to that for *shi*, or Korean *meju*, preparation, which is a moldy starter prepared from soybeans, for soysauce fermentation.

The use of *chu* for rice-wine making was commonly practiced in the periods of Spring and Fall and Warrior Periods of China and the beginning of the Three Nation's Periods in Korea (B.C. 1-A.D. 2 centuries). This process was transferred from Korea to Japan in the 3rd century by Inborn, according to *Kojiki*, or Chin, whose memorial tablet is kept in a shrine, Matsuo Taisha, in Kyoto, Japan [7].

As mentioned above *chu* has been commonly used in the Asia-Pacific region as an enzyme source for the degradation of complex plant tissue to produce cereal-wines, soysauce, fish and meat sauce, sour bread, and fermented porridges and snacks. Table 1 summarizes the names of *chu* in different countries and their ingredients. According to Chi-Min-Yao-Su (530-550 A.D.), *chu* was made from barley, rice and wheat, and can be classified as described in Figure 1 (6). Ten different types of *chu* were described in Chi-Min-Yao-Su, all of which were used for the fermentation of alcoholic beverages. Cake type *ping-chu* is identical to *nuruk* in Korea, and the granular type *san-chu* is similar to Japanese *koji*.

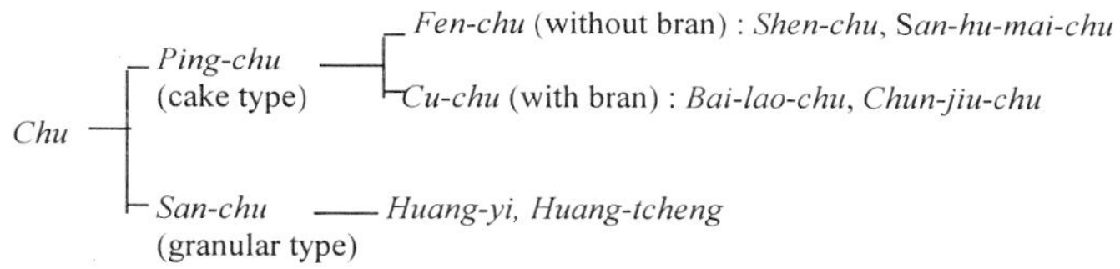

Fig 1. Classification of fermentation starters described in Chi MinYao Su.

Table 1. Fermentation starters and major ingredients used in different countries.

Country	Name	Ingredients commonly used	Shape	Microorganisms
China	*Chu*	Wheat, barley, millet, rice (whole grain, grits or flour)	Granular or cake	*Rhizopus* *Amylomyces*
Korea	*Nuruk*	Wheat, rice barley (whole grain, grits or flour)	Large cake	*Aspergillus* *Rhizopus*, Yeast
Korea	*Meju*	Soybean (whole seed)	Large ball	*Aspergillus* *Bacillus*
Japan	*Koji*	Wheat, rice . (whole grain, grits or flour)	Granular	*Aspergillus*
Indonesia	*Ragi*	Rice (flour)	Small cake	*Amylomyces* *Endomycopsis*
Malaysia	*Ragi*	Rice (flour)	Small cake	
Philippines	*Bubod*	Rice, glutinous rice (flour)	Small cake	*Mucor*, *Rhizopus* *Saccharomyces*
Thailand	*Loogpang*	Bran	powder	*Amylomyces* *Aspergillus*
India	*Marchaa*	Rice	Flat cake	*Hansenular*, *Mucor*, *Rhizopus*

According to Yokotsuka [5], *chu* may either be yellow (*huang*) possibly due to *Aspergillus oryzae* or white probably due to *Rhizopus* and *Mucor*. *Huang-chu* was widely used for alcoholic fermentations as well as for fermentation of soybean foods. Three types of *huang-chu* were described, *huang-yi, huang-tcheng* and *nu-chu*. *Huang-yi* is prepared from crushed wheat, which is first washed, soaked in water until sour and then drained and steamed. After cooling, the steamed wheat is piled and covered with leaves for 7 days, following covering or mycelia and spores. The *chu* thus prepared is then sun dried. In the preparation of *huang-cheng,* wheat flour and some water are shaped into a ball or cake, which is steamed, cooled, and then covered with leaves until it developed cultures of molds. *Nu-chu* is prepared from cooked rice, which is shaped into a cake and then cultured with molds (5). Wheat *chu* originated in the Northern part of China and the Korean peninsular, while rice *chu* originated in the South. This is reflected by the main ingredients of the fermentation starters prepared today in the countries of South-Pacific region as shown in Table 1. In Korea, the fermentation starters for cereal alcoholic fermentation and soybean fermentation have been made differently by using raw materials. Cereal alcoholic fermentation starters are made from raw wheat grits, while soybean fermentation starters are predominantly made form cooked soybean.

2.2. Cereal Alcoholic Fermentation Starters

Figure 2 compares the preparation processes of Korean *nuruk*, Japanese *koji,* Indonesian *ragi,* and Philippine *bubod* [1,16]. *Nuruk, ragi* and *bubod* are similar in that they are prepared by the natural fermentation of raw cereal powders which are molded into the shape of a cake or ball, while *koji* is prepared by controlled fermentation of cooked cereals in a granular form, which are commonly inoculated with the molds.

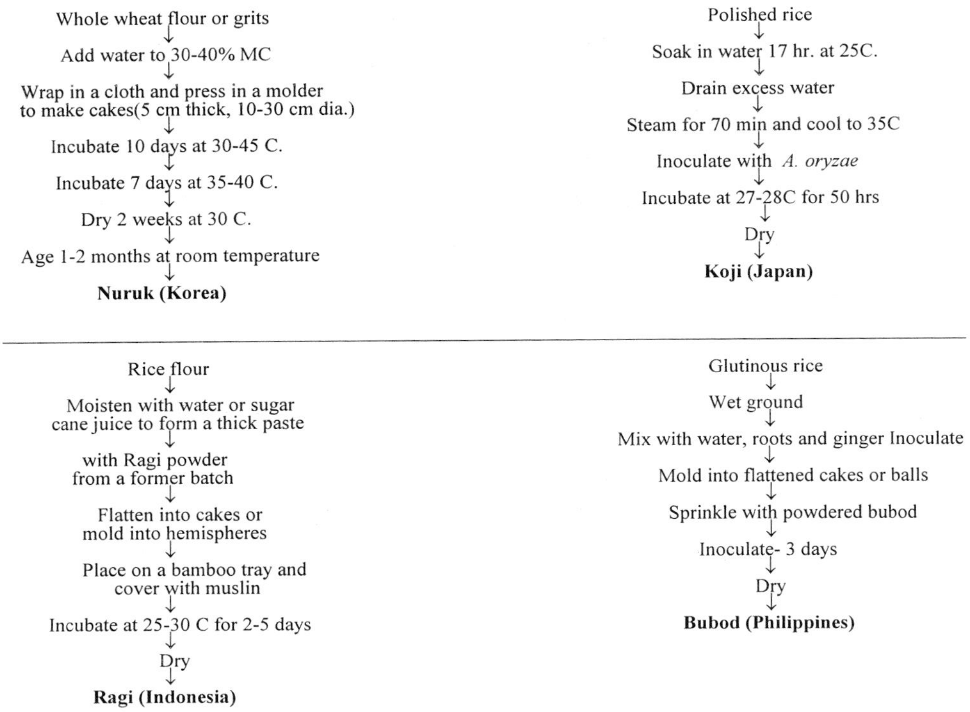
Whole wheat flour or grits
Add water to 30-40% MC
Wrap in a cloth and press in a molder to make cakes(5 cm thick, 10-30 cm dia.)
Incubate 10 days at 30-45 C.
Incubate 7 days at 35-40 C.
Dry 2 weeks at 30 C.
Age 1-2 months at room temperature
Nuruk (Korea)
Polished rice
Soak in water 17 hr. at 25C.
Drain excess water
Steam for 70 min and cool to 35C
Inoculate with A. oryzae
Incubate at 27-28C for 50 hrs
Dry
Koji (Japan)
Rice flour
Moisten with water or sugar cane juice to form a thick paste
with Ragi powder from a former batch
Flatten into cakes or mold into hemispheres
Place on a bamboo tray and cover with muslin
Incubate at 25-30 C for 2-5 days
Dry
Ragi (Indonesia)
Glutinous rice
Wet ground
Mix with water, roots and ginger Inoculate
Mold into flattened cakes or balls
Sprinkle with powdered bubod
Inoculate- 3 days
Dry
Bubod (Philippines)

Fig 2. Flow charts for the preparation of solid-fermented starters made in different countries of Asia-Pacific region.

Numerous types of microorganisms, molds, bacteria and yeasts, are found in the naturally fermented starters. In *nuruk, Aspergillus oryzae* (1×10^7 cfu/g), *Aspergillus niger* (1×10^7 cfu/g), *Rhizopus* (1×10^6 cfu/g), bacteria (1×10^7 cfu/g) and yeasts (1×10^5 cfu/g) were found [8]. In bubod, the number of molds (1×10^3-10^7/g), yeasts (1×10^5-10^7/g) and lactic acid bacteria (1×10^5-10^7/g) varied with the source and district of collection [9]. The important microorganism in *ragi* were *Amylomyces rouxii* and *Endomycopsis burtonii* [10], while those in *loog-pang Amylomyces, Aspergillus, Rhizopus, Mucor*, and Absidia were dominant molds. [11]. Of 41 yeast strains isolated from Indonesian *ragi* and *tape* by Saono and co-workers [12], 19 were amylolytic, none were proteolytic, and 14 were lipolytic. All mold isolates were amylolytic and lipolytic, and 89% of them also exhibited proteolytic activity.

The enzyme activities of *koji* are generally higher than those of *nuruk* [13, 8]. This may be due to the fact that the pure culture of *A. oryzae* on loose cereal granules allows maximum growth during the preparation of *koji.* On the other hand, in *nuruk* mixed culture of *Aspergillus, Rhizopus, Mucor* and other fungi imperfects grow on compact structure of *nuruk* cake resulting lower enzyme activity [7]. However, the yeasts and lactic acid bacteria grow simultaneously in nuruk play important role in the later stage of alcoholic fermentation, and the mixed culture contributes to the deeper flavor of Korean rice-wine.

The growth of mold and enzyme activity varies with the degree of gelatinization (DG) of the substrate grain, cultivation time and the type of mold [14]. Figure 3 shows the microscopic view of the hyphae and conidial spore of *A. oryzae* grown on rice of different DG. Figure 4 compares amylase and protease activities of two strains of *A. oryzae* grown on rice for 5 days. In general, maximum enzyme activity is attained after 4 days of cultivation, and the mold with long hyphae gives higher amylase activity, while short hyphae mold produce *koji* with strong protease activity.

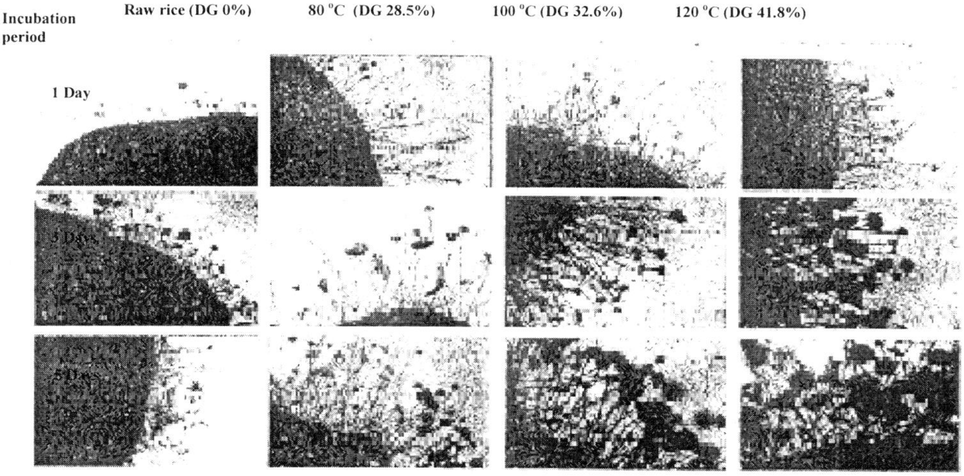

Fig 3. Microscopic views of *Aspergillus oryzae* grown on rice of various degrees of gelatinization at 30 °C (x10)

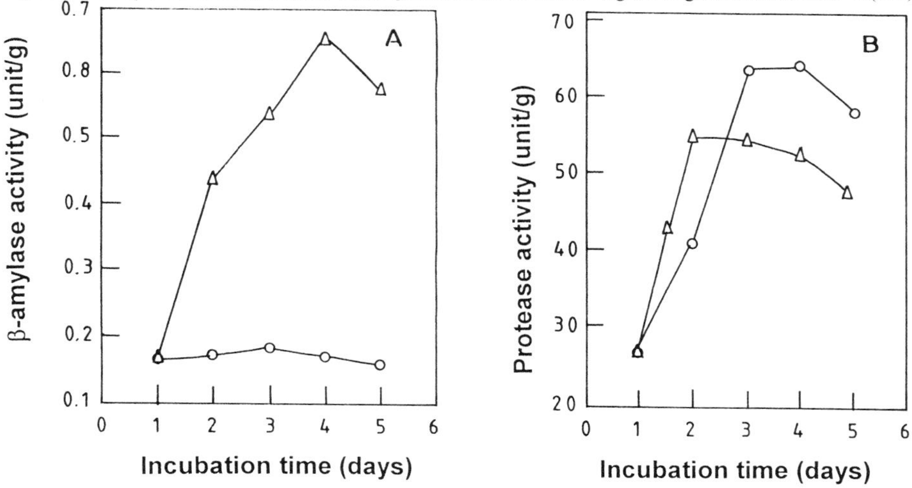

Fig 4. Changes in amylase (A) and protease (B) activities of two different strains of *Aspergillus oryzae* grown on cooked rice (DG 41.8%) for 5 days. - - : *A. oryzae* MJW001, - - : *A. oryzae* MHK001

2.3. Soybean Fermentation Starters

Figure 5 shows the processing procedures of soysauce fermentation starters made in Korea and Japan. Korean *meju* is prepared from cooked whole soybean, while Japanese *koji* is prepared from a mixture of roasted wheat and defatted soybean cake.

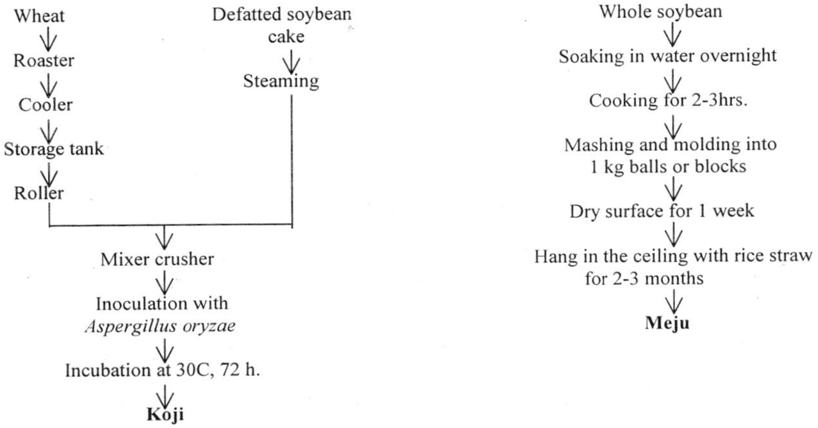

Fig 5. Flow chart for the preparation of solid-state fermented starters for soybean Fermentation (2, 13)

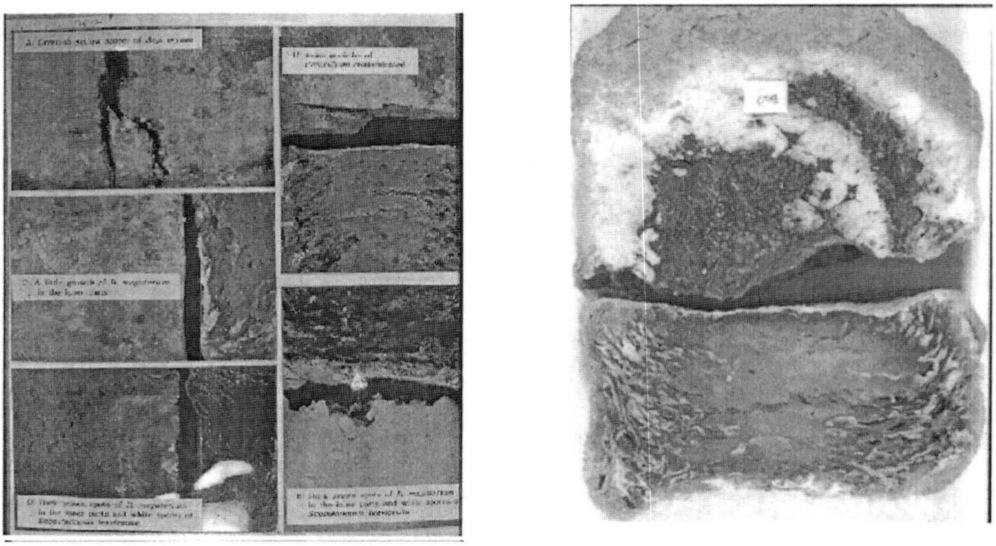

Fig 6. Typical pictures of the surface and cut profile of *meju;* Left - A: Greenish yellow spores of *A. oryzae*, B: Some growth of *Penicillium* contamination, C: A little growth of *B. megaterium* [26] in the inner parts, D and E: Dark brown spots of *B. megaterium* in the inner parts a and white spores of *Scopulariopsis brevicaulis*, Right - Top: Doksan-meju, grown with Mucor hiemalis and *Penicillium citrium*, Bottom: *Sunchang-meju*, grown with *Rhizopus* and *Absidia*, no brown inner spots of bacteria.

The latter is prepared by the inoculation of *A. oryzae* in a controlled fermentation, while the former is prepared by uncontrolled natural fermentation. The outer layer of the *meju* ball is over grown with molds, while on the inside, bacteria, mainly *Bacillus subtilis* grow [36]. Figure 6 shows some typical pictures of *meju* collected from different regions in Korea.

Table 2. Colors, smell and protease activities of the *meju* made with the single inoculum of the fungi and the features of *kanjang* made from the *meju* [24]

Microorganisms isolated[a]	*Meju* made with the single isolated		*Kanjang*[c]
	Colors or Smell	Protease activity (U)	
Zygomycetes			
M. isabellina	Black-rotten	389	Soysauce
M. hiemalis f. hiemalis	Black-savory	6780	Sweet soysauce
M. circinelloides f. *griseo-cyanus*	Black-savory	7640	Soysauce
M. circinelloides f. *circinelloides*	White-acidic	300	Salty
M. jansseni	Black-acidic	400	Salty
M. racemosus f. *racemosus*	Gray-brown	1929	Salty
R. stolonifer	Dark Brown-acidic	5504	Bitter
R. oryzae	Black-savory	14313	Salty
A. corymbifera	Black	1223	Soysauce
Hyphomycetes			
A. flavus	Gray-rotten	722	Soysauce
A. oryzae	Green-rotten	5990	sweet soysauce
A. oryzae	Gray-green	7830	Soysauce
A. terreus	Cinnamon	375	Soysauce
A. flavus v. colum	Dark-green	111	Soysauce
P. botryosoum	Green-rotten	111	Soysauce
P. gorlenkoanum	White-green	Nd	Rotten
P. griseo-purpureum	Green-rotten	202	Salty
P. citrinum	Green	12621	Soysauce
P. miczynskii	Green-trooten	11175	Salty
P. gaditanum	Green-rotten	4948	Salty
P. turolense	Green	8340	Sweet soysauce
P. funiculosum	Green	10452	Sweet soysauce
P. rubicundum	Green-rotten	2780	Salty
P. godlewskii	Green-rotten	112	Salty
P. jensenii	Green-rotten	722	Salty
P. roqueforti	Green-rotten	8173	Salty
P. volgaense	Green-rotten	4670	Salty
P. verrucosum v. corymbiferum	Gray-rotten	3447	Salty
S. brevicaulis	Cinnamon, Savory	8775	Sweet soysauce
Bacterium			
B. megaterium	Brown	12754	Tasteless

[a]Most microorganisms directly isolated from the traditional Korean home made *meju* cakes [15-23]. [b] The *meju* made from the autoclaved soybeans with the inoculum of a single isolate and cultured at 28 C for two weeks; [c]Soysauces made with the saline solution for two months and evaluated by the experts, having more than 20 years'experiences for Korean traditional home made *meju*s. The protease produced by *M. circinelloides* f. *griseo-cyanus* would be similar to that by *Bacillus megaterium* in its reaction of protein [26]

Numerous species of molds including *Mucor*, *Rhizopus*, *Penicillium*, and *Aspergillus*, have been identified from *meju* [15-23]. The contribution of these molds on the quality of soybean sauce was examined [24]. Table 2 shows the color/smell and protease activity of *meju* made by the controlled fermentation inoculated with the strains isolated from *meju* and the features of *kanjang* (Korean soysauce) made from them. There have been identified many strains of mold which play important role in the protein hydrolysis and flavor formation of Korean soybean sauce making [15-23].

3. FUNGAL FERMENTED FOODS

3.1. Cereal Alcoholic Products

Unlike in Europe and the Middle East, where most of the indigenous alcoholic beverages are produced from fruits, in the Asia-Pacific region alcoholic beverages are produced from cereals and serve as an important source of nutrients. European beer uses barley malt as the major raw material, while Asian beer utilizes rice with molded starters as the raw material. Beverages vary from crystal-clear products to turbid thick gruels and pastes. Clear products which are generally called *shaosingjiu* in China, *chongju* in Korea and *sake* in Japan, contain around 15% alcohol and are designated as rice-wine. Turbid beverages *takju* in Korea and *tapuy* in the Philippines, contain less than 8% alcohol along with suspended insoluble solids and live yeasts, and are referred to as rice-beer [1]. Examples of alcoholic beverages prepared from cereals in Asia-Pacific region are listed in Table 3.

Table 3. Examples of cereal alcoholic beverages in Asia-Pacific region.

Product name	Country	Major ingredients	Microorganisms	Appearance
Rice wine				
Shaosingjiu	China	Rice	*S. cerevisiae*	Clear liquid
Chongju	Korea	Rice	*S cerevisiae*	Clear liquid
Sake	Japan	Rice	*S. sake*	Clear liquid
Rice - beer				
Takju	Korea	Rice, wheat	*Lactic acid bacteria, S. cerevisiae*	Turbid drink
Tapuy	Philippine	Rice, glutinous rice	*Sacchromyces, Mucor Rhizopus Aspergillus, Leyconostoc, L. plantarum*	Sour, sweet liquid, Paste
Brem bali	Indonesia	Glutinous rice	*Mucor indicus Candida*	Dark brown liquid Sour, alcoholic
Alcoholic rice paste				
Khaomak	Thailand	Glutinous rice	*Rhizopus, Mucor Saccharomyces, Hnasenula*	Semisolid, Sweet alcoholic
Tapai pulut	Malaysia	Glutinous rice	*Chlamydomucor, Endomycopsis Hansenula*	Semisolid, Sweet alcoholic
Tape-ketan	Indonesia	Glutinous rice	*A.rouxii, E. burtonii, E. finulinger*	Sweet/sour alcoholic Paste
Lao-chao	China	Rice	*Rhizopus, A. rouxii*	
Alcoholic rice seasoning				
Mirin	Japan	Rice, alcohol	*A. ouyzae, A. usamii*	Clear liquid seasoning

3.1.1. Rice-wine

The process of cereal alcohol fermentation using molded starter was well established in the years of 1000 B.C, and 43 different types of cereal wines and beers were described with detailed processing procedures in Chi-Min-Yao-Su (530-550 A.D.). Millet appeared to be the main ingredient for alcohol fermentation. Among the 43 kinds of products described, 16 were red from millet, 11 from rice and 12 from Indian millet. The dried and powdered starter was mixed with water and steamed grains, and fermented for 2-3 weeks or up to 5-7 months depending on the brewing method. Multiple brews prepared by adding newly cooked grains to the fermenting mash for 2, 3, 4 and up to nine times were described [6]. Figure 7 compares processing procedures for the preparation of *quing-chu-jiu* as appeared in Chi-Min-Yao-Su with traditional *chong-ju* prepared in Korea.

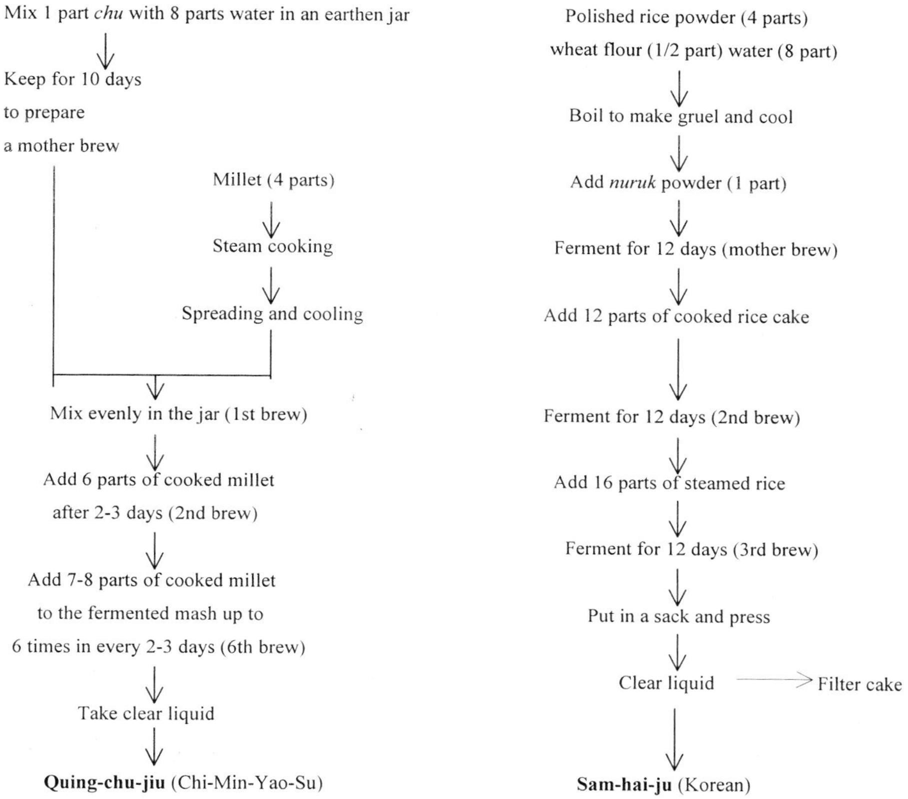

Fig 7. Flow chart for the production of rice-wines in China and Korea.

Newly cooked cereals are added at the end of each step of the fermentation process. The incubation period for each step of the brewing process varied from 2 days to one month depending on the fermentation temperature. Low temperatures (ca. 10^0C) are better for improving taste and keeping good quality of the wine. Traditionally these wines are prepared in late autumn or early spring, when ambient temperatures are below 10^0C in the far Eastern region. The volume of wine produced is approximately the same as that of raw grain used [27].

The traditional method of rice-wine brewing was industrialized by Japanese brewers in the late 19th century, who adopted pure starter culture manufacturing technology from Europe and transferred it to Korea and China. Figure 8 shows the Japanese process for preparing of rice-wine [28].

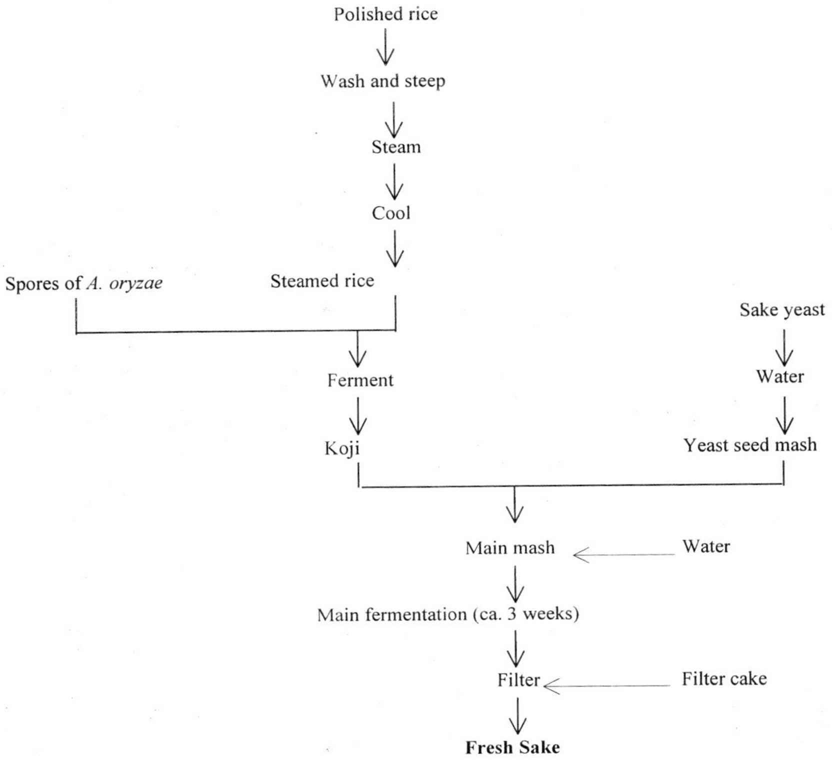

Fig. 8. Flow chart for Japanese Sake brewing process.

3.1.2. Rice-beer

Cereal beers are produced at a higher temperature (ca. 20^{0}C) of fermentation. The ground fermentation starter is mixed with cooked cereals incubated at approximately 20^{0}C for 2-3 days, following which it is filtered through a fine mesh sieve or cloth. These beers are usually prepared by either single or double brew. Figure 9 compares the preparation of Korean *takju* with that of Philippine *tapuy* [28]. Similar products are also prepared in other Southeast Asian countries, such as *brem bali* in Indonesia [12] and *jaanr* and *bhatte jaanr* in Sikkim, India [29, 30]. Figure 10 shows the biochemical changes occurring during the fermentation of *takju* [8].

Cereal-beers are abundant in micro-nutrients, such as vitamin B groups formed during the fermentation, and provide rapid energy supplements with the ethyl alcohol and partially hydrolyzed polysaccharides. Korean *takju* contains 7% alcohol, 1.9% protein and 1.2% carbohydrate (21), while Indonesian *brem bali* contains 16-23% reducing sugar and 6-14% ethanol [12].

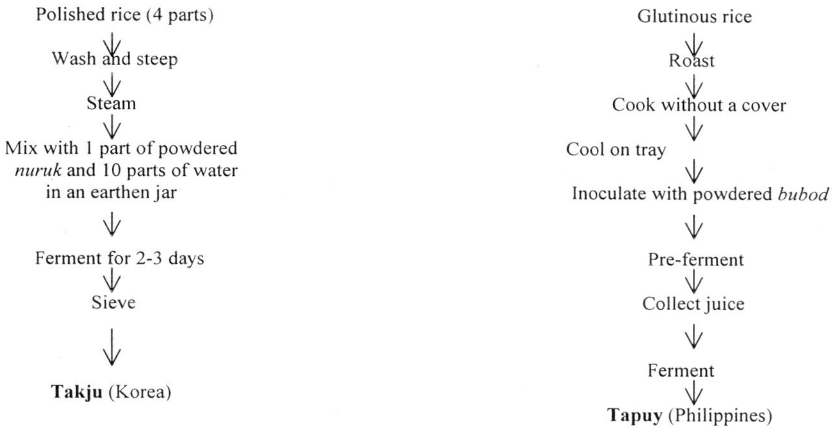

Fig. 9 Flow chart for the processing of *takju* and *tapuy*.

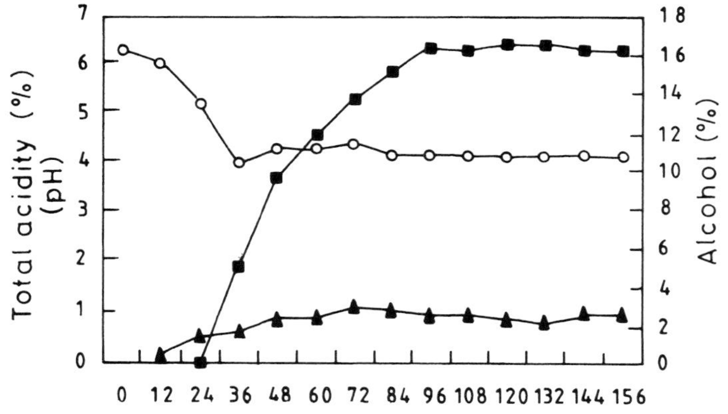

Fig. 10 Changes in alcohol (%), pH and total acid (%) during *takju* fermentation [8].

3.1.3. Alcoholic Rice Paste

In the Southeast Asian countries, alcoholic fermentation has been used for the preparation of paste-type food products. Figure 11 outlines the processing procedure for the manufacture of *tape ketan* in Indonesia [12] and *bhatte jaanr* in East Sikkim, India [29]. The major contributing microorganisms for tape ketan fermentation are *Amylomyces rouxii* and *Endomycopsis burtonii*. During the course of *tape kekan* fermentation, the pH decreases to 4.0 and ethanol content increases to 7% within 48-72 hr of fermentation, while starch and reducing sugars are decreased [28].

3.1.4. Alcoholic rice seasoning

Mirin is a traditional Japanese alcoholic seasoning prepared from rice and *koji*. The digestion of rice by native enzymes in *koji* is carried out in ethanol in order to prevent microbial contamination. *Mirin*, the final product, is a clear liquid containing approximately 40% sugar, which is produced as a result of starch hydrolysis by *koji* enzymes and approximately 80% of the sugar is glucose [32].

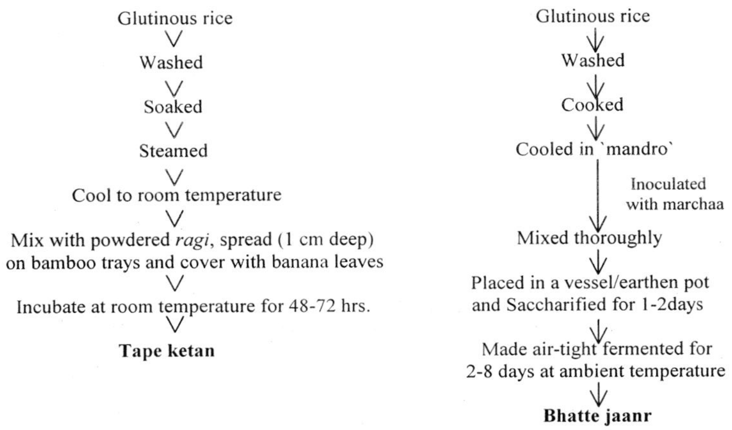

Fig. 11 Flow chart for the manufacture of Indonesian *tape ketan* and Indian *Bhatte jaanr* [12, 29].

3.2. Fermented Soybean Products

The first invention of soybean fermentation is not known. However, soybean sauce, *kanjang*, and soybean paste, *doenjang*, have been used in Korea for more than 2000 years and formed the characteristic flavor of Korean cuisine. The term "Shi", the Chinese letter indicating *meju*, first appears in *Jijiupian* written in Han period (206 B.C.-208 A.D.) of China. *Bowuzhi* of Jin (265-420 A.D.) of China describes that *Shi* was originated from foreign country, and the letter is a dialect. *Xintangshu* of Tang period (618-807 A.D.) in China names Shi as a special product of Balhai or Bohai (688 - 826 A.D.), a nation founded by the refugees from defeated Kokuryo (37 B.C.-668 A.D.) [33].

It is generally recognized that Koreans were the first to experiment with soybean fermentation, sparking the beginning of the soybean sauce culture of the Orient [1]. Their traditional fermentation technology was so advanced that they taught their techniques to neighboring countries. The history book of Wei, the Chinese dynasty of the third century, praises the fermentation skill of Korean people. The techniques were transferred to Japan around that period, 2-3 Century A.D. from Kokuryo, located in the northern part of Korean peninsula. The Japanese term of soybean paste, *miso* was originated in Korean language *maldo* or *maljang*, an old term indicating fermented soybean products. *Masakurainbunsyo*, written in Japan in 752 A.D., describes *miso* as a dialect from Koryo, the name of old Korea, and often called as Koryojang, Korean soy sauce.[34]

Japanese modified *meju* preparation method in their own way. They grow mold, *A. oryzae* on cooked rice or cooked wheat grits to make *koji*. It is mixed with cooked soybean for further fermentation, and then ripened in the brine. They separated the production of soybean paste from soy sauce making; *koji* is mixed with cooked soybean and salt, and then mashed into paste and ripened. This process makes easy the industrialization of soybean sauce production. However, Korean people prefer the strong flavor of traditional soybean sauce and soybean paste, same as European people distinguish Roquefort from processed Cheddar cheese. The differences start from the preparation of the fermentation starters, *meju* and *koji*, as discussed above. Both are made by the typical solid state fermentation, but the materials and fermentation conditions are different.

3.2.1. Korean *kanjang* and *doenjang*

Meju, the fermentation starter for Korean soysauce, *kanjang*, is made from soybean soaked in water overnight and cooked for 2-3 hours until the whole kernel is easily crushed by finger press. The cooked soybean is poured in a sack and crushed by stepping on the sack or pounded in a mortar. The crushed soybean is molded into a brick, a cone or a ball shape. It is placed on rice straw, dried in the sun and stacked under a cover during the night for a week. When the outside is dried, it is tided with rice straw and hung under the ceiling of warm living rooms. Normally *meju* preparation starts in October, the beginning of Korean autumn. and the slow drying process undergoes for 10 weeks in the rooms. During this periods, molds grow on the surface of *meju* ball, and bacteria, especially *Bacillus* species [26], grow in the inside of the ball. When fully dried, the *meju* ball hung in the room is taken into a sack made by rice straw and kept in the storage room. It is periodically taken out in the sun for a half day drying. In the next spring, *meju* balls are taken out from the straw sack and brushed out the surface mold and straw dirts, and put in a large earthen jar washed with boiling water and filled with brine. The mixing ratio of *meju*, water and salt is 2:4:1.2 in volume, which makes the salt concentration of the mash around 20%. Small amounts of charcoal. dried whole red-pepper and dried date are added to the jar, which is placed in the sunny place and covered with fine net in order to protect from flies. The ripening of *kanjang* mash in the brine is ended in 1-2 months. During this period, the amino acids and sugars formed by the enzymatic hydrolysis of soybean proteins and carbohydrates exude out to the brine and they undergoes Maillard reaction to produce dark brown *kanjang*. The halophilic yeasts, mainly *Saccharomyces rouxi*, grow in the mash and produce alcohols and other organic compounds to add the flavor of the sauce. At the end of the ripening the mash is filtered and the liquid part, *kanjang*, is boiled and kept in an earthen jar for longer periods of storage. The residue remained after filtration is mixed with small amounts of *meju* powder and rice porridge, packed in a small earthen jar with additional salt spread on the top, and kept in the sunny place for several weeks to make soybean paste, *doenjang*. Figure 12 shows the processing procedure of Korean *kanjang* and *doenjang* [1].

3.2.2. Japanese *shoyu* and *miso*

The fermentation starter for Japanese style soysauce, *koji*, is made by controlled fermentation of a inoculated mold, *A. oryzae*, on the mixture of roasted broken wheat and cooked soybean or defatted soybean meal. *koji* is mixed with brine to make soysauce mash for ripening. The mixing ratio of *koji*, water and salt is 1:7:1.8 and the higher soybean content in the mash results in stronger meaty flavor but less sweet taste. The mash is stirred once in a day, and ripening continues normally for 1-2 months, but the longer period of ripening, for example 6 months or one year, produces better quality of soysauce. After ripening the mash is filtered to separate clear liquid and the residue which are discarded. The clear liquid is sterilized by boiling and then bottled for sale. Soybean paste, *miso*, is made separately in Japanese style soybean fermentation. Rice *koji* is mixed with cooked soybean and salt, and ground to make paste. It is packed in a container and ripened for several weeks. The pure culture of mold excluding the growth of *Bacillus* and the use of cereals in the starter render the products milder aroma and sweeter taste.

3.2.3. Tempe

Tempe is found in all parts of Indonesia but is particularly important in Java and Bali. It is also produced in some Malaysian villages and in Singapore. *Tempe* is a white, mold-covered cake produced by fungal fermentation of dehulled, soaked in water and partially cooked

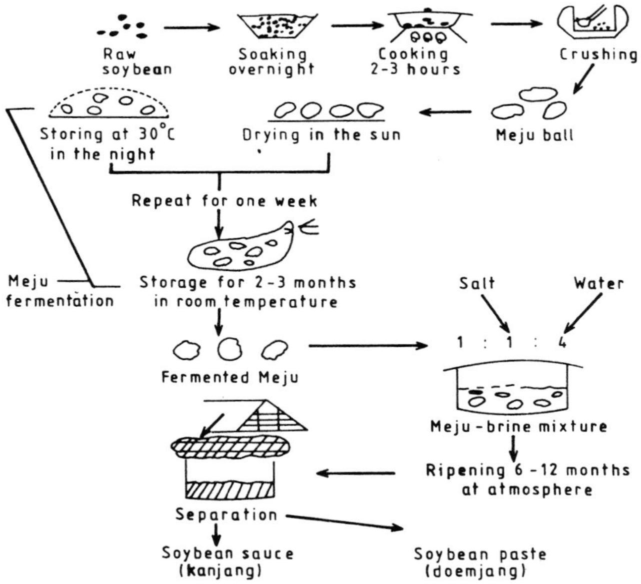

Fig. 12. The processing procedure of Korean *kanjang* and *doenjang* [1].

soybean cotyledons [28]. It is packed in wilted banana leaves and sold in the market. Essential steps in the preparation of tempe include cleaning the beans, soaking in water, dehulling and partial cooking of the dehulled beans. Dehulling is important for the growth of mold on the surface of cotyledon. Soybean is not necessarily being cooked fully, because subsequent mold growth is able to softening the texture. Under the natural conditions in the tropics, *tempe* production involves two distinct fermentations; bacterial acidification of the beans during soaking and fungal overgrowth of the cooked bean cotyledon by the mold mycelium. A previous batch of sporulated *tempe* or sun-dried pulverized *tempe* powder(1-3g) is sprinkled over the cooked and drained soybean cotyledon(1kg) and thoroughly mixed to distribute the mold spores over the surfaces of all the beans. *Rhyzopus oligosporus* is known to *tempe* mold, and the pure culture of strain NRRL2710 or CBS 338.62 can be used for the inoculum.

A handful of inoculated beans are placed on wilted banana leaves or other large leaves and packed. The leaf keeps the soybean cotyledon moist during the fermentation and allows for gaseous exchange. Incubation can be at temperatures from 25 to 37°C. The higher the incubation temperature, the more rapidly the *tempe* molds will grow. For example, 80 h of incubation is required at 25°C, 26 h at 28°C, and 22 hrs at 37°C. The *tempe* should be harvested as soon as the bean cotyledons have been completely overgrown and knitted into a compact cake. The cotyledons should be soft and pasty (not rubbery), and the pH should have risen to about 6.5.

Tempe should be consumed immediate after harvest. It can be stored for one or two days without refrigeration. If the *tempe* is not going to be consumed immediately, it should be deep-fried, in which form it remains stable for a considerable time, or it should be blanched by steaming and refrigerated. It can be stored after dehydration, either by sun-drying or hot-air drying and keeping in plastic bags. Subsequent keeping quality is excellent because *tempe* contains a strong antioxidant produced by the mold and is resistant to development of rancidity. *Tempe* is consumed as fresh or in deep-fried form.

3.2.4. Chinese *Sufu*

Chinese *sufu* (*tosufu, toufuru, fuyu* or *tauhuyi*) is a highly flavored, creamy bean paste made by overgrowing soybean curd with a mold belonging a genus *Actinomucor, Rhizopus*, or *Mucor* and fermenting the curd in a salt brine/rice wine mixture [18]. In the West, *sufu* has been referred to as Chinese cheese, which is made from soybean curd. *Sufu* is usually of red or white blocks 2-4 cm square by 1-2 cm thick, and the white *sufu* is untreated, while the red variety is colored with Chinese red rice, *hung chu*. The procedure for making *sufu* consists of five steps: preparation of soybean curd (tofu), preparation of molded *tofu* (*pehtze*), salting, fermenting in salt brine/rice wine, processing and packaging. Soybeans are cleaned, soaked in water and ground to make soybean milk slurry. The slurry is heated to boil and filtered through cloth, and the residue is discarded. The filtrate soymilk is added with coagulants (Calcium chloride/calcium sulfate mixture or sea salt brine) to make soybean curd. The amount of coagulant used to produce *tofu* for sufu manufacture is 20% higher than that used for regular *tofu*. Moreover, after the coagulate are mixed with soybean milk, the mixture needs to be agitated vigorously in order to break the coagulated protein into small pieces, after which it is set aside for 10 min to complete the process of coagulation. This process reduces the water content of the curd and makes the texture harder. If the water content is more than 60%, the inoculation of fungi is deferred until the water remaining on the curd surface is reduced by drying.

Pehtze is the soybean curd overgrown with the grayish hairlike mycelium of molds belonging to genera *Actinomucor*, *Rizopus*, or *Mucor*. These fungi are normal contaminants in rice straw. Traditionally inoculation was performed by placing the *tofu* on the rice straw, but this method does not always yield a high quality because of undesirable contaminating microorganisms. In spring or autumn when the ambient temperature is 10-20°C, white fungal mycelium is visible on the surface of the cubes after 3 to 7 days, at which point they are taken out and immediately salted in a large earthenware jars. Each layer of *pehtze* is sprinkled with a layer of salt, and after 3-4 days when the salt is absorbed, the *pehtze* is removed, washed with water, and put into another jar for processing.

For processing, a dressing mixture, which varies for each type of *sufu*, is placed on the jar. To make red *sufu*, *anka koji* and soy mash are added: to make fermented rice (*tsao*) *sufu*, fermented rice mash is added; to make Kwantung *sufu*, red pepper and anise are added in addition to salt and red *koji*. Alternate layers of *pehtze* and dressing mixture are packed into the jar until it is filled to 80% of its volume, then a brine with a concentration of approximately 20% NaCl is added. Finally, the mouth of the jar is wrapped with the sheath leaves of bamboo shoots and sealed with clay. After 3 to 6 months fermenting and aging, the *sufu* is ready to consumption. Table 4 shows the chemical composition of several types of *sufu* [28].

3.3. Other Fermented Products

3.3.1. Chinese red rice (*Anka*)

Anka, also known as *ang-kak*, *beni-koji* and red rice is used in China, Taiwan, the Philippines, Thailand and Indonesia to color foods which include fish, rice wine, red soybean cheese, pickled vegetables and salted meats [50]. It is a product of fermentation of rice with various strains of *Monascus purpureus*. A number of countries are gradually adopting this natural pigment to replace coal-tar dyes, as the latter have been implicated as carcinogens. The advantages of using *Anka* are that its raw materials are readily available, the yield is good, the color of pigment produced is consistent and stable, the pigment is water soluble, and there is no evidence of any toxicity or carcinogenicity [35]. *Anka* is produced in industrial scale in Taiwan. Non-glutinous rice (1450 kg) is washed and steamed for 60 min.

Water (1.8 hectaliters) is sprayed on the rice, which is again steamed for 30 min. The steamed rice is mixed with 32 liters of *chu chong tsaw*, a special variety of red rice inoculum in Taiwan, after cooling to 36oC, and heaped in a bamboo chamber. When the temperature of the rice rises to 42°C, it is spread on plates and shelved. *Anka* is produced by moistening the rice three times during incubation, followed by final drying; 700 kg of *Anka* are produced from 1450 kg of rice [41].

Table 4. Analytical data on several types of *sufu* [18] (unit : %).

	Tsao sufu	Red Sufu	Kwantung sufu	Rose sufu
Water content	69.03	61.25	74.46	59.99
Crude protein	12.87	14.89	12.42	16.72
Ether extract	12.89	14.31	6.39	13.74
Crude fiber	0.13	0.42	0.11	0.14
Ash	5.08	9.13	6.61	9.41
Total nitrogen	2.06	2.38	1.99	2.68
Protein nitrogen	1.30	1.56	1.27	1.81
Ammonia nitrogen	0.20	0.16	0.18	0.18
Amino acid nitrogen	0.27	0.27	0.24	0.31
Other nitrogen	0.34	0.33	0.30	0.38

An uncommon phenomena of the mold *Monascus purpureus* is the extrusion of granular fluid through the tips of the hyphae. When the culture is still young, the freshly exuded fluid is colorless, but gradually changes to reddish-yellow and purple-red. The red coloring matter diffuses throughout the substrate [35]. The dark red color consists of two pigments, the red monascorubrin ($C_{22}H_{24}O_5$) and the yellow monascoflavin ($C_{17}H_{22}O_4$). The strains of *Monascus purpureus* that are adopted to the production of red rice are only those which are capable of impregnating rice with a dark red color in the presence of water concentrations sufficiently low so that no distortion is produced in the hydrated grains [36].

3.3.2. Enzyme foods

Enzyme food is a new food item classified in health food category in Korea and Japan. It is made from cereals, vegetables and herbs grown with inoculated mold, especially *A. oryzae*. It has relatively high enzyme activities, mainly protease and amylase, produced by the mold together with the health beneficial effect of the raw materials. Recently, many people in East Asia use this item for helping digestion and for alternative and complimentary medicine (CAM) purposes. Figure 16 shows the processing procedure of enzyme food generally applied in the health food industry [26].

4. HYGENIC ASPECTS OF FUNGAL FERMENTED FOODS

Fungal fermented foods are always exposed to the risk of mycotoxin contamination. There are two possible ways of toxin contamination: (1) toxin from the contaminated cereals used for the raw materials [16,38-40], and (2) toxin produced by the mold contaminated during fermentation [41]. Table 5 summarizes mycotoxins produced by the fungi that are often encountered in food and animal feed [42].

4.1. Mycotoxin in fermentation raw materials

Cereals are often contaminated with mycotoxin by the toxin producing fungi parasitizing plants during cultivation [38,39] and those grown on the grain during storage and transportation [41,43]. Species of *Fusarium*, *Alternaria* and *Acremonium* are typical parasites dwelling on roots and stem of plants and produce zearalenones, alternarion and tremprgenic toxins [40,44]. Contamination of mold on grain during storage and transportation occurs often in the inter-continental trade of cereals. Wheat, barley, corn and soybeans are controlled for their mycotoxin contamination by the quarantine service of export and import harbors. However, the risk of contamination of these toxins in fermented food is not much as far as the harvest is normal and the quarantine control is properly conducted.

Raw materials
↓
Heating at 100oC for 50 min and cooling
↓
Inoculation with starter culture (*A. oryzae*)
↓
Cultivation at 28-35oC, relative humidity 80%, for 3-12 h
↓
Drying and blending with other sub-ingredients
↓
Granulation and packaging
↓
Enzyme food

Fig. 16. An example of flow chart for enzyme food production

Table 5. Mycotoxins produced by the fungi.

Toxins	**Fungi produced**	**Symptoms**
Aflatoxin	*Aspergillus flavus*, *A. paraciticus*	Heptotoxic, heptocarcinogenic, mutagenic, teratogenic
Zearalenones	*Fusarium roseum*, other *Fusarium* spp.	Estrogenic, infertility in cattle, swine, and poultry
Vomitoxin	*Fusarium*	Weight loss in most livestock, feed refusal, vomiting, immunosuppression
T-2 toxin	Pink *Fusarium* molds	Feed refusal, vomiting, Nervous system dysfunction, immunosuppression
Trichothecenes	*Fusarium* spp.	Gastrointestinal tract herorrhaging, alimentary toxic aleukia, feed rejection, vomiting, and immunotoxicity
Fumonisins	*F. moniliforme*	Esophageal carcinogenic in human (ELEM) in equines, geptocarcinogenic in rats
Orchatoxins	*Aspergillus* spp., *Penicillium* spp.	Nephrotoxic, teratogenic, immunotoxic, Suspected carcinogen
Sterigmatocystin	*A. versicolor*, *other Aspergillus* spp.	Heptotoxic, heptocarcinogenic
Citrinin	*Aspergillus* spp. *Penicillium* spp.	Nephrotoxic
Rubratoxin	*Penicillium* spp.	Heptotoxic

4.2. Mycotoxin Formed During Fermentation

The traditional fermentation starters used in Asia-Pacific region have high risk of mycotoxin contamination. The uncontrolled natural fermentation of rice, wheat and soybeans

with molds may cause mycotoxin problems. *Penicillium* and *Aspergillus* are the most important for the toxin production on cereals during storage and natural fermentation.
Penicillium glandicola produces ochratoxin A on cereals having over 19% moisture, *Penicillium expansum* produces patulin on apples and pear, and *Penicillium loquefoti* produces citrinin on blue cheese [45]. *Aspergillus flavus* and *Aspergillus parasiticus* produce aflatoxin B and G on corn, peanut and rice, and *Aspergillus ochraceus* produces ochratoxin. *Aspergillus flavus* can also produce mycotoxin cyclopiazonic acid [46]. *Aspergillus flavus* is the most important mycotoxin producing organism in the soysauce culture region including northern China, Korea and Japan, where the solid-state fermentation of cereals with *A. oryzae* has been used traditionally. There are two possible ways of contamination by this organism in fermentation process: (1) contamination from air during the uncontrolled solid-state fermentation of traditional starters, and (2) contamination in starter culture prepared for the controlled fermentation like *koji*. Since the morphology and physiological characteristics of *Aspergillus flavus* is very similar to those of *A. oryzae*, the risk of contaminating *A. flavus* [47].

Korean *nuruk*, *meju* and other traditional fermentation starters are exposed to the contamination of mycotoxin producing fungi. Several species of *Penicillium* and *Aspergillus* including *A. flavus* are detected in *meju* (Table 2). Some speculate that the high incidence of stomach and liver cancer in Korean and Japanese may have relation with the use of traditional fermented soybean products. However, it was reported that most of the anti-nutritional and toxic effects of traditional *meju* disappeared during the long period of aging process in the brine [48]. It is also believed that soybean is not a good substrate for *A. flavus* to grow and to produce mycotoxin. This idea is supported by some survey data on the mycotoxin content in *meju* and traditional fermented soybean products in Korea [30]. Mycotoxins, both aflatoxin and ochratoxin A, were not detected over the concentration range of 0.01 to 1.0 ng per gram of *mejus*, which were collected from different regions of Korea. It was speculated that the mycotoxins produced by *meju*-fermenting fungi might be degraded under the conditions of fermentation and storage of traditional home-made *meju* [30].

5. CONCLUSIONS

Fermentation of food by fungi is an old food processing technology dating back to several thousand years ago in the Northeast Asian region, where cereals are the main ingredients for alcoholic fermentation. The production of alcohol from cereals requires the saccharification of starch by molds to produce sugar, which is then converted to alcohol by the action of starch. This two-step fermentation process was probably established as early as five thousand years ago. The development of this alcoholic fermentation technology contributed greatly to the start of soybean fermentation technology in the later days, probably around B.C.1500, the Bronze Age of Korean Peninsula. From then on development, techniques of the fermentation were mainly applied for food preservation and flavor which were essential elements for the survival and social development of the people in this region. Fungi have been used for major source of enzyme in Northeast Asia, while malt has been used in Europe for the same purpose. *Aspergillus*, *Rhizopus* and *Mucor* are most frequently used species. Although the Japanese type of fermentation that uses starters made from controlled fermentation with inoculated single strain gives consistent product quality. In case of uncontrolled fungal fermentation, the risk of contamination with mycotoxin always exists. Because of the morphological and biochemical similarities of *A. flavus* with food grade molds including *A. oryzae*, there is possibility of contaminating with this mold on the various starter cultures used in Asia and Pacific regions.

6. REFERENCES

1. Lee, C.H. (2001). Fermentation Technology in Korea, Korea University Press, Seoul.
2. Lee, C.H. and Jul, M. (1982). The effect of Korean soysauce fermentation on the protein quality of soybean, in Traditional Food Fermentation as Industrial Resources in ASCA Counties. Eds. S. Saono, F.F. Winarno and D. Karjadi, LIPI, Indonesia. pp. 209-220.
3. Min, S.K. and Morashima, M. (1997). Samsung Economics Research Institute, Seou.l
4. Batra, L.R. and Millner, P.D. (1974). Some Asian fermented foods and beverages and associated fungi. Mycologia 66: 942-950.
5. Yokozuka, T. (1985). Fermented protein foods in the Orient, with emphasis on *shoyu* and *miso* in Japan, in Microbiology of Fermented Foods, vol. 1. (Ed) B.B. Wood. Elsevier Appl. Sci., London, 197-247.
6. Yoon, S.S. (1985). Cheminyosul, A Translation of Chiminyaosu in Korean, MinEumSa, Seoul.
7. Lee, C.H. (1995). An introduction to Korean Food Culture. Korean American Studies Bulletin, 6: 6-10.
8. Kim, C.J. (1968). Microbiological and enzymological studies on Takju Brewing. J. Korean Agric. Chem. Soc., 60: 66-69.
9. Tanimura, W, Sanches, P.C. and M. Kozaki. M (1978). The fermented food in the Philippines (Part 1) Tapuy (rice wine). J. Agric. Soc. (Japan) 22: 118-133.
10. Ko, S.D. (1972). Tape fermentation. Appl. Microbiol., 23: 976-978.
11. Pichiyangkura, S. and Kulprecha, S. (1977). Survey of mycelial molds in loogpang from various source in Thailand, Symposium on Indigenous Fermented Foods, Bangkok, Thailand
12. Saono, S., Hull, R.R. and Dhamchree, B. (1986). A Concise Handbook of Indigenous Fermented Food in ASCA Countries. LIPI, Jakarta, Indonesia.
13. Nunikawa, Y. and Ouchi, K. (1973). Kagaku to seibutsu. 11: 216.
14. Lee, B.I., Yoon, S.W. and Lee, C.H. (1999). Determination of the degree of gelatinization of cooked rice and its effect on the enzyme activity of the Korean gokja grown with *Aspergillus oryzae*. Food Sci. Biotech, 8: 162-167.
15. Barron, G.L. (1977). The Genera of Hyphomycetes from Soil. Robert E Krieger Publishing Co.
16. Choi, C, Choi, K.S., Lee, S.H., Son, J. H., Choi, H. J. and An, B.J. (1996). Characteristics and Ambient Pattern of protease from *Scopulariopsis brevicaulis* in Korean traditional soy source. Korean J. Microbiol. Biotechol. 25: 56-61
17. Gilman, J. C. (1968). A Manual of Soil Fungi. The Iowa State University Press. Ames, Iowa, USA, pp. 8-70.
18. Lee, S. S., Park, K. H., Choi, K. J., and Won, S.A. (1993). A Study on Hyphomycetous fungi found on Maejus, a raw material of Korean traditional soysauce. Korean J. Mycology 21: 242-272.
19. Lee, S. S., Park, K. H., Choi, K. J., and Won, S. A. (1993). Identification and isolation of Zygomycete fungi found on *Maeju*- a raw material of Korean traditional soysources. Korean J. Mycology 21: 172-187.
20. Pitt, J. I. (1979). The Genus *Penicillium* and its teleomorphic states *Eupenicillium* and *Talaromycis.* Academic Press, pp. 634.
21. Ramirez, C. and Martinez, A. T. (1982). Manual and Atlas of the Penicillia. Elsevier Biomedical Press, 847.
22. Raper, K. B. and Fennell, D. I. (1973). The genus *Aspergillus*. Robert E. Krieger Publishing Company, Huntington, New York, 686,
23. Zycha, H. and Siepmann, P. (1969). Mucorales. Verlag von J. Cramer. pp. 355,
24. Lee, S.S. (1995). *Meju* fermentation for a raw material of Korean traditional soy products, Korean J. Mycology 23: 161-175.
25. Lee, S.S., Yoon, Y. S., Yoo, J. Y. (1997). The fungal isolates of *Scopulariopsis* collected from Korean homemade Mejus. Korean J. Mycology 25: 35-45.
26. Sneath, P. A. (1986). Bergey's Manual of Systematic Bacteriology. Vol. 2, Williams & Wilkins. pp. 1134.
27. Lee, C.H. and Kim, G.M. (1993). Korean rice-wine, the types and processing methods in old Korean literature. Bioindustry 6: 8-25.
28. Steinkraus, K.H. (1983). Handbook of Indigenous Fermented Foods. Marcel Dekker, Inc., New York.
29. Tamang, J.P., Thapa, S., Tamang, N., Rai, B. (1996). Indigenous fermented food beverages of Darjeeling Hills and Sikkim: process and product characterization. J. Hill Res., 9: 401-411.
30. Korean Rural Nutrition Institute (1991). Food Composition Table (4th edition). RDA, Seoul, Korea.
31. Takayama, T., Oyashlki, H. and Sakai, T.(1997). Immobilized glucose isomerase use in Mirin production, J. Food Sci., 62: 237.
32. Lee, S.W. (1984). Korean Dietary Culture. Kyomunsa, Seoul, Korea
33. Lee, S.W. (1990). A study on the origin and interchange of Dujang (also known as soybean sauce) in ancient Asia, Korean J. Dietary Culture 5: 313.
34. Sooksan, R. and Gongsakdi, S. (1977). The red rice-anka. Symposium on Indigenous Fermented Foods.

Bangkok, Thailand.
35. Chen, J. W. (1997). Study of characteristic of water soluble Monascus pigments Food Additive of China 5: 15-19.
36. Lee, E.J. and Lee, C.H. (2001). Effective components of commercial Enzyme Food products and their HACCP scheme. Korean J. Food Sci. Technol., 33 (In press).
37. Bullerman, L.B. (1986). Mycotoxins and food safety. Food Technol., 40: 59-66.
38. Jone, B.D. (1972). G70 Methods of aflatoxin analysis. Tropical Products Institute. Ministry of Overseas Development, U.K.
39. Kasasian, R. and Dendy, D.A.V. (1979). G117 Grain processing losses bibliography: covering threshing, shelling, hulling, milling, grinding etc and excluding harvesting and storage. Tropical Products Institute. Ministry of Overseas Development, U.K.
40. McGee, D. S. (1991). Soybean disease. A reference source for seed technologists. APS press. The American Phytopathological Society, St. Paul, Minnesota , pp,151.
41. Pitt, J.I. and Hocking, A. D. (1985). Fungi and Food Spoilage. Academic Press, N.Y.
42. Betina, V. (1984). Mycotoxins: Production, Isolation, Separation and Purification. Elsevier, Amsterdam.
43. Bullerman, L. B. (1986). Mycotoxins and Food Safety. Food Technol, 40: 56-59.
44. Collins, G. J. and Rosin, J. D. (1981). Distribution of T-2 Toxin in wet milled corn products. J. Food Science 46: 877-1433.
45. Lee, C.H. (1976). The effect of Korean soysauce and soypaste making on soybean protein quality. Part III, Changes in the lysine availability. Korean J. Food Sci. Technol. 8: 63-69.
46. Han J., Kim, H. J., Lee, S. S. and Lee, I. S. (1999). Inhibitive effects of *Meju* extracts made with a single inoculum of the fungi isolated from the traditional *meju* on the human leukemia cell line. Korean Mycology 27: 312-317.
47. Jennifer, Y. L., Hagler, W. H. Jr. (1991). Aflatoxin and Cyclopiazonic acid production by *Aspergillus flavus* isolated from contaminated maize. J. Food Sci., 56:871-872.
48. Lee, S.S. (1998). Detection of mycotoxins on the Korean traditional home made *meju*. Korean J. Mycology, 26: 487-495.
49. Lee, S.S. , Pard, K.H., Choi, K.J. and Won, S.A. (1993). Identification and isolation of Zygomycetous fungi found on *Maeju*, a raw material of Korean traditional soysauce. Korean J. Mycology, 21: 172-187.
50. Blanc, P. J., Hajjaj, H. Loret, M.O., and Goma, G. (1998). Control of the production of citrinin by *Monascus*. Proceedings of Symposium on *Monascus* culture and applications. Toulouse, France.

Applied Mycology and Biotechnology
Volume 2. Agriculture and Food Production
Published by Elsevier Science B.V.

Mycotoxins Contaminating Cereal Grain Crops: Their Occurrence and Toxicity

Deepak Bhatnagar , Robert Brown, Kenneth Ehrlich and Thomas E. Cleveland
Food and Feed Safety Research Unit, U. S. Department of Agriculture, Agricultural Research Service, Southern Regional Research Center, New Orleans, Louisiana 70124, USA (E mail: dbhatnag@nola.srrc.usda.gov).

Mycotoxins are natural products of fungi that are capable of producing acute toxic or chronic carcinogenic, mutagenic, teratogenic or estrogenic response in higher vertebrates and other animals. Mycotoxins, however, are not grouped with antibiotics, compounds which also may be produced by fungi, but have microorganisms as the target species. Mycotoxins are not required for the growth of the producing fungus and, therefore, are considered secondary metabolites. Presumably these compounds play some role in the ecology of the fungus, but their function has not been clearly defined. Many mycotoxins have been characterized and they show significant diversity in their chemical structures and biological activity. Further, many fungi produce multiple families of mycotoxins, but the three genera *Aspergillus*, *Penicillium* and *Fusarium* comprise the largest number of mycotoxin producing species, particularly in cereal grains. But not all species within these genera produce toxins. The incidence and extent of mycotoxin contamination is dependent primarily on geographic and seasonal factors, as well as cultivation, harvesting, storage and transportation conditions. The toxic effect of mycotoxins is referred to as mycotoxicosis, and its severity depends on the toxicity of the mycotoxin, the extent of exposure, age health and nutritional status of the affected animal or human, and syngeristic effects of other compounds to which the affected individual has been exposed.

1. INTRODUCTION AND HISTORICAL PERSPECTIVE

Fungi produce a wide variety of compounds that are not required for their growth, hence these compounds are called secondary metabolites. Many of these compounds have adverse and toxic effects in animals and humans, and these compounds are called mycotoxins. Mycotoxins are extremely diverse in their chemical structure. They are low molecular weight, non-proteinaceous compounds derived primarily from amino acids, shikimic acid or acetyl and malonyl CoA. Fungal species that produce mycotoxins are also very diverse. A single mycotoxin may be produced by a number of fungal species or genera, or by a single fungal species.

The toxic effects of mycotoxins are as diverse as the fungal species that make these toxins. The toxicity of mycotoxins can be acute by short-term ingestion of higher quantities of the toxic compound, or chronic after long-term exposure to low levels of the toxin. The toxic effects of mycotoxins are termed "mycotoxicoses", and the severity of the effect depends on the type of mycotoxin, the extent of the exposure [duration and dose], age, nutritional status and health of the affected individual, and the synergistic effects of the mycotoxin with other chemical exposure of the individual. Mycotoxicosis is usually mediated by damage to cells of all the major organs, such as liver, kidney, lungs and the nervous, endocrine and immune systems. Although a number of mycotoxins have been implicated in human illnesses, for only a very few has a direct connection between the mycotoxin and a corresponding mycotoxicosis has been demonstrated. A very comprehensive review on mycotoxins has been recently published [1].

Table 1. Mycotoxin contamination of cereal grain commodities and the producing fungi

Mycotoxin	Commodities contaminated	Producing Fungi
Aflatoxins	corn, wheat, rice	*Aspergillus flavus* *Aspergillus parasiticus*
Citrinin	wheat, barley, corn, rice	*Penicillium verrucosum*
Citreoviridin	rice	*Penicillium* *Aspergillus*
Cyclopiazonic acid	corn, kodo millet	*Penicillium cyclopium* *Aspergillus flavus*
Fumonisin	corn	*Fusarium verticilliodes*
Ochratoxin A	wheat, barley, oats, corn	*Aspergillus ochraceus* *Penicillium verrucosum*
Patulin	wheat straw residue	*Aspergillus* sp.
Penicillic acid	stored corn, cereal grains	*Penicillium roqueforti* *P. caseicolum*
Sterigmatocystin	wheat	*Aspergillus spp.* *Penicillium luteum*
Trichothecenes	corn, wheat, barley	*Fusarium graminearum*
Zearalenone	corn, hay	*Fusarium graminearum*

* Compiled from CAST report [1].

condiments, and other items such as fodder, hay, silage etc [9]. Factors affecting mycotoxin production include temperature, humidity, extent of physical damage to the commodity i.e., factors that allow fungi to colonize seeds pre- and post-harvest.

The history of mycotoxicoses can be traced back nearly 3000 years [10, 11, 12, 13, 14] with the plagues in Egypt, Greece and other parts of Europe attributed to intake of fungal-contaminated food. For almost 1000 years, since the ninth century, ergotism from ergot-infected rye has afflicted large populations in Europe [15]. Ergotism was also called *ignis sacer* [sacred fire] or St. Anthony's fire because it was believed that a pilgrimage to the shrine of St. Anthony would bring relief from the intense burning sensation caused by the toxin [16,17]. In 1850's, ergotism was finally found to be caused by alkaloids from the sclerotia of *Claviceps purpurea* [13,18,19]. Modern mycotoxicology, however, began in 1960, after 10,000 turkeys died when fed a mycotoxin-contaminated peanut meal ("Turkey-X" disease). It was the discovery of aflatoxins (produced by *Aspergillus flavus*) as the causative agent in the peanut meal causing the "Turkey-X" disease that provided the impetus to chemically isolate and associate fungal toxins with illness in animals [20]. In the last 40 years, many mycotoxins have been identified and characterized and their biosynthetic origin in various fungi elucidated [21]. Mycotoxins have recently been the subject of many reviews [1, 22, 23, 24]. In this chapter, a comprehensive review has been provided on the occurrence and toxicity of mycotoxins contaminating cereal grain crops.

2. POTENTIAL ECONOMIC COSTS OF MYCOTOXINS

The most obvious cost of mycotoxin contamination is the inability to sell contaminated commodities for human food or animal feed due to the regulatory guidelines for sale of these commodities within a country's border or as export products [Table 2]. In countries where there are no regulations, or the defined regulations are not followed rigidly, the economic impact, from risks to toxin exposure, of higher human health costs and animal losses are extremely high. Loss of crops due to mycotoxin contamination can reduce the supply and raise the price of cereal grains. Additional losses are from enhanced expenditures on testing, and development of better detection methods, regulatory enforcement and quality control, litigation, and research to prevent contamination. The exact economic costs of mycotoxin contamination are difficult to ascertain because of uncertainty about various factors such as the extent and level of contamination, variability in contamination analysis, the variability of the price and quantity of the affected commodities, the costs of efforts to mitigate the contamination, the loss in livestock value from contaminated feed, and the variability in health care costs of affected humans. In spite of this variability, direct economic costs of mycotoxin contamination of cereal grains are estimated in the billions of U. S. dollars worldwide. The recently published CAST report [1] provides "first simulated estimates of the economic effects of mycotoxins" at $932 million for U. S. A. alone.

Table 2. Limits for Aflatoxins in Foodstuffs*

Countries	Maximum Allowed Total aflatoxins (ppb or μg/kg])
Austria	0.02 - 1
Germany	5
India	30
Japan	10
Malaysia	35
Mexico	20
The Netherlands	0. 02 - 5
United Kingdom	10
United States (food and feed)	20
(milk and eggs)	

*Adapted from van Egmond [25].

3. DETECTION AND SCREENING OF MYCOTOXINS

Surveillance programs for establishing and reducing the risk of mycotoxin consumption by humans and animals include monitoring, managing and controlling mycotoxin levels in food and feed. Analytical testing methods for rapid analyses of a large number of samples of foodstuffs have been developed for several mycotoxins [26, 27]. Traditional analytical techniques for accurate characterization and quantitation of mycotoxins include chemical methods such as Thin Layer Chromatography [TLC], High Pressure Liquid Chromatography (HPLC) and Gas Chromatography (GC). Most chemical methods require extensive sample clean up, could be time consuming, and often require specialized expensive equipment. Radioimmunoassay [RIA] and Enzyme-linked immunosorbent assay [ELISA] tests are now readily available for a large number of most commonly encountered mycotoxins [reviewed in 28, 29, 30]. The application of mycotoxin immunoassay technique is not limited to food or feed but has also been effectively used for sensitive monitoring of mycotoxins in body fluids as well as tissues or organs of humans and animals which have been exposed to these toxins [reviewed in 28, 29, 30]. The commercial ELISA kits [27, 28, 29, 30] are also available for field testing of mycotoxins in agricultural commodities such as grain. These kits can rapidly detect mycotoxins in foods at levels as low as one tenth of a nanogram per milliliter [9].

4. MYCOTOXIGENIC FUNGAL CONTAMINANTS

The major fungal genera that contaminate cereal grains and produce toxins are *Aspergillus, Fusarium* and *Penicillium*.

4.1. *ASPERGILLUS*

The fungi of the genus *Aspergillus* occupy diverse ecological niches, but are found most commonly in sub-tropical and warm temperate climates [31]; most abundantly between latitudes 26^0 to 35^0 north or south of the equator. These saprophytic fungi grow on a variety of substrates and are considered significant in nutrient cycling in nature. The ability of these fungi to survive in high temperatures with extremely low water activity makes them well suited to colonize a number of grains. The most common agronomically significant, mycotoxigenic contaminants of grain amongst *Aspergillus* are *A. flavus* and *A. parasiticus* because they produce the toxic secondary metabolities aflatoxins, incidently the most well studied mycotoxins [32, 33]. *A. flavus* can contaminate commodities before harvest. Even though reports of the presence of this fungus in corn ear have appeared since 1920 [34], *A. flavus* has never been considered a serious pathogen of corn ears.

4.2. FUSARIUM

Fusarium is a very large genus and its species are found in a wide variety of habitats throughout the world, specially a group of plant pathogenic fungi of grain crops. However, out of dozens of *Fusarium* species, very few are responsible for most mycotoxin contamination of food and feed [35]. Many *Fusarium* species can infect heads of wheat and other small grain cereals under field conditions, causing premature bleaching symptoms known as head scab or blight [36]. *F. graminearum* (the sexual stage is called *Gibberella zeae*) is the major causal agent of head scab of small grains, including wheat, barley, rye, and oats, whereas other species, e.g., *F. avenaceum, F. culmorum, F. poae, F. sporotrichioides,* can predominate in small grains in some regions and environmental conditions. *Fusarium graminearum* is also known to cause red ear rot of maize, because the fungus appears as a pink to red mold of the kernels and, in severe infections, of the cob, husks, and shank. Similar ear rots of maize can be caused by a group of closely related *Fusarium* species, predominately *F. verticillioides, F. proliferatum,* and *F. subglutinans*. *F. graminearum and F. fujikuroi* can cause ear blights of rice, but the grain within the ears is not usually infected. The *Fusarium* species of most concern are those that produce mycotoxins in wheat, maize, rice, barley, and other cereal grains that are significant commodities for worldwide human diet and important animal feed ingredients. *Fusarium graminearum* produces trichothecenes and zearalenone, but these toxins are not produced by *F. verticillioides* and related species. Fumonisins can be produced by all of the toxigenic *Fusarium* species. However, strains of *F. subglutinans* are poor fumonisin producers and strains of *F. proliferatum* vary from low to high production. *Fusarium verticillioides,* a consistent, high-level fumonisin producer, is the predominant fumonisin-producing *Fusarium* species in maize grown in the United States and many other regions of the world [37].

4.3. *PENICILLIUM*

Unlike *Aspergillus* spp. that prefer warmer climates, *Penicillium* species are more abundant in temperate climatic zones. Moisture requirements for *Penicillium spp.* vary widely among species, with some species colonizing substrates with no free water, and others invading grain and their products with moisture contents in equilibrium with relative humidities of 90% or above, at temperatures as low as -2° to +5°C [38]. The production of the mycotoxin patulin by *P. expansum* can occur between 0°C and 24°C. *Penicillium* spp. can grow and produce mycotoxins over a wider range of temperatures than *Aspergillus* spp. [39]. *Penicillium* species are more commonly associated with storage than with preharvest contamination of grain [38]. However, these fungi can cause preharvest contamination of corn with *P. funiculosum and P. oxalicum* in the mid-western United States; *P. purpurogenum, P. funiculosum,* and *P. citrinum* in the southeastern United States; and *P. chrysogenum, P. steckii,* and *P. purpurogenum*, in Spain [40]. *Penicillium oxalicum* infection can be detected by lesions on husks and kernels and resulting ear rot whereas *Penicillium funiculosum* causes streaking of the pericarp without affecting seed viability [40]. *Penicillium* spp. can occasionally be found on small grains, with infection favored by prolonged wet weather or lodging of grain [40].

5. SELECTED MYCOTOXINS

Some of the most prevalant mycotoxins that are found associated with cereal grains, their producing fungi, and the major toxic effects of these toxins in human and animals are discussed below. Although specific mycotoxicoses have been attributed to various mycotoxins, and the mechanisms of toxicity have been established *in vitro* [41], there are many pitfalls in interpreting the results of *in vitro* studies and extrapolating them to *in vivo* effects, and suggesting a natural disease associated with consumption of a mycotoxin. Riley [42], in describing these difficulties, suggested, ".. the factors that tend to confound *in vitro* mechanisms studied include: [1] failure to differentiate secondary effects from the primary biochemical lesion; [2] failure to relate the effective intracellular concentration *in vitro* to the tissue concentration of toxin which causes disease *in vivo*; [3] choice of an *in vitro* model which is either deficient in the biochemical target or unresponsive due to other inadequacies of the model system; and [4] failure to adequately model the complexity of the *in vivo* exposure with regard to potential interactions with other toxins, drugs, environmental, and/or nutritional factors".

5.1. AFLATOXINS

Aflatoxins are by far the best characterized class of mycotoxin. They are a group of polyketid-derived bis-furan containing dihydrofuranofuran and tetrahydrofuran moieties [rings] fused with a substituted coumarin (Figure 1). At least 16 structurally related toxins have been characterized [43]. These toxins are primarily produced by *Aspergillus flavus* and *A. parasiticus* on agricultural commodities, and infrequently by *A. tamarii* and *A. nomius* [44]. There are 4 major aflatoxins B_1, B_2, G_1 and G_2. *A. flavus* produces only the B aflatoxins, and *A. parasiticus* produces both B and G toxins. The aflatoxins were originally isolated from *A. flavus* hence the name *A-fla-toxin*. The B toxins fluoresce blue under U.V. light and the G toxins fluoresce green. Other significant members of the aflatoxin family, M_1 and M_2, are oxidative metabolites of aflatoxin B_1 produced by animals, and isolated from milk, urine and faeces.

Attempts to decipher the aflatoxin biosynthetic pathway began with the discovery of the structure of these toxins [43]. However, the major biochemical steps and the corresponding molecular genetics of AFB_1 biosynthesis have been elucidated only in the last decade [45, 46, 47, 48, 49, 50, 51]. Starting with the polyketide precursor, acetate, there are at least 23 enzymatic steps in the AFB_1 biosynthetic pathway.

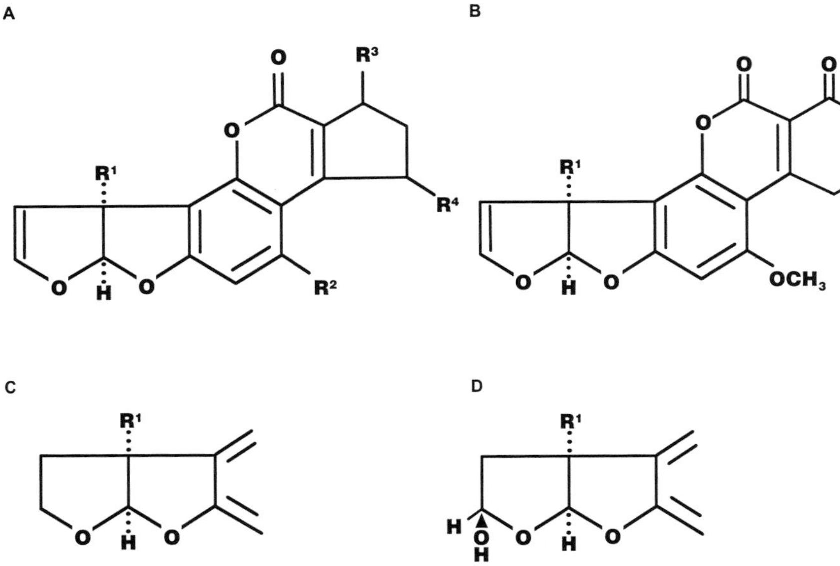

Aflatoxins	Structure	R^1	R^2	R^3	R^4
B_1	A	H	OCH_3	=O	H
M_1	A	OH	OCH_3	=O	H
P_1	A	H	OH	=O	H
Q_1	A	H	OCH_3	=O	OH
R_0	A	H	OCH_3	OH	H
R_0H_1	A	H	OCH_3	OH	OH
B_2	AC	H	OCH_3	=O	H
B_{2a}	AD	H	OCH_3	=O	H
M_2	AC	OH	OCH_3	=O	H
G_1	B	H	-	-	-
G_2	BC	H	-	-	-
G_{2a}	BD	H	-	-	-
GM	BC	OH	-	-	-

Figure 1. Chemical structures of aflatoxins. (A) The B-type aflatoxins are characterized by a cyclopentane E-ring and fluoresce blue under long-wavelength UV-light. (B) The G-type aflatoxins have a xanthone ring in place of the cyclopentane, and fluoresce green under UV-light. (C) Aflatoxins of the B_2 and G_2 type have a saturated bis-furanyl ring. (D) Aflatoxins of the B_{1a} and G_{1a} type have a hydrated bis-furanyl structure.

toxins. The aflatoxins were originally isolated from *A. flavus* hence the name *A-fla-toxin*. The B toxins fluoresce blue under U.V. light and the G toxins fluoresce green. Other significant members of the aflatoxin family, M_1 and M_2, are oxidative metabolites of aflatoxin B_1 produced by animals, and isolated from milk, urine and faeces.

Attempts to decipher the aflatoxin biosynthetic pathway began with the discovery of the structure of these toxins [43]. However, the major biochemical steps and the corresponding molecular genetics of AFB_1 biosynthesis have been elucidated only in the last decade [45, 46, 47, 48, 49, 50, 51]. Starting with the polyketide precursor, acetate, there are at least 23 enzymatic steps in the AFB_1 biosynthetic pathway. AFB_2, G_1, and G_2 are synthesized from pathways that diverge from the B_1 pathway [52, 53, 54]. The genes for almost all the enzymes have been cloned and a regulatory gene (*aflR*) coding for a DNA-binding, Gal 4-type 47 kDa protein has been shown to be required for transcriptional activation of all the structural genes [46, 48, 49, 51, 55, 56, 57, 58, 59, 60, 61, 62, 63]. It has also been shown by restriction mapping of cosmid and phage DNA libraries of *A. flavus* and *A. parasiticus* that all the AFB_1 pathway genes are clustered within a 75-kb region of the fungal genome [48, 49, 51, 62].

Since the discovery of the Turkey X disease in 1960, aflatoxins have been established as immunosuppressive mutagenic, tetratogenic and hepatocarcinogenic in experimental animals. Aflatoxin B_1 [AFB_1] is the most toxic of this group of toxins and the order of toxicity is $B_1 > G_1 > B_2 > G_2$. Aflatoxin M_1 is 10-fold less toxic that B_1, but its presence in milk is of concern in human health [64, 65, 66]. AFB_1 is also one of the most carcinogenic natural compound known; therefore, extensive research has been done on its synthesis, toxicity and biological effects [51, 67, 68, 69, 70].

The liver is the target organ for aflatoxins, and aflatoxicosis leads to proliferation of the bile duct, centrilobular necrosis and fatty infiltration of the liver, hepatomas and hepatic lesions. The susceptibility of animals to AFB_1 varies with species [reviewed in 70]. In addition to the liver, AFB_1 also affects other organs and tissues, such as the lungs and the entire respiratory system [71, 72, 73]. Several outbreaks of aflatoxicosis have occurred in tropical countries. These occurrences were mostly found among adults in rural populations with inadequate nutrition, and with corn being the staple food[13].

In animal models, activation of AFB_1 by microsomal cytochrome P-450 is required for carcinogenicity [70, 74]. The enzyme cytochrome P-450 monooxygenase in the liver converts AFB_1 to a variety of metabolites of increased polarity including AFB_1-8, 9-epoxide, the ultimate carcinogen which binds covalently to N7 position of guanine in DNA, resulting in defective repair and DNA damage, mutations and ultimately carcinomas in many animal species [68, 75, 76]. Epidemiological studies suggest that aflatoxins can be responsible for cancer of the liver in humans in different parts of African and Asia, in combination with hepatitis [70, 76, 77, 78, 79, 80]. Exposure to aflatoxin results in the detection of DNA-aflatoxin adducts found in urine samples of affected humans. These adducts are the results of the modification of a guanine residue at a mutation hot spot, the transversion of guanine to thiamine at the third base of codon 249 in the tumor suppressor gene p53. This gene [p53] is a transcription factor involved in the regulation of the cell cycle and which is commonly mutated in human cancers [81]. Since multiple factors may be involved in carcinogenesis [82], aflatoxins in combination with hepatitis B virus may be significant dual etiological factors in hepatocellular carcinoma [83, 84]. There are also indications that kwashiorkor, a disease caused by protein deficiency, can be at least partially, the result of exposure to aflatoxins [reviewed in 1].

5.2. OCHRATOXINS

Ochratoxins, a group of dihydroisocoumarins (Figure 2), are secondary metabolites produced by a number of *Penicillium* and *Aspergillus* species [85]; the largest amounts of ochratoxins are made by *A. ochraceus* and *P. verruculosum* [86, 87,88]. Other fungi such as *Petromyces alliaceus* (alternate name *Aspergillus alliaceus*) [89, 90] and *A. niger* have also been reported to produce ochratoxins. Ochratoxins are produced primarily in cereal grains [barley, oats, rye, corn, wheat] and mixed feed during storage in temperate climatic conditions; but are also found to contaminate beans, coffee, nuts, olives, cheese, fish, pork, milk powder, wine, beer and bread [91, 92]. Ochratoxins bind tightly to serum albumin and are carried in animal tissues and body fluids. Consequently, the occurrence of this toxin has been reported in pork sausage made from contaminated organ meats, a product widely consumed in northern Europe and the Balkans.

The most frequently occurring member of this group of mycotoxins is Ochratoxin A, which is also the most toxic, and has been found to be a potent nephrotoxin causing kidney damage in many animal species as well as liver necrosis and enteritis [23, 87, 93, 94, 95]. This toxin in experimental animals, has been shown to be a immunosuppessor [95, 96, 97], also demonstrating teratogenic [98], mutagenic [99], weak genotoxic [100] effects on test animals. Some studies have shown that ochratoxins may be weakly carcinogenic, affecting both the kidney and the liver [96, 101]. In vitro studies have shown that ochratoxin is an inhibitor of transfer ribonucleic acid (tRNA) synthetase and protein synthesis. Human exposure to ochratoxins can come from ingestion of contaminated sausage meats or cereals, cereal products, coffee, beer and even pulses [23, 102].

Although the role of ochratoxins in human pathogenesis is still speculative, owing to the similarity of morphological and functional lesions in ochratoxin A induced porcine nephropathy and endemic nephropathy, this mycotoxin has been proposed as the causative agent of endemic nephropathy [16, 103]. The pathological legions of nephropathy in humans were reported to be similar to those observed in porcine nephropathy [104]. Outbreaks of kidney disease [Balkan endemic nephropathy] in rural populations in Bulgaria, Romania, Tunisia and the former Yugoslavia was associated with ochratoxin A [103, 104, 105]. These correlations were based on detection of ochratoxin A in human serum, human milk and kidneys in these countries. Among 77 countries which have regulations for different mycotoxins, 8 have specific regulations for ochratoxin A, with limits ranging from 1 to 20 g/kg in different foods [23]. Administration of phenylalanine can reverse some of the toxic effects of ochratoxins. But other efforts to eliminate the presence of this toxin in food have not been successful.

5.3. CYCLOPIAZONIC ACID

This fungal metabolite is produced by several species of the genus *Penicillium, P. aurantiogriseum, P. crustosum, P. griseofulvum, P. camemberti,* as well as by *A. flavus*, *A. tamarii, and A. versicolor,* but not *A. parasiticus* [44, 106]. Natural occurrence of cyclopiazonic acid [CPA] has been reported in corn, peanuts and cheese, along with the presence of aflatoxins. The association of CPA with aflatoxins together with a lack of well-developed analytical methods for its determination in foods has prevented an assessment of the possible health effects of CPA. Cyclopiazonic acid apparently causes hyperesthesia and convulsions as well as liver, spleen, pancreas, kidney, salivary gland and myocardial damage [69, 107]. Kodua poisoning resulting from ingestion in India of Kodo millet seeds contaminated with *Aspergillus*, has been attributed to cyclopiazonic acid. CPA exhibits an ability to chelate metal ions [108]; and chelation of calcium,

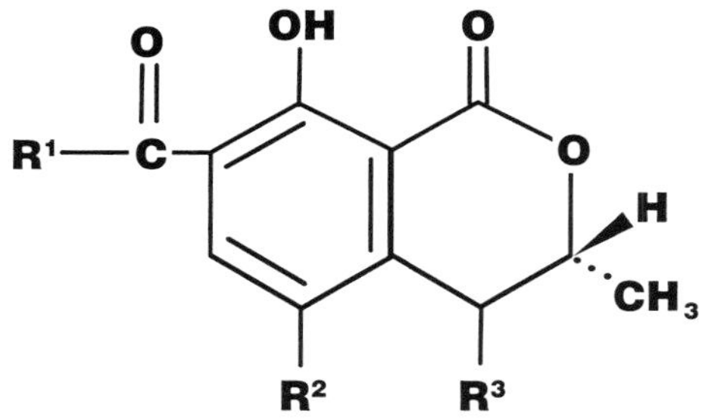

Ochratoxins	R^1	R^2	R^3
A	a*	Cl	H
B	a	H	H
C	b	Cl	H
A Methyl ester	c	Cl	H
B Methyl ester	c	H	H
B Ethyl ester	b	H	H
4 - hydroxy ochratoxin A	a	Cl	OH
α	OH	Cl	H
β	OH	H	H

* a = $C_6H_5CH_2CH(COOH)NH$-

b = $C_6H_5CH_2CH(COOEt)NH$-

c = $C_6H_5CH_2CH(COOMe)NH$-

Figure 2. Structure of different classes of ochratoxins depending on the nature of the amide group (a-c), and the presence or absence of a chlorine moiety at R_2 in the phenyl.

magnesium and iron may be an important mechanism of its toxicity[109].

5.4. PATULIN

Patulin was originally considered desirable for its antibacterial properties. However its toxicity to mammals precluded its use as an antibiotic. Patulin, a metabolite of many species of *Penicillium* and *Aspergillus*, has been detected in damaged apples, apple juice and apple cider made from partially decaying apples and in other fruit juices as well as in wheat. A heat-resistant fungus *Byssochlamys nivea*, frequently found in foods also produces patulin [109]. Sorenson et al [110] found that protein synthesis was inhibited in rat alveolar macrophages exposed to patulin in vitro and that cell membrane function was compromised. Due to highly reactive double bonds that readily react with sulfhydral groups in foods, patulin is not very stable in foods containing these groups [111]. Toxic effects include contact edema and hemorrhage upon ingestion. Patulin appears to be carcinogenic in experimental animals, but no instances of human cancers are known [69, 112]. At least 10 countries have regulatory guidelines that limit the presence of patulin to 50 : g/kg in various foods and juices.

5.5. PENICILLIC ACID

Penicillic acid, also a metabolite of many *Penicillium and Aspergillus* species, is a contaminant of corn ("blue eye corn"), beans and even meat. Even though penicillic acid has been shown to be a liver toxin, no instances of animal or human toxicity have been reported. In spite of this the potential for these toxins to act synergistically with other more potent toxins is a concern since the species that produce these toxins usually produce other toxins as well [69].

5.6. CITRININ

Frequently associated with the natural occurrence of ochratoxin A, citrinin is produced by *P. citrinum* and several other *Penicillia* and *Aspergilli* [113] and *Monascus ruber* and *M. purpureus* [114]. As with patulin, citrinin was originally investigated for its antibacterial properties. But, this toxin has been reported to be a nephrotoxin in chickens fed corn contaminated with *P. citrinum*. Citrinin has been found to be potentially an immunostimulator rather than an immunosuppressor. This golden-yellow compound is easily detected as a free compound, but upon binding to proteins particularly during the drying of corn, the color is lost, while the toxicity remains. Therefore, significant quantities of citrinin, while part of the diet, could lead to chronic and hard to diagnose kidney disease in susceptible individuals and animals, particularly ones that were fed products made from low quality corn [69].

5.7. STERIGMATOCYSTIN

Sterigmatocystin [ST] is produced by various *Aspergilli.* But there are reports which indicate that fungi of the genera *Bipolaris* and *Chaetomium*, and *Talaromyces luteus* also produce ST [113]. Structurally related to aflatoxins, and a precursor of AFB_1 [69, 115, 116], sterigmatocystin is a hepatotoxic and carcinogenic mycotoxin. But its carcinogenicity is several order of magnitudes less than that of aflatoxin B_1 in test animals [117, 118]. Sterigmatocystin is a contaminant of cereal grains (barley, rice and corn), coffee beans and cheese [119]. It has been reported as a contaminant in foods from regions of the world with high incidence of cancer e.g., esophageal cancer in Linxian province in the People's Republic of China [120] and liver cancer in Mozambique [118]. Although ST producing fungi have been isolated from patients with esophageal cancer, the role of ST in human carcinogenesis appears to be indirect and inconclusive [23].

5.8 ZEARALENONE

This toxin is produced by several *Fusarium spp.*, in particular by the scabby wheat fungus *F. graminearum* (roseum) which also makes deoxynivalenol, DON [69]. It has an unusual macrocyclic structure (6 10-hydroxy-6-oxo-trans-1-undecenyl]-beta-resorcyclic acid-: -lactone]. Zearalenone [also called F-2] is found in the gluten of contaminated wet-milled wheat, and also in corn and corn products. Zearalenone has estrogenic properties and can cause premature onset of puberty in female animals, as well as hyperestrogenic effects and reproductive problems in animals, especially swine. This toxin has also been associated with increased incidence of cervical cancer. Zearalenone can be toxic to plants; it can inhibit seed germination and embryo growth at low concentrations. Contamination of feed with zearalenone as well as with deoxynivalenol (DON) may result in severe economic losses to the swine industry due to both feed refusal and adverse estrogenic effects.

5.9. FUMONISINS

Fumonisins are a group of toxins produced primarily by *Fusarium verticillioides* [formerly called *F. moniliforme*], *Fusarium proliferatum* and other related species which readily colonize corn all over the world [121, 122, 123, 124, 125]. Nine structurally related fumonisins including B_1, B_2, B_3, B_4, A_1 and A_2, have been described (Figure 3). Chemically, fumonisin B_1 is a derivative (diester) of propane-1,2,3-tricarboxylic acid of 2-amino-12,16-dimethyl-3,5,10,14,15-pentahydroxyicosane [122, 123, 124, 125, 126, 127, 128, 129, 130, 131]. The other fumonisins lack the tricarballylic acid or other ester groups [122, 132, 133]. Fumonisins are chemically similar in structure to toxins (AAL) produced by *Alternaria alternata* [134]. Production of fumonisins by *Alternaria* has also been reported [135, 136], and some fumonisin-producing Fusaria have been known to produce AAL toxins [137].

Fumonisins B_1, B_2 and B_3 are the major mycotoxin contaminants in corn and corn-based foods and other grains such as sorghum and rice. The level of fumonisins varies from negligible to greater than 100 ppm, but is generally reported to be between 1 to 2 ppm. Corn frequently contains detectable but low levels of fumonisins, even though there is no visible sign of fungal contamination. In laboratory cultures, highest levels of this toxin are produced on corn, and less on rice; whereas peanuts and soybeans are very poor substrates for fumonisin production [reviewed in 23].

Like most other mycotoxins, fumonisins are quite stable and are not removed from grains by changes in temperature, pH and salt concentration during food processing. Therefore, significant amounts of fumonisins are usually found in foods and feeds. Stability of fumonisins is of concern because some products, such as gluten from wet milling and distillers' dried grains from fermentation of contaminated grains, contain higher levels of fumonisins than found in the starting raw material. Acid hydrolysis of fumonisins causes the loss of the tricarballylic acid moiety. Significant amounts of fumonisins are not transmitted to milk, meat or eggs because animals rapidly excrete these toxins [138, 139].

Fumonisin biosynthesis has been of significant research interest in the last few years [reviewed in 140]. A 20- carbon chain is the backbone of fumonisins which structurally resembles fatty acids and linear polyketides [141]. Labelling studies with ^{13}C- and ^{14}C- labeled acetate, has suggested that a polyketide synthase, alanine, methionine and other amino acids are involved in fumonisin synthesis. Four loci (designated *fum 1, fum 2, fum 3 and fum 4*) involved in fumonisin biosynthesis have been identified by classical genetic analyses utilizing *Gibberella fujikuroi*, the sexual stage of *F. verticillioides* [140]. These genes are clustered on chromosome 1 [141, 142].

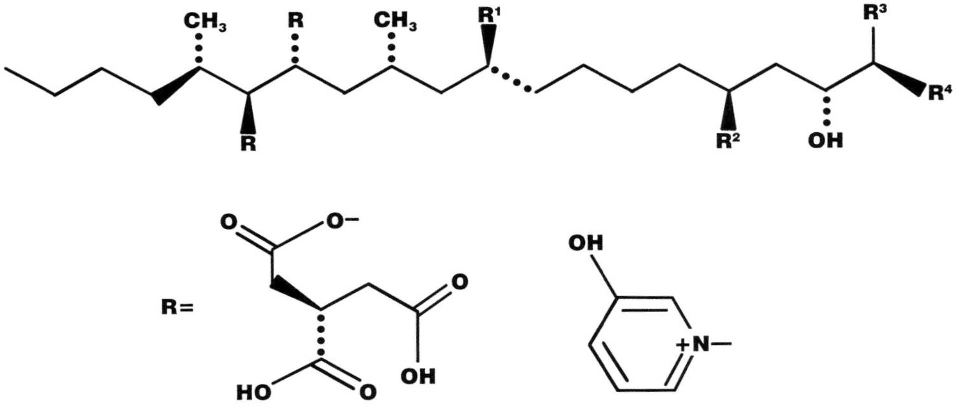

Tricarballylic acid **3-Hydroxypyridinium (3HP)**

Fumonisins	R^1	R^2	R^3	R^4
B_1	OH	OH	NH_2	CH_3
B_2	H	OH	NH_2	CH_3
B_3	OH	H	NH_2	CH_3
B_4	H	H	NH_2	CH_3
A_1	OH	OH	$NHCOCH_3$	CH_3
A_2	H	OH	$NHCOCH_3$	CH_3
C_1	OH	OH	H	H
C_4	H	H	H	H
P_1	OH	OH	3HP	CH_3
P_2	H	OH	3HP	CH_3
P_3	OH	H	3HP	CH_3

Figure 3. Chemical structures of various fumonisins. The R residue is 3-hydroxy-pyridinium in fumonisins P1 to P3 while the others are tri-carballyl esters.

Fumonisin B_1, first isolated in South Africa, is regarded as the most significant of the fumonisins. Fumonisins have been shown to have diverse biological and toxicological effects as discussed here. A correlation between moldy corn and sporadic outbreaks of equine leukoencephalomalacia (ELEM), a neutrotoxic response, has been established for many years. "ELEM" was shown to be caused by fumonsins [143, 144]. Leukoencephalomalacia in horses and other Equidae [donkeys and ponies] is a seasonal disease occurring in late fall and early spring. The disease is manifested by neurological changes in animals, including uncoordinated movement and apparent blindness. Clinically, the disease is characterized by edema in the cerebrum of the brain. The levels of fumonisins B_1 and B_2 in feeds associated with ELEM ranges from 1.3 to 27 ppm [23]. The mechanism of fumonisin toxicity is not well understood. Studies have shown that fumonisins are inhibitors of ceramide synthase; their effect thus appears to be related to interference with sphingolipid biosynthesis in multiple organs, such as brain, lung, liver and kidney, of the susceptible animals. Fumonisin B_1 has also been shown to be a hepatotoxin and a carcinogen in rats resulting in liver cirrhosis and hepatic nodules, adenofibrosis, hepatocellular carcinoma, ductular carcinoma and cholangio-carcinoma [126, 145, 146, 147, 148]. Because of its potential role in hepatocarcinoma formation, fumonisin B_1 has been classified as a class II carcinogen by the International Agency for Research on Cancer, 1993. In cell culture systems, fumonisin B_1 has been demonstrated to be mitogenic and cytotoxic, without genotoxic effects [147, 148]. Kidney cells have also been shown to be targeted by these toxins. Porcine pulmonary edema caused by ingestion of fumonisin contaminated wheat and corn is one of the recorded toxic effect of this group of mycotoxins. Reports have indicated a possible role of fumonisin B_1 in the etiology of human esophageal cancer in the regions of South Africa, China and northeastern Italy where *Fusarium* species are common contaminants[reviewed in 1].

Current data suggest that fumonisins may have greater effect on the health of farm animals than on humans because studies on the impact of levels of fumonisins in human food have not yielded significant correlations [128, 149]. The Mycotoxin Committee of the American Association of Veterinary Laboratory Diagnostics recommended that permissible levels of fumonisins (B_1) in feeds be limited to 5 ppm for horses, 10 ppm for swine and 50 ppm for cattle and poultry.

5.10. TRICHOTHECENES

Several species of *Fusarium* infect corn, wheat, barley, and rice. Under favorable conditions they elaborate a number of different types of tetracyclic sesquiterpenoid mycotoxins that are composed of the epoxytrichothecen skeleton and an olefinic bond with different side chain substitutions (Figure 4). Based on the presence of a macrocyclic ester or ester-ether bridge between C-4 and C-15, trichothecenes are generally classified as macrocyclic (Type C) or non-macrocyclic (Types A and B). Other fungal genera producing trichothecenes are *Myrothecium*, *Trichoderma*, *Trichothecium*, *Acremonium*, *Verticimonosporium* and *Stachybotrys*. The term trichothecenes is derived from trichothecin, the first compound isolated in this group [116, 143, 144, 150, 151, 152, 153, 154].

The biosynthesis of the trichothecenes involves the incorporation of three molecules of mevalonate into the trichothecane nucleus of trichothecin. The precursor farnesyl pyrophosphate can fold in different ways depending on the position and stereochemistry of the three double bonds. The 6, 7-trans olefin is favored to fold to give trichodiene, the first stable intermendiate in the biosynthetic pathway, a process catalysed by the enzyme trichodiene synthase. Trichodiene synthases have a high degree of similarity in different trichothecene-producing fungi. Subsequent conversions of the trichothecane nucleus involve a series of specific oxidations of the three rings,

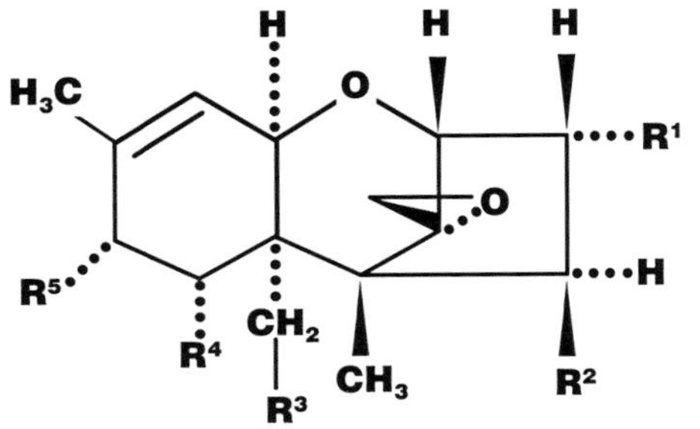

Trichothecenes	R[1]	R[2]	R[3]	R[4]	R[5]
Diacetoxyscirpenol	OH	OAc[1]	OAc	H	H
4-Monoacetoxyscirpenol	OH	OAc	OH	H	H
15-Monoacetoxyscirpenol[a]	OH	OH	OAc	H	H
Scirpentriol	OH	OH	OH	H	H
Deoxynivalenol	OH	H	OH	OH	=O
Nivalenol	OH	OH	OH	OH	=O
Fusarenon-x	OH	OAc	OH	OH	=O
T-2	OH	OAc	OAc	H	ISV[2]
HT-2	OH	OH	OAc	H	ISV
T-2 triol	OH	OH	OH	H	ISV
3'-OH-T-2	OH	OAc	OAc	H	OH-ISV[3]
T-2 tetraol	OH	OH	OH	H	OH
Neosolaniol	OH	OAc	OAc	H	OH
Roridin A,B,E,H,J	H	MC[a]	MC	H	H
Roridin K	H	MC	MC	H	OH
Verrucarin A,B,J,K	H	MC	MC	H	H
Verrucarin L	H	MC	MC	H	OH
Satratoxin F,G,H	H	MC	MC	H	H
Verrucarol	H	OH	OH	H	H

[a] Macrocyclic
[1] OAc $OCOCH_3$
[2] ISV $OCOCH_2CH(CH_3)_2$
[3] OH-ISV $OCOCH_2C(OH)(CH_3)_2$

Figure 4. Chemical structures of various trichothecenes. MC, macrocyclic; ISV, isovalerate; OH-ISV, hydroxysovalerate.

mediated by cytochrome P450 oxidases encoded by genes in a gene cluster, as has been found for biosynthesis of other fungal toxic secondary metabolites. The most significant oxidation is provided by a cytochrome P450 monooxygenase which introduces the C12, 13 epoxide. At least 12 genes (*Tri1* to *Tri 12*) are involved in the biosynthesis of trichothecenes, These genes have been cloned, and are clustered on a 25 kb region of the chromosome. *Tri 6* is a regulatory gene [140, 155, 156, 157, 158].

The trichothecenes constitute a class of over 80 different members that elicit different toxic effects in animals and humans. The trichothecenes causing the most concern for food and feed consumption are nivalenol, deoxynivalenol (DON), and T2 toxin, and the macrocyclic trichothecenes. The most potent trichothecenes are dermal toxins and cause severe blistering and necrosis of the target tissue. Their cytotoxicity parallels their acute toxicity in animals, with T-2 toxin being more potent than nivalenol. Nivalenol is much more potent as an acute toxin than is deoxynivalenol. DON is of considerable importance to agricultural economies because it is more frequently found in corn and cereal grains and animals refuse to eat contaminated grain. And economic losses from both rejected feed stocks and underfed animals are significant. Unlike most mycotoxins, T-2 toxin is produced optimally at 15°C instead of 25-30°C for other toxins [153, 159]. Contamination of cereal grains such as barley, corn, oats, and wheat with T-2 toxin is less frequent than with DON. However, T-2 toxin is at least 20-fold more toxic to animals, and possibly to humans, than DON [23].

The trichothecene mycotoxicoses are difficult to distinguish because they affect many organs, including gastrointestinal tract, hematopoietic, nervous, immune, hepatobiliary, and cardiovascular systems [reviewed in 150]. Ingestion of trichothecene contaminated food and feed causes many types of mycotoxicoses in humans and animals, for example, moldy corn toxicosis, scabby wheat toxicosis (Akakabi-byo disease), vomiting and feed refusal, fusaritoxicosis, and hemorrhagic syndrome. The trichothecenes have now been found to be responsible for the human disease alimentary toxic aleukia in western Siberia prior to World War II which was then attributed to ingestion of moldy millet and wheat. These contaminated grain contained certain macrocyclic epoxytrichothecenes and the severe blistering agent T-2 toxin. Feed refusal by pigs exposed to moldy corn containing deoxynivalenol (also called vomitoxin) was a severe economic problem in the 1970s in midwest United States. Head scab of wheat caused by *F. graminearum* was also a serious problem for wheat and barley growers from 1991-1997 in the eastern and midwestern U. S., and in the Canadian provinces of Ontario and Manitoba [160, 161, 162]. The major mycotoxin associated with head scab is deoxynivalenol [153, 163]. Red-mold disease caused by contamination of wheat and barley with *F. graminearum,* among other Fusaria, was a sporadic problem in Japan and Korea in the 1970s. The combined result of *Fusarium* mycotoxin contamination is world-wide economic losses in the millions of dollars annually.

Another disease, termed Stachybotryotoxicosis, caused the death of thousands of horses in the Soviet Union prior to WWII and was attributed to contamination of straw and hay with the fungus *Stachybotrys atra*. More recently, spores from *S. atra* growing on moist wood or building materials in houses have been associated with pulmonary hemosiderosis in young infants. Trichothecenes may also be involved in this "sick building" syndrome in humans. Symptoms such as headaches, sore throats, hair loss, diarrhea, fatigue and general malaise have been reported in humans living in water-damaged houses [151, 164, 165].

The unusual 12,13-spiroepoxy ring of the trichothecenes has been shown to be necessary for their toxicity (reviewed in 150). Their toxicity is mainly due to the inhibition of initiation of polypeptide synthesis or polypeptide chain elongation during protein synthesis. All types of trichothecenes interfere with the action of peptidyl transferase by binding to the 60S subunits of the eukaryotic ribosome. The higher molecular weight trichothecenes inhibit initiation by binding to the P site of the ribosome and preventing translocation. The lower molecular weight species inhibit both the initiation and elongation steps of protein synthesis.

6. MANAGEMENT OF MYCOTOXIN CONTAMINATION

The significant negative impact of mycotoxin contamination on economics of agriculture, as well as human and animal health has created the need to eliminate or at least minimize this contamination of food and feed worldwide. The strategies for achieving this goal include both pre and post-harvest control measures; and decontamination proceduces, if contamination does occur in the commodities. Traditional methods to reduce mycotoxin contamination in field grain crops include appropriate cultural practices such as irrigation, control of insect pests etc. [reviewed in166, 167, 168, 169]. However, these practices may not be feasible for every environmental condition or geographic location because of economic or physical difficulties in administering them. Significant control of pre-harvest toxin contamination is expected to be dependent on a detailed understanding of the physiological and environmental factors that affect the biosynthesis of the toxins, the biology and ecology of the fungus, and the parameters of the host plant- fungal interactions. For examples, insect damage to seed is a significant source of entry of toxigenic fungi into the crop or commodity, and control of insect- population could have a significant impact. Research efforts have been made worldwide to elucidate these factors for at least the most agriculturally significant toxins, namely aflatoxins, fumonisins and trichothecenes [156, 170, 171, 172]. The strategies to manage mycotoxin contamination both pre-harvest or past-harvest can be summarized as follows:

6.1. Reduction in levels of toxigenic fungi

The presence of fungi can be achieved by chemical means by use of fungicides or insecticides (to control insect pests as fungal vectors), use of natural products as antifungal or anti-insect agents. Transgenic crops containing genes coding for anti-insect gene products can reduce levels of toxigenic fungi in the crops. Use of specific atoxigenic fungi to outcompete the toxin producing species appears to be an effective strategy. Native atoxigenic *A. flavus* strains as biocompetitive agents has been effective in reducing aflatoxin contamination [173]. Breakthroughs in the understanding of the molecular biology of toxin bio-synthesis has enabled the selective designing of effective biocompetitive agents whenever a native biocontrol strain is not available [170]. Biocompetition may be more feasible in cropping areas that are subject to a chronic toxin contamination problem, as well as in crops where natural resistance against fungal infections is not available due to limited genetic diversity.

6.2. Development of Resistant Germplasm

Germplasm, that offers resistance to toxigenic fungal infection or toxin production, can be developed through plant breeding or genetic engineering. While fumonisins may not be essential for virulence, they may play a significant role in fungal infection of corn [174]. And, since *F. verticilloides* hyphae are found to be confined to the pedical of the corn kernel, introduction by genetic engineering or plant breeding and expression of antifungal proteins in the pedicel region of the corn kernel could impart resistance in corn against preharvest fumonisin contamination [174].

F. graminearum is the causative organism of head blight [scab] of wheat, and trichothecenes produced by this fungus appear to be involved in fungal virulence [158]. This suggests that resistance in wheat against trichothecene production may provide a means of controlling not only head blight but preharvest contamination of crops with trichothecenes such as deoxynivalenol and nivalenol. In case of aflatoxins, the problem is that *A. flavus* is not a typical plant pathogen, and identification of resistant cultivars in crops such as corn is difficult. However, some naturally resistant germplasm has been commercially released (reviewed in 175). Additionally, research is underway [175] to identify specific biochemical factors from both host and non-host crops that cause inhibition of fungal growth or toxin production. Several of these factors are being examined for use in plant breeding or genetic engineering of crops [175] to assess their efficacy in preventing pre-harvest toxin formation. However, the aflatoxin contamination problem in field crops is so complex [167] that a combination of approaches will be required to eliminate or even control the preharvest toxin contamination problem [46].

6.3. Post-harvest Management

Post-harvest mycotoxin contamination is prevalent in most tropical countries due to hot, wet climate coupled with sub-adequate methods of harvesting, handling and storage practices, which often lead to severe fungal growth and mycotoxin contamination of food and feed [168, 175, 176, 177, 178]. Fungal growth and toxin production largely depend on physical factors such as water activity (a_w) and temperature. Therefore, storage conditions with water activity below 0.7 and at temperatures below 15°C could reduce fungal growth and toxin production in food (reviewed in 1). However, mycotoxins are sometimes produced before or immediately after harvest, and good storage practices may not prevent contamination of stored food grain.

If preventive measures are ineffective, the only recourse left for salvaging the crop or commodity is to eliminate the contaminating toxins by detoxifying the commodity. Detoxification can be achieved by removal (segregation) or inactivation (degradation) of toxins present in the commodities through physical, chemical or biological means. A successful detoxification procedure should be economical, relatively simple and rapid, and should not pose any health hazard or impair nutritional quality [169, 179,180]. Various detoxification processes can be categorized as follows: *(i) Removal of mycotoxins:* Since most of the mycotoxin burden in contaminated commodities is localized to a relatively small number or seeds or kernels (reviewed in 179, 181, 182), removal of these contaminated seeds/kernels is effective in detoxifying the commodity. Methods currently used include (a) physical separation by identification and removal of damaged seed, mechanical or electronic sorting, flotation and density separation of damaged or contaminated seed; (b) removal by filtration and adsorption of toxins onto filter pads, clays, activated charcoal etc.; (c) removal of the toxin by milling processes, or by solvent extraction. *(ii) Inactivation of mycotoxin:* When removal of mycotoxins is not possible, mycotoxins can be inactivated by (a) physical treatments such as heat, microwave, photochemical or gamma irradiation; (b) chemical methods such as treatment of commodities with acids, alkalis, aldehydes, oxidizing agents, and gases like chlorine, sulfur dioxide, ozone and ammonia [183]; (c) biological methods such as fermentations or enzymatic digestions that cause the breakdown of mycotoxins [184].

6.4. Dietary Modification

Modifications to diets, such as use of agents for absorption, distribution and metabolism of mycotoxins, can affect the toxicity of mycotoxins. Dietary additives including anti-carcinogenic substances [185, 186, 187] and chemoprotective agents [188, 189, 190,191, 192] have been found to inhibit the carcinogenic effects of aflatoxins in test animals by preventing formation of aflatoxin-DNA adducts [193]. Antioxidants such as vitamin C and E reduced the toxic effects of ochratoxin A [194, 195], whereas ascorbic acid provided protection against aflatoxins [196]. The toxic effects of ochratoxin have also been reduced by aspartame because it competitively prevents the binding of ochratoxin to serum albumin [197]. The introduction of hydrated sodium calcium aluminasilicates in diets of animals has reduced the toxicity of various mycotoxins [such as aflatoxins] to animals because these and related compounds (HSCAS or NOVASIL) have a high affinity for mycotoxins [198, 199,200, 201, 202] that prevents the absorption of mycotoxins by animals. Other absorbants that have been tested for mycotoxins such as T-2 toxin included zeolite, bentonite and superactive charcoal.

7. CONCLUSION

Mycotoxins are natural products (secondary metabolites) produced by fungi that evoke a toxic response in higher vertibrates and other animals when fed at low concentrations; some mycotoxins may also be phytotoxic or antimicrobial. Mycotoxins, however, are not grouped with antibiotics, compounds which also may be produced by fungi with microorganisms as the target species. Also, poisonous metabolites produced by mushrooms and yeasts are not included in this group. Mycotoxin production is associated with fungal development, even though, often there is no positive correlation between fungal growth in a commodity and the toxin levels detected in that commodity. Presumably these compounds play some role in the ecology of the fungus, but their function has not been clearly defined. Many mycotoxin-producing fungi can cause plant diseases in the field, but it has been difficult to establish a causal role of mycotoxins in plant disease. Critical analysis needs to be done to demonstrate that production of a mycotoxin increases the ability of the fungus to cause a plant disease. Laboratory and field tests have indicated that trichothecenes play an important role in wheat head blight and maize ear rot caused by *Fusarium graminearum* [203, 204]; but fumonisins, produced by another *Fusarium* species (*F. verticilioides*) is not required for maize ear rot [205]. However, recent understanding of the genetic regulation of mycotoxin biosynthesis has revealed that the synthesis of several of these toxins occurs through a number of complex steps. A cascade of regulatory elements that govern processes expressed by gene clusters specific for secondary metabolism indicates that fungi must have had an evolutionary or physiological advantage in producing mycotoxins. In spite of this dilema, one thing is certain that mycotoxin production causes a very significant negative impact on agricultural economics throughout the world due to the perceived toxic effects of mycotoxins. In case of mycotoxins such as aflatoxins, the toxic effects and human mycotoxicoses have been fairly well established by detailed epidemiological studies. Very few reports, however, have demonstrated a direct correlation between a mycotoxin and its corresponding mycotoxicosis, although significant circumstantial and research information implicates mycotoxins in human illnesses. Therefore, research on mycotoxins should continue to focus on development of effective and economical strategies to control and ultimately eliminate mycotoxin contamination of food and feed.

8. REFERENCES

1. Richard, J. L. and Payne, G. A. (2001). Mycotoxins: Risks in plant and animal systems. Council for Agricultural Science and Technology (U.S.A.) Task Force Report No. 138, pp. 303.
2. D'Mello, J. P. F. and Macdonald, A. M.C. (1997). Mycotoxins. Animal Feed Science Technology. 69: 155-166.

3. Scott, P.M. (1989). The natural occurrence of trichothecenes. In Trichothecene Mycotoxicosis. Pathophysiologic Effects, vol. I. (Beasley, V.D. Ed.), pp. 1-26. CRC Press, Boca Raton, FL.
4. Strange, R.N. (1991). Natural Occurrence of mycotoxins in groundnuts, cottonseed, soys and cassava. In Mycotoxins
and Animal Foods (Smith, J. E.. and Henderson, R.S. Eds.), pp. 341-362. CRC Press, Boca Raton, FL.
5. Shotwell, O. L. (1991). Natural occurrence of mycotoxins in corn. In Mycotoxins and Animal Foods (Smith, J. E. Henderson, R.S. Eds.) pp. 325-340. CRC Press, Boca Raton, FL.
6. Marasas, W. F.O. (1995). Fumonisins: their implications for human and animal health. Nat. Toxins. 3193-198.
7. Yoshizawa, T.(1991). Natural occurrence of mycotoxins is small grain cereals (wheat, barley, rye, oats, sorghum, millet, rice). In Mycotoxins and Animal Foods (Smith, J. E. and Henderson, R.S. Eds.), pp.. 301-324. CRC Press, Boca Raton, FL.
8. Mannon, J. and Johnson, E. (1985). Fungi down on the farm. New Scientist, 1445: 12-16.
9. Malloy, C.D. and Marr, J. S. (1997). Mycotoxins and Public Health: A review. J. Public Health Management Practices 3: 61-69.
10. Schoental, R. (1984). Mycotoxins and the Bible, Perspectives in Biology and Medicine, no 1, 28: 117-120.
11. Marr, J. S. and Malloy, C.D. (1996). An epidemiologic analysis of the ten plagues of Egypt. Caduceus. 12:7-24.
12. Schoental, R. (1994) Mycotoxins in food and the plague of Athens, Journal of Nutritional Medicine, 4: 83-85.
13. Peraica, M., Radic, B., Lucic, A. and Pavlovic, M. (1999). Toxic effects of mycotoxins in humans, Bulletin of the World Health Organization. 77: 754-766.
14. Matossian, M. K. (1989). Poisons of the past, *Molds, Epidemics, and History,* New Haven, CT, Yale University Press.
15. Castiglioni, A. (1947). A history of medicine, New York, Alfred A. Knopf.
16. Cartwright, F. F. (1975). Disease and history, New York, Thomas Y, Crowell Co.
17. Van Dongen, P.W. J.and DeGroot, A.N.J.A. (1995). History of ergot alkaloids from ergotism to ergometrine, European Journal of Obstetrics, gynecology and reproductive biology. 60:109-116.
18. Maxcy, K. F. (1956). Rosenau preventive medicine and public health. In (Ed.) New York: Appleton-Century-Crofts.
19. Ackerknocht, E. H. (1965). History and Geography of the most important diseases, New York, Hafner Publishing.
20. Bradburn, N., Coker, R. D. and Blunden, G. (1994). The aetiology of Turkey X Disease. Phytochemistry. 35:817.
21. Sharma, R.P. and Salunkhe, D.K. (1985) Animal and Plant Toxins. In Modern Toxicology, vol 1, (Gupta, P.K., Salunkhe, D.K. Eds.) pp. 252-316, Metropolitan Book Co.
22. Scudmore, K. A. (2000). Mycotoxins. In Natural Toxicants in Food (D Watson Ed.) 147-181. CRC Press Boca Raton, FL USA.
23. Chu, F. S. (2000). Mycotoxins and mycotoxicosis. In Foodborne Infections and Intoxications (Third edition) (Reimann, D. Cliver, DO Eds.) Academic Press, New York (in press).
24. Bhatnagar, D., Ehrlich, K.C. and Chang, P-K. (2000). Mycotoxins. In Agriculture Encyclopedia of Life Sciences, Nature Publishing Group, London, (in press).
25. Van Egmond, H.P. (1993). Rationale for regulatory programmes for mycotoxins in human foods and animal feeds. Food Additives and Contaminants. 10, 29-36.
26. Coker, R. D. (1998). Design of Sampling plans for determination of mycotoxin in food and feed. In Mycotoxin in Agriculture and Food Safety (Sinha, K.K. and Bhatnagar, D. Eds.), pp. 109-134 Marcel Dekker, New York.
27. Wilson, D. M., Sydenham, E. W., Lombaert, G.A., Trucksess, M.W., Abramson, D. and Bennett, G.A. (1998). Mycotoxin analytical techniques. In Mycotoxin in Agriculture and Food Safety (Sinha, K.K. and Bhatnagar, D. Eds.), pp. 135-182. Marcel Dekker, New York.
28. Chu, F.S. (1991). Development and use of immunoassays in detection of the ecologically important mycotoxins. In Mycotoxins in Ecological Systems, vol. V. (Bhatnagar, D. Lillehoj, E. B. Arora, D.K. Eds.), pp. 87-136. Handbook of Applied Mycology. Marcel Dekker, Inc., New York.
29. Chu, F.S. (1995). Mycotoxin analysis. In Analyzing Food for Nutrition Labeling and Hazardous Contaminants (Jeon, I. J. and Ikins, W.G. Eds.), pp. 283-332. Marcel Dekker, New York.
30. Chu, F.S. (1996). Recent studies on immunoassays for mycotoxins. In Immunoassays for Residue Analysis (Beier, R. C. and Stanker, L. H. Eds.) 294-313. Food Safety ACS Symposium Series Book no. 621. American Chemical Society, Washington, DC.
31. Klich, M. A., Tiffany, L. H. and Knaphus, G. (1994). Ecology of the aspergilli of soils and litter. In (Klich, M. A. and Bennett, J. W. Eds.), pp. 329-353. *Aspergillus* Biology and Industrial Applications. Butterworth-Heineman, Boston.
32. Payne, G. A. (1992). Aflatoxin in maize. Crit Rev. Plant Sci. 10: 423-440.

33. Payne, G. A. (1998). Process of contamination by aflatoxin-producing fungi and their impact on crops. In (Sinha, K. K. S. and Bhatnagar, D. Eds.), pp. 279-306. Mycotoxins in Agriculture and Food Safety. Marcel Dekker, Inc., NewYork.
34. Taubenhaus, J. J. (1920). A Study of the Black and Yellow Molds of Ear Corn. Texas Agricultural Experiment Station Bulletin No. 270. College Station.
35. Marasas, W. F. O., Nelson, P. E. and Toussoun, T. A. (1984b). Toxigenic Fusarium Species: Identity and Mycotoxicology. The Pennylvania State University Press, University Park. Pennsylvania.
36. Sutton, J. C. (1982). Epidemiology of wheat heat blight and maize ear rot caused by Fusarium graminearum. Can. J. Plant Pathol. 4:195-209.
37. Munkvold, G. P. and Desjardins. (1997). Fumonisins in maize: Can we reduce their occurrence? Plant Dis. 81: 556-565.
38. Sauer, D. B., Meronuck, R. A. and Christensen, C. M. (1992). Microflora. In (Sauer, D. B. Ed.), pp. 313-340. Storage of Cereal Grain and Their Products. 4th ed. American Association of Cereal Chemists, Inc., St. Paul, Minnesota.
39. Ominski, K. H., Marquardt, R. R., Sinha, R. N. and Abramson, D. (1994). Ecological aspects of growth and mycotoxin production by storage fungi. In (Miller, J. D. and Trenholm, H. L. Eds.), pp. 287-312. Mycotoxins in Grain: Compounds Other than Aflatoxin. Eagan Press, St.Paul, Minnesota.
40. Lacey, J., Ramakrishna, N., Hamer, A., Magan, N. and Marfleet, I. C. (1991). Grain Fungi. In (Aurora, D. K., Mukerji, K. G. and Marth, E. H. Eds.), pp. 121-177. Handbook of Applied Mycology. Vol. 3. Marcel Dekker, Inc., New York.
41. Riley, R. T. and Norred, W.P. (1996). Mechanism of mycotoxicity. In The Mycota Vol VI.(Howard, D.H. and Miller J.D. Ed.), 193-211. Berlin, Springer.
42. Riley, RT. (1998). Mechanistic interactions of mycotoxins. Theoretical considerations. In (Sinha, K. K. and Bhatnagar D. Eds.), pp 266-254. Mycotoxins in Agriculture and Food Safety. Marcel Dekker, New York.
43. Goldblatt, L.A. (1969). Aflatoxin-Scientific Background, Control and Implications, Academic Press, New York.
44. Goto, T., Wicklow, D.T. and Ito, Y. (1996). Aflatoxin and cyclopiazonic acid production by a sclerotium-producing *Aspergillus tamarii* strain. Appl. Environ. Microbiol. 62: 4036-4038.
45. Bhatnagar, D., Lillehoj, E. B. and Arora D. K. (1992). (Eds.) Mycotoxins in Ecological Systems. Handbook of Applied Mycology, vol V, Marcel Dekker, Inc., New York.
46. Bhatnagar, D., Payne, G., Linz, J. E. and Cleveland T. E. (1995). Molecular biology to eliminate aflatoxin. International News on Fats, Oils and Related Materials. 6: 262-271.
47. Dutton. M.F. (1988). Enzymes and aflatoxin biosynthesis. Microbiol. Rev. 52: 274-295.
48. Townsend, C.A. (1997). Progress towards a biosynthetic rationale of the aflatoxin pathway. Pure and Appl. Chem. 58: 227-238.
49. Trail, F., Mahanti, N. and Linz, J. (1995). Molecular biology of aflatoxin biosynthesis. Microbiology, 141: 755-765.
50. Payne, G.A. and Brown, M.P. (1998). Genetics and physiology of aflatoxin bioxynthesis. Annu. Rev. Phytopathol. 36: 329-362.
51. Cary, J. W., Bhatnagar, D. and Linz, J. E. Aflatoxin: Biological significance and regulation of biosynthesis. In Microbial Foodborne Diseases. Cary, J. W., Linz, J. E. and Bhatnagar, D. Eds.), (2000) pp. 317-361. Mechanism of pathogenesis and toxin synthesis. Tecchnomic Publishing Co. Inc., Lancaster, PA.
52. Bhatnagar, D., Cleveland, T. E. and Kingston, D.G.I. (1991). Enzymological evidence for separate pathways for aflatoxin B_1 and B_2 biosynthesis. Biochemistry 30: 4343-4350.
53. Yu, J., Chang, P-K., Ehrlich, K.E., Cary, J.W., Montalbano, B., Dyer, J. M., Bhatnagar, D. and Cleveland, T. E. (1998). Characterization of the critical amino acids of an *Aspergillus parasiticus* cytochrome P450 mono-oxygenase encoded by *ordA* that is involved in the biosynthesis of aflatoxins B_1, G_1, B_2 and G_2. Appl. Environ. Microbiol. 64: 4834-4841.
54. Yabe, K., Nakamura, M. and Hamasaki, T. Enzymatic formation of G-group aflatoxin and biosynthetic relationship between G-and B-group aflatoxins. Appl. Environ. Microbiol. 65: 3867-3872.
55. Cary, J. W., Wright, M., Bhatnagar, D., Lee, R. and Chu, F.S. (1996). Molecular characterization of an *Aspergillus parasiticus* dehydrogenase gene, *nor*A, located on the aflatoxin biosynthesis gene cluster. Appl. Environ. Microbiol. 62: 360-366.
56. Chang, P-K., Cary, J. W., Yu, J., Bhatnagar, D. and Cleveland, T.E. (1995). The *Aspergillus parasiticus* polyketide synthase gene *pksA*, a homolog of *Aspergillus nidulans wA,* is required for aflatoxin B_1 biosynthesis. Mol. Gen. Genetics. 248: 270-277.

57. Chang, P-K., Ehrlich, K.C., Yu, J., Bhatnagar, D. and Cleveland, T.E. (1995). Increased expression of *Aspergillus parasiticus aflR,* encoding a sequence-specific DNA-binding protein, relieves nitrate inhibition of aflatoxin biosynthesis. Appl. Environ. Microbiol. 61: 2372-2377.
58. Chang, P.K., Ehrlich, K.C., Linz, J. E., Bhatnagar, D., Cleveland, T.E. and Bennett, J.W. (1996). Characterization of the *Aspergillus parasiticus niaD* and *nii*A gene cluster. Curr. Genet. 30: 68-75.
59. Payne, G.A., Nystrom, G. J., Bhatnagar, D., Cleveland, T. E. and C.P. (1993). Cloning of the *afl*-2 gene involved in aflatoxin biosynthesis for *Aspergillus flavus.* Appl. Environ. Microbiol. 59: 156-162.
60. Prieto, R. and Woloshuk, C. P. (1997). *ord1*, an oxidoreductase gene responsible for conversion of O-methylsterigmatocystin to aflatoxin in *Aspergillus flavus.* Appl. Environ. Microbiol. 63: 1661-1666.
61. Woloshuk, C. P., Yousibova, G. L., Rollins, J. A., Bhatnagar, D. and Payne, G.A. (1995). Molecular characterization of the *afl*-1 locus in *Aspergillus flavus.* Appl. Environ. Microbiol. 61:3019-3023.
62. Yu, J., Chang, P.K., Cary, J.W., Wright, M., Bhatnagar, D., Cleveland, T.E., Payne, G.A. and Linz, J.E. (1995). Comparative mapping of aflatoxin pathway gene cluster in *Aspergillus parasiticus* and *Aspergillus flavus.* Appl. Environ. Microbiol. 61: 2365-2371.
63. Yu, J. H., Butchko, R. A. E., Fernandes, M., Keller, N.P., Leonard, T. J. and Adams, T. H. (1996). Conservation of structure and function of the aflatoxin regulatory gene *aflR* from *Aspergillus nidulans* and *A. flavus.* Curr. Genet. 29: 549-555.
64. Cullen, J. M., Ruebrier, B. H., Hsieh, L.S., Hyde, D.M. and Hsieh, D.P. (1987). Carcinogenicity of dietary aflatoxin M_1 in male Fischer rats compared to aflatoxin B_1. Cancer Res. 47: 1913-1917.
65. Galvano, F., Galofaro, V. and Galvano, G. (1996). Occurrence and stability of aflatoxin M_1 in milk and milk products: A worldwide review. J. Food Protect. 59: 1079-1090.
66. Van Egmond, H.P. (1989). Current situation on regulations for mycotoxins: Overview of tolerances and status of standard methods of sampling and analysis. Food Addit. Contam. 6 :134-188.
67. Busby, W.F. Jr. and Wogan, G.N. (1979). Food-borne mycotoxins and alimentary mycotoxicoses. In Food-borne Infections and Intoxications (Rieman, H. and Byran, F. L. Ed.), New York, Academic Press.
68. Busby, W.F. Jr. and Wogan, G. N. (1981). Aflatoxins. In Mycotoxins and N-nitrosocompounds: Environmental Risks, vol. II (Shank, RC Ed..), pp. 3-45. CRC Press, Boca Raton, FL.
69. CAST. (1989). Mycotoxins, Economics and Health Risks. Council of Agricultural Science and Technology (CAST), Task force report No. 116. p. 92. CAST, Ames, IA.
70. Eaton, D.L. and Groopman, J. D. (1994). The toxicology of aflatoxins-Human health, Veterinary and Agricultural significance. (Eds.), Academic Press, New York.
71. Massey, T. E. (1996). The 1995 Pharmacological Society of Canada Merck Frosst award - Cellular and molecular targets, in pulmonary chemical carcinogenesis - studies with aflatoxin B_1. Can. J. Physiol. Pharmacol. 74: 621-628.
72. Heinonen, J. T., Fisher, R., Brendel, K. and Eaton, D. L. (1996). Determination of aflatoxin B_1 biotransformation and binding to hepatic macromolecules in human precision liver slices. Toxicol. Appl. Pharmaco*l.* 136: 1-7.
73. Kelly, J.D., Eaton, D.L., Guengerich, F. P. and Coulombe, R.A. (1997). Aflatoxin B_1 activation in human lung. Toxicol. Appl. Pharmocol. 144: 88-95.
74. Newberne, P. M. and Rogers, A. E. (1981) Animal toxicity of major environmental mycotoxins. In Mycotoxins and Nitroso Compounds, Environmental Risks, vol. I, (Shank, R. C. Ed.), pp. 51-106. CRC Press, Inc., Boca Raton, FL.
75. Wild, C.P. and Kleihues, P. (1996). Etiology of cancer in human and animals. Exp. Toxic Pathol. 48: 95-100.
76. Wogan, G.N. (1992). Aflatoxins as risk factors for hepatocellular carcinoma in humans. Cancer Res.(Suppl). 52: 2114s-2118s.
77. Hsieh, D. P. H. (1989). Potential human health hazards of mycotoxins. In (Natori, S. Hashimoto, K. and Ueno Y. Eds.), pp. 69-80. Mycotoxins and Phycotoxins 88. Elsevier, Amsterdam.
78. Peers, F., Bosch, X., Kaldor, J., Linsell, A. and Pluijiman, M. (1987). Aflatoxin exposure, hepatitis B virus infection and liver cancer in Swaziland. *Intl.* J. Cancer. 39: 545-553.
79. Wild, C.P., Shresma, S.M., Anwar, W.A. and Montesano, R. (1992). Field studies of aflatoxin exposure, metabolism and induction of genetic alterations in relation to HBV infection and hepatocellular carcinoma in The Gambia and Thailand. Toxicol. Lett. 64/65: 455-461.
80. Zhu, J.Q., L.S. Zhang, X. Hu, Y. Xiao, J.S. Chen, Y.C. Xu, J. Fremy, F.S. Chu. (1987). Correlation of dietary aflatoxin B_1 levels with excretion of aflatoxin M_1 in human urine. Cancer Res. 47: 1848-1852.
81. Groopman, J.D., Wogan, G.N., Roebuck, B.D. and Kensler, T.W. (1994). Molecular biomarkers for aflatoxins and their application to human cancer prevention. Cancer Res. 54: (Suppl.), 1907S-1911S.
82. Harris, C.C. and Sun, T.T. (1986). Interactive effects of chemical carcinogens and hepatitis B virus in the pathogenesis of hepatocellular carcinoma, Cancer Surv. 5: 765-780.

83. Chen, C. J., Wang, L. Y., Lu, S. N., Wu, M. H., You, S.L., Zhang, Y. L., Wang, L.W., and Santella, R.M. (1996). Elevated aflatoxin exposure and increased risk of hepatocellular carcinoma. Hepatology. 24: 38-42.
84. Chen, C.J., Yu, M. W., Liaw, Y. F., Wang, L.W., Chiamprasert, S., Matin, F., Hirvonen, A., Bell, D.A. and Santella, R.M. (1996). Chronic hepatitis B carriers with null genotypes of glutathione S-transferase M1 and T1 polymorphisms who are exposed to aflatoxin are at increased risk of hepatocellular carcinoma. Am. J. Human Genetics. 59: 128-134.
85. Chu, F.S. (2000). Mycotoxins. In (Cliver, D.O. Ed.), Foodborne Diseases (second edition). Academic Press, New York, (in press).
86. Chu, F.S. (1974). Studies on ochratoxins. Crit. Rev. Toxicol. 499-524.
87. Kuiper-Goodman, T. and Scott, P.M. (1989). Risk assessment of the mycotoxin ochratoxin. A. Biomed. Environ. Sci. 2: 179-248.
88. Pohland, A. E., Nesheirn, S. and Friedman, L. (1992). Ochratoxin A: a review. Pure Appl. Chem. 64: 1029-1046.
89. Abarca, M. K., Bragulat, M. R., Castella, G. and Cobanes, F. J. (1994). Ochratoxin A production by strains of *Aspergillus niger* var. niger. Appl. Environ. Microbiol. 60: 2650-2652.
90. Ono, H., Kataoka, A., Koakutsu, M., Tanaka, K., Kawasugi, S. , Wakazawa, M., Ueno, Y. and Manabe, M. (1995). Ochratoxin producibility by strain *of Aspergillus niger* group stored in IFO culture collection. Mycotoxins. 41: 47-51.
91. Van Egmond, H.P. and Speijers, G. J. A. *(*1994). Survey of data on the incidence and levels of ochratoxin A in food and animal feed worldwide, Natural Toxins, 3: 125 -144.
92. Studer-Rohr, I., Dietrich, D. R., Schlatter, J. and Schlatter, C. (1995). The occurrence of ochratoxin A in coffee. Food Chem. Toxicol. 33: 341-355.
93. Chu, F. S. (1974). Studies on ochratoxins. Crit. Rev. Toxicol. 2: 499-524.
94. Fink-Gremmels, J. , Jahn, A. and Blom, M.J. (1995). Toxicity and metabolism of ochratoxin A. Nat. Toxins. 3: 214-220.
95 . Simon, P. (1996). Ochratoxin and kidney disease in the human. J. Toxicol.-Toxin Rev. 15: 239-249.
96. Boorman, G. (1988). NTP technical report on the toxicology and carcinogenesis studies of ochratoxin A (CAS no. 303-47-9) in F344/N rats (gavage studies). NIH publication No., 89-2813. U.S. Dept. of Health and Human Services, NIH, Bethesda, MD, USA.
97. Boorman, G. A., McDonald, M. R., Imoto, S. and Persing, R. (1992). Renal lesions induced by ochratoxin A exposure in the F344 rat. Toxicol. Pathol. 20: 236-245.
98. Hayes, A.W. (1981). Mycotoxin Teratogenicity and Mutagenicicty. CRC Press, Inc. Boca Raton, FL.
99. deGroene, E. M., Jahn, A., Horbach, G. J. and Fink-Gremmels, J. (1996). Mutagenicity and genotoxicity of the mycotoxin ochratoxin A. Environ. Toxicol. Pharmacol. 1: 21-26.
100. Dirheimer, G. (1996). Mechanistic approaches to ochratoxin toxicity. Food Additives Contam. (Suppl.), 13: 45S-48S.
101. Schlatter, C., Studerrohr, J. and Rasonyi, T. (1996). Carcinogenicity and kinetic aspects of ochratoxin A. *Food Addit. Contam.* (Suppl.),13: 43S-44S.c
102. Stegen, G.V.D. Jorissen, U., Pittet, A., Saccon, A. M., Steiner, W., Vincenzi, M., Winkler, M., Zapp, J. and Schlatter, C. (1997). Screening of European coffee final products for occurrence of ochratoxin A (OTA). Food Addit. Contamin. 14:211-216.
103. Maaroufi, K., Achour, A., Zakharna, F., Ellouz, M., ELMay, E.E. and Creppy, Bacha, H. (1996). Human nephropathy related to ochratoxin A in Tunisia. J. Toxicol.-Toxin Reviews. 15: 223-237.
104. Krogh, P. (1976). Epidemiology of mycotoxic porcine nephropathy. Nord. Vet. Med. 28: 452-458.
105. Petkova-Bocharova, T. and Castegnaro, M. (1985). Ochratoxin A contamination of cereals in an area of high incidence of Balkan endemic nephropathy in Bulgaria. Food Addit. Contam. 2:267-270.
106. Huang, X. Dorner, J. W. and Chu, F. S. (1994). Production of aflatoxin and cyclopiazonic acid by various aspergilli: An ELISA analysis. Mycotoxin Res. 10: 101-106.
107. Riley, R. T. and Goeger, D. E. (1991). Cyclopiazonic acid, Speculation on its function in fungi. In (Bhatnagar,D., Lillehoj, E.B., Arora, D.K. Eds.), pp. 385-402. Handbook of Applied Mycology: Vol. V, Mycotoxins in Ecological Systems. Marcel Dekker, Inc., New York.
108. Gallagher, R. T., Richard, J. L., Stahr, H. M. and Cole, R. J. (1978). Cyclopiazonic acid production by aflatoxigenic and non-aflatoxigenic strains of Aspergillus flavus. Mycopathologia. 66: 31-36.
109. Tournas, V. (1994). Heat-resistant fungi of importance to the food and beverage industry. Crit. Rev. Microbiol.. 20: 243-263.

110. Sorenson, W. G., Simpson, J. and Castranova, V. (1985). Toxicity of the mycotoxin patulin for rat peritoneal macrophages. Envir. Res. 38: 407-416.
111. Scott, P.M. (1975). Patulin. In (Purchase, I.F.H. Ed.), p. 383-403. Mycotoxins, Elsevier Scientific Publishing Co., New York.
112. Wilson, B.J. and Hayes, A.W. (1973). Microbial toxins. In Toxicants Occurring Naturally in Foods. pp. 372-423. NRC, U. S. Natl. Acad. Sci., Washington, DC.
113. Cole, R.J. and Cox, E. H. (1981). Handbook of Toxic Fungal Metabolites. Academic Press, New York.
114. Pastrana, L., Loret, M.O., Blanc, P. J. and Goma, G. (1996). Production of citrinin by *Monascus ruber* submerged culture in chemical defined media. Acta Biotech. Environ. Microbiol. 16: 315-319.
115. Betina, V. (1989). Mycotoxins: Chemical Biological and Environmental Aspects. Elsevier Publisher, Amsterdam-Oxford-NY.-Tokyo.
116. Chu, F.S. (1991). Mycotoxins: food contamination, mechanism, carcinogenic potential and preventive measures. Mutat. Res. 259: 291-306.
117. Mori, H. and Kawai, K. (1989). Genotoxicity in rodent hepatocytes and carcinogenicity of mycotoxins and related chemicals. In (Natori, S., Hashimoto, K. and Ueno, Y. Eds.), pp. 81-90. Mycotoxins and Phycotoxins 88, Elsevier, Amsterdam.
118. Van der Watt, J.J. (1977). Sterigmatocystin. In (Purchase, I.F.H. Ed.), pp. 368-382. Mycotoxins, Elsevier Scientific Publishing Co., New York.
119. Jelinek, C.F., Pohland, A.E. and Wood, G.E. (1989). Worldwide occurrence of mycotoxins in foods and feeds-an update. J. Assoc. Off. Anal. Chem. 72: 223-230.
120. Zhang, R.F., Chen, C.S., Yu, L., Sun, H.L., Fu, C.G. and Xu, H.D. (1988). *Aspergillus versicolor* and sterigmatocystin might be related to genesis of gastric cancer. Abstract 7th International Union of Pure and Applied Chemistry (IUPAC), International Symposium on Mycotoxins and Phytotoxins, Tokyo, Japan.
121. Dutton, M.F. (1996). Fumonisins, mycotoxins of increasing importance: Their nature and their effects. Pharmacol. Therap. 70:137-161.
122. Jackson, L., DeVries, J.W. and Bullerman, L.B. (1996). (Eds.), Fumonisins in Food. Plenum, New York.
123. Riley, R.T. and Richard J. L. (1992) (Eds.), Fumonisins: A current perspective and view to the future. Mycopathologia 117: 1-124.
124. Riley, R.T. , Norred, W. P. and Bacon, C.W. (1993). Fungal toxins in foods: recent concerns. Ann. Rev. Nutr. 13:167-189.
125. Scott, P.M. (1993). Fumonisins. Internatl. J. Food Microbiol. 18: 257-270.
126. Gelderblom, W.C.A., Jaskiewicz, K., Marasas, W.F.O., Thiel, P.G., Horak, R.M. Vleggar, R. and Krek, N.P.J. (1988). Fumonisins-novel mycotoxins with cancer-promoting activity produced by *Fusarium moniliforme.* Appl. Environ. Microbiol. 54: 1806-1811.
127. Gelderblom, W.C., Marasas, W.F.O., Vleggaar, R., Thiel, P.G. and Cawood, M.E. (1992). Fumonisins: isolation, chemical characterization and biological effects. Mycopathologia. 117: 11-16.
128. Marasas, W.F.O., Nelson, P.E. and Toussoun, T.A. Toxigenic *Fusarium* species: identity and mycotoxicology. Penn. State Univ. Press, University Park, PA.
129. Nelson, P.E., Desjardins, A.E. and Plattner, R.D. (1993). Fumonisins: mycotoxins produced by *Fusarium* species: biology, chemistry, and significance. Ann. Rev. Phytopathol. 31: 233-252.
130. Norred, W.P. (1993). Fumonisins - Mycotoxins produced by *Fusarium moniliforme.* J. Toxicol. Environ. Health. 38: 309-328.
131. Shier, W.T. (1992). Sphingosine analogs: an emerging new class of toxins that includes the fumonisins. J. Toxicol.-Toxin Rev. 11: 241-257.
132. Seo, J.A., Kim, J.C. and Lee, Y.W. (1996). Isolation and characterization of two new type C fumonisins produced by *Fusarium oxysporum.* J. Nat. Prod. 59: 1003-1005.
133. Musser, S.M., Gay, M.L., Mazzola, E.P. and Plattner, R.D. (1996). Identification of a new series of furmonisins containing 3-hydroxypyridine. J. Pat. Prod. 59: 970-972.
134. Abbas, H.K., Tanaka, T. and Shier, W.T. (1995). Biological activities of synthetic analogues of *Alternaria alternata* toxin (AAL) and fumonisin in plant and mammalian cells. Phytochemistry 40: 1681-1689.
135. Chen, J., Mirocha, C.J., Xie, W., Hogge, L. and Olson, D. (1992). Production of the mycotoxin fumonisin B1 by *Alternaria alternata f. sp lycopersici.* Appl. Environ. Microbiol. 58. 3928-3931.
136. Abbas, H.K. and Riley, R.T. (1996). The presence and phytotoxicity of fumonisins and AAL-toxin in *Alternaria alternata.* Toxicon. 34: 133-136.

137. Mirocha, C.J., Gilchrist, D.G., Shier, W.T., Abbas, H.K., Wen,Y. and Vesonder, R.F. (1992). AAL toxins, fumonisins (biology and chemistry) and host-specificity concepts. Mycopathologia 117: 47-56.
138. Hammer, P., Bluthgen, A. and Walte, H.G. (1996). Carry-over of fumonisin B_1 into milk of lactating cows. Milchwissenschaft 51: 691-695.
139. Richard, J.L., Meerdink, G., Maragos, C.M., Tumbleson, M., Bordson, G., Rice, L.G. and Ross, P.F. (1996). Absence of detectable furnonisins in the milk of cows fed *Fusarium proliferatum* (Matsushima) Nirenberg culture material. Mycopathologia 133: 123-126.
140. Proctor, R.H. (2000). Fusarium toxins: Trichothecenes and Fumonisins. In (Cary, J. W., Linz, J.E., Bhatnagar, D. Eds.), pp. 363-381. Mcicrobial Foodborne Diseases: Mechanism of pathogenesis and toxin synthesis. Technomic Publishing Co. Inc., Lancaster, PA.
141. Desjardins, A.E., Plattner, R.D. and Proctor, R.H. (1996). Linkage among genes responsible for fumonisin biosynthesis in *Gibberella fujikuroi* mating population A. Appl. Environ. Microbiol. 62: 2571-2576.
142. Xu, J.R. and Leslie, J.F. (1996). A genetic map of *Gibberella fujikuroi* mating population A *(Fusarium moniliforme).* Genetics 143: 175-189.
143. Beasley, V.R. (1989). Trichothecene Mycotoxicosis, Pathophysiologic Effects. 175 (Vol I) and 198 (Vol II). CRC Press, Inc, Boca Raton, FL.
144. Miller, J.D. and Trenholm, H.L. (1994). Mycotoxins in grain: compounds other than aflatoxin. Eagan Press. St. Paul, MN.
145. Gelderblom, W.C.A., Krick, N.P.J., Marasas, W.F.O. and Thiel, P.G. (1991). Toxicity and carcinogenicity of the *Fusarium moniliforme* metabolite, fumonisin B_1 in rats. Carcinogenesis 12: 1247-1251.
146. Gelderblom, W.C., Semple, A.E., Marasas, W.F.O. and Farber, E. (1992). The cancer-initiating potential of the fumonisin B mycotoxins. Carcinogenesis 13: 433-437.
147. Gelderblom, W.C.A., Cawood, M.E. Snyman, S.D. and Marasas, W.F.O. (1993). Structure-activity relationships of fumonisins in short-term carcinogenesis and cytotoxicity assays. Food Chem. Toxicol. 31: 407-414.
148. Gelderblom, W.C.A., Cawood, M.E., Snyman, S.D. and Marasas, W.F.O. (1994). Fumonisin B_1 dosimetry in relation to cancer initiation in rat liver. Carcinogenesis 15: 209-214.
149. Norred, W.P., Voss, K.A. (1994). Toxicity and role of fumonisins in animal diseases and human esophageal cancer. J. Food Prot. 57: 522-527.
150. Chu, F.S. (1997). Trichothecene mycotoxicosis. In (Dulbecco, R. Ed.), pp. 511-522. Encyclopedia of Human Biology, second edition, Vol. 8. Academic Press, New York.
151. Jarvis, B.B., Salenime, J. and Morais, A. (1995). *Stachybotrys toxins.* 1. Nat. Toxins 3: 10-16.
152. Marasas, W.F.O., Nelson, P.E. and Toussoun, T.A. (1986). Toxigenic *Fusarium* species: Identity and Mycotoxicology. Penn. State Univ. Press. University Park, PA.
153. Ueno, Y. (1983). Trichothecenes: Chemical, Biological, and Toxicological aspects. Elsevier Publisher, Amsterdam.
154. Vesonder, R.F. and Golinski, P. (1989). In (Chelkowski, J. Ed.), pp. 1-39. Fusarium mycotoxins, Taxonomy, and Pathogenicity, Elsevier, New York.
155 Desjardins, A.E., Hohn, T.M. and McCormick, S.P. (1993). Trichothecene biosynthesis in *Fusarium* species: Chemistry, genetics and significance. Microbiol. Rev. 57: 595-604.
156. Hohn, T.M., McCormick, S.P. and Desjardins, A. E. (1993). Evidence for a gene cluster involving trichothecene-pathway biosynthetic gene in *Fusarium sporotrichioides.* Curr. Genet. 24: 291-295.
157. Keller, N.P. and Hohn, T.M. (1997). Metabolic pathway gene clusters in filamentous fungi. Fungal Genetics Biology 21: 17-29.
158. Proctor, R.H., Desjardins, A.E., McCormick, S.P. and Hohn, T.M. (1995). Trichothecene toxins and wheat head scab. Proceedings of the USDA-ARS *Fusarium*/Fumonisin Workshop, Beltsville, MD, 30.
159. Park, J., Smalley, E. B. and Chu, F.S. (1996). Natural occurrence of *Fusarium* mycotoxins of the 1992 corn crop in the field. Appl. Environ. Microbiol. 62: 1642-1648.
160. Anon. (1983). Protection against trichothecenes. Reported by Committee on Protection Against Mycotoxins, National Research Council, National Academy of Science, National Academy Press, Washington, DC, 17-166.
161. Rotter, B.A., Prelusky, D.B. and Pestka, J.J. (1996). Toxicology of deoxynivalenol (vomitoxin). J. Toxicol. Environ. Health 48: 1-34.
162. Trenholm, H.L., Prelusky, D.B.,Young, J.C. and Miller, J.D. (1988). Reducing mycotoxins in animal feeds, Agriculture Canada Publication 1827E. Agriculture Canada, Ottawa, Canada.
163. Luo, X.Y. (1988). Fusarium toxin contamination of cereals in China. Proc. Jpn. Assoc. Mycotoxicol. Special issue No. 1, 97-98.

164. Hendry, K.W. and Cole, E.C. (1993). A review of mycotoxins in indoor air. Journal of Toxicological Environmental Health, 38: 183-198.
165. Nikulin, M., Pasanen, A. L., Berg, S. and Hintikka, E.L. (1994). *Stachybotrys atra* growth and toxin production in some building materials and fodder under different relative humidities. Appl. Environ. Microbiol. 60: 3421-3424.
166. Lisker, N. and Lillehoj, E. B. (1991). Prevention of mycotoxin contamination (principally aflatoxins and *Fusarium* toxins) at the preharvest stage. In (Smith, J.E. and Henderson, R.S. Eds.), pp. 689-719. Mycotoxins and Animal Foods. Boca Raton: CRC Press.
167. Sinha, K.K. and Bhatnagar, D. (1988). (Eds.), Mycotoxin in Agriculture and Food Safety. Marcel Dekker, New York.
168. Payne, G.A. (1998). Process of contamination by aflatoxin - producing fungi and their impact on crops. In (Sinha, K.K. and Bhatnagar, D. Eds.), pp 278-306. Mycotoxins in Agriculture and Food Safety. Marcel Dekker, New York.
169. Sinha, K.K. (1998). Detoxification of mycotoxins and food safety. In (Sinha, K.K. and Bhatnagar, D. Eds.), pp. 381-406. Mycotoxin in Agriculture and Food Safety. Marcel Dekker, New York.
170. Brown, R.L., Bhatnagar, D., Cleveland, T.E. and Cary, J.W. (1998). Recent advances in preharvest prevention of mycotoxin contamination. In (Sinha, K.K. and Bhatnagar, D. Eds.), pp. 351-380. Mycotoxins in Agriculture and Food Safety. Marcel Dekker, New York.
171. Desjardins, A.E., Plattner, R.D., Nelson, T.C. and Leslie, J. (1995). Genetic analysis of fumonisin production and virulence of *Gibberella fujikuroi* mating population A (*Fusarium moniliforme)* on maize (*Zea mays)* seedlings. Appl Environ Microbiol. 61: 79-86.
172. Hohn, T.M., Desjardins, A. E., McCormick, S.P. and Proctor, R.H. (1995). Biosynthesis of trichothecenes, genetic and molecular aspects. In (Eklund, M., Richard, J.L. and Mise, K. Eds.), pp. 239-248. Molecular Approaches to Food Safety: Issues Involving Toxic Microorganisms. Alaken, Inc., Ft. Collins, CO.
173. Cotty, P.J. and Bhatnagar, D. (1994). Variability among atoxigenic *Aspergillus flavus* strains in ability to prevent aflatoxin contamination and production of aflatoxin biosynthetic pathway enzymes. Appl. Environ. Microbiol. 2248-2251.
174. Muhitch, M.J. (1995). A genetic engineering approach to lowering fumonisin levels in maize kernels. Proceedings of the USDA-ARS *Fusarium*/Fumonisin Workshop, Beltsville, MD, 27.
175. Chen, Z.-Y., Cleveland, T.E., Brown, R.L., Bhatnagar, D., Cary, J.W. and Rajasekaran, K. (2000). Corn as a source of antifungal genes for genetic engineering of crops for resistance to aflatoxin contamination. American Chemical Society Publication (in Press).
176. FAO. (1979). Prevention of mycotoxins. FAO Food and Nutrition Paper 10, 71.
177. Phillips, T.D., Clement, B.A. and Park, D.L. (1994). Approaches to reduction of aflatoxin in foods and feeds. In (Eaton, D.L. and Groopman, J. D. Eds.), pp. 383-406. The Toxicology of Aflatoxins: Human Health, Veterinary and Agricultural Significance. Academic Press, San Diego.
178. Haumann, F. (1995). Eradicating mycotoxins in food and feeds. INFORM. 6 248-257.
179. Sharma, A. (1998). Mycotoxins-Risk evaluation and management in radiation-processed food. In (Sinha, KK. and Bhatnagar, D. (eds), pp. 435-458. Mycotoxin in Agriculture and Food Safety. Marcel Dekker, New York.
180. Goldblatt, L.A. and Dollear, F.G. (1977). Review of prevention, elimination and detoxification of aflatoxins. Pure Appl. Chem. 49: 1759-1764.
181. J.G. Heathcote, J.R. Hibbert, Biochemical effects, structure activity relationships: *in* L.A. Goldblatt (ed.), Aflatoxin: Chemical and Biological Aspects. Amsterdam, Elsevier (1978) 112-130.
182. Dickens, J.W. (1977). Aflatoxin control programme for peanuts. J. Am. Oil Chem. Soc. 54:225A-228A.
183. Lopez-Garcia, R. and Park, D.L. (1998). Effectiveness of post-harvest procedures in management of mycotoxin hazards. In (Sinha, K.K. and Bhatnagar, D. Eds.), pp. 407-434. Mcycotoxin in Agriculture and Food Safety. Marcel Dekker, New York.
184. Bhatnagar, D., Lillehoj, E.B. and Bennett, J.W. (1991). Biological detoxification of mycotoxin. In (Smith, J.E. and Henderson, R.S. Eds.), pp. 815-826. Mycotoxins and Animal Foods. Boca Raton, CRC Press.
185. Newberne, P.M. (1987). Interaction of nutrients and other factors with mycotoxins. In (Krogh, P. Ed.), pp. 177-216. Mycotoxins in Food, Academic Press, New York.
186. Dashwood, R.H., Arbogast, A., Fong, T., Perieira, C., Hendricks, J.D. and Bailey, G.S. (1989). Quantitative inter-relationships between aflatoxin B_1 carcinogen dose, indole-3-arbinol anti-carcinogen dose, target organ DNA adduction and final tumor response. Carcinogenesis 10: 175-181.
187. Whitty, J.P. and Bjeldanes, LF. (1987). The effects of dietary cabbage on xenobiotic-metabolizing enzymes and the binding of aflatoxin B_1 to hepatic DNA in rats. Food Chem. Toxicol. 25: 581-587.

188. Wattenberg, L.W. (1986). Protective effects of 2(3)-tert-butyl 4-hydroxanisole on chemical carcinogensis. Flood Chem. Toxicol. 24: 1099-1102.
189. Williams, G. M., Tanaka, T. and Maeura, Y.(1986). Dose-related inhibition of aflatoxin B_1 induced hepatocarcinogenesis by the phenolic antioxidants, butylated hydroxyanisole and butylated hydroxtoluene, Carcinogenesis 7: 1043-1050.
190. Cabral, J.R.P. and Neal, G.E. (1983). The inhibitory effects of ethoxyquin on the carcinogenic action of aflatoxin B_1 in rats. Cancer Lett. 19: 125-132.
191. Buetler, T.M., Bammler, T.K., Hayes, J.D. and Eaton, D.L. (1996). Oltipraz-mediated changes in aflatoxin B_1 biotransformation in rat liver: implication for human chemo intervention. Cancer Research 56: 2306-2313.
192. Qin, G., Gopalan-Kirczky, P., Su, J., Ning, Y. and Lotlikar, P.D. (1997). Inhibition of aflatoxin B_1 -induced initiation of hepatocarcinogenesis in the rat by green tea. Cancer Lett. 112: 149-154.
193. Jhee, E.C., Ho, L.L., Tsuji, K., Gopalan, P. and Lotlikar, P.D. (1989). Effect of butylated hydroxyanisole pretreatment on aflatoxin B_1-DNA binding and aflatoxin B_1-glutathione conjugation in isolated hepatocytes from rats. Cancer Res. 49: 1357-1360.
194. Bose, S. and Sinha, S.P. (1994). Modulation of ochratoxin produced genotoxicity in mice by vitamin C. Food Chem. Toxicol. 32:533-537.
195. Hoehler, D. and Marquardt, R.R. (1996). Influence of vitamins E and C on the toxic effects of ochratoxm A and T-2 toxin in chicks. Poultry Sci. 75: 1508-1515.
196. Netke, S.P., Roomi, M. W. , Tsao, C. A. (1997). Niedzwiecki, Ascorbic acid protects guinea pigs from acute aflatoxin toxicity. Toxicol. Appl. Pharmacol. 143: 429-435.
197. Creppy, E.E., Baudrimont, I., Belmadani, A. and Betbeder, A.M. (1996). Aspartame as a preventive agent of chronic toxic effects of ochratoxin A in experimental animals. J. Toxicol. -Toxin Reviews, 15: 207-221.
198. Beaver, R.W., Wilson, D.M., James, M.A., Haydon, K.D., Colvin, B.M., Sangster, L.T., Pikul, A.H. and Groopman, J.D. (1990). Distribution of aflatoxin in tissues of growing pigs fed an aflatoxin contaminated diet amended with a high affinity aluminosilicate sorbent. Vet. Hum. Toxicol. 32: 16-18.
199. Colvin, B.M., Sangster, L.T., Haydon, K.D., Beaver, R.W. and Wilson, DM. (1989). Effect of high affinity aluminosilicate sorbent on prevention of aflatoxicosis in growing pigs. Vet. Hum. Toxicol. 31: 46-48.
200. Harvey, R.B., Kubena, L.F., Philips, T.D., Huff, W.E. and Corrier, D.E. (1989). Prevention of aflatoxicosis by addition of hydrated sodium calcium aluminosilicate to the diets of growing barrows. Am. J. Vet. Res. 50: 416-420.
201. Kubena, L.F., Harvey, R.B., Huff, W.E., Elissalde, M.H., Yersin, A.G., Philips, T.D. (1993). Efficacy of a hydrated sodium calcium aluminosilicate to reduce the toxicity of aflatoxin and diacetoxyscirpenol. Poultry Sci. 72: 51-59.
202. Smith, E.E., Phillips, T.D., Ellis, J.A., Harvey, R.B., Kubena, L.F., Thompson, J. and Newton, G. (1994). Dietary hydrated sodium calcium aluminosilicate reduction of aflatoxin M_1 residue in dairy goat milk and effects on milk production and components. J. Anim. Sci. 72: 677-682.
203. Desjardins, A. E. and Hohn. (1997). Mycotoxins in plant pathogenesis. Molec. Plant-Microbe Interact 10: 147-152.
204. Harris, L. J., Desjardins, Ac. E., Plattner, R. D., Nicholson, P., Butler, G., Young, J. C., Weston, G., Proctor, R. H. and Hohn, T. H. (1999). Possible role of trichothecene mycotoxins in virulence of Fusarium graminearum on maize. Plant Dis. 83: 954-960.
205. Desjardins, A. E., Plattner, R. D. and Proctor, R. H. (2000). Maize ear rot and systemic infection by fumonisin B_1 -nonproducing mutants of Gibberella fujikuroi mating population A. Phytopathology (Abstract).

Applied Mycology and Biotechnology
Volume 2. Agriculture and Food Production

Emerging Strategies to Control Fungal Diseases in Vegetables

Padma K. Pandey and Koshlendra K. Pandey
Department of Plant Pathology, Indian Institute of Vegetable Research, 1 Gandhi Nagar (Naria), P.B.No. 5002, P.O. BHU, Varanasi 221 005, India (e-mail: pdveg@up.nic.in).

Plant biotechnology has made significant achievements in the past fifteen years encompassing within its fold the spectacular developments in plant molecular biology and genetic engineering. Biotechnological approaches are now being considered as the best options to develop disease resistant varieties in vegetable crops for effective management of fungal diseases. The molecular techniques allow the transfer of resistance traits into plants without altering their intrinsic properties. The major approaches focus on the production of antifungal compounds such as antifungal proteins, hydrolytic enzymes and toxins that directly affect the growth and development of the fungus in transgenic vegetable crops. Strategies to circumvent the problems of gene transfer from transgenic plants via pollen are being developed. The another approach is to develop a hypersensitive plant response leading to cell death around the pathogen-infected area and thereby isolating the infected tissue from the healthy tissues. Several antifungal polypeptides have been described and found to inhibit fungal growth *in vitro* and *in vivo*. Combinations of more than two compatible PR proteins act in a synergistic way and displayed a tremendous reduction in fungal growth. However, the degree of resistance was not very high and limited for certain fungi only. Long term durability and overall performance of disease resistant transgenic plants could be developed by utilizing the multiple cloned resistance genes or new genes encoding proteins. Many transgenic plants in potato against *Phytophthora infestans* and tomato against *Cladospirum fulvum* have been developed successfully showing severe localized necrosis after inoculation of the respective fungi. Pathogen induced cell death at site of infection and hypersensitive cell death through oxidative burst will have more potential for pathogen resistance in future. Developing world must adopts these fast-changing technologies and harness their unprecedented potential for the benefit of the mankind.

1. INTRODUCTION

Vegetables are important source of dietary proteins, minerals and vitamins. Shifting from a non-vegetarian diet to vegetarian, global recognition of the importance of vegetables for human health and their medicinal and nutritional value have contributed to a steady upward trend in vegetable production. China is ranked first in the world and currently produces 237 million tons of vegetable [1]. The total vegetable production of India during the year 1999-2000 was approximately 90 million tons [2]. India has made a quantum leap in vegetable production, securing second position in the world [2]. Other major vegetable producing countries are Japan, the Republic of Korea and Iran. The rest of Asia and the Pacific produced only around 36 million tons in 1997 [1].

Fungi, bacteria, viruses and phytoplasma are responsible for causing diseases that result in reduction of yield and quality. The abiotic factors like sunscald, frost injury and nutritional disorders are also indirectly associated with the vegetable diseases. Introduction of hybrid vegetable crops has created several new disease problems like *Pseudocercospora* leaf blight in tomato and brinjal in India [3]. Intensive culturing and use of monocultures of vegetables have also increased the disease pressure [3]. Moreover, the off-season vegetable growing practices have prolonged the survival period of pathogens in the field [3]. Seed borne diseases and the inoculum of pathogen in vegetable crops have increased throughout the South-East Asian countries due to Open Government License (OGL) and General Agreement of Tariff and Trade (GATT) agreements [3]. Disease pressure in the standing crop from the seedling stage to harvest, spoilage caused by microorganisms during transit, storage and marketing are one of the main constraints in vegetable production. It has been estimated that a total of 65 million tons of pre-harvest losses of vegetables (10.5%) are due to disease alone [4]. Pre-harvest losses mean that the quality of produce was affected by indirect physiological interference of the pathogen or it was directly destroyed by disease before harvesting. A maximum loss of 44 million tons has been reported for developing countries [4]. Post-harvest losses are also of much concern and are estimated to be about 20-40% of the total fruit and vegetable production, costing more than 652 million US dollars annually. A very high percentage of water, relatively high metabolic activity and inherent short self-life of vegetables make them highly perishable commodities. These characteristics propose a greatest problem in the successful transportation, storage and marketing. Among pathogens, fungi are the main reason for the extensive damage of vegetable crops especially in tropical and subtropical countries.

Presently, the control of fungal vegetable diseases is mainly based on the application of agro-chemicals, crop rotation and the use of pathogen-resistant varieties. Despite the proven efficiency of agro-chemicals in the prevention and spread of the fungal diseases, their high cost makes the vegetable production very expensive for the marginal farmers of the developing countries of South East Asia [5]. Moreover, chemical residues remaining in the vegetables, especially in celery, cucumber, lettuce, leafy vegetables, melons, tomato and okra form an important health hazard. Development of disease resistant and tolerant varieties of vegetable crops is one of the best options to minimize the losses. However, breeding of disease resistant varieties of vegetable crops is time-consuming and new races of pathogens evolve continuously. The process of co-evolution of plants and pathogens has led to incompatibility in many cases determined by a single dominant gene for resistance in the host. Incompatibility is associated with the defence response genes where products may include, (i) hydrolytic enzymes such as chitinase, β -1,3 glucanase and other pathogenesis related (PR) proteins, (ii) ribosome inactivating proteins (RIP), (iii) antifungal proteins (AFP), (iv) biosynthetic enzymes for the production of antimicrobial phytoalexins, (v) wall bound phenolics, osmotins, thionions, lectins etc, and (vi) hydrogen peroxide. Plant genetic manipulation through expression of either novel proteins from foreign organisms or over expression of a part of their own defensive arsenal for disease resistance has become a reality. Therefore, new strategies based on the biotechnological approaches should be developed to protect vegetable crops from diseases.

Revolutionary methods based on biotechnology to protect plants were first applied to tobacco by introducing genes encoding viral coat proteins to produce plants resistant to virus infection [6]. Development of tobacco varieties resistant to fungal disease for the first time involved introduction of *Phaseolus vulgaris* genes encoding PR proteins into tobacco, making it resistant to the fungal pathogen *Rhizoctonia solani* [7]. Biotechnology has been also successfully applied to tomato for its protection against tomato mosaic virus. Today, biotechnological methods are frequently applied to develop disease resistance in vegetables.

Molecular techniques in general have proven a great success in developing vegetable varieties resistant to virus diseases. Significant results have also been achieved in improving bacterial disease resistance but very limited research has been carried into fungal disease resistance in vegetable crops. In this chapter, we will review and discuss the role of biotechnological tools to the control of fungal diseases of vegetable crops with major emphasis on disease resistance.

2. BIOTECHNOLOGICAL APPLICATIONS IN FUNGAL DISEASE RESISTANCE

Since new races of fungal pathogens evolve continuously, so traditional breeding for developing disease resistance in vegetable crops is neither possible nor feasible. Therefore, new strategies based on the application of molecular biological techniques have been introduced for the evaluation of fungal disease resistance varieties. These include restriction length polymorphism (RFLP), randomly amplified polymorphic DNA (RAPD) and amplified fragment length polymorphism (AFLP). These techniques are being explained in brief as below:

2.1 RFLP

Restriction Fragment Length Polymorphism (RFLP) analysis is based on two techniques that are widely used in modern molecular biology, the restriction endonuclear digestion of DNA and the transfer of DNA fragments to a filter into fragment. Type II restriction endonucleases of bacteria recognize and cut specific nucleotide notify in a DNA sequence (the enzymes commonly used for RFLP analysis recognize 4-6 base pair sequences). They are, therefore, capable of population of fragments with discrete sizes. In practice, fragments range in size from a few to more than several thousand base pair. A polymorphism in a restriction pattern occurs when the mutation creation, of a new restriction site, or when by insertion/deletion, the size of a restriction fragment is altered. When the RFLP patterns of nuclear DNA are examined in segregating populations they behave like classical co-dominant genetic markers and can be used to create RFLP linkage maps. In vegetable, linkage analysis of DNA markers has been accomplished in tomato [8].

2.2 RAPD

Towards the end of 1990 two teams simultaneously reported the development of novel molecular genetic techniques based upon the polymerase chain reaction (PCR). These were capable of producing diagnostic fingerprints of any genomic DNA sample, by non-radioactive means, within a short period of time [9]. Although the two techniques were virtually identified the researcher coined different terms for the procedure, hence both "Arbitrary Primed PCR" (AP-PCR) and "Randomly Amplified Polymorphic DNAs" (RAPDs) were chosen to describe essentially the same technique. Following introduction, the RAPDs technique found immediate favour with molecular and population biologists due to its simple and easy protocol. The technique relies on the fact that whereas the standard PCR requires two different oligonucleotides where base composition is fixed by the sequence of the fragment to be amplified, RAPDs require only the presence of a single randomly chosen oligonucleotides. RAPDs technology is being used to screen genomic DNA from individuals for variation in sequence. Their sequence changes may result in a change in the pattern of amplification products following agarose gel electrophoresis. This makes RAPDs a very useful technique not only for screening populations for sequence diversity but also, when used in conjunction with bulk segregate analysis for generating molecular markers from specific regions of the genome without any requirement for a detailed genetic map.

2.3 AFLP

The AFLP technology [10] provides a new powerful tool for the detection and evaluation of genetic variation, and in tagging of resistant genes. The AFLP technique is based on the principle of selectively amplifying a subset of restriction fragments from a complex mixture of DNA fragments obtained after digestion of genomic DNA with restriction endonuleases. Polymorphisms are then detected by differences in the length of the amplified fragments by polyacrylamide gel electrophoresis (PAGE).

Using these techniques, linkage of a particular resistant gene to a specific chromosome fragment can be established and a specific fungal disease resistance trait can be selected. Molecular technology allows the transfer of DNA encoding disease resistance into various vegetables without altering their intrinsic properties. Therefore, the knowledge of the molecular mechanisms of fungal pathogenicity and disease resistance combined with molecular biology and genetic engineering methods offer a promising alternative to produce new varieties of vegetable crops resistant to fungal diseases. It may be possible to develop resistance to fungal diseases using resistance genes isolated from other microorganisms. For example, genes responsible for the production of phenazine and phloroglucinol antibiotics have been expressed as part of the plant genome to create disease resistance [11]. An important approach is to introduce genes encoding proteins with antifungal activity into the plant genome therefore enabling the plant to protect itself against fungal attack. The use of multiple genetic markers is necessary for the vegetable breeders to follow the inheritance of several resistance genes. For example, there are at least 15 genes that can result in resistance to various strains of downy mildew in lettuce. In order to develop resistant cultivars, which possess several of these genes to give a broad-spectrum resistance, the rare individuals who inherit all of the desired genes from both parents must be identified. In order to accomplish this, screening of each seedling against all pathogenic strains is very difficult through conventional methods. Alternatively, seedlings can be screened by using techniques such as RFLP to establish linkage of a specific molecular marker (resistance gene) with a particular plant variant. Only those seedlings possessing all of the desired genes will then be grown for the next generation. Examples of some markers linked to fungal disease resistance genes in vegetable crops are summarized in table-1.

Table 1. Linkage of resistance genes to plant varieties resistant to fungal disease

Host	Pathogens	Genes	References
Tomato	*Cladosporium fulvum*	*Cf9, Cf8, Cf4*	[12]
	Leveillula taurica	*Lv*	[13]
	Oidium lycopersicon	*Ol*	[14]
	Phytophthora infestans	*Ph 2*	[15]
	Verticillium dahliae	*Ve*	[16]
Potato	*Phytophthora infestans*	*R1 & R3*	[17]
Pea	*Erysiphe polygoni*	*Er*	[18]
	Fusarium oxysporum	*Fw*	[18]
French bean	*Uromyces appendiculatus*	*Up2*	[14]
Lettuce	*Bremia lactuceae*	*DM 17 & DM 18*	[19]
		DM 8 and DM 10	[20]
	Plasmopara lactuceae	*Plr*	[21]
Chinese cabbage	*Plasmodiophora brassicae*		[22]

Gene mapping of *I-2*, a resistance locus to *Fusarium oxysporum* in tomato has been carried out successfully [23]. Genes conferring resistance to downy mildew and leaf blight in onion were cloned through its identification by RFLP and RAPD [24]. Specific genes that are associated with particular DNA separated by gel electrophoresis are these markers. In general, two major approaches have been followed to introduce resistance to fungal diseases into plants. The first focuses on the production of antifungal compounds such as antifungal proteins and toxins that directly affect the growth and development of the fungus in transgenic plants. The second approach aims for generation and improvement of plant responses leading to cell death around the pathogen-infected area and thereby isolation of the infected tissue from the healthy cells [25].

3. METHODS OF GENE TRANSFER

Plant genetic engineering involves techniques and approaches to introduce foreign genes into plants. This requires availability of suitable procedures to regenerate plant species of interest *in vitro*. Fortunately, advances made in tissue culture in the past thirty years have led to the development of techniques to regenerate a wide range of plant/crop species [26]. Based on such procedures, three major gene transfer techniques have been employed to transform various plant species. These are (i) *Agrobacterium* mediated approach, (ii) protoplast based approach and (iii) Biolistic approach. Many other methods have also been developed for gene transfers to plants which, are listed in table-2.

Table 2. Various approaches for gene transfer to plants

	Major Approaches	Minor Approaches
1.	*Agrobacterium tumefaciens*	*Agrobacterium rhizogenes*
2.	Gene transfer to protoplasts	Agroinfection Agrolistics
	PEG - mediated	
	Electroporation	Macro injection
	Liposome function	Pollen tube pathway
	Micro injection	Seed incubation Microlasers
3.	Biolistic method	Electrophoresis
	Microprojectiles	Liposome injection
	Agrolistic method	Silicone whislcers
	Microtargeting	

3.1 *Agrobacterium*-mediated Gene Transfer

A. tumefaciens is a gram-negative plant pathogenic bacterium, which causes crown gall disease in certain plant species. The tumors produced by the plants upon infection by the bacterium are because of the transfer of a sequent of oncogenic DNA into the plant cells at wound sites. This DNA segment (transfer DNA or T-DNA) is carried on large plasmids called Tumor inducing (Ti) plasmids in the bacterium. The T-DNA caries many genes encoding biosynthetic enzymes of hormones and nitrogenous metabolites called opines. The T-DNA integrates into the plant chromosomes by illegitimate recombination. Upon integration, the plant cells start proliferating and producing opines that are metabolized by the bacterium [27]. Thus, *Agrobacterium* creates a niche for its multiplication by genetically engineering the plant genome. This property is utilized by molecular biologists to manipulate plant genomes. "Superfluous" genes in T-DNA are replaced by genes of interest. The T-DNA is cloned in Ti-plasmid that are reduced in size and made to replicate in *Escherechia coli* as well, to facilitate molecular manipulations. These vectors are mobilized into *Agrobacterium* host strains and used to infect the plant tissues. The infected plant tissues are grown on media containing plant hormones to facilitate regeneration of transformed cells. The selection of transformed cells is aided by the presence of a suitable gene in T-DNA, which encodes an enzyme that detoxifies and antibiotic/herbicide/metabolite present in the tissue culture medium. The antibiotic resistant shoots are further grown, tested for foreign gene integration and agronomic performance. *Agrobacterium* has been successfully used to transform a wide variety of plant species including legumes, cereals, forest and tree species [28].

3.2 Direct Gene Transfer to Protoplasts

Protoplasts are cells devoid of the cell walls by enzymatic digestion. For many years, direct uptake of naked DNA by plant protoplasts has been practiced till *Agrobacterium* approach came into vogue. Co-incubation of protoplasts with the nucleic acids (plasmid DNA) and polyethylene glycol (PEG) leads to reversible permeabilization of the plasma membranes and allows the nucleic acid to enter the protoplasts [29, 30]. Electroporation (making apertures in plasma membranes by electrical pulses) of protoplasts or plant cells that are still surrounded by parts of or intact walls to facilitate DNA uptake is also feasible [31]. The production of transgenic plants by direct gene transfer to protoplasts depends on protoplast to plant regeneration protocol and an efficient selection system for transgenic clones.

3.3 Biolistic Gene Transfer

In many important crop species the limiting factors for gene transfer are the host specificity of *Agrobacterium* or the protoplast to plant regeneration protocols. Alternative approaches like bombardment of regenerable tissues with DNA coated microprojectiles at a very high velocity have resulted in considerable success. Many cereals, legumes and forest tree species have been transformed by adopting this strategy. The microprojectiles are usually made of gold or tungsten and are coated with DNA. Inside the cell, DNA molecules are released from the particles and eventually become integrated into the nuclear or organelle genome of the host cell. The tissue explants are later cultured on media containing hormones and antibiotics to select the transformed shoot [26]. In the future, the techniques to transform plant species will be more precise and direct. For instance, the penetration and distribution of the microprojectiles in biolistic gene transfer can be controlled by microbiolistic targeting approach. The instrument can direct microprojectiles to very small areas, requiring a stereomicroscope to aim at targets smaller than 0.15 nm in diameter. This approach has the potential for overcoming the genotype dependent problems of tissue culture.

4. TRANSGENIC VEGETABLES CONTAINING ANTIFUNGAL COMPOUNDS

Antifungal compounds are either proteins or secondary metabolites. Antifungal proteins include polypeptides, ribosome inhibiting proteins and plant-produced antibodies raised against essential fungal product [32]. Polypeptides affecting the integrity of the fungal cell wall include hydrolytic enzymes such as chitinases and antifungal metabolites include compounds such as phytoalexins. Antifungal proteins with the ability to inhibit the growth of fungi *in vitro* or to kill them are abundant in nature. Numerous antifungal polypeptides have been described to date and many of them seem to inhibit fungal growth in a synergistic manner both *in vitro* and *in planta* [32]. Genes coding for such antifungal proteins can be used to render the plant fungus-resistant. Genes encoding antifungal proteins like osmotin, glucanase, thionin, peroxidase and chitinase have been successfully introduced into vegetables, although the exact involvement of these genes in the defence response under *in vivo* conditions is not always known [32]. Studies into constitutive expression of genes encoding antifungal proteins in transgenic plants are in progress [32].

4.1 Pathogensis Related Proteins

Among the antifungal proteins, pathogenesis related (PR) proteins have been studied in more detail. PR proteins are plant proteins, which are induced upon pathogen attack or related situations, including wounding and application of elicitors or chemicals that induce host response [33]. PR proteins are believed to have a role in resistance to fungal diseases and defence responses [32]. They are effectively induced during a hypersensitive response (HR) in resistant or near immune cultivars of vegetables, concomitantly with the induction of systemically acquired resistance (SAR, discussed in section 8). Several PR proteins have been purified and found inhibitory to fungal growth on *in vitro* bioassay. Over expression of individual PR protein genes coding for chitinase and glucanase enzymes has been used to engineer disease-resistant varieties of vegetable crops. In this respect, successful field trials have been reported with carrot [34]. The best source for genes encoding PR proteins is tobacco (*Nicotiana tabacum*). Constitutive expression of the PR *Ia* gene of tobacco has resulted in enhanced resistance to *Peronospora* sp. Activity and biological function of some of the PR proteins have been characterized in detail. PR 2 (β-1, 3-glucanase) and PR 3 (chitinase) proteins hydrolyze glucan and chitin respectively, which are important structural components in the cell wall of most of the fungal pathogens [32]. β-glucanase and chitinase, when added to fungi *in vitro*, cause lysis of hyphal tips. PR 5 proteins, which interfere with permeability of the fungal cell wall also, cause lysis of fungal hyphae [35]. Genetically engineered plants expressing PR 8 protein genes show enhanced fungal resistance, however, antifungal activity of the proteins has not been shown *in vitro*. Despite the ability of certain PR proteins to inhibit growth of pathogenic fungi *in vitro*, transgenic plants constitutively express in these PR proteins did not show enhanced resistance against those fungi [32]. Some molecular approaches into fungal disease resistance of vegetables using single PR genes are listed in table-3.

Table 3. Single PR protein genes used in vegetable crops to achieve increased fungal resistance

PR genes	Source	Recipient	Fungus tested	References
PR 2 (Class I glucanase)	Tobacco	Tomato	*F.o.* f.sp. *lycopersici*	[36]
PR 3 (Class I chitinase)	Tobacco	Tomato	*F.o.* f.sp. *lycopersici*	[36]
PR 5	Tobacco	Potato	*P. infestans*	[35]
PR5	Tobacco	Carrot	*E. heraclei*	[34]
PR8 (Class III chitinase)	Cucumber	Tobacco	*R. solani*	[37]
PR 3 (Class I chitinase)	Bean	Tobacco	*R. solani*	[7]
PR 3(Class I chitinase)	Tomato, Tobacco	Rape seed	*Cylindrosporium concentricum* *Phoma lingum* *Sclerotinia sclerotiorum*	[38]

The combined effect of two or more PR protein genes has been studied in relation to the inhibition of fungal pathogens. In an experiment carried out combinations of different PR proteins displayed remarkable increase in the inhibition of fungal growth when compared to the growth inhibition by only a single PR protein gene product [39]. Consequently, there are a number of published reports in which constitutive co-expression of PR 2 and PR 3 genes resulted in an increase in the resistance against several fungi [34, 36]. This can be explained by a synergistic effect of the two gene products. Synergy between the PR protein genes encoding enzymes degrading the fungal cell wall occurs both *in vitro* and *in planta*. Some examples of synergy between gene products are given in table-4. A tomato variety Arka Vikas, susceptible to *Alternaria solani* was transformed with *A. tumefaciens* carrying an alfalfa glucanase gene and a rice chitinase gene in a *Ti* plasmid [40]. The presence of *npt* II as a selectable marker under the constitutive CaMV 35S promoter facilitated the selection and regeneration of a transformed plant from the cotyledon and hypocotyl explants [40]. Out of 37 transformants, only six could be established *in vivo* successfully. All the transformed tomato plants were morphologically normal and self-fertile. Two of them did not set fruit on their own. Manual pollination was required for one, while another had very low frequency of flowers with high pollen sterility. However, all the transformants expressed vigorous vegetative growth and delayed senescence [40]. In addition, osmotin gene from tobacco has been successfully used to enhance resistance to fungal diseases when constitutively expressed in tobacco or potato [35].

Table 4. Synergy of tobacco-derived PR genes encoding chitinase enzymes in transgenic tomato and carrot plants for increased resistance to pathogenic fungi

Gene 1	Source	Gene 2	Source	Recipient	Fungus tested	References
PR 3 (Class I Chitinase)	Tobacco	PR 2 (Class II glucanase)	Tobacco	Tomato	*F.o.* f.sp. *lycopersici*	[36]
PR 3 (Class I Chitinase)	Tobacco	PR 2 (Class I glucanase)	Tobacco	Carrot	*Alternaria dauci* *Cercospora carotae* *A. radicina* *E. heraclei*	[34]

4.2 Fungal Resistance to Cell Wall-degrading Enzymes

It has long been recognized that extracellular hydrolytic enzymes synthesized and released by soil microbes might play a role in the antagonism of phytopathogenic fungi. Expression of chitinases and glucanases, like other PR proteins are induced in host plants in response to pathogen attack to degrade the invading fungal cell wall [41]. Therefore, different clones of chitinases and glucanases of diverse source have been engineered into transgenic plants with varied degree of success against different fungal pathogen [7]. In particular, numerous correlations between fungal antagonism and bacterial production of chitinases and/or β-1, 3-glucanases have been noted [42, 43]. Chitin and β-1, 3-glucans are major constituents of many fungal cell walls and various workers have demonstrated *in vitro* lysis of fungal cell walls either by bacterial chitinase or β-1, 3-glucanases alone or by a combination of both enzymes [44, 45, 46]. Such studies have lent support to the hypothesis that these hydrolytic enzymes contribute to biocontrol efficacy.

Several genes, which encode cell wall degrading enzymes, mostly chitinases, have been isolated from plants and microorganisms. Chitinases have been shown to exhibit fungal growth inhibition *in vitro* [32]. Those pathogenic fungi that successfully attack a given vegetable have developed a way to resist or avoid the defence enzymes produced by the plant. A way around this evolutionary mechanism of host pathogen adoption may be the use of chitinases from bacterial origin. For example, strains of *Serratia marcescens* have been isolated, which have shown to be effective in the biocontrol of a number of plant pathogenic fungi, particularly *S. rolfsii* [45]. It was shown that this biocontrol relied on the production and excretion of a chitinase enzyme. This chitinase (*chi A*) was very stable, resistant to heat and inhibited fungal growth *in vitro*. Chitinase encoded by the *S. marcescens chi A* produced in *Escherichia coli* showed antifungal activity *in vitro* against *R. solani* [45]. The gene product was also shown to be active in transgenic plant [45]. These plants exhibited a markedly increased tolerance towards attack by the fungal pathogen *R. solani*.

The cloning of chitinase genes from *S. marcescens* QMB1466 allowed the disruption of the *chiA* locus and the recombinational exchange of the disrupted gene into the *S.* chromosome [47]. The *ChiA* deficient mutant demonstrated reduced chitinase production, reduced inhibition of fungal germ tube elongation and reduced biological control of *F.o.* f.sp. *pisi* infection of pea seedlings. It was proposed that chitinase most likely was acting in concert with an array of other compounds produced by *Serratia* to show fungal growth. Sudhinam and coworkers [48] cloned chitinase genes from *S. marcescens* BJL 200 and introduced one such chitinase gene into a *P. fluorescens* strain on a broad-host-range plasmid. Culture filtrates of a *Pseudomonas* transconjugant that expressed chitinase activities were reported to inhibit *in vitro* the germ tube growth of two *F. oxysporum* strains. Enhanced biocontrol of *F. o.* f.sp. *redolens* by the *Pseudomonas* transconjugant on radish in a growth chamber experiment was also observed. One isolate of *P. stutzeri* with native chitinase and β-1, 3-glucanases activities was identified recently [43] and demonstrated to inhibit growth of *F. solani in vitro.* Shapira [45] constructed an *E. coli* strain harbouring the *Serratia chiA* gene on a multicopy plasmid. Greenhouse experiments indicated that daily application of the chitinase producing *E. coli* strain reduced disease incidence caused by *S. rolfsii* on bean plants. The combinative effect between chitinase and glucanase has also been tested and found to increase the efficacy as it gives synergistic effect [49].

4.3 Cystein-rich Proteins

Another example of antifungal polypeptides is small cystein-rich proteins in which hevein, a chitin-binding protein of 43 amino acids has been derived from latex. The hevein precursor is homologous to the tobacco PR 4 protein [39]. Hevein and *Urtica dioica* agglutinin (UDA) are the only two chitin-binding plant lectins known to show growth inhibiting activity on several fungi *in vitro.* The hevein gene has been expressed in transgenic tomato plants [50]. However, despite poor processing of the hevein precursor, the transgenic tomato plant exhibited less disease symptoms. Another class of small cystein-rich protein is plant defensins. Plant defensins originally isolated from *Raphanus sativus* seeds are a small group of cysteine rich proteins that accumulate systemically at high levels after localized fungal infection have been proven to have antifungal activity *in vitro* [51]. At least two groups of plant defensins show *in vitro* antifungal activity against various fungi by inducing morphological changes such as increase in hyphal branching in fungi. Some defensins cause no morphological change in the fungal pathogen. Introduction of plant defensins genes, Rs-AFP2, under the control of CaMV 35S promoter in tobacco conferred enhanced resistance to the foliar pathogen *Alternaria longipes* [52].

4.4 RIPs and other Antifungal Proteins

Plant ribosome inactivating protein (RIPs) are antifungal proteins that inhibit protein elongation. 28S RNA is modified by the N-glycosidase activity of RIP, resulting in the elimination of an adenine residue and the inability of the 60S eukaryotic subunit to bind elongation factor 2. Plant RIPs inactivate foreign ribosomes of distantly related species and of other eukaryotes including fungi. A purified RIP from barley inhibits growth of several fungi *in vitro* [53]. Tobacco plants transgenic for the cDNA encoding the barley type 1 *RIP* gene under the control of potato *won1* promoter showed enhanced resistance to *R. solani* [27]. This resistance was increased due to synergism between RIP and either PR2 or PR3.

Production of proteinaceous antifungal toxins may render plant fungus resistance. KP1, KP4 and KP6 are killer toxins, which are secreted by three subtypes (P1, P4 and P6 respectively) of *Ustilago maydis,* a fungal pathogen of corn [54]. Each of these subtypes harbour a specific double-stranded RNA virus encoding the antifungal killer toxin. Fungal strains producing the toxins are naturally resistant to their own toxins, but susceptible to killer toxins produced by other strains. *U. maydis* strains are resistant to the toxin produced by themselves, but sensitive to the killer toxins of other strains. Recently, active killer toxins were produced in plants transgenic for either the KP4 or KP6 toxin gene [54]. In another approach, a hen egg white lysozyme (HEWL) gene has been expressed in transgenic potato and tobacco plants. The HEWL recovered from the transgenic tobacco plants exhibited antifungal activity against fungi containing chitin in their cell wall [55]. This approach has been proved effective against fungi like *B. cinerea*, *V. albo-atrum* and *R. solani* [55]. The cellulose containing fungal cell wall was not inhibited by HEWL [56].

4.5 Phytoalexins

Phytoalexins, a group of antimicrobial compounds of low molecular weight belonging predominantly to the families of the phenylpropanoids, isoprenoids and acetylenes are synthesized and accumulated in plants in response to microbial infections [56]. Phytoalexins are non-specific, multisite toxicants. When fungi are treated with phytoalexins their cytoplasm rapidly become granular and cell contents are lost, suggesting that membranes are one site of attack. For example, the carrot phytoalexin 6-methoxymellein interacted with the membranes of *Candida albicans* and caused leakage of cell contents. Phytoalexin biosynthesis in plant is complex and involves many biochemical pathways. Stilbenes are typical phytoalexins of *Vitis vinifera* and *Picea sitehensis* involved in the defence of there plant species against fungal attack [57]. The precursors for the biosynthesis of stilbenes are present in most plant species, including tobacco, but they lack the enzyme stilbene synthase. Hain and coworkers [58] introduced the gene enconding stilbene synthase from grapevine (*V. vinifera*) into tobacco plants. In tobacco the substrate for stilbene synthase is present, but stilbene synthase itself is absent. The expression of the stilbene synthase gene resulted in the production of resveratrol, a stilbene-type phytoalexin as a result of which organic plants showed enhanced resistance to *Bortrytis cinerea*. These results demonstrate that constitutive production of phytoalexins in heterologous plant systems may result in increased pathogen resistance.

Many transgenic plants developed through the introduction of genes encoding antifungal compounds have been successful in controlled laboratory conditions in many cases. However, this does not guarantee success in the field. Constitutive production of phytoalexins, toxins or antifungal proteins does not render fungus resistant plants to an absolute level and resistance observed for limited number of fungi only. Transgenic crops overexpressing a chitinase gene and found to be resistant to a number of fungal pathogens containing chitin in their cell walls. However, this is not effective against the fungi containing cellulose in their cell wall. A fungus may become adapted by modifying its cell wall composition leading it insensitive to chitinase produced by plant. It is also possible that some fungal strains, particularly sexually reproducing fungi, may evolve mechanism to detoxify certain phytoalexin [32].

5. TRANSGENIC IN VEGETABLES BY PATHOGEN-INDUCED CELL DEATH RESISTANCE

Other strategies were also developed to provide engineered plant with more durable resistance against broad-spectrum fungal pathogens infecting the same host. The strategy involves pathogen induced cell death and mechanisms based on defence reaction at site of infection on the host during incompatible host-pathogen interaction [32]. The strategy of hypersensitive cell death employing genetic engineering in response to fungal pathogen attack at the site of infection has been employed successfully. Under this approach, a bacterial ribonuclease gene (barnase) from *Bacillus amyloliquefeciens* and an inhibitor of barnase, barstar were introduced into potato under the control of fungal infection specific (prp 1-1) promoter and CaMV 35S promoter, respectively [59]. It has been postulated that the level of barstar is exceeded by the level of barnase only in the close vicinity of infection sites, thereby leading to cell death specifically in infected host tissue. On detached leaves from transgenic potato lines having the component systems, sporulation of the late blight fungus, *P. infestans* was considerably reduced. Many transgenic potato plants expressing both genes showed severe localized necrosis of tissue after inoculation by *P. infestans* resulting reduction in symptom developments.

Table 5. Transgenic plants generated in various vegetable crops against fungal resistance

Plants	Gene/protein	Target Pathogens	References
Brassica napus	Bean chitinase	*R. solani*	[7]
	Chimeric endocutinase by fusion of tomato and tobacco chitinase gene	*C. concentricum*, *S. sclerotiorum*, *P. lingam*	[38]
Potato	β-1, 3-endoglucanase from soybean antifungal proteins	*P. infestans*	[60]
Potato	Osmotin from tobacco	*P. infestans*	[35]
	Genes/proteins that induce artificial cell death	*P. infestans*	
Potato	Barnase and barstar	*P. infestans*	[59]
Potato	Glucose oxidase from *Aspergillus niger*	*P. infestans*	[61]

Many sexually produced fungus-plant interactions fit in the gene for gene model. In this case, transgenic plants are also produced with resistance gene (R) and for corresponding avirulence (*Avr*) gene under the control of pathogen inducible promoter [62]. It will impart resistance reaction upon pathogen attack provided all components necessary to complete the signal transduction pathway leading to resistance are present [63]. The constitutive expression of the gene encoding *Avr 9* elicitor from *C. fulvum* in tomato carrying the corresponding resistance gene *(cf 9)* resulted in delayed growth, necrosis and even complete cell death [64]. Introduction of resistance gene R from a plant variety resistant to a certain pathogen into susceptible varieties or even in other species is another strategy [64]. Development of a tomato transgenic plant lacking *cf 9* gene with this recently isolated gene might render them resistant to all races of *C. fulvum* producing *Avr 9* [64].

Hypersensitive cell death through oxidative burst (as a result of accumulation of H_2O_2 and other reactive oxygen species) is another potential tool to achieve broad range pathogen resistance [61]. Hydrogen peroxide shows direct inhibitory effect on microbial growth [61]. It also activates the production of phytoalexins and cross-linking of the hydroxyproline rich cell wall protein [61]. Active oxygen as H_2O_2 is produced during the oxidative burst, which shows incompatible pathogen-host interaction [61]. Glucose oxidase (GO) is an enzyme found in some fungi and catalyses the oxidation of glucose, thus producing H_2O_2 and gluconic acid [61]. Transformation of potato plants with a glucose oxidase gene from *A. niger* under CaMV 35S promoter led to resistance development against *P. infestans* [61]. Transgenic potato showed increased concentration of H_2O_2 with normal growth and less susceptible to *P. infestans* and *V. dahliae*. This strategy can be used to develop transgenic plants resistant to many fungal pathogens.

Induction of hypersensitive reaction with the production of an elicitor in response to infection by pathogen was adopted recently [46]. Transgenic tobacco plants were generated harbouring a fusion between the pathogen inducible tobacco *hsr* 203J gene promoter and a *P. cryptogea* gene encoding the highly active elicitor cryptogein [65]. Under non-induced conditions the transgene was silent but upon infection by virulent fungal pathogen, cryptogram production was stimulated which coincided with the fast induction of several defence genes at an around the infection sites. Induced elicitor production resulted in a localized necrosis that restricted further growth of the pathogen. The transgenic plants displayed enhanced resistance to fungal pathogen that were unrelated to *Phytophthora* species such as *T. basicola*, *E. cichoracearum* and *B. cinerea* [65].

Plant polygalacturonase inhibiting proteins (PGIPs) inhibit fungal endo-polygalacturonidases, thereby creating mixtures of oligogalacturonides of different sizes, among which larger molecular may be active as nonspecific elicitors, which shows defence response [55]. Transgenic plants could be developed which express PGIPs upon pathogen attack with increased resistance to fungal pathogen [55].

6. MECHANISMS OF FUNGAL DISEASE RESISTANCE IN TRANSGENIC PLANTS

Cell wall degrading enzymes like chitinases and β-1, 3-glucanases are known to be induced during fungal infection, treatment with fungal elicitors and play an active role in plant defence response [62,66]. The levels of translatable chitinase mRNAs increased in melon plants when infection takes place by the anthracnose fungus *C. lagenarium* [67]. Hypocotyls of the melons were locally inoculated by this pathogen showed increased activity of chitinase throughout the plant. These fungal elicitor treated plants were subsequently exposed to *C. lagenarium* resulted in increased resistance to infections. Necrosis of tissues was more limited in these plants after infection. This indicated that the increased level of chitinase activity was partially responsible for the induced systemic resistance against *C. lagenarium* [68]. Two distinct enzymes, β-1, 3-glucanases and six chitinases accumulate after infection of *P. infestans* or treatment of detached potato leaves with *P. infestans* elicitor. Infection of *C. fulvum* on tomato results in the accumulation of several pathogenesis related proteins in the apoplast [11]. *F. oxysporum* infected tomato roots showed induction of chitinase in the incompatible interaction [11]. In compatible and incompatible systems, the presence of chitinase was highly correlated with the distribution of the pathogen. The preferential association of the enzyme with altered segments of the fungal cell wall suggested that the action of chitinase might be preceded by coincident with hydrolytic enzyme β-1, 3-glucanases [69]. *In vitro* activity of chitinase and glucanase on many fungi by using protein extracts from pea pods infected with *F. s.* f.sp. *phaseoli* as well as purified enzymes and also observed growth inhibition of the pathogen by β-1, 3-glucanases [70]. It has been further reported that inhibitory effect of fungus mycelium is due to enzyme-catalyzed hydrolysis of chitin and glucans polymers in the cell wall of the growing hyphal tips [70]. In bean plants, chitinase activity is associated with a basic 30 KD protein, which is localized in the vacuole [71]. Infection of chitinase *E. coli* mediated β-glucuronidase (GUS) plants with fungal pathogen also resulted β-glucuronidase activity [11]. *B. cinerea* infection indicated that GUS activity was localized at the site of inoculation and a mild induction of chitinase promoter was also evident in close vicinity of the lesion. The fungal mycelia were found to be concentrated within the necrotic lesion. The induction pattern for the bean chitinase promoter in *B. cinerea* infected plants was not restricted to this pathogen [72]. Similar results were obtained in the *R. solani* and *S. rolfsii* infected transgenic plant [72]. Maximum GUS activity was near the infection site and decreased as distance increases from lesions.

The enzymes chitinase and β-1, 3-glucanases alone or in combination of both have been inhibited the growth of some pathogenic fungi *in vitro* [70]. The activation of the 5B promoter during fungal infection has indicated that chitinase activity is intimately associated with the response of the plant to pathogen infection [73]. The enhanced resistance of 35S-chitinase tobacco plants to *R. solani* infection appeared to be correlated with the level of bean chitinase expression. Plants containing higher levels of the bean polypeptide displayed a greater survival rate in fungal infested soil. When the 35S-chitinase plants were grown in the presence of *P. aphanidermatum* that lacks a chitin containing cell wall, no differences in survival was detected compared to control plants [71]. *R. solani* upon prolonged exposure to purified bean chitinase resulted structural modification in mycelium including retraction of plasmalemma, cell wall disruption and leakage of cytoplasm. In some transgenic plant reduction in fungal biomass, chitin breakdown and fungal lysis was observed [22]. Thus it is clear that transgenic plants containing high constitutive leaves of bean endo chitinase are more resistant to infection by the soilborne pathogen particularly *R. solani* than the plants which do not contain chimeric chitinase gene [22]. The resistance level in these plants against fungal resistance varies with fungal inoculum load and indicates quantitative resistance [22].

7. RACE IDENTIFICATION AND DISEASE RESISTANT GENES

In most of the fungal diseases of vegetable crops, resistant genes and existence of races were not identified at molecular level. If it could be properly characterized and cloned that would be a potential tool for effective disease control. Resistance to *F. o.* f.sp. *lycopersici* race 3 is controlled by one major locus *I3* tightly linked to *Got 2* and also resistance to race 1 and 2. Thus the *Got 2* locus can be used as a selective marker for resistance to all 3 races of tomato wilt [60]. In the case of cucumber anthracnose caused by *C. lagenarium* (*C. orbaculare*) cDNA was successfully equalized by differential hybridization of 1900 cDNA clones in the equalized cDNA library and RNA blot analysis of candidate clones, 11 independent cDNA clones were identified [74]. Total 124 markers were analyzed for linkage in 138 F_2 plants of cabbage and identified Quantitative Trait Loci (QTL) for resistance against club root disease caused by *Plasmodiophora brassicae* [75]. A number of dissociation tagged *cf-4* mutants, identified on the basis of their insensitivity to *Avr-4* were resistant to infection by *C. fulvum* (*Fluvia fulva*) race 5. Molecular analysis of 16 *cf*-4 mutants of tomato indicated that additional resistance specificity was encoded by *Hcr-9-4E* a novel *F. fulva* avirulence determinant and designated as *Avr 4E* [71]. By utilizing restriction enzyme mediated integration (REMI) method AAL toxin deficient mutants were developed from a library of transformants obtained with the addition of enzymes. These mutants were unable to cause symptoms on susceptible tomato [76]. This technique can be utilized for converting the pathogenic isolate of *A. a.* f.sp. *lycopersici* into non-pathogenic isolates without host specific toxin producing ability. Taxonomic status of tomato crown and root rot pathogen was confirmed on the basis of DNA -DNA reassociation kinetics, show high (96-100%) homology and indicated that *F. o.* f.sp. *radicis-lycopersici* is correct name of the pathogen rather than *F. o.* f.sp. *lycopersici* race J3 in Japan [14]. *A. a.* f.sp. *lycopersici* produces AAL-toxins that function as chemical determinants of the *Alternaria* stem canker disease in the tomato. In resistant cultivars, the disease is controlled by the *Asc* locus on chromosome 3L [77]. RFLP mapping and allelic testing indicated that high intensity to AAL toxin from *Lycopersicon pennellii* is conferred by *Asc* locus [77]. Tomato strain 32 of *F. o.* f.sp. *lycopersici* showed an elevated genetic uniformity and 2 haplotype by contrast with 10 haplotypes in strain 15 of *F. o.* f.sp. *radicis-lycopersici.*

8. SAR AND LAR IN FUNGAL DISEASE RESISTANCE

Many vegetable crops could be protected against infection from virulent strain of pathogen by prior inoculation with another pathogen, avirulent strain of the same pathogen (Table-6). Induced/acquired resistance, if expressed locally on the tissue, known as localized acquired resistance (LAR) and when expressed systemically in whole plant, called systemic acquired resistance (SAR). LAR and SAR are similar in that they are effective against various types of pathogens in many vegetables. However, a signal that propagates enhanced defence capacity throughout the plant in SAR is lacking in LAR [78]. SAR is characterized by the accumulation of PR proteins and salicylic acid (SA) [79, 80].

Significant interpopulation genetic distance (phist-0.2275) was observed between isolates of *A. solani* collected from potatoes and tomatoes, suggesting the possibility of pathogenic specialization. There was significant difference between isolates of *A. solani* and *A. alternata* [81]. The susceptible cultivar Red River to *F. o.* f.sp. *lycopersici* acquires a stable competence for active defence particularly when the phytohormones equilibrium is altered in favour of cytokinins. The expressions of PR-protein genes involved in the defence response against the above pathogen. Fungal cell wall components, glutathione, salicylic acid and the ethylene forming ethaphon are used as probes for the induction of defence processes including ethylene production. Results indicated that expression of the intracellular PR proteins was constitutive in all the transgenic tissues and control while extracellular proteins was inducible only in control tissue of resistant cultivar [82]. In *C. fulvum* tomato systems the isolated R genes *cf-2, cf-4, cf-5, cf-9* etc. have been found in cluster and *cf-5* gene with its six homologous are characterized by pronounced allelic variation in leucine rich report proteins (LRR) [25]. Induced resistance is associated with the induction of several PRs in which β-1, 3-glucanases and chitinases are important and responsible for hydrolyzing fungal cell wall. Induced resistance to anthracnose fungus in cucumber showed inhibition of penetration by *C. lagenarium* in the host and direct correlation was recorded between failure of pathogen penetration and deposition of lignin like compounds in the outer epidermal cell wall [83].

Table 6. Examples of induced resistance in vegetable crops.

Crop/LAR/SAR	Prior inoculation	Induced resistance against	References
Cucumber	Anthracnose (*C. lindemuthianum*)	Scab (*C. cucumerinum*)	[83]
	Scab (*C. cucumerinum*)	Anthracnose (*C. lagenarium*)	[83]
	F. oxysporum (avirulent form)	*F. oxysporum* (virulent strain)	[85]
	F. oxysporum (avirulent form)	*F. oxysporum* (virulent strain)	[86]
Watermelon	Anthracnose (*C. lagenarium*)	*C. lagenarium* (virulent strain)	[87, 88]
	Maize pathogen (*Cochlibolus carbonum*)	Wilt (*F. oxysporum*)	[89]
Muskmelon	Fungal elicitors	Anthracnose (*C. lagenarium*)	[73]
	C. lagenarium	*C. lagenarium* (virulent strain)	[87, 88]
	Non pathogens of bean	Anthracnose (*C. lindemuthianum*)	[90, 91]
	Anthracnose (*C. lindemuthianum*) (conidia in lower leaf)	*C. lindemuthianum* (upper leaf)	[92, 93]
Green bean	Anthracnose (*C. lindemuthianum*)	*C. lindemuthianum* *U. appendiculatus*	[94]
	2, 6-dichloroisonicotinic acid (INA)	*C. lindemuthianum* *U. appendiculatus*	[95]
	U. appendiculatus	Tobacco necrosis virus	[96]
	C. lindemuthianum	*U. appendiculatus*	[97]
Tomato	*P. infestans*	*P. infestans*	[98, 99]
	Salicylic acid	*A. solani*	[100]

Strong silicon signal has also been observed in the same induced tissues where lignin was present [84]. Thus lignification and silicon strengthen the cell wall and prevent penetration. Induced resistance of tomato to *P. infestans* revealed in histological study that number of germinated cytospores decreased, infection process stopped at the site of penetration on epidermis and hypersensitive cell death. At molecular level, the set of genes are induced during SAR may differ among the species. In cucumber, class II chitins are the most highly induced SAR genes, albeit cucumber possesses PR-1 homologue in its genome. Genetic engineering may play a greater role in developing the mild strain an avirulent strain of fungus, SAR producing PR protein genes, phytoalexins etc. Colonization with beneficial isolates of *P. putida* increases the defensive potential of bean plants against *F. s.* f. sp. *phaseoli* and noted that lignin increased in root tissue in response to this bacterial colonization.

9. USE OF CLONED RESISTANCE GENES : ADVANTAGES AND PROBLEMS

The molecular cloning of resistance genes targeted against specific pathogen has major impact on vegetable production. If many cloned disease resistance genes are deployed against *P. infestans* better stability in resistance would be expected as compared to only one gene introduced in potato plants. The use of genetic transformation to introduce resistance genes into desirable cultivars avoids introduction of undesirable traits and repeated back crossing because resistance genes are often derived from germplasm sources, which are inferior for many other qualitative and quantitative traits. Several different cloned disease resistance genes could be introduced simultaneously into a particular vegetable by a single transformation. Cloned disease resistance genes can be successfully utilized for disease management but it has several limitations apart from the useful effect.

9.1 Advantages

1. Once the cloned resistance genes against any fungal pathogen introduced in a particular vegetable crop then it inherited and become permanent till the outbreak of new race.
2. The method is simple, cost effective, no need of any application systems, no need to give attention for pathogen population, alternate hosts etc.
3. The method is safe and ecofriendly since no chemical and bioagents are involved in field.
4. Resistance expressed only when the plant is invaded by the pathogens.
5. Genetic resistance is generally less affected by soil type, climate, pathogen inoculum, previous year disease pressure and crop rotation.
6. The probability is high of an introduced disease resistance gene working in a new genetic background.

9.2 Disadvantages

1. In sexually produced fungi often virulent pathogenic races may appear resulting in breakdown of the single gene resistance.
2. Resistance genes against some fungal pathogen are unavailable in the germplasms of some vegetable crops.
3. Introduction, cloning and deployment of resistance genes are difficult or impossible in vegetatively propagated vegetables where same material has to be routinely utilized.
4. Introduction of resistance gene into elite cultivars requires extensive back crossing to re-establish desired traits.
5. In some vegetable crops cloned resistance genes from wild taxa are not compatible with cultivated variety and very difficult to regenerate transgenic resistance plants against some fungal pathogen.

9.3 New approaches on availability of general cloned resistance genes

1. Multiple resistance genes introduced into plants by transformation would avoid negative qualitative and quantitative characteristics of the plants, which could come through crossing and save time in several back crossing.
2. On availability of several cloned resistance genes against a pathogen only those should be introduced in plant that are difficult to over come so that disability can be prolonged.
3. New combinations of cloned resistance genes could be released periodically to overcome the emergence of virulent pathogenic races.
4. The availability of cloned resistance genes might permit the construction of chimeric or synthetic resistance genes encoding none recognition specifications.
5. The overall performance and durability in terms of disease, yield and cost-economics will be more in the vegetable crops developed by utilizing the multiple cloned resistance genes against the fungal pathogen.

10. CONCLUSIONS

The advent of biotechnology has opened up a lot of exciting possibilities in the improvement of vegetable crops. Genetic engineering can be successfully utilized for the development of disease resistant transgenic against many pathogens. Contrary to the overwhelming success in the production of commercially useful virus and insect resistant plants, the development of fungal resistant in vegetable has been more limited. However, several new advances in the field of molecular biology are available at present for the development of fungus resistant vegetable variety for effective disease management. In this regard, one of the major approaches is the use of antifungal compounds in engineered transgenic plants. The most widely exploited antifungal compounds are pathogenesis-related proteins, chitinase β-1, 3-glucanases, ribosome inactivating proteins, cystein-rich proteins, phytoalexin etc. Many genes encoding antifungal polypeptides have been described and transgenic plants were developed particularly in tomato but the resistance was not absolute in field condition. New strategy for long term success, effective against many races of a fungal pathogen and multiple resistance genes is being utilized. More emphasis should also be given to develop transgenic for pathogen induced cell death at the site of infection by a fungal pathogen. However, future prospects of transgenic against fungal pathogen resistance in vegetable will depend upon patent issues, safety concerns, regulations, consumer's acceptance, social and economic factors.

11. REFERENCES

1. Annonymous (1995). Report on the Expert Consultation of the Regional Network on Vegetable Crops. F.A.O./RAPA Publication.
2. Annonymous (1997) Vision 2020. Project Directorate of Vegetable Research, Prospective Plan (1997) ICAR, New Delhi.
3. Pandey, K. K. (2000). Important fungal diseases of vegetable crops and their integrated management. National Training course on vegetable production. Ministry of Agriculture, Krishi Bhawan, New Delhi, pp. 176-184.
4. Snowden, A. L. (1991). Post harvest diseases of fruits and vegetables. Wolfe Scientific Ltd., London, Vol. I, II.
5. Premnath (2000). Salient Issue of Vegetable Production in Asia and the Pacific. *In:* Emerging Scenario in Vegetable Research and Development (G. Kalloo and Kriti Singh, eds.), Research Periodicals and Book Publishing House, New Delhi, pp. 11-31.
6. Powell, P. A., Nelson, R. S., De Hoffman, B. N., Rogers, S. G. N., Fraly, R. T. and Beachy, R. N. (1986). Delay of disease development in transgenic plants that express the tobacco mosaic virus coat protein gene. Science 232: 738.

7. Broglie, K., Chet, I., Holliday, M., Cressman, R., Biddle, P., Knowlton, S., Mauvais, C. J. and Broglie, R. (1991). Transgenic plants with enhanced resistance to the fungal pathogen *Rhizoctonia solani*. Science 245: 1194-1197.

8. Helentjaris, T., King, G., Slocum, M., Siedestrang, C. and Wegman, S. (1985). Restriction fragment polymorpholismas probes for plant diversity and their development as tool for applied plant breeding. Pl. Mol. Bio. 5:109-118.

9. Williams, J. G. K., Kubelik, A. R., Livak, K. J., Rafalski, J. A. and Tingey, S. V. (1990). DNA polymorphisms amplified by arbitrary primers are useful as genetic markers. Nucl. Acids. Res. 18: 6531-6535.

10. Vos, P., Hogers, R., Bleeker, M., Rijans, M., Van de Lee, T., Hornes, M., Frifters, A., Pot, J., Peleman, J., Kuiper, M. and Zabeau, M. (1995). AFLP : A new technique for DNA fingerprinting. Nucl. Acids Res. 23: 4407-4414.

11. Broglie, K., Broglie, R., Benhamou, N. and Chet, I. (1993). The role of cell wall degrading enzymes in fungal disease resistance. *In*: Biotechnology in Plant Disease Control (I. Chet, ed.), Wiley-Liss Inc., New York, pp. 139-156.

12. Thomas, C. M., Jones, D. A., Parniske, M., Harrison, K., Balint Kurti, P. J., Hatzixanthis, K. and Jones, J. D. (1997). Characterization of the tomato *cf-4* gene for *Cladosporium fulvum* identifies sequences that determine recognitional specificity in *cf-4* and *cf-9*. Plant Cell 9: 2209-2224.

13. Chunwongse, J., Bunn, T. B., Crossman, C., Jiang, J. and Tanksley, S. D. (1994). Chromosomal location and molecular marker tagging of the powdery mildew resistance gene (*Lv*) in tomato. Theor. Appl. Genet. 89: 76-79.

14. Michelmore, R. W. (1995). Molecular approaches to manipulation of disease resistance genes. Ann. Rev. Phytopathol. 33: 301-309.

15. Moreau, P., Thoquet, P., Olivier, J., Laterrot, H. and Grimsley, N. (1998). Genetic mapping of Ph-2 a single locus controlling partial resistance to *Phytophthora infestans* in tomato. Mol. Pl. Microb. Interac. 11: 259-269.

16. Kwachuk, L. M., Lynch, D. R., Hachey, J., Bains, P. S. and Kulcsar, F. (1994). Identification of a codominant amplified polymorphic DNA marker linked to the *Verticillium* wilt resistance gene in tomato. Theor. Appl. Genet. 89: 661-664.

17. El-Kharbotly, A., Leonards-Schippers, C., Huigen, D. J., Jacobsen, E., Pereira, A., Stiekema, W. J., Salamini, F. and Gabhardt, C. (1994). Segregation analysis and RFLP mapping of the R1 and R3 alleles conferring race specific resistance to *Phytophthora infestans* in progeny of dihaploid potato parents. Mol. Gen. Genet. 242: 749-754.

18. Dirlewanger, E., Isaac, P. G., Ranade, S., Belajonza, M., Cousin, R. and Veinn, D. D. (1994). Restriction fragment length polymorphisim analysis of loci associated with disease resistance genes and developmental traits in *Pisum sativum* L. Theor. Appl. Genet. 88: 17-27.

19. Maisonneuve, B., Bellec, Y., Anderson, P. and Michelmore, R. W. (1994). Rapid mapping of two genes for resistance to downy mildew from *Lactuca serriola* to existing clusters of resistance genes. Theor. App. Genet. 89: 96-104.

20. Witsenboer, H., Kasseli, R. V., Fortin, M. G., Stanghellini, M. and Michelmore, R. W. (1995). Sources of genetic structure of cluster of genes for resistance to three pathogen in lettuce. Theor. Appl. Genet. 91: 178-188.

21. Kesseli, R. V., Paran, I. and Michelmore, R. W. (1994). Analysis of a detailed genetic linkage map of *Lactuca sativa* (lettuce) constructed from RFLP and RAPD markers. Genetics 136: 1435-46.

22. Kuginuiki, Y., Nakamura, K., Hida, K. I. and Yosikawa, H. (1997). Varietal differences in embryogenic and regenerative ability in microspore culture of chinese cabbage (*Brassica rapa* L. ssp. *pekinensis*). Breeding Science 47: 341-346.

23. Segal, G., Sarfatti, M., Schaffer, M. A., Ori, N., Zamir, D. and Fluhr, R. (1992). Correlation of genetic and physical structure in the region surrounding the I-2 *Fusarium oxysporum* resistance locus in tomato. Mol. Gen. Genet. 231: 179-185.

24. Kik, C., Verbeck, W. H. J., Wietsma, W. A. and Van Ooijen (1995). Towards an AFLP linkage map of onion (*Allium cepa*). Plant Genome IV Conference, San Diego, CA.

25. Dixon, M. S., Hatzixanthis, K., Jones, D. A., Harrison, K. and Jones, J. D. G. (1998). The tomato *cf-5* disease resistance gene and six homologus show pronounced allelic variation in leucine rich repeat copy number. Plant Cell 10: 1915-1925.

26. Birch, R. G. (1997). Plant transformation: Problems and strategies for practical application. Ann. Rev. Pl. Physiol. 48: 297-326.

27. Hooykaas and Beijersbergen, A. G. M. (1994). The virulence of *Agrobacterium tumefaciens*. Ann. Rev. Phytopathol. 32: 157-179.

28. Potrykus, I., Bilang, R., Fuetterer, J., Sautter, C., Schrott, M. and Spangenberg, G. (1998). Genetic engineering of crop plants. *In*: Agricultural Biotechnology. (A. Altman, ed.), Marcel Dekker, New York, pp. 119-157.
29. Krens, F. A., Molendij, K. L., Wullems, G. J. and Schilperoort, R. A. (1982). *In vitro* transformation of plant protoplats with plasmid DNA. Nature 296: 72-74.
30. Paszkowski, J., Shillito, R. D., Saul, M. W., Mandak, V., Hohn, T., Hohn, B. and Potrykus, I. (1984). Direct gene transfer to plants. EMBO J. 3: 2712-2722.
31. Kloti, A., Iglesias, V. A., Wunn, J., Burkhard, P. K., Dalta, S. K. and Potrykus, I. (1993). Gene transfer by electroporation into intact scutellum cells of wheat embryos. Plant Cell Rep. 12: 671-675.
32. Does, M. P. and Cornelissen, B. J. C. (1998). Emerging strategies to control fungal diseases using transgenic plants. *In*: 2nd Int. Crop Science Cong. (V.L. Chopra, R.B. Singh and A. Varma, eds.), Proc. Oxford IBH Pub. New Delhi, pp. 233-244.
33. Van Loon, L. C., Pierpoint, W. J., Boller, T. H. and Conejero, V. (1994). Recommendation for naming plant pathogensis related proteins. Mol. Biol. Rep. 12: 245-265.
34. Stuiver, M. H., Tigelaar, H., Molendijk, L., Troost-van Deventer, E., Sela-Buurlage, M. B., Storms, J., Plooster, L., Sijbolts, F., Custers, J., Apotheker-de Groot, M. and Melchers, L. S. (1996). Broad-spectrum resistance in transgenic carrot plants. *In:* 8th International Congress Molecular Plant-Microbe Interactions (G. Stacey, B. Mullin and P. M. Greshoff, eds.), Knoxville, USA, p. B-93.
35. Liu, D., Raghothama, K. G., Hasegawa, P. M. and Bressan, R. A. (1994). Osmotin over expression in potato delays development of disease symptoms. Proc. Natl. Acad. Sci. USA 91: 1888-1892.
36. Jongedijk, E., Tigelaar, H., Van Roekel, J. S. C., Bres-Vloemans, S. A., Dekker, I., Van den Elzen, P. J. M., Cornelissen, B. J. C. and Melchers, L. S. (1995). Synergistic activity of chitinases and β-1, 3-glucanases enhances fungal resistance in transgenic tomato plants. Euphytica 85: 173-180.
37. Lawton, K., Uknes, S., Friedrich, L., Gaffney, T., Alexander, D., Goodman, R., Metraux, J. P., Kessmann, H., Ahl Goy, P., Gut Rella, M., Ward, E. and Ryals, J. (1993). The molecular biology of systemic acquired resistance. *In:* Mechanisms of Plant Defence Responses. (B. Fritig and M. Legrand, eds.), Kluwer Academic Publishers, Dordrecht, The Netherlands, pp. 422-432.
38. Grison, R., Grezes-Besset, B., Schneider, M., Lucante, N., Olsen, L., Leguay, J. J. and Toppan, A. (1996). Field tolerance to fungal pathogens of *Brassica napus* constitutively expressing a chimeric chitinase gene. Nature Biotechnol. 4: 643-646.
39. Cornelissen, B. J. C., Does, M. P. and Melchers, L. S. (1996). Strategies for molecular resistance breeding and transgenic plant. *In: Rhizoctonia Species*: Taxonomy, molecular biology, ecology, pathology and control. (S. Sneh, S. Jabaji-Hare, Neate and G. Deist,eds.), Kluwer Academic Publishers, Dordrecht, The Netherlands, pp. 529-536.
40. Mythili, J. B., Anand, L. and Ravishankar, K. V. (2000). Transformation of tomato with chitinase and glucanase genes for disease resistance. Symposium on Biotechnology of plant protection application and technology development, Varanasi, pp. 49-50.
41. Strittmatter, G., Goethals, K. and Van Montagu, M. (1998). Strategies to engineer plants resistant to bacterial and fungal diseases. *In*: Subcellular Biochemistry. Vol. 29 (B. B. Biswas and H. K. Das, eds.), Plenum Press, New York, pp. 191-213.
42. Inbar, J. and Chet, I. (1991). Evidence that chitinase produced by *Aeromonas caviae* is involved in the biological control of soil borne pathogens by this bacterium. Soil Biol. Biochem. 23: 239-242.
43. Lim, H, Kim, Y. and Kim, S. (1991). *Pseudomonas stutzeri* YPL-1 genetic transformation and antifungal mechanism against *Fusarium solani,* agents of plant root rot. Appl. Environ. Microbiol. 57: 510-516.
44. Ordentlich, A., Elad, Y. and Chet, I. (1988). The role of *Serratia marcescens* in biocontrol of *Sclerotium rolfsii*. Phytopathol. 78: 84-88.
45. Shapira, R., Ordentlich, A., Chet, I. and Oppenheim, A. B. (1989). Control of plant diseases by chitinase expressed from cloned DNA in *Escherichia coli*. Phytopathol. 79: 1246-1249.
46. Enkerli, J., Gisi, U. and Mosinger, E. (1993). Systemic acquired resistance to *Phytophthora infestans* in tomato and the role of pathogensis related proteins. Physiol. Mol. Plant Pathol. 43: 161-171.
47. Jones, J. D. G., Grady, K. L., Suslow, T. V. and Bedbrook, J. R. (1986). Isolation and characterization of genes encoding two chitinase enzymes from *Serratia marcescens*. EMBO J. 5: 467-473.
48. Sundheim, L., Poplawsky, A. R. and Ellingboe, A. H. (1988). Molecular cloning of two chitinase genes from *Serratia marcescens* and their expression in *Pseudomonas* species. Physiol. Mol. Plant Pathol. 33: 483-491.
49. Jach, G., Longemann, S., Wolf, G., Oppenheim, A., Chet, I., Scheel J. and Logemann, J. (1992). Expression of a bacterial chitinase leads to improved resistance of transgenic tobacco plants against fungal infection. Biopract. 1: 33-40.

50. Lee, H. I. and Raikhel, N. V. (1995). Prohevein is poorly processed but shows enhanced resistance to a chitin-binding fungus in transgenic tomato plants. Braz. J. Med. Biol. Res. 28: 743-750.
51. Terras, F. R. G., Schoots, H. M. E., deBolle, M. F. C., van Leunen, F., Rees, S. B., Vanderleyden, J., Cammue, B. P. A. and Broekaert, W. F. (1992). Analysis of two novel classes of antifungal proteins from radish (*Raphanus sativus* L.) seeds. J. Biol. Chem. 267: 15301-15309.
52. Terras, F. R. G., Eggermont, K., Kovaleva, V., Raikhel, N. V., Osborn, R. W., Kester, A., Rees, S. B., Torrekens, S., van Leunen, F., Vanderleyden, J., Cammue, B. P. A. and Broekert, W. F. (1995). Small cysteine-rich antifungal proteins from radish: their role in host defence. Plant Cell 7: 573-588.
53. Leah, R., Tommerup, H., Svendsen, I. and Mundy, J. (1991). Biochemical and molecular characterization of three barley seed proteins with antifungal properties. J. Biol. Chem. 266: 1464-1573.
54. Park, C. M., Berry, J. O. and Bruenn, J. A. (1996). High level secretion of a virally encoded anti fungal toxin in transgenic tobacco plants. Plant Mol. Biol. 30: 359-366.
55. Trudel, J., Potvin, C. and Asselin, A. (1995). Secreted hen lysozyme in transgenic tobacco: recovery of bound enzyme and *in vitro* growth inhibition of plant pathogens. Plant Science 55-62.
56. Herrera-Estrella, L. and Simpson, J. (1995). Genetically engineered resistance to bacterial and fungal pathogens. World J. Microbiol. Biotechnol. 11: 383-392.
57. Woodward, S. and Perce, R. B. (1988). The role of stilbenes in resistance to Sitka spruce (*Picea sitchensis* Bong. Carr.) to entry of fungal pathogens. Physiol. Mol. Plant Pathol. 33: 127-149.
58. Hain, R., Reif, H. J., Krause, E., Langebartels, R., Kindl, H., Vornam, B., Wiese, W., Schmelzer, E., Schreier, P. H., Stocker, R. H. and Stenzel, K. (1993). Disease resistance results from foreign phytoalexin expression in a novel plant. Nature 361: 153-156.
59. Strittmatter, G., Janssens, J., Opsomer, C. and Botterman, J. (1995). Inhibition of fungal disease development in plants by engineering controlled cell death. Bio/Technol. 13: 1085-1089.
60. Bournival, B. L., Vallejos, C. E. and Scott, J. W. (1990). Genetic analysis of resistance to race 1 and 2 of *Fusarium oxysporium* f.sp. *lycopersici* from the wild tomato *Lycopersicon pennellii*. Theor.Appl. Genet. 79: 641-645.
61. Wu, G., Shortt, B. J., Lawrence, E. B., Levine, E. B., Fitzsimmons, K. C. and Shah, D. M. (1995). Disease resistance conferred by expression of a gene encoding H_2O_2 generating glucose oxidase in transgenic potato plant. Plant Cell 7: 1357-1368.
62. Kumar, S., Singh, A. K. and Kalloo, G. (2000). Disease resistance mechanisms and breeding in vegetable Crops. *In:* Emerging Scenario in Vegetable Research and Development (G. Kalloo and Kriti Singh, eds.), Research Periodicals and Book Publishing House, New Delhi, pp. 118-143.
63. De Wit, P. J. G. M. (1992). Molecular characterization of gene for gene systems in plant fungus interactions and the application of avirulence genes in control of plant pathogens. Annu. Rev. Phytopath. 30: 391-481.
64. Honee, G., Melchers, L. S., Vleeshouwers, A. A., Van Roekel, J. S. C. and De Wit, P. J. G. M. (1995). Production of the AVR 9 elicitor from the fungal pathogen *Cladosporium fulvum* in transgenic tobacco and tomato plant. Plant Mol. Biol. 29: 909-920.
65. Keller, H., Pambukdjian, N., Ponchet, M., Poupet, A., Delon, R., Verrir, J. L., Roby, D. and Ricci, P. (1999). Pathogen-induced elicitin production in transgenic tobacco generates a hypersensitive response and non-specific disease resistance. Plant Cell 11: 223-235.
66. Carr, J. P. and Klessing, D. F. (1989). The pathogens related proteins of plants. *In:* Genetic Engineering, Principles and Methods (J. K. Setlow, ed.), Plenum Press, New York, Vol. II, pp. 65-89.
67. Roby, D., Toppan, A. and Esquerre Tugaye, M. T. (1987). Cell surface in plant microorganism interactions, VII: increased proteinase inhibitor activity in melon in response to infection by *Colletotrichum lagenarium* or to treatment with an elicitor from this fungus. Physiol. Mol. Plant Pathol. 30: 453-460.
68. Kombrink, E., Schroder, M. and Hahlbrock, K. (1988). Several pathogensis related proteins in potato 1, 3-β-glucanases and chitinases. Proc. Natl. Acad. Sci. USA 85: 782-786.
69. Benhamou, N., Joosten, M. H. A. and De Wit, P. J. G. M. (1990). Subcellular localization of chitinase and of its potential substrate in tomato root tissues infected by *Fusarium oxysporum* f.sp. *radicis-lycopersici*. Plant Physiol. 92: 1109-1120.
70. Mauch, F., Mauch-mani, B. and Boller, T. (1988). Antifungal hydrolases in pea tissue. Plant Physiol. 88: 936-942.
71. Boller, T., Gehri, A., Mauch, F. and Vogeli, U. (1983). Chitinase in bean leaves: induction by ethylene, purification, properties and possible function. Planta. 157: 22-31.
72. Broglie, K., Broglie, R., Benhamou, N. and Chet, I.(1993). The role of cell wall degrading enzymes in fungal disease resistance. *In*: Biotechnology in plant disease control. (I. Chet, ed.), John Wiley & Sons, USA, pp. 139-156.

73. Roby, D. and Esquerre-Tugaye, M. (1987). Induction of chitinases and translatable mRNA for these enzymes in melon plants infected with *Colletotrichum lagenarium*. Plant 52: 175-185.
74. Inagaki, A., Takano, Y., Kubo, Y., Misc, K. and Fureyawa, I. (2000). Construction of an equalized cDNA library from *Colletotrichum lagenarium* and its application to the isolation of differentially expressed genes. Can. J. Microbiol. 46: 150-158.
75. Moriguchi, K., Kimizuka Takagi, C., Ishii, K. and Nomura, K. (1999). A genetic map based on RAPD, isozyme, morphological markers and QTL analysis for clubroot resistance in *Brassica oleracea*. Breeding Science 49: 257-265.
76. Kodama, M., Akamatsu, H., Itoh, Y., Narusaka, Y., Sanekata, T., Otani, H., Kohmoto, K. and Yoder, O. C. (1997). Host specific toxin deficient mutants of the tomato pathotypes of *Alternaria alternata* obtained by restriction enzyme mediated integration. Molecular genetics of host specific toxins in plant disease, Proceedings of the 3rd Tottori International Symposium Daisen, Tottori, pp. 35-42.
77. Biezen, E. A., Van der, Glagotskaya, T., Overduin, B., Nijkamp, H. J. J. and Hille, H. (1995). Inheritance and genetic mapping of resistance to *Alternaria alternata* f. sp. *lycopersici* in *Lycopersicon pennellii*. Mol. Gen. Genet. 247: 453-461.
78. Van Loon, L. C., Bakker, P. A. H. M. and Pieterse, C. M. J. (1998). Systemic resistance induced by rhizosphere bacteria. Ann. Rev. Phytopathol. 36: 453-483.
79. Kessmann, H., Staub, T., Ligon, J., Oostendrop, M. and Ryals, J. (1994). Activation of systemic acquired disease resistance in plants. Eur. J. Plant Pathol. 100: 359-369.
80. Ryals, J. A., Neuenschwander, U. H., Willits, M. G., Molina, A. and Steiner, H. Y. (1996). Systemic acquired resistance. Plant Cell 3: 1085-1094.
81. Weir, T. L., Huff, D. R., Christ, B. J. and Romaine, C. P. (1998). RAPD-PCR analysis of genetic variation among isolates of *Alternaria solani* and *Alternaria alternata* from potato and tomato. Mycologia 90: 813-821.
82. Bettini, P., Cosi, E., Pellegrini, M. G., Turbanti, L., Vendramin, G. G. and Buiatti, M. (1998). Modification of competence for *in vitro* response to *Fusarium oxysporum* in tomato cells, III. PR protein gene expression and ethylene evolution in tomato cell lines transgenic for phytohormone related bacterial genes. Theor. Appl. Genet. 97: 575-583.
83. Hammerschmidt, R., Acres, S. and Kuc, J. (1976). Protection of cucumber against *Colletotrichum lagenarium* and *Cladosporium cucumerinum*. Phytopathology 66: 790-793.
84. Stein, B. D. (1991). The ultrastructure and histochemistry of infection by *Colletotrichum lagenarium* in cucumbers induced for resistance. Ph.D. Thesis, Michigan State University, East Lansing.
85. Michail, S. H., Sheir, H. M. and Rasmy, M. R. (1989). Cross protection of watermelon and cucumber plants against wilt by prior inoculation with an irrespective form of *Fusarium oxysporum*. Acta. Phytopathol. Entomol. Hung. 24: 301-309.
86. Mandeel, Q. and Baker, R. (1991). Mechanisms involved in biological control of *Fusarium* wilt of cucumber with strains of non-pathogenic *Fusarium oxysporum*. Phytopathol. 81 462-469.
87. Caruso, F. L. and Kuc, J. (1977a). Field protection of cucumber, watermelon and muskmelon against *Collectotrichum lagenarium*. Phytopathol. 67: 1290-1292.
88. Caruso, F. L. and Kuc, J. (1977b). Protection of watermelon and muskmelon against *Collectotrichum lagenarium*. Phytopathol. 67: 1285-1289.
89. Shimotsuma, M. J., Kuc, J. and Jones, C. M. (1972). The effect of prior inoculation with non-pathogenic fungi on *Fusarium* wilt of watermelon. HortSicence 7: 72-73.
90. Rahe, J. E., Kuc, J., Chuang, C. M. and Williams, E. B. (1969). Induced resistance in *Phaseolus vulgaris* to bean anthracnose. Phytopathol. 59: 1641-1645.
91. Skipp, R. A. and Deverall, B. J. (1973). Studies on cross protection in the anthracnose disease of bean. Physiol. Plant Pathol. 3: 299-314.
92. Sutton, D. C. (1979). Systemic cross protection of bean against *Colletotrichum lindemuthianum*. Austral Plant Pathol. 8: 4-5.
93. Cloud A. M. E. and Deverall, B. J. (1987). Induction and expression of systemic resistance to the anthracnose disease in bean. Plant Pathol. 36: 551-557.
94. Dann, E. K. and Deverall, B. J. (1995). Effectiveness of systemic resistance in bean against foliar and soilborne pathogens as induced by biological and chemical means. Plant Pathol. 44: 458-466.
95. Metraux, J. P., Ahl Goy, P., Staub, T. H., Speich, J., Steinemann, A., Ryals, J. and Ward, E. (1991). Induced systemic resistance in cucumber in response to 2, 6-dichloroisonictinic acid and pathogens. *In:* Adv. Mol. Genet. of Plant Microbe. Interac. (H. Hennecke and D. P. S. Verma, eds.), Vol. 1, Kluwer Academic, Amsterdam, pp. 432-439.
96. Kutzner, B., Hellwald K. H. and Buchenauer, H. (1993). Systemic induction of resistance in *Phaseolus vulgaris* L. to tobacco necrosis virus *Uromyces appendiculatis*. J. Phytopathol. 138: 9-20.

97. Takahashi, K., Inaba, T. and Morinaka, T. (1985). Systemic induction to bean rust induced in *Phaseolus vulgaris* by the preinoculation with *Colletotrichum lindemuthianum*. Ann. Rev. Phytopathol. Soc. Japan 51: 399-404.
98. Enkerli, J., Gisi, U. and Mosinger, E. (1993). Systemic acquired resistance to *Phytophthora infestans* in tomato and the role of pathogensis related proteins. Physiol. Mol. Plant Pathol. 43: 161-171.
99. Kovats, K., Binder, A. and Hohl, H. R. (1991). Cytology of induced systemic resistance of cucumber to *Collectotrichum lagenarium*. Planta. 183: 484-490.
100. Spletzer, M. E. and Enyedi, A. J. (1999). Salicylic acid induces resistance to *Alternaria solani* in hydroponically grown tomato. Phytopathol. 89: 722-727.

Biological Control of Postharvest Diseases of Fruits and Vegetables

Ahmed El Ghaouth[1,2], Charles Wilson[2], Michael Wisniewski[2], Samir Droby[3], Joseph L. Smilanick[4] and Lise Korsten[5]

[1]Universite De Nouakchott, Faculte des Sciences et Techniques, B.P.5026, Nouakchott, Mauritanie [2]Appalachian Fruit Research Station, USDA-ARS, 45 Wiltshire Road, Kearneysville, WV, 25430, [3]Dept of Postharvest Science, ARO, The Volcani Center, P.O. Box 6, Bet Dagan 5250, Israel, [4]USDA-ARS, 2021 South Peach Avenue, Fresno, CA, 93727, [5]Dept of Microbiology and Plant Pathology, University of Pretoria, Pretoria 0002, South Africa (E Mail: AelGhaouth@afrs.ars.usda.gov).

Worldwide, postharvest losses of fruits and vegetables are estimated at more than 25% and much of this is due to postharvest decay. Currently, fungicides are the primary means of controlling postharvest diseases of fruits and vegetables. Public concern over food safety has enhanced interest to find effective alternatives to pesticides to control postharvest diseases of fruits and vegetables. Several promising biological approaches that include antagonistic microorganisms, compounds of natural origin, and induced resistance have been proposed as potential alternatives to synthetic fungicides for postharvest disease control. Among the proposed alternatives, the use of antagonistic microorganisms as a pre-storage treatment of fruit and vegetables have been the most studied and substantial progress has been made in this area. In this chapter, we will present an overview of the potential application, and limitations of microbial antagonists to prolong shelf-life, and control of decay of harvested commodities.

1. INTRODUCTION

A considerable amount of fruits and vegetables is lost to spoilage after harvest. This loss can range from 10-50% depending on the commodity and country. Developing countries experience the greatest losses due to inadequate storage and shipping facilities. Any reasonable reduction in the postharvest losses would ease the pressure on the production resources and generate much needed revenue in developing countries. Postharvest losses of fruits and vegetables are particularly devastating in that the total investment in the production, harvesting, processing, and distribution of a food commodity is lost, as well as any potential profits from this investment. The major cause of postharvest losses are usually due to the biochemical changes associated with respiratory metabolism, biosynthesis and action of ethylene, transpiration, and decay. Although manipulating environmental storage conditions to delay senescence can reduce losses associated with accelerated metabolic turnover, these beneficial practices are usually not sufficient to protect postharvest pome fruits, grape, and tomato, against postharvest decay [14, 18-25]. Although the fruit surfaces and vegetative tissue have been a rich source of microbial antagonists, the superiority of this method has not been demonstrated and the selection of candidates by other reasoning cannot be excluded.

3. BIOCONTROL ACTIVITY OF ANTAGONISTS

Considerable success has been realized utilizing antagonistic microorganisms to control postharvest diseases and a large body of information regarding postharvest biocontrol antagonists is now available (see ref. 5, 6, 11). Several species of bacteria and yeast have been isolated and shown to control decay on a variety of fruits. Gutter and Littauer (26) demonstrated the potential use of microbial antagonists to control postharvest decay as early as 1953 where they showed that *Bacillus subtilis* was capable of reducing decay of citrus fruits. As a follow up to this early work, Singh and Deverall [27] demonstrated that dipping citrus fruit in a cell suspension of *B. subtilis* delayed decay caused by *Alternaria citri*, *Geotrichum candidum* and *Penicilluim digitatum*. Partial control of decay of strawberry and apple was also reported with *Trichoderma viride* and *T. harzianum*, respectively, when applied as a preharvest treatment [28, 29]. Continuing this approach, several potential antagonistic bacteria and yeasts were isolated and evaluated on stone, citrus, and pome fruits [12, 16, 30]. Biological control of postharvest decay was demonstrated on peaches using an isolate of *B. subtilis* [30, 31]. After the initial screening of 122 potential antagonists on citrus, Wilson and Chalutz [16] showed that four antagonists, two bacteria (*Pseudomonas cepacia* and *Pseudomonas syringae*) and two yeasts (*Debaryomyces hansenii* and *Aureobasidium pullulans*), were the most effective in reducing decay caused by either *P. digitatum* or *P. italicum*. These four isolates reduced decay of grapefruit by more than 50 % after 11 days of storage at room temperature [16]. Effective control of decay of grapefruit was also observed with yeasts *Candida guilliermondii*, *C. oleophila*, and *C. sake*, while *C. guilliermondii* was the most predominant species found on fruit surfaces [18, 32]. These same antagonistic yeasts were also effective against decay of oranges and lemons caused by *P. digitatum* and *P. italicum* [18, 19, 33, 34] and against sour rot of lemon caused by *G. candidum* [19].

A reduction in decay of oranges caused by *P. digitatum* has also been reported with other antagonistic yeasts such as *Saccharomyces cerevisiae* [35], *C. sake* [36], *C. famata* [37], and *C. saitoana* [38]. A comparable level of control rot on oranges and lemons caused by *P. digitatum* was also demonstrated with bacterial antagonists *Pseudomonas* spp. [39, 40, 41], *B. subtilis* [27, 42], and fungal antagonists including *Myrothecium rorodum* [43], and *T. viride* [44]. When used alone on oranges and lemons, *C. saitoana* resulted in 40% less decay than in the water-treated control [38]. A slightly better level of control was obtained with *C. oleophila* on oranges inoculated with *P. digitatum* [45, 46]. The biocontrol potential of microbial antagonists was also reported on pome and stone fruits. On apples and pears, bacterial antagonists (*P. cepacia* and *P. syringae*) were shown to be effective in controlling decay caused by *B. cinerea* and *P. expansum* [12, 47]. The combination of the bacterium *P. syringae* and the pink yeast *Sporobolomyces roseus*, against blue mold on apple controlled this disease more effectively when combined at approximately equal biomass (50:50 of the same turbidity) than in individual applications [48]. Nutritional profiles of these antagonists, based on utilization of 35 carbons and 33 nitrogen sources, revealed significant differences in carbon catabolism [48]. Control of apple decay was also shown with *C. sake* [49], *C. oleophila* [50], *C. saitoana* [24], *Cryptococcus infirmominiatus*, and *C. laurentii* [23]. McLaughlin et al., [51] showed that lesion size and frequency of Botrytis rot of apple were significantly less in fruit pretreated with aqueous suspensions of *Candida* sp. strains 87 and 101 as compared with controls pretreated with water. Teixido et al., [49] reported that *C. sake* when applied as a wound treatment reduced lesion diameter by more than 80 % and resulted in 50 % reduction in the incidence of lesions. In apples wounds inoculated with *P. expansum*, *C. infirmominiatus* combined with *C. laurentii* reduced decay by 84 % and the combination of *C. infirmominiatus* with a low dose of thiabendazole (264 mg/ml) resulted in 91% reduction of decay [23]. *C. infirmominiatus* and *C. laurentii* were also effective in controlling decay of pears caused by *B. cinerea* and *P. expansum* [23].

Antagonistic bacteria [30], yeast, and fungi [51, 52] have also been shown to reduce brown rot of stone fruit caused by *Monilinia fructicola*. *Bacillus subtilis* reduced the incidence of brown rot of peach in laboratory studies and in semi-commercial trials [30]. Reduction of peach and plum decay was also reported with *Trichoderma atroviride* isolates, *T. viride* isolate 23-E-6, and *Rhodotorula* sp. BI-54 [52]. These isolates reduced brown rot on peaches by 63-98% and plums by 67-100%, when fruits were inoculated with *M. fructicola* following the application of the biological control agent. The control of brown rot on plums was also obtained when the antagonists were applied 12 h earlier than inoculation with *M. fructicola* [52].

The biocontrol potential of microbial antagonists has been demonstrated on kiwi fruit [53], potato [54], strawberry fruit [55], avocado [56, 57], feed grain [58], table, and cherry tomato [59]. On kiwi fruit, topical application of five different yeast to pedicel wounds conferred a significant level of biocontrol of *Botrytis* rot following applications made simultaneously with, or up to 96 h after inoculation with *B. cinerea* [53]. Biocontrol activity of the yeast was further increased by fruit curing but only when yeast application was made after 96 h of fruit curing at 10 C. On avocado, pre-inoculation of fruit with an isolate of *Colletotrichum gloeosporioides* generated by insertional mutagenesis delayed symptom development by the wild type isolate of *C. gloeosporioides* without causing any symptoms on the pericarp [57]. In an artificial inoculation study on avocado, *B. subtilis* was effective against *C. gloeosporioides*, *Fusarium solani*, and *Trichoderma pseudotrichia* [56]. Scheena et al. [59] showed that isolates of *A. pullulans* isolates were effective against *B. cinerea*, *Rhizopus stolonifer* and *Aspergillus niger* on table grape and *B. cinerea* and *R. stolonifer* on cherry tomato. The preharvest application of *A. pullulans* isolate L47 on table grape resulted in a significant 27.1 to 49.5% reduction of gray mold compared to the untreated control. Microbial antagonists also have potential as preharvest and postharvest treatment for grain crops to prevent disease during plant development and suppress mycotoxin production during seed storage when kernels are not dried adequately [58, 60]. On corn kernels, a strain of *T. viride* isolated from corn root was found to reduce fumonisin B-1 production by *Fusarium moniliforme*. Levels of the mycotoxin decreased by 85% when the pathogen and the antagonist were inoculated onto kernel the same day and by 72% when inoculation of *T. viride* was delayed by 7 days after inoculation with *F. moniliforme* [60]. Control of grain spoilage was also reported with *Pichia anomala* on wheat kernels inoculated with *Penicillium roqueforti* [58].

4. COMMERCIAL APPLICATIONS

The success of some of these microbial antagonists in laboratory studies has generated interest by several agrochemical companies in the development and promotion of microbial antagonists for control of rots of fruits and vegetables. Several microbial antagonists have been patented and evaluated for commercial use as a postharvest treatment. From the industry perspective, the selected antagonists must meet certain criteria to be successfully developed for commercial use on harvested crops. Selected microbial antagonists should: (1) pose no risk to human health and environment; (2) be genetically stable; (3) provide effective and consistent control of a wide range of pathogens on a variety of fruits and vegetables; (4) be effective at low concentrations; (5) not be fastidious in its nutrient requirements; (6) be able to survive and remain active under various environmental storage conditions; (7) be amenable to production on an inexpensive growth medium; (8) be amenable to a formulation with a long shelf life; (9) be easy to dispense; (10) be compatible with commercial processing procedures; and (11) be nonpathogenic to the host commodity. Presently, four antagonistic microorganisms, yeasts *C. oleophila* and *C. albidus,* and two strains of the bacterium, *P. syringae,* are commercially available under the trade names ASPIRE, YieldPlus, and BIOSAVE-110, respectively. In commercial packinghouse tests on oranges conducted in Israel, the combination of Aspire with 200-micrograms/ml thiabendazole reduced the incidence of green and blue molds to a level comparable to that obtained with the conventional fungicide treatments containing sodium O-phenyl-phenate, thiabendazole, imazalil, and metalaxyl [61]. Aspire treatment also was highly effective against sour rot caused by *G. candidum*, a decay not controlled by the conventional fungicide treatments. On pears, packinghouse trials showed that the level of control by Bio-Save 110 +100 ppm thiabendazole and Aspire +100 ppm thiabendazole was comparable to that obtained with the maximum label rate of thiabendazole [61]. Both products, however, in laboratory tests were less effective than the antagonistic yeasts *R. glutinis*, *C. infirmo-miniatus*, and *C. laurentii* in controlling decay caused by thiabendazole-resistant and thiabendazole-sensitive *P. expansum* [62]. These latter yeasts, in semi-commercial trials on apple and pears, were as effective as the commercial recommended rate of thiabendazole (528 micrograms /ml), when used in combination with a reduced rate of thiabendazole (264 micrograms /ml) or in combination with one another [23]. The effect of Aspire and Bio-Save was evaluated on several varieties of Florida oranges. Both biological products significantly reduced green mold of citrus fruit, but the level and consistency of control of decay was usually less compared to thiabendazole and imazalil [45]. Aspire and Bio-Save equaled the efficacy of the standard rate of thiabendazole and imazalil only when combined with a reduced rate of thiabendazole.

Acceptance of the biological control of postharvest diseases, like biological control approaches to manage other pests such as weeds or insects, is greeted with caution and skepticism by many in the agricultural community. Pests and diseases are in most cases usefully reduced by biological control product applications, while the impressive but often temporary near-elimination of pest or pathogen populations that typically transpires when a new chemical pesticide is introduced is not likely to occur. Like any new technology, the adoption of these products is greatly facilitated by successful commercial tests. Unfortunately, the efficacy of some postharvest biological control antagonists evaluated under simulated and actual commercial conditions has been irregular [45] unless they are combined with other treatments, such as low rates of fungicides.

Considerable progress has been made to examine each phase of fresh fruit handling and processing to determine its influence on the efficacy of postharvest antagonists, but more work is needed so strategies to improve their reliability and effectiveness can be developed. The efficacy of antagonists is influenced by: (1) interval between application of the antagonist and inoculation by the pathogen; (2) the temperature of storage and transportation; (3) the season of harvest and maturity of the fruit; (4) mortality of the antagonist by common sanitizers and fungicides; (5) characteristics of the wound site; (6) formulations and fruit wax composition.

4.1. Application of the Antagonist and Inoculation by the Pathogen

Unlike synthetic fungicides, microbial antagonists have no eradicant activity against established infections [38, 63] and act primarily as wound protectants. Antagonists are effective only when applied into a wound before or within few hours of infection by the pathogen. This prerequisite is difficult to meet since in most commercial packinghouses delays of up to 24 hours between infection during harvesting and the time of treatment with decay control products on a commercial line are quite common. Eradicant activity is particularly needed for pathogens that infect wounds made during harvest and subsequent handling, such as *Penicillium* decay and sour rot of citrus [63, 64]. Intervals of 24 hours or more between harvest and the first opportunities for packingline antagonist application are common because at many facilities the fruits are harvested, trucked overnight to the packinghouse, and processed the next day. However, efforts to manage harvest and processing schedules to minimize this interval are feasible in many operations, particularly where the fruit processed are harvested locally. The lack of eradicant activity by microbial antagonists may in part explain their reduced efficacy and/or inconsistency in comparison to synthetic fungicides. Recent efforts to overcome the lack of eradicant activity and improve the efficacy of existing biological products have involved the use of cold storage, physical treatments such as curing, heat, and ultraviolet light [22, 40, 65, 66, 67], and the use of various naturally-occurring, bioactive compounds [21, 38, 50, 63, 68].

On pome and citrus fruit, the addition of $CaCl_2$ increased the protective effect of some antagonistic yeast [50, 69]. Control of green mold of oranges and lemons was maximized when fruits were pretreated with sodium carbonate prior to treatment with Bio-Save [67] or *C. saitoana* [38, 63]. Biocontrol activity of microbial antagonists on citrus and pome fruits was also shown to be augmented by addition of amino acids [21], 2-deoxy-D-glucose, a sugar analog [21, 32], and chitosan [63]. The level of control obtained from the combination of *C. saitoana* with 0.2 % 2-deoxy-D-glucose on lemon and orange fruit was similar to that of imazalil, a common fungicide with worldwide usage. On Gala apple, the protection provided by curing fruit at 38 C for 4 days was further enhanced by treating fruit wounds with the antagonist *P. syringae*, or one of two yeast antagonists [66].

4.2. The Temperature of Storage and Transportation

Temperatures of fruit after harvest can impact the efficacy of the antagonists [70, 71, 72]. Brown et al. [70] reported *C. oleophila* applied in the Aspire formulation, colonized wounds on oranges after 2 days at 21 and 30 C, but not at 13 C. In harvest operations in California and elsewhere, some facilities cool oranges to 10 C or less soon after harvest. This is thought to reduce the efficacy of *C. oleophila*. Manipulation of the postharvest temperature to favor the yeast and retard green mold development, which occurs at temperatures as low as 30 C, was discussed by Brown [70]. However, the higher temperature favors other pathogens, such as *G. citri-aurantii*, and can aggravate physiological disorders such as rind pitting of citrus [73]. With pear [71], apple [71] and cherry [72] fruits, the storage temperature influenced the effectiveness of most yeast antagonists, with some superior at warm temperatures and others at cool temperatures [74].

4.3. The Season of Harvest and Maturity of the Fruit

Season of harvest and maturity of the fruit also influences efficacy; efficacy declines with later harvest or advanced maturity of pear [75], apple [76] and citrus fruit [38]. Therefore, in a citrus processing facility where fruit are stored for later marketing, application of an antagonist to freshly harvested fruits entering storage would probably be more effective than that to fruits shipped after storage, and better effectiveness would be anticipated early in the season rather than later.

4.4. Mortality of the Antagonist by Common Sanitizers and Fungicides

Sanitation has long been an important component in the management of many postharvest pathogens, and more recently, it has become an important element to minimize the contamination of fruit with human pathogens. Chlorine is in widespread uses to sanitize the fruit and packingline process water, such as that in dump tanks, pressure washers, and flumes. Chlorine will kill antagonists; bacterial antagonists, such as Pseudomonas spp., are killed by chlorine dosages about 50 times lower than those that kill most fungi [77]. Little has been published on the impact of chlorine use on practical postharvest biological control, but it is obvious care must be taken that processes where chlorine is in use must be separated from and precedes antagonist application. Some have discussed the use of chlorine oxidants, such as calcium thiosulfate, within the antagonist formulations or in rinse water applied before the antagonists are applied, to minimize chlorine-caused mortality of the antagonists. The influence of fungicides and their formulations on the effectiveness of the fungicides must be determined. Some, such as thiabendazole, have been compatible with many types of yeast and have no antibacterial activity, so they can be used with many antagonists. Others, such as fludioxonil, inhibit a wide range of yeasts and other fungi, so its use with antagonists may be more limited.

4.5. Characteristics of the Wound Site

The nature of the wound site significantly impacts antagonist behavior and efficacy. Mercier and Wilson [78] showed the application of *C. oleophila* should follow the occurrence of wounding as soon as possible in order to obtain optimal disease control during storage, because wound drying significantly reduced the rate of colonization of apples. Conversely, on citrus fruit Brown and coworkers [70] reported peel oil was toxic to cells of *C. oleophila* but not to spores of *P. digitatum*, and that ruptured oil glands were colonized more effectively if treated 7 h after injury rather than immediately. They showed *C. oleophila* colonized punctures more uniformly than individually damaged oil glands, and provided more effective control of green mold originating at punctures that avoided oil glands. Eckert [79] showed that if oils accompanied the *C. oleophila* application, populations of the yeasts subsequently recovered from wounds on the lemons were far higher. Deep wounds that penetrate into the juice sacs of citrus fruit, such as those caused during the harvest of 'Lisbon' lemon, which has by the long, sharp thorns, greatly facilitate pathogen infection and cannot be controlled by any treatment [64]. These wounds are less frequent on cultivars that lack thorns, such as 'Eureka' lemon.

4.6. Formulations and the Influence of Fruit Waxes

The lower performance of formulated biocontrol products in comparison to synthetic fungicides may be attributed to inherent limitation of antagonists or to the reduction of performance of antagonists as result of the industrial mass production of active dried yeast.

Downstream processing and formulation involves various steps, such as drying, addition of inert ingredients, adhesives, emulsifiers, and adjuvant. All these actions may adversely affect the characteristics important to the biocontrol activity. In many instances, the dried formulation of antagonistic yeasts was shown to be less effective than the laboratory produced yeast paste. The rapid uptake of water by dried yeast was shown to cause plasmolysis unless certain precautions, such as using Luke warm instead of cold water. The type of the delivery system used to apply a biocontrol agent such as on-line drenchers and sprayers are also important to its performance.

In addition to efficacy, other issues influence to the adoption of this technology, such as its price, ease of use, acceptance of treated products by regulators and buyers, and the spectrum of pathogens that require management. Food safety issues, primarily addressing the accidental contamination of fresh fruit with human pathogens, are of increasing concern to buyers, regulators and consumers, and biological control agents have an impact on this issue as well. Janisiewicz and coworkers [80] reported the antagonist *P. syringae* in wounds of 'Golden Delicious' apple prevented *Escherichia coli* 0157:117 from growing in the wounds. These results indicate that biocontrol agents that control the decay of fruits may have the additional benefit of preventing the growth of foodborne pathogens in freshly wounded tissue of intact and fresh-cut fruits. Conversely, solutions of antagonists at packinghouses can be potential food safety hazard points, because the multiplication of microbes of food safety concern could conceivably occur within them, and they cannot be chlorinated or heat pasteurized. To manage this hazard, the antagonists solution should not be repeatedly reused, such as in a fruit dip tank or re-circulating spray, and the useful life of the diluted solution used in the packinghouse should be carefully monitored and relatively brief.

Stem-end rot of citrus, the primary pathogen of citrus in Florida and other humid citrus growing areas of the world, has been poorly controlled to date by antagonists [38, 70]. Therefore, reliance on antagonists alone at this time in Florida is not feasible, while in California and other arid growing areas this issue is of little importance because stem-end rot is less prevalent.

5. MODE OF ACTION OF ANTAGONISTIC MICROORGANISMS

Although the biocontrol potential of microbial antagonists has been demonstrated with several commodities, the mode action of antagonists has not been fully elucidated. The reported data indicate that it may involve antibiosis, nutrient competition, site exclusion, direct parasitism, and induced resistance [81, 82]. If we are to maximize the biocontrol potential of microbial antagonists, a more fundamental understanding of their mode of action is imperative. As we learn more about the mechanism by which antagonistic yeasts protect fruits from rotting, more effective methods of formulating, applying and selecting antagonists will ultimately emerge.

5.1. Antibiosis

Antibiosis occurs when the production of antibiotics or toxic metabolites by one microorganism has a direct effect on another microorganism. The selection of antibiotic-producing microorganisms is generally based on crude plate assays in which the antagonist is shown to produce an inhibition zone restricting the radial growth of the target pathogen. Antibiosis is a commonly assumed mechanism for the biocontrol activity of bacterial species on leaf surfaces and in other habitats. Several types of antibiotics including bacteriocin-like compounds and siderophores have been suggested as inhibitory agents responsible for biocontrol of certain foliar and soil-borne diseases [83, 84]. For instance the amendment of sterilized soil with the pyoverdin produced by *Pseudomonas* spp. was shown to reduce the rate of germination of *Fusarium* chlamydospores. Also isogenic mutant of *P. putida*, disenabled in its capacity to produce siderphores, provided further evidence for the involvement of competition for iron as a mechanism of the antagonism expressed by fluorescent *Pseudomonas* spp against *F. oxysporum* [83].

Involvement of antibiotics in the biocontrol of antagonistic bacteria against postharvest decay of fruit has also been inferred [31, 41, 47]. Several bacterial antagonists have been shown to reduce postharvest decay and their protectant effect was in part associated with the production of antibiotics. *B. subtilis* (strain B-3) an effective antagonist against brown rot of peaches produces an antibiotic, iturin, which is capable of protecting wounds in the absence of the bacterium [31]. Topical treatment of fruit wounds with *Iturin* alone-reduced decay caused by *M. fructicola* [31]. Similarly, reduction of decay was also reported in apples, pears, and citrus treated with *Burkolderia* (*Pseudomonas*) *cepacia* or its antibiotic, pryrrolnitrin [41, 47]. On citrus fruit, a pryrrolnitrin-resistant strain of *P. digitatum* was partially controlled on citrus fruit by *B. cepacia* [41]. Also, in other host-pathogen combinations, some pyrrolnitrin production-deficient mutant strains of *B. cepacia* retained all of their biocontrol capacity [85] or lost some or all of their effectiveness [86]. Thus it is suggested that other mechanisms beside antibiosis may be involved in the biocontrol activity of *B. cepacia*.

The bacterial antagonist *P. syringae*, which controls *Penicillium* molds of citrus fruit and gray molds of pome fruit, also produces an antibiotic, syringomycin, that is inhibitory to *P. digitatum in vitro* and in fruit wounds [87, 88]. Characterization of syringomycin production by syringomycin-deficient mutants strains of *P. syringae* suggest that syringomycin is secreted in the infection court and may have an active role in control of green mold [88]. Although antibiotic-producing microorganisms have a potential to be used as biocontrol agents of postharvest diseases, more effort is being deployed toward the development of non-antibiotic-producing, antagonistic yeasts for the biological control of postharvest diseases [11].

5.2. Direct Interaction with Pathogens

Mycoparasitism, the phenomenon of one fungus parasitizing another, is well documented in biological control of soil-borne and foliar diseases by *Trichoderma* spp. [89, 90]. Parasitism depends on close contact and recognition between antagonist and host, on the secretion of enzymes, and on the active growth of the hyperparasite into the host [90]. Direct parasitism of fungal pathogens was also reported with several antagonistic yeasts [24, 81, 91]. For instance, *P. guilliermondii* and *D. hansenii* when co-cultured with *B. cinerea* strongly attach to hyphae of *B. cinerea* and cause swelling and, in extreme cases, the complete disruption of hyphal wall structure. When yeast cells were dislodged from the hyphae, a concave appearance of the hyphal surface and partial degradation of the cell wall of *B. cinerea* was observed at the attachment sites [81]. The partial degradation of *B. cinerea* cell walls by *P. guilliermondii* was attributed to its tenacious attachment to hyphal walls in conjunction with its production of ß-1,3-glucanase [81]. The close contact of the yeast cells with the fungal cell wall could also facilitate the efficient uptake of nutrients from the immediate vicinity surrounding the fungus and the intramural release of lytic enzymes by the yeast. *Pichia guilliermondii* produced high levels of ß-1,3-glucanase activities when co-cultured on various carbon sources or on cell walls of several fungal pathogens [81]. The glucanase secreted by *P. guilliermondii* was identified as an exo-ß-1,3-glucanase with molecular weight of 45 KDa and its production kinetics during growth of the yeast revealed that its activity in culture filtrate reached its maximum after 36-48 h growth [92].

A similar attachment pattern has been observed with the antagonistic yeast *C. saitoana* when co-cultured with *B. cinerea* [24]. When *B. cinerea* was allowed to grow for 16 h before adding *C. saitoana*, the yeast cells attached tightly to the hyphae despite extensive rinsing with distilled water. Closer examination by scanning electron microscopy showed that the yeast cells become embedded in the hyphal walls. In apple wounds, *C. saitoana* prevented the proliferation of *B. cinerea* and attached to the hyphae of *Botrytis* [24]. *Botrytis* hyphae in contact with yeast cells displayed severe cellular alterations that included cell wall swelling, extensive vacuolation, and cytoplasmic degeneration. Like *P. guilliermondii*, *C. saitoana* is also capable of producing fungal cell wall degrading enzymes, namely chitinase and ß-1,3 glucanase. This, in part, may explain the severe alterations observed in fungal wall areas in contact with *C. saitoana* cells. The cellular damage evident in *Botrytis* cells was similar in many respects to that reported in aged and nutrient-deprived fungal cells [93]. It is quite possible that the observed cellular alterations of *B. cinerea* cells may be the result of nutritional starvation caused by rapidly multiplying *C. saitoana*.

Attachment of antagonistic yeasts to fungal cell walls was blocked when the yeast cells or the pathogen hyphae were exposed to compounds that affect protein integrity or respiratory metabolism, thus indicating a lectin-type of recognition system [81]. Recently, a gene ALA1 from *C. albicans* that encode for cell surface molecules, adhesins, have been isolated and shown to promote both attachment and aggregation in *Saccharomyces cerevisiae* transformants [94]. The adherence mediated by the ALA1 gene appear to consist of two sequential steps: an initial attachment phase to protein-coated surface that occur rapidly followed by a rather slower aggregation phase in which cell-to-cell interaction predominate. The aggregation step is believed to assist in the colonization of the host by recruiting more cells. The adherence of *S. cerevisiae* overexpressing ALA1 gene to protein ligands was unaffected by shear force and the presence of several competing molecules including salts, sugar, and detergent, but was inhibited by agents that disrupt hydrogen bonds [94].

As to the role played by the attachment capability of antagonistic yeasts in the observed control of lesion development, it has been suggested that attachment may enhance nutrient competition as well as interfere with the ability of the pathogen to initiate infection. Analysis *in planta* of the effect of treatments that negate attachment may provide further insight regarding the extent of the role played by attachment in the biocontrol activity of antagonistic yeasts. *In vitro* studies have shown that yeast attachment to fungal hyphae occurs within 12 to 16 hours of incubation [81], while in fruit tissue yeast attachment to Botrytis hyphae was detected within 48 to 72 hours after inoculation [24]. Thus it is unlikely that attachment might play a decisive role in local protection, considering that most fungal pathogen start penetrating fruit tissue within 6 to 10 hours. This does not, however, exclude a supporting role for attachment in local protection provided by antagonistic yeast.

5.3. Nutrient and Space Competition

Competition is probably one of the most studied modes of action of biocontrol microorganisms. Competition occurs when there is demand by two or more microorganisms for the same resource. According to Lockwood [95], microbial competition is based on exploitation, during which one organism depletes resources without inhibiting access of another organism, or on interference, during which such access is inhibited by the presence of the first colonizer. Therefore, reduction in available nutrients by actively growing antagonistic microorganisms appears to be a suitable method to reduce the incidence of infection by plant pathogens. Nutrient competition is major component of the mode action of microbial antagonists in the phylloplane and rhizosphere. Nutrient stress is only a suitable mechanism of antagonism for pathogens which are sensitive to it and for which infection can be stimulated by nutrient addition. Blakeman and Brodie, [96] showed that in phyllosphere bacterial activity similar to soil fungistasis [97] inhibited spore germination of *B. cinerea* by competition for exogenous nutrient, namely amino acids, and by utilizing nutrient leached from the ungerminated spores during water uptake. In an amino acids-glucose mixture, *Pseudomonas* sp. removed more than 80% of amino acids in 5 h and inhibited germination of *B. cinerea* conidia [96]. Using ^{14}C-labeled glucose and amino acids, Edwards and Blakeman [98] showed that in presence of *Sporobolomyces* sp. or *P. fluorescens* and *B. cinerea*, nutrients were partitioned in favor of the antagonists.

Competition for nutrients and space is also believed to constitute a major component of the complex mode of action of antagonistic microorganisms against postharvest diseases [18, 82]. Reports on the interaction between epiphytic microorganisms showed that antagonistic bacteria and yeasts take up nutrients from dilute solutions more rapidly than germ tubes of filamentous fungal pathogens [18, 61, 82, 91]. For instance, *P. guilliermondii*, a non-antibiotic- producing yeast, when co-cultured with *P. expansum* in minimal synthetic medium or in wound leachate solutions inhibited the spore germination and growth of *P. expansum* [18]. Similarly, *Enterobacter cloacae*, a bacterial antagonist, inhibited germination of *R. stolonifer* spores through nutrient competition [91]. Recent studies *in vitro* partitioning of ^{14}C glucose between cells of antagonist yeast *S. roseus* and conidia of *B. cinerea* indicated much greater removal of glucose by the antagonist, which prevented germination of the conidia by nutrient deprivation [99]. This competition most likely extends to other nutrients in fruit wounds, particularly nitrogenous compounds that are at low concentrations. In deed, in apple wounds nitrogen rather than carbon was shown to be the principal limiting factor for both the antagonists and the pathogen [100]. The yeast population was unaffected by addition of glucose but increased when amino acids were added to the wounds.

Competition for nutrients and space was also indirectly demonstrated in fruit wounds and shown to play a major role in the mode of action of P. *guilliermondii* against *P. digitatum* [18], *E. cloacae* against *R. stolonifer* [91], *C. laurentii* against *B. cinerea* [14], and *S. roseus* against *B. cinerea* and *P. expansum* [21]. Droby et al. [18] showed that the biocontrol potential of *P. guilliermondii* was highly dependent on the ability of the antagonist to prevent the establishment of a nutritional relationship between the pathogen and the host. In fruit wounds, the biocontrol activity of *P. guilliermondii* was easily reversed by the addition of low amount of glucose [18]. Negation of biocontrol activity by the addition of exogenous nutrients into the wound site was also demonstrated with *E. cloacae* [91], *C. laurentii* [14], and *S. roseus* [21]. Involvement of nutrient competition in the mode of action of antagonists is also suggested by the close relationship between the performance of the antagonist and stage of fruit ripeness. In general, the level of control provided by antagonistic bacteria and yeasts often diminishes with an increase in fruit maturity [38, 41]. This is believed to be due to an increase in nutrient availability concomitant with the onset of ripening as a result of the biochemical changes associated with ripening.

In addition to nutrient competition, the biocontrol activity of antagonistic microorganisms also seems to be in part due to their ability to outcompete pathogens for space [82]. The level of control offered by antagonistic microorganisms highly depends on the initial concentration of antagonist applied in the wound and the ability of the introduced antagonists to rapidly colonize the wound site [12 18, 51, 91]. In most reports on the biological control of postharvest diseases of fruits, a quantitative relationship has been demonstrated between the concentration of the antagonist applied in the wound and the efficacy of the biocontrol agent [6, 18, 51, 91]. In most commodities, the biological activity of antagonistic yeasts and bacteria increases with increasing concentration of antagonist applied in the wound [6, 14, 18, 20, 51]. In general, with most harvested commodities, biocontrol activity increased with increasing dose of the antagonist and decreasing dose of the pathogen. Microbial antagonists are most effective in controlling postharvest decay when applied at 10^7 to 10^8 CFU/ml [6, 18, 24]. Often, little to no control of decay has been observed when antagonistic yeasts were applied at 10^5 CFU/ml.

The performance of antagonists is also dependent on their ability to proliferate rapidly in fruit wounds in order to prevent the establishment of the pathogen in the fruit tissue. This was demonstrated by using a mutant of *P. guilliermondii*, which lost its biocontrol activity against *P. digitatum* on grapefruit and against *B. cinerea* on apples, even when applied to the wound at concentrations as high as 10^{10} cells/ml [101]. The cell population of this mutant remained constant at the wound sites during the incubation period, while that of the wild type increased 10 to 20-fold, within 24 hours. With apple and citrus fruit, examination of fruit wounds showed that introduced antagonistic yeasts often formed a matrix of yeast cells that covered the cell layers ruptured during wounding [18, 24, 102]. Similar tissue colonization pattern was also displayed by bacterial antagonists [6]. The study of the population dynamics of antagonists showed that antagonists started multiplying in the wound site within three to six hours after application. Within 24 h, the population of introduced antagonists usually increases by nearly ten-fold, stabilizing thereafter [18, 21, 62, 87, 102, 103]. This may result in a reduction in available nutrients and thus a reduction in pathogen spore germination and hyphal development.

Although competition for limiting nutrients appears to be a more general mechanism of antagonism in fruit wounds, the extent to which it reduces infection may vary with the infection strategy of the pathogen involved and commodity used. Even though the same concentration antagonistic yeast such as *C. oleophila* and *C. saitoana* is applied to citrus, pome, and stone fruits, the level of control offered by the yeast appear to differ considerably from one commodity to the other and within the same commodity depending on the target pathogens tested [5, 7, 8, 51]. Thus, indicating not only a differential sensitivity of the pathogens to the antagonism of the yeast but also differences in the commodities in term of efficiency of yeast biocontrol activity. In pome fruits, *Candida* sp. was shown to be more effective in controlling lesion development caused by *B. cinerea* than that caused *P. expansum* [8, 51]. Also antagonistic yeasts attached more tenaciously to *B cinerea* than to *P. expansum* hyphae [81]. The deferential sensitivity of the pathogens to *Candida* sp. is likely the result of differences in biochemical and genetic traits whereby *B. cinerea* shows probably greater similarity in ecological niche to *Candida* sp. than *P. expansum*. A high degree of similarity in term of nutrient requirements for growth and pathogenesis should increase competition between the saprophytic antagonist and the pathogen, provided the resources for which antagonist outcompete the target pathogen in fruit wounds are a limiting factor. Presently, however, there are no clear information regarding the specificity of competition between antagonistic yeasts and various pathogens in fruit wounds. Antagonistic yeasts especially *Candida* sp. tend to be more effective in controlling decay of citrus and apple than that of stone fruits [5, 7, 8, 51]. Also different varieties of a given crop may affect the performance of microbial antagonist differently. Within the same crop, the efficacy of microbial antagonist may vary from one variety to another. If one considers that competition for nutrient is the primary mechanism of microbial antagonists, then it is quite possible that the resources for which antagonist competes successfully on one commodity may not be limiting on different variety or a different commodity. Currently, there is only fragmented data regarding the host-antagonist interaction and antagonist-pathogen interaction in term of competition for limiting nutrient essential for pathogenesis. Once this information is available and genes responses of antagonism of biocontrol agents have been characterized, it will be possible to develop antagonistic strains with a higher rate of transport and/or metabolism of limiting nutrient essential for pathogenesis.

5.4. Induced Resistance

In recent years, considerable attention has been placed on induced resistance in vegetative crops as an important form of plant protection. This interest has been generated by the large volume of basic and applied research which demonstrates that plants can be rendered resistant by artificially turning on their natural defense mechanisms following treatment with microbial and biochemical elicitors [104-106]. Upon treatment with microbial and chemical elicitors, plant tissue usually reacts by producing highly coordinated local and systemic defense responses that are orchestrated in a complex spatial and temporal manner [107, 108]. Localized responses are expressed in tissue around the site of interaction and may include a hypersensitive reaction (HR), callose formation, lignification, and phytoalexin accumulation. Whereas, systemic defense responses develop in cells distant from the site of the stimulus and involve the activation of antifungal hydrolases such as chitinase, ß-1,3-glucanases, and peroxidases. The effectiveness of both localized and systemic responses against a pathogen lies both in their magnitude and rapidity of their onset.

Recently, antagonistic yeasts have been shown to induce several biochemical defense responses in harvested commodities and the involvement of induced resistance as part of their mode of action has been inferred [24, 25, 82]. In apple fruit, *C. saitoana* induced a transient increase in chitinase activity and the formation of papillae along host cell walls that appeared to interfere with pathogen ingress [24]. Activation of defense responses was also reported with *A. pullulans* on apple fruits [25]. When applied in apple wounds, *A. pullulans* caused a transient increase in □-1,3-glucanase, chitinase, and peroxidase activities starting 24 h after treatment and reaching maximum levels 48 and 96 h after treatment. Induction of disease resistance responses were also reported in pineapple, avocado and citrus fruits [37, 109-111]. With avocado fruits, the protection provided by a reduced-pathogenicity isolate of *C. gloeosporioides* coincided with an increase in the levels of preformed antifungal diene [57, 110]. Induction of antifungal secondary metabolites was also seen in citrus fruit following treatment with two antagonistic yeasts, *P. guilliermondii* and *C. famata* [37, 111]. Arras [37] showed that accumulation of the phytoalexin scoparone could be 19 times higher when the antagonist *C. famata* was applied to fruit 24 h prior to *P. digitatum*, and only four times higher if applied 24 h after the pathogen. A comparable accumulation kinetic of scoparone was reported in lemon fruit treated with *C. guilliermondii* [111]. Although the ability of various antagonistic yeasts to activate of fruit defense responses is well established, the identity and structure of the signal molecules or stimuli generated by antagonistic yeast has not been elucidated. Oligosaccharide fragments of yeast cell wall polysaccharides composed of 3-, 6-, and 3,6-linked b-glucosyl residues are known to be active elicitors of host defense responses [112] and their release at the plant-interface is mediated by host-cell wall degrading enzymes. Induction of disease resistance responses was also reported in cultured tomato cells treated with a high mannose-containing glycopeptide obtained from yeast glycoprotein [113-116]. Both the glycan and the peptide portion of the elicitor were shown to be essential for activity. The removal of the peptide portion caused a loss of elicitor activity, thus indicating either that the peptide interacts with the binding proteins or stabilizes elicitor-active conformation [113, 114]. On the other hand, the free glycans released from the yeast glycopeptides appear to suppress the eliciting ability of the glycopeptide in tomato cells [113, 114], presumably by competing for identical binding site.

The normal ultrastracture appearance of antagonistic yeasts in fruit wounds suggests that the active elicitor may not be a product of host lytic enzymes. In fruit wounds, yeast cells appeared normal and showed no sign cell wall alteration. Yeast cells were surrounded by compact cell walls and the only discernible structures in their dense cytoplasm were lipid bodies [24]. Antagonistic yeasts often produce a large amount of extra cellular mucilage along host cell walls and in many instances, the yeast cells appeared to be embedded in the fibrillar material [24, 81, 91]. It is quite possible that the extracellular mycilage may be implicated in cell adhesion and could contain active chemical elicitor that provides signals for recognition and subsequent responses. The physical contact mediated by adhesion could be expected to facilitate communication between the antagonist and host tissue. In pea endocarp, the firm adhesion of fungal conidia to host tissue was require for the released fungal elicitor to induce pisatin synthesis [117]. Implication of microbial mucilage and its content in fungal adhesion to plant surface and activation of host defense responses have been suggested in other plant-microbe system [118]. Mucilage extracted from the surface of the yeast cells was shown to be inhibitory to several postharvest pathogens and prevented the development of green mold on oranges [119]. The ability of extracellular mucilage to activate host defense responses, however, was clearly demonstrated. Furthermore, the yeast adhesive compounds that mediate attachment to fruit tissue and fungal walls has not been isolated or characterized and the molecular bases of antagonistic yeast-substratum binding have not been elucidated.

In most of these studies, however, the induction of disease-resistance by microbial antagonists has been inferred but not clearly established since other putative modes of action of the antagonist could not be ruled out. While a causal connection between the accumulation of chitinase, ø-1,3-glucanase and phytoalexin and bioprotection by antagonistic yeasts has not yet been established, the occurrence of high levels of these compounds in protected tissue suggest their involvement in disease resistance. In several plant-pathogen interactions, the induction and accumulation of PR proteins including chitinase and ø -1,3-glucanase is often correlated with the onset of induced resistance [104, 105, 106, 120]. Chitinase and ø -1,3-glucanase are known to hydrolyze fungal cell walls and in combination they have been shown to inhibit the *in vitro* growth of several pathogenic fungi [121, 122]. Considering that antagonistic yeasts are effective only when introduced into a wound before or within few hours of inoculation with the pathogen [24], the expression of disease-resistance, 24 to 96 h after yeast treatment, may not have a decisive role in local protection. This does not preclude a supporting role for induced resistance in combination with the antagonistic activity of the yeast. Nutritional-starvation of pathogen cells caused by rapidly multiplying antagonistic yeast can be expected to increase the susceptibility of pathogen cells to host antifungal enzymes namely chitinase and ø-1,3-glucanase.

Antagonistic yeasts, also possess ø-1,3-glucan and chitin in their cell walls [123], this raises the question of its sensitivity to chitinases and ø-1,3-glucanases produced by the fruit as well as to phytoalexins. In apple fruits, the growth of *A. pullulans* in wound cavities was not impaired by the accumulation of chitinase and ø-1,3-glucanase [25]. While in kiwi fruit, host-resistance mechanisms triggered by curing did not appear be specific to the pathogen, *B. cinerea*, since the biocontrol activity of epiphytic yeasts appear to be also reduced by factors induced in the first 24-48 h of curing [53]. It is quite possible that lytic enzymes of host origin may act differently on antagonistic yeast confined to wound cavities than on necrophitic pathogens. In ectomycorrhizal interactions plant-secreted chitinases and ø-1,3-glucanases inhibitory toward pathogenic fungi showed no effect on ectomycorrhizal fungi, thus allowing the establishment of symbiotic interactions [124]. Also, filamentous fungi, despite having similar cell wall constituents [123] are known to display differential sensivity to chitinases and ø -1,3-glucanases [122].

6. CONCLUSIONS

Biocontrol products are expected to protect wounds from being infected during processing, deactivate wound infection occurring during the harvesting process, and this protection must last during storage, shipping and retailing phase. In recent years, the development of microbial antagonists has received considerable attention, and presently a number of microbial antagonists are commercially available for the control of postharvest decay. If we are to maximize the biocontrol potential of microbial antagonists, a more fundamental understanding of their mode of action, their ecology, their compatibility with postharvest commercial practices, and the effect of host physiology on their biological activity is needed. Currently, several other promising biological control approaches that include the use of bioactive compounds and induced resistance are available and could be used in combination with microbial antagonists to provide a greater stability and effectiveness than the approach of utilizing a single microbial biocontrol agent. As we learn more about the fundamental bases underlying the protective effect of microbial antagonists more effective methods of formulating combinations of complementary biological approaches for additive and/or synergistic effects will emerge. So far, the results obtained with the different combinations of biological products demonstrate the potential of this multifaceted approach as a viable alternative to synthetic fungicides. The complexity of the mode of action displayed by combined alternatives should make the development of pathogen resistance more difficult and present a more highly complex disease deterrent barrier than an approach relying on a single biological agent. Therefore, such biological control strategies can also be expected to have greater stability and effectiveness than approaches utilizing single biological agents.

7. REFERENCES

1. Eckert, J.W. and Raynayake, M. (1983). Host-pathogen interaction in postharvest diseases. *In*: Postharvest Physiology and Crop Preservation (M. Liberman, Ed.), Plenum Press, New York, pp. 247-260.
2. Swinburne, T. R. (1983). Quiescent infections in postharvest diseases. *In* : Postharvest Pathology of Fruits and Vegetables *In*: Postharvest Pathology of fruits and vegetables (C. Dennis Ed.), Academy Press, London, pp. 1-25.
3. Eckert, J. W. and Ogawa, J. M. (1988). The chemical control of postharvest diseases: Deciduous fruits, berries, vegetables and roots/tuber crops. Annu. Rev. Phytopath. 26: 433-469.
4. Spotts, R. A. and Cervantes, L. (1986). Populations, pathogenicity, and benomyl resistance of

Botrytis spp., and *Mucor piriformis* in packinghouses. Plant Dis. 70: 106-108.
5. Wilson, C. L. and Wisniewski, M. E. (eds.). (1998). Biological Control of Postharvest Diseases of Fruits and Vegetables-Theory and Practice, CRC Press, Boca Raton, Florida.
6. Janisiewicz, W. (1998). Biocontrol of postharvest diseases of temperate fruits: Challenges and opportunities. *In*: Plant-microbe interactions and biological control (J. Boland and L.D. Kuykendall, eds.), Marcel Dekker, New York, pp 171.
7. El Ghaouth, A. and Wilson, C. (1995). Biologically-based technologies for the control of postharvest diseases. Postharvest. News. Inf. 6: 5-11.
8. El Ghaouth, A. (1997). Biologically based alternatives to synthetic fungicides for the control of postharvest diseases. J. Indust. Microbiol. Biotechnol. 19: 160-162.
9. Korsten, L. De Villiers, E. E. Wehner, F. C. and Kotze, J. M. (1994). A review of biological control of postharvest diseases of subtropical fruits. In: BR Champ, E Highley and GI Johnson, Eds. Postharvest handling of tropical fruits. ACIAR Proceedings, 50: 172-185.
10. Wilson, C. L. Wisniewski, M. E. El Ghaouth, A. Droby, S. and Chalutz, E. (1996). Commercialization of Antagonistic Yeasts for the Biological Control of Postharvest Diseases of Fruits and Vegetables. J. Indust. Microbiol. Biotechnol. 46 237-242.
11. Wisniewski, M.E. and Wilson, C.L. (1992). . Biological control of postharvest diseases of fruits and vegetables Recent. Adv. Hort. 27: 94-98.
12. Janisiewicz, W. and Roitman, J. (1988). Biological control of blue mold and gray mold on apple and pear with *Pseudomonas cepacia*. Phytopathology 78:1697-1700.
13. Redmond, J. C Maroi, J. J. and MacDonald, J. D. (1987). Biological control of *Botrytis cinerea* on roses with epiphytic microorganisms. Plant Dis. 71: 799-802.
14. Roberts, R. G. (1990). Biological control of gray mold of apple by *Cryptococcus laurentii*. Phytopathology 80:526-530.
15. Wilson, C. L. Wisniewski, M. E. Droby, S. and Chalutz, E. (1993). A selection strategy for microbial antagonists to control postharvest diseases of fruits and vegetables. Sci. Hort. 40:105-112.
16. Wilson, C.L. and Chalutz, E. (1989). Postharvest biological control of *Penicillium* rots of citrus with antagonistic yeasts and bacteria. Sci. Hort. 40:105-112.
17. Tronsmo, A. (1989). Effect of fungicides and insecticides on growth of *Botryis cinerea*, *Trichoderma viride*, and *T. harzianum*. Norwegian. J. Agric. Soc. 3: 151-156.
18. Droby, S. Chalutz, E. Wilson, C. L. and Wisniewski, M. E. (1989). Characterization of the biocontrol activity of *Debaryomyces hansenii* in the control of *Penicillium digitatum* on grapefruit. Can. J. Microbiol. 35: 794-800.
19. Chalutz, E. and Wilson, C. L. (1990). Postharvest biocontrol of green and blue mold and sour rot of citrus by *Debaryomyces hansenii*. Plant Dis. 74:134-137.
20. Gullino, M. L. Aloi, C. Palitto, M. Benzi, D. and Garibaldi, A. (1991). Attempts at biocontrol of postharvest diseases of apple. Med. Fac. Landbouw. Rijksuiv. Gent. 56: 195-200.
21. Janisiewicz, W. (1994). Enhancement of biocontrol of blue mold with nutrient analog 2-deoxy-D-glucose on apples and pears. Appl. Environ. Microbiol. 60: 2671-2676.
22. Lurie, S. Droby, S. Chalupowicz, L. and Chalutz, E. (1995). Efficacy of *Candida oleophila* strain 182 in preventing *Penicillium expansum* infection of nectarine fruits. Phytoparasitica 23:231-234.
23. Chand-Goyal, T. and Spotts, R. A. (1997). Biological control of postharvest diseases of apple and pear under semi-commercial and commercial conditions using three saprophytic yeasts. Biol. Control. 10: 199-206.
24. El Ghaouth, A. Wilson, C. and Wisniewski, M. (1998). Ultrastructural and cytochemical aspect of the biocontrol activity of *Candida saitoana* in apple fruit. Phytopathology 88: 282-291.
25. Ippolito, A. El Ghaouth, A. Wisniewski, M. and Wilson, C. (2000). Control of postharvest decay of apple fruit by *Aurobasidium pullulans* and induction of defense responses. Postharvest. Biol. Technol. 19:265-272.
26. Gutter Y. and Littauer, F. (1953). Antagonistic action of *Bacillus subtilis* against citrus fruit pathogens. Bull. Res. Counc. Israel. 33: 192-197.

27. Singh, V. and Deverall, B. J. (1984). *Bacillus subtilis* as a control agent against fungal pathogens of citrus fruit. Trans. Br. Mycol. Soc. 83: 487-490.
28. Tronsmo, A. and Ystaas, J. (1980). Biological control of *Botrytis cinerea* on apple. Plant Dis. 64: 1009.
29. Tronsmo, A. (1986). Trichoderma used as a biocontrol agent against *Botrytis cinerea* rots on strawberry and apple. Sci. Rep. Agric. Univ. Norway. 65: 1-22.
30. Pusey, P. L. and Wilson, C. L. (1984). Postharvest biological control of stone fruit brown rot. by *Bacillus subtilis*. Plant Dis. 68: 753-756.
31. Pusey, P. L. (1989). Use of *Bacillus subtilis* and related organisms as biofungicides. Pestic. Sci. 27: 133-140.
32. Droby, S.Lischinski, S. Cohen, L. Weiss, B. Daus, A. Chand-Goyal, T. Echert, J. W. and Manulis, S. (1999). Characterization of an epiphytic yeast population of grapefruit capable of suppression of green mold decay caused by *Penicillium digitatum*. Biol Control. 16: 27-34.
33. Droby, S. Hofstein, R. Wilson, C. Wisniewski, M. Fridlender, B. Cohen, L. Weiss, B. Daus, A. Timar, D. and Chalutz, E. (1993). Pilot testing of *Pichia guilliermondii*: A biocontrol agent of postharvest diseases of citrus fruit. Biol. Control. 3:47-52.
34. McGuire, R. G. (1994). Application of *Candida guilliermondii* in commercial citrus coating for biocontrol of *Penicillium digitatum* on grapefruits. Biol Control. 3: 1-7.
35. Cheah, L. H. and Tran, T. B. (1995). Postharvest biocontrol of *Penicillium* rot of lemons with industrial yeasts. Proceedings of 48th New Zealand, Plant Protection, pp. 155-157.
36. Arras, G. Sanna, P. and Astone, V. (1997). Biological control of *Penicillium italicum* of citrus fruits by *Candida sake* and calcium salt. Proceedings of 49th International Symposium, Crop Protection, pp. 1071-1078.
37. Arras, G. (1996). Mode of action of an isolate of *Candida famata* in biological control of *Penicillium digitatum* in orange fruits. Postharvest. Biol . Technol. 8:191-198.
38. El Ghaouth, A. Smilanick, J. G. Eldon Brown, E. Ippolito, A. and Charles L. Wilson, C. (2000). Application of *Candida saitoana* and glycolchitosan for the control of postharvest diseases of apple and citrus fruit under semi-commercial conditions. Plant Dis. 84: 243-248.
39. Bull, C. T. Stack, J. P. and Smilanick, J. L. (1997). *Pseudomonas syringae* strains ESC-10 and ESC-11 survive in wounds on citrus and control green and blue molds of citrus. Biol. Control. 8: 81-88.
40. Huang, Y. Deverall, B. J. and Morris, S. C. (1995). Postharvest control of green mold on oranges by a strain of *Pseudomonas glathei* and enhancement of its biocontrol by heat treatment. Postharvest. Biol. Technol. 3: 129-137.
41. Smilanick, J. L. and Dennis-Arrue, R. (1992). Control of green mold of lemons with *Pseudomonas* species. Plant Dis. 76:481-485.
42. Arras, G. and D'Hallewin, G. (1994). *In vitro* and *in vivo* control of *Penicillium digitatum* and *Botrytis cinerea* in citrus fruit by *Bacillus subtilis* strains. Agric. Mediter. 124: 56-61.
43. Appel, D. J. Gees, R. and Coffey, M. D. (1988). Biological control of the postharvest pathogen *Penicillium digitatum* on Eureka lemons. Phytopathology 12:1595
44. Borras, A. D. and Aguilar, R. V. (1990). Biological control of *Penicillium digitatum* by *Trichoderma viride* on postharvest citrus fruits. Intl. J. Food Microbiol. 11: 179-184.
45. Brown, G. E. and Chambers, M. (1996). Evaluation of biological products for the control of postharvest diseases of Florida citrus. Proc. Fla. State. Hort Soc. 109: 278-282.
46. Katz, M. (1996). Bio-fungicides at postharvest. Citrograph, 81: 3-5.
47. Janisiewicz, W. J. Yourman, L. Roitman, J. and Mahoney, N. (1991). Postharvest control of blue mold and gray mold of apples and pears by dip treatment with pyrrolnitrin, a metabolite of *Pseudomonas cepacea*. Plant Dis. 75 : 490-494.
48. Janisiewicz, W. J. and Bors, B. (1995). Development of a microbial community of bacterial and yeast antagonists to control wound-invading postharvest pathogens of fruits. Appl. Environ. Microbial. 61: 3261-3267.

49. Teixido, N. Usall, J. and Vinas, I. 1999. Efficacy of preharvest and postharvest *Candida sake* biocontrol treatments to prevent blue mould on apples during cold storage Int. J. Food. Microbiol. 50: 203-210.
50. Wisniewski, M. Droby, S. Chalutz, E. and Eilam, Y. (1995). Effect of Ca^{+2} and Mg^{+2} on *Botrytis cinerea* and *Penicillium expansum in vitro* and on the biocontrol activity of *Candida oleophila*. Plant Path. 44: 1016-1024.
51. McLaughlin, R.J. Wisniewski, M.E. Wilson, C.L. and Chalutz, E. (1990). Effect of inoculum concentration and salt solutions on biocontrol of postharvest diseases of apple with *Candida* sp. Phytopathology 80: 456-461.
52. Hong, C. Michailides, T.J. and Holtz, B.A. (1998). Effects of wounding, inoculum density, and biological control agents on postharvest brown rot of stone fruits. Plant Dis. 82:1210-1216.
53. Cook, D. W. M. Long, P. G. and Ganesh, S. (1999). The combined effect of delayed application of yeast biocontrol agents and fruit curing for the inhibition of the postharvest pathogen *Botrytis cinerea* in kiwifruit. Postharvest. Biol. Technol. 16: 233-243.
54. Burkhead, K. D., Schisler, D. A. and Slininger, P. J. (1995). Bioautography shows antibiotic production by soil bacterial isolates antagonistic to fungal dry rot of potatoes. Soil. Biol. Biochem. 26: 1611-1616.
55. Guinebretiere, M. H. Nguyen-the, C. Morrison, N. Reich, M. and Nicot, P. (2000). Isolation and characterization of antagonists for the biocontrol of the postharvest wound pathogen *Botrytis cinerea* on strawberry fruits. J. food Prot.63: 386-394.
56. Korsten, L. De Jager, E. S. De Villers, E. E. Lourens, A. Kotze, J. M. and Wehner, F. C. (1995). Evaluation of bacterial epiphytes isolated from avocado leaf and fruit surfaces for biocontrol of avocado postharvest diseases. Plant Dis. 79: 1149-1156.
57. Yakoby, N. Zhou, R. Kobiler I., Dinoor, A. and Prusky, D. (2001). Development of *Colletotrichum gloeosporioides* restriction enzyme-mediated integration mutants as biocontrol agents against anthracnose disease in avocado fruits. Phytopathology 91: 143-148.
58. Peterson, S. Johnson, N. and Schnurer, J. (1999). *Pichia anomala* as a biocontrol agent during storage of high-moisture feed grain under airtight conditions. Postharvest. Biol. Technol. 15: 175-184.
59. Schena, L. Ippolito, A. Zahavi, T. Cohen, L. Nigro, F. and Droby, S. (1999). Genetic diversity and biocontrol activity of *Aureobasidium pullulans* isolates against postharvest rots. Postharvest. Biol. Technol. 17: 189-199.
60. Yates, I. E. Meredith, F. Smart, W. Bacon, C. W. and Jaworski, A. J. (1999). *Trichoderma viride* suppresses fumonisin B1 production by *Fusarium moniliforme*. J. Food. Prot. 62: 1326-1332.
61. Droby, S. Cohen, A. Weiss, B. Horev, B. Chalutz, E. Katz, H. Keren-Tzur, M. and Shachnai, A. (1998). Commercial testing of Aspire: A yeast preparation for the biological control of postharvest decay of citrus. Biol. Control. 12 : 97-100.
62. Sugar, D. and Spotts, R.A. (1999). Control of postharvest decay in pear by four laboratory-grown yeasts and two-registered biocontrol products. Plant Dis. 83: 155-158.
63. El Ghaouth, A. Smilanick, J. Wisniewski, M. and Wilson, C. (2000). Improved control of apple and citrus fruit decay with a combination of *Candida saitoana* with 2-deoxy-D-glucose. Plant Dis. 84: 249-253.
64. Eckert , J. W. and Eaks, I. L. (1989). Postharvest disorders and diseases of citrus fruits. *In*: The Citrus Industry, Vol. 4, (W. Reuther, E. Calavan, G.E. Carman eds.) University of California, Division of Natural Resources, Oakland, California, pp.179.
65. Barkai-Golan, R. and Douglas, J. P. (1991). Postharvest heat treatment of fresh fruits and vegetables for decay control. Plant Dis. 75: 1085.
66. Leverentz, B. Janisiewicz, W. J. Conway, W. S. Saftner, R. A. Fuchs, Y. Sams, C. E., and Camp, M. J. (2000). Combining yeasts or a bacterial biocontrol agent and heat treatment to reduce postharvest decay of 'Gala' apples. Biol. Technol. 21: 87-94.

67. Stevens, C. Kahn, V. A. Lu, J. Y. Wilson, C. El Ghaouth, A. Chalutz, E. and Droby, S. (1996). Low dose UV-C light as a new approach to control decay of harvested commodities. Rec. Res. Develop. Plant Pathol. 1: 155-169.
68. Smilanick, J. L. Margosan, D. A. Mlikota, F. Usall, J. and Michael, I. F. (1999). Control of citrus green mold by carbonate and bicarbonate salts and the influence of commercial postharvest practices on their efficacy. Plant Dis. 83:139-145.
69. Janisiewicz, W. J. Conway, W. S. Glenn, D. M. and Sams, C. E. (1998). Integrating biological control and calcium treatment for controlling postharvest decay of apples. Hort Science, 33: 105-109.
70. Brown, G. E. Davis, C. and Chambers, M. (2000). Control of citrus green mold with Aspire is impacted by the type of injury. Postharvest Biol. Technol. 18: 57-65.
71. Chand-Goyal T. and Spotts, R. A. (1996). Postharvest Biological Control of Blue Mold of Apple and Brown Rot of Sweet Cherry by Natural Saprophytic Yeasts Alone or in Combination with Low Doses of Fungicides. Postharvest Biol. Technol. 7: 51-64.
72. Chand-Goyal T. and Spotts, R. A. (1996). Postharvest Biological Control of Blue Mold of Apple and Brown Rot of Sweet Cherry by Natural Saprophytic Yeasts Alone or in Combination with Low Doses of Fungicides. Biol Control. 6: 253-259.
73. Petracek, P. D. Wardowski, W. F. and Brown, G. E. (1995). Pitting of Grapefruit that Resembles Chilling Injury. HortScience 30:1422-1426.
74. Kampp, J. (1994). Biological control of postharvest diseases of apples and pears. Acta. Hortic., 36: 69-77.
75. Sugar, D. Roberts, R. G. Hilton, R. J. Reghetti, T. L. and Sanchez, E. E. (1994). Integration of cultural methods with yeast treatment for control of postharvest decay in pear. Plant Dis. 78: 791-795.
76. Sobiczewski, P. and Bryk, H. (1996). Biocontrol of *Botrytis cinerea* and *Penicillium* on postharvest apples by antagonistic bacteria. Acta Hortic. 422: 344-345.
77. White, G. C. (Ed.) Handbook of chlorination and alternative disinfectants, Fourth Ed., Wiley-Interscience, John Wiley & Sons, New York, 1999.
78. Mercier, J. and Wilson, C. (1995). Effect of wound moisture on the biocontrol by *Candida oleophila* of gray mold rot (*Botrytis cinerea*) of apple. Postharvest. Biol.Technol., 6: 9-15.
79. Eckert, J. W. (1999). Investigation of new fruit fungicides, sanitizing agents and biological treatments for control of postharvest decays. Annual Report of California Citrus Research Board, p.42.
80. Janisiewicz, W. J. Conway, W. S. and Leverentz, B. (1999). Biological control of postharvest decays of apple can prevent growth of *Escherichia coli* 0157:1-17 in apple wounds. J. Food Protect. 62: 1372-1380.
81. Wisniewski, M. Biles, C. Droby, S. McLaughlin, R. Wilson, C. and Chalutz, E. (1991). Mode of action of the postharvest biocontrol yeast, *Pichia guilliermondii*. I. Characterization of attachment to *Botrytis cinerea*. Physiol. Mol. Plant. Pathol., 39: 245-258.
82. Droby, S. and Chalutz, E. (1994). Mode of action of biocontrol agents for postharvest diseases. *In*: Biological Control of Postharvest Diseases of Fruits and Vegetables-Theory and Practice (C.L. Wilson and M.E. Wisniewski eds.), CRC Press, Boca Raton, Florida, pp. 63.
83. Raaijmakers, J. M. Leeman, M. Oorschot, M. M. P. Van der Sluis, I. Schippers, B. and Baker, P. A. H. (1995). Dose-response relationship in biological control of fusarium wilt of radish by *Pseudomonas* spp. Phytopathology 85: 1075-1081.
84. Lindow, S. E. and Windels C. E. (eds.) (1985). Biological control on the phylloplane, American Phytopathology Society, St. Paul, Minnesota.
85. Kraus, J. and Loper, J. E. (1992). Lack of evidence for a role of antifungal metabolite production by *Pseudomonas fluorescens* Pf-5 in biological control of *Pythium* damping-off of cucumber. Phytopathology 82: 264-271.

86. Pfender, W. F. Kraus, J. and Loper, J. E. (1993). A genomic region from *Pseudomonas fluorescens* Pf-5 required for pyrrolnitrin production and inhibition of *Pyrenphora tritici-repentis* in wheat straw. Phytopathology 83: 1223-1228.
87. Bull, C. T. Wadsworth, M. L. K. Sorenson, K. N. Takemoto, J. Austin, R. and Smilanick, J. L. (1998). Syringomycin E produced by biological agents controls green mold on lemons. Biol. Control. 12: 89-95.
88. Bull, C. T. Wadsworth, M. L. Pogge, T. D. Le, T. T. Wallace, S. K. and Smilanick, J. L. (1998). Molecular investigations into the mechanisms in the biological control of postharvest diseases of citrus. Biological Control of Fungal and Bacterial Plant Pathogens. IOBC Bulletin, 21: 1-6.
89. Elad, Y. Chet, I. and Henis, Y. (1982). Degradation of plant pathogenic fungi by *Trichoderma harzianum*. Can. J. Microbiol. 28: 719-725.
90. Elad, Y. (1995). Mycoparasitism *In*: Pathogenesis and Host-Parasite Specificity in Plant Diseases: Histopathological, Biochemical, Genetic and Molecular Basis (K. Kohomo, R.P. Singh, U.S. Singh, R. Zeigler Eds.), Pergamon Press Oxford, pp. 152.
91. Wisniewski, M. Wilson, C. and Hershberger, W. (1989). Characterization of inhibition of *Rhizopus stolonifer* germination and growth by *Enterobacter cloacae*. Can. J. Bot. 67:2317-2323.
92. Avraham, A. M.Sc. Isolation and Biochemical Characterization of Exo-□-1,3-glucanasefrom *Pichia guilliermondii*. Thesis, Faculty of Agriculture, Hebrew University of Jerusalem, Israel, 1994.
93. 93. Campbell, R. (1970). An electron microscope study of exogenously dormant spores, spore germination, hyphae and conidiophores of *Alternaria brassicola*. New Phytol. 69: 287-293.
94. Gaur, N. K. Klotz, S. A. and Henderson, R. L. (1999). Overexpression of the *Candida albicans* ALA1 gene in *Saccharomyces cerevisiae* results in aggregation following attachment of yeast cells to extracellular matrix proteins, adherence properties similar to those of *Candida albicans*. Infect. Immun., 67: 6040-6047.
95. Lockwood, J. L. (1992). Exploitation competition *In*: The Fungal Community (G.C. Carroll and D.T. Wicklow, Eds.), Marcel Dekker, New York, pp. 243.
96. Blakeman, J. P. and Brodie, I. D. S. (1977). Competition for nutrients between epiphytic microorganisms and germination of spores of plant pathogens on beetroot leaves. Physiol. Plant Pathol. 10: 29-35.
97. Lockwood, J. L. and Filonow, A. B. (1981). Responses of fungi to nutrient-limiting conditions and to inhibitory substances in natural habitats. Adv. Microbiol. Ecol. 5: 1-10.
98. Edwards, M. C. and Blakeman, J. P. (19840. J. Microscopy. 133: 205.
99. Filonow, A. B. Vishniac, H. S. Anderson, J. A. and Janisiewicz, W. J. (1996). Biological control of *Botrytis cinerea* in apple by yeasts from various habitats and their putative mechanisms of antagonism. Biol. Control. 7: 212-220.
100. Vero, S and Wisniewski, M. (2001). Characterization of biocontrol of two yeast strains from Uruguay against blue mould of apple, Postharvest. Biol. Technol. (2001) In Press.
101. Droby, S. Chalutz, E. and Wilson, C.L. (1991). Antagonistic microorganisms as biological control agents of postharvest diseases of fruits and vegetables. Postharvest. News. Inf. 2: 169-173.
102. Mercier, J. and Wilson, C. (1994). Colonization of apple wounds by naturally occurring microflora and introduced *Candida oleophila* and their effect on infection by *Botrytis cinerea* during storage. Biol Control. 4: 138-144.
103. Shefelbine, P.A. and Roberts, R.G. (1990). Population dynamics of *Cryptococcus laurentii* in wounds in apple and pear fruit stored under ambient or controlled atmospheric conditions. Phytopathology 80:1020.
104. Kuc, J. and Strobel, N. (1992). Induced resistance using pathogens and nonpathogens. *In*: Biological Control of Plant Diseases (E. Tjamos and G. Papavisas, Eds.), Plenum Press, New York, pp. 295.
105. Ryalls, J. Neuenschwander, U. Willits, M. Molina, A. Steiner, H. Y. and Hunt, M. (1996). Systemic acquired resistance. Plant Cell. 8: 1809-1819.

106. Van Loon, L. C. Bakker, P. A. H. M. and Pieterse, M. J. (1998). Systemic resistance induced by rhizosphere bacteria. Annu. Rev. Phytopathol. 36: 453-483.
107. Bowles, D. J. (1990). Defense-related proteins in higher plants. Annu. Rev. Biochem. 59: 873-907.
108. Dixon, R. A. and Harrison, M. (1990). Activation, structure, and organization of genes involved in microbial defense in plants. Adv. Genet. 28: 165-180.
109. Tong-Kwee, L. and Rohrbock, K. G. (1980). Role of *Penicillium funiculosum* strains in the development of pineapple fruit diseases. Phytopathology 70: 663-665.
110. Prusky, D. Freeman, S. Rodrigues, R. J. and Keen, N. T. (1994). A nonpathogenic mutant strain of *Colletotrichium magna* induces resistance to *C. gloesporiodes* in avocado fruits. Mol. Plant. Microbe Interact. 7: 326-333.
111. Rodov, V. Ben-Yehoshua, S. Albaglis, R. and Fang, D. (1994). Accumulation of phytoalexins scoparone and scopoletin in citrus fruits subjected to various postharvest treatments. Acta Hort . 381: 517-523.
112. Ayers, A. R. Ebel, J. Finelli, F. Berger, N. and Albersheim, P. (1976). Host-pathogen interactions. Quantitative assays of elicitor activity and characterization of the elicitor present in the extracellular medium of cultures of *Phytophthora megasperma* var. *sojae*. Plant Physiol. 57: 751-759.
113. Base, C. W. Bock, K. and Boller, T. (1992). Elicitors and suppressors of the defense response in tomato cells. Purification and characterization of glycopeptide elicitors and glycan suppressors generated by enzymatic cleavage of yeast invertase. J. Biol. Chem. 267: 10258-10265.
114. Base, C. W. and Boller, T. (1992). Glycopeptide elicitors of stress responses in tomato cells. N-linked glycans are essential for activity but act as suppressors of the same activity when released from the glycopeptides. Plant Physiol. 98: 1239-1247.
115. Felix, G. Grosskopf, D. G. Regenass, M. Basse, C. W. and Boller, T. Elicitor-induced ethylene biosynthesis in tomato cells. Characterization and use as a bioassay for elicitor action. Plant Physiol. 97: 19-25.
116. Grosskopf, D. G. Felix, G. and Boller, T. (1991). A yeast-derived glycopeptide elicitor and chitosan or digitonin differentially induce ethylene biosynthesis, phenylalanine ammonia-lyase and callose formation in suspension-cultured tomato cells. J. Plant Physiol. 138: 741-746.
117. Smith, M. M. and Cruickshank, I. A. M. (1987). Dynamics of conidial adhesion and elicitor uptake in relation to pisatin accumulation in aqueous droplets on the endocarp of pea. Physiol. Mol. Plant Pathol. 31: 315-324.
118. Pascholati, S. F. Yoshioka, Y. Kunok, H. and Nicholson, R. L. (1992). Preparation of the infection court by *Erysiphe graminis* f. sp. *Hordei* : cutinase is a component of the conidial exudate. Physiol. Mol. Plant Pathol. 41: 53-59.
119. Droby, S. Chalupovicz, L. Chalutz, E. Wisniewski, M.E. and Wilson, C.L. (1995). Inhibitory activity of yeast cell wall materials against postharvest fungal pathogens. Phytopathology 85:1123.
120. Sticher, L. Mauch-Mani, B. and Metraux, J. P. (1997). Systemic acquired resistance. Annu. Rev. Phytopathol. 35: 235-270.
121. Schlumbaum, A. Mauch, F. Vogeli, U. and Boller, T. (1986). Plant chitinases are potent inhibitors of fungal growth. Nature 324: 365-367.
122. Sela-Buurlage, M.B. Ponstein, A.S. Bres-Vloemans, B. Melchers, L.O. Van den ELzen, P. Cornelissen, B.J.C. (1993). Only specific tobacco chitinases and ß-1,3-glucanases exhibit antifungal activity. Plant Physiol. 101:857-863.
123. Bartnicki-Garcia, S. (1968). Cell wall chemistry, morphogenesis, and taxonomy of fungi. Ann. Rev. Microbiol. 22: 87-108
124. Peter, S., Hubner, B., Sirrenberg, A., Hager, A. (1997). Differential effect of purified spruce chitinases and Ø -1,3-glucanases on the activity of elicitors from ectomycorrhizal fungi. Plant Physiol. 114: 957-968.

Applied Mycology and Biotechnology
Volume 2. Agriculture and Food Production

Biological Weed Control with Pathogens: Search for Candidates to Applications

S.M. Boyetchko[a], E.N. Rosskopf[b], A.J. Caesar[c], R. Charudattan[d]
[a]Agriculture and Agri-Food Canada, Research Centre, 107 Science Place, Saskatoon, Saskatchewan Canada S7N 0X2; [b] USDA/ARS, 2199 S. Rock Rd, Fort Pierce, Florida 34945, USA; [c]USDA/ARS, 1500 N. Central Ave., Sidney, Montana, 59270, USA; [d]University of Florida, Plant Pathology Department, 1453 Fifield Hall, Gainesville, Florida 32611-0680, USA (E Mail: boyetchkos@EM.AGR.CA).

Biological control of weeds with microbial agents has had a long and successful history worldwide. While weeds contribute significantly to crop yield losses, some of the reported negative impacts of chemical pesticides such as residues in soil and water, spray drift, and weed species exhibiting herbicide resistance underscore the importance of exploring alternatives to chemical herbicides. There have been a number of successful introductions of classical biological control agents, including the more notable *Puccinia chondrillina* for control of skeleton weed. In the early 1980s, DeVine® was introduced as the first registered bioherbicide in the U.S. to control stranglervine in citrus. This was followed by Collego®, which is commercially available for northern jointvetch control and several other bioherbicides which are registered or are in precommercial evaluation stages. Several reviews published in the last 10 years have described the various biocontrol approaches and provide examples [1-11]. This review provides an update on some of the biological control agents, but also focuses on new approaches for implementation of biocontrol, including the integration of multiple pathogens and insects and biotechnological approaches such as development of effective formulations and DNA technologies to enhance virulence and pathogenicity and monitor the dispersal and eventual risk of introducing these microbial agents into the environment. Other considerations include understanding the weed population structure and selection of candidate organisms. More importantly, efforts to fit biologicals into production agriculture and their role in integrated weed management systems is discussed.

1. INTRODUCTION

Weeds continue to be one of the major contributing factors to yield loss in agricultural production in developing countries. They contribute to crop losses through competition for water and nutrients, prolific seed production that significantly increases the weed seed bank, lower crop grades, dockage, and costs associated with seed cleaning. Fundamental weed management tactics include cultural, mechanical, chemical, and biological methods, but there appears to be a continued reliance on herbicides as the main weed control method. Globally, the agrochemical market was almost $27 billion in 1991, with herbicides accounting for 44% of all pesticide sales [12]. Chemical herbicides account for 60% and 70% of the total pesticide sales in the U.S. and Canada, respectively [13,14]. They are considered to be relatively inexpensive, reliable, and consistent in their performance and are therefore often a preferred form of weed control by pest managers [15]. Moreover, progress in agricultural biotechnology over the past few decades has resulted in significant improvements in crop yields (more than double) and these have been attributed to crop protection products and fertilizers [16]. In more recent years, however, crop protection practices have come under greater scrutiny because of reported negative impacts on the environment. Off-target spray drift, residues in the soil and water, and development of herbicide-resistant weed populations are only a few issues which have prompted government legislators, researchers, industry, and consumers to re-evaluate the benefits and risks of herbicide applications [17, 18]. For example, over 100 weed species have been reported to exhibit resistance to chemical herbicides [19]. In view of these factors, it is becoming essential for weed managers to provide alternatives to chemical pesticides. The use of plant pathogens for biological weed control offers potential opportunities as nonchemical alternatives, and has been gaining in acceptance as an environmentally safe and viable weed management option.

2. BIOLOGICAL CONTROL OF WEEDS WITH PATHOGENS

Two strategies by which microbial agents are often used for biological control of weeds are the classical approach and inundative approach. Classical biological control is characterized by the importation, introduction and release of a natural enemy from the same geographic origin as the weed into an area where the weed is a problem [10, 11, 20]. Following release, the natural enemy (insect or fungal pathogen) is allowed to self-perpetuate, survive, and establish, thus providing long-term weed control over a period of several years. Classical biological control is often more appropriate in pasture or rangeland, where often site disturbance is minimal. The inundative approach, also known as the bioherbicide approach, focuses on the application of high doses (i.e. inoculum levels) of a host-specific pathogen to a target weed [1, 4, 9]. The pathogens are often indigenous, artificially mass-produced and applied during the growing season. Weed control is short-term, compared to classical biocontrol agents, and the microbes are not expected to persist in the environment. The majority of microbes used as bioherbicides are predominantly fungal pathogens, while there are a growing number of examples of foliar and soil-applied bacterial agents being explored and developed as bioherbicides as well.

2.1. Classical approach in natural and rangeland areas

There are a number of examples of introductions of fungal pathogens for classical biological control (Table 1). The first successful use of the classical approach employing a plant pathogen was that of *Puccinia chondrillina*, which reduced stand density of *Chondrilla juncea* (skeleton-weed) both in Australia and the U. S. [20-22]. This finding might have stimulated an increased effort in the use of fungal agents that could effect a similar level of impact against perennial weeds of uncultivated areas, but this has not occurred. The failure to advance beyond this successful use of a rust in North America has been attributed in the case of another rust, *Puccinia carduorum*, to undue delay in obtaining approval for its release. The control of *C. juncea* was attributed to the combined effects of the rust disease caused by *P. chondrillina* and water stress [20, 21]. Based on this simple mechanism, testing for the interaction of a candidate rust with moisture stress for increased impact against the target weed probably should have been a criterion for new candidate agents for perennial weeds in uncultivated situations. All four rust fungi in highly successful or prominent ongoing classical biological control programs are autoecious: *P. chondrillina* against *C. juncea*, *Phragmidium violaceum* against *Rubus* spp. (blackberry), *Uromycladium tepperianum* against *Acacia saligna* and *P. carduorum* to control

Carduus thoermeri (=*Carduus nutans*) (musk thistle). This commonality could constitute another criterion when considering possible agents against perennial weed species. It remains to be determined whether this trend might predict any eventual success of *Puccinia thlaspeos*, an autoecious rust that is a candidate for biological control of Dyer's Woad (*Isatis tinctoria*) [23].

Table 1. Examples of Classical Biological Control Agents.

Target Weed	Pathogen	Country of Importation
Acacia saligna (Port Jackson willow)	*Uromycladium tepperianum* (Acacia gall rust)	South Africa
Ageratina riparia (Mistflower)	Entyloma compositarum (smut fungus)	Hawaii, New Zealand
Chondrilla juncea (Skeleton Weed)	Puccinia chondrillina (rust fungus)	Australia, U.S.
Carduus thoermeri (*C. nutans*) (Musk thistle, Nodding thistle)	*Puccinia carduorum* (rust fungus)	Canada, U.S.
Centaurea diffusa (Diffuse knapweed)	*Puccinia jaceae* (Centaurea rusts)	Canada (B.C.), U.S. (Washington, Oregon, Idaho,Montana, South Dakota)
Passiflora tripartita var. *tripartita* (Banana poka)	*Septoria passiflorae* (Septoria leaf spot)	Hawaii
Rubus spp. (Blackberry)	*Phragmidium violaceum* (Blackberry rust)	Chile, Australia

Another agent that was effective against the same target in widely different regions was the smut fungus *Entyloma compositarum* against *Ageratina riparia* (mistflower). Highly effective in Hawaii [3], *E. compositarum* was similarly successful in New Zealand [24]. Such pathogen/target weed systems should probably receive closer study to elucidate mechanisms that can serve as criteria to use for selecting future agents. The involvement of *Entyloma* as a widely successful agent in the case just described may indicate the promise of another smut fungus as a biocontrol agent of *Hieracium* spp., a perennial that is a highly invasive exotic weed both in New Zealand and the northwestern U.S. An *Entyloma* spp. was found in Europe in the 1990s on *Hieracium* (Hasan, personal communication) but, unfortunately, progress was arrested by

Figure 1. Weed suppression of green foxtail using a DRB. Untreated control plots (upper) compared to 85% control using a pesta-formulated *Pseudomonas* spp. (lower) 8 weeks after application.

Figure 2. Application of multiple pathogens (*Drechslera gigantea, Exserohilum rostratum,* and *E. longirostratum*) on green foxtail. Untreated control (far left), multiple pathogens in water and 18 h dew (middle), multiple pathogens in psyllium (Metamucil®) and 18 h dew.

Table 2. Examples of inundative biological control

Target Weed	Pathogen (Registered or Trade Mark name)	Country	Status[a]
Annual bluegrass	*Xanthomonas campestris* pv. *poae* (Camperico®)	Japan	4
Barnyardgrass	*Exserohilum monoceras*	Australia, Japan	3
		Phillippines-Canada	
Black and golden wattle	*Cylindrobasidium laeve* (Stumpout®)	South Africa	5
Bracken	*Ascochyta pteridis*	Scotland, UK	3
Cocklebur, Noogoora burr	*Colletotrichum orbiculare*	Australia	4*
Cogongrass	*Colletotrichum caudatum*	Malaysia	1
Common groundsel	*Puccinia lagenophorae*	UK-Switzerland	4
Dandelion	*Sclerotinia sclerotiorum*	Canada	4
Dodder	*Alternaria destruens*	USA, Florida	4
Gooseweed	*Alternaria* sp.	Philippines	2
	Colletotrichum gloeosporioides	Malaysia	2
Grass weeds	*Drechslera* spp., *Exserohilum* spp.	Vietnam-Australia	2
		USA, Florida	3
	Pseudomonas spp.	Canada, USA	3
Hakea	*Colletotrichum gloeosporioides*	South Africa	4
Hemp Sesbania	*Colletotrichum truncatum*	USA, Mississippi	4
Lambsquarters	*Ascochyta caulina*	Netherlands	4
Northern joint	*Colletotrichum gloeosporioides* f.sp. *aeschynomene* (Collego®)	USA, Arkansas	5
Nutsedges	*Puccinia canaliculata* (Dr. Biosedge®)	USA, Georgia	5*
	Dactylaria higginsii	USA, Florida	3
	Cercospora caricis	Brazil, Israel	3
Pigweeds and Amaranths	*Phomopsis amaranthicola*	USA, Florida	3
Round-leaved mallow	*Colletotrichum gloeosporioides* f.sp. *malvae* (Mallet WP)	Canada/USA	5*
Stranglervine	*Phytophthora palmivora* (DeVine®)	USA, Florida	5
Various annual weeds	*Myrothecium verrucaria*	USA, Maryland	3
Velvetleaf	*Colletotrichum coccodes*	Canada	4*
Water hyacinth	*Cercospora rodmannii*	USA, Florida	4*
		South Africa	3
	Alternaria eichhorniae	Egypt, India, SE Asia	3
Weedy hardwood	*Chondrostereum purpureum*		
	(BioChon™)	Netherlands	5
	(CHONTROL)	Canada	4

Note: Compiled from published and unpublished reports; [a] Status: 1 = exploratory phase; 2 = laboratory and/or greenhouse testing underway; 3 = field trials in progress; 4 = under early commercial or practical development; 4* = commercial development tried but registration uncertain; 5 = available for commercial or practical use; and 5* = available registered as a microbial herbicide but currently unavailable for use due to economic reasons.

personnel changes at both the European weed labs of the U.S. and Australia. Meanwhile, a promising insect agent for biocontrol of *Hieracium* spp. has yet to be identified after several years of effort. The same situation exists in the case of the weed *Acroptilon repens*, which still awaits a suitably host-specific and effective insect to control it, again despite much previous effort. An emphasis on plant pathogens would now seem justified in these cases and in several others. This situation may be due to the small number of personnel involved in classical weed who are trained as plant pathologists [6], a situation that exists within each of the major organizations conducting this type of research.

While some deuteromycetous fungi have been found that cause foliar disease on exotic, invasive perennial weeds of rangelands, such as common crupina (*Crupina vulgaris*) and St. John's wort (*Hypericum perforatum*) [25, 26], enthusiasm should be tempered by such findings as the failure even of a highly destructive necrotrophic pathogen of above-ground plant parts, *Sclerotium rolfsii*, to cause more than negligible overall impact against *Euphorbia esula/virgata* [27]. The plants of this deep, extensively rooted perennial recovered from complete kill-down of tops by *S. rolfsii* and reestablished within a single season from the unscathed root system.

2.2. Bioherbicide Approach

Bioherbicides are applied in a similar fashion to chemical herbicides, with the fungal or bacterial agent comprising the active ingredient. Although the majority of research has focused on foliar fungal pathogens as bioherbicides, greater research efforts have begun to explore foliar and soil-applied bacterial agents. To date, eight bioherbicides have been registered in various countries, and several other microbial agents are either in the early to late stages of evaluation and development (Table 2).

2.2.1. Fungal agents

The first registered bioherbicide in the U.S. was DeVine®, a liquid formulation comprised of chlamydospores of the soilborne fungus *Phytophthora palmivora* for control of stranglervine (*Morrenia odorata*) in citrus [4, 28, 29]. Although the market is limited, Encore Technologies (Minnetonka, MN) has recently acquired the rights to sell the product (David Goulet, Encore Technologies, personal communication). The product provides consistent weed control of over 90% and control can persist for at least 2 years. Collego® is another bioherbicide comprised of a wettable powder formulation of the fungus *Colletotrichum gloeosporioides* f.sp. *aeschynomene* registered for the control of northern jointvetch (*Aeschynomene virginica*), a leguminous weed in rice and soybean [29]. The pathogen is host-specific and causes anthracnose symptoms to the foliage and stem. Originally developed at the University of Arkansas, U.S. Department of Agriculture and the Upjohn Company [30], Collego® is now commercially available through Encore Technologies.

The first registered bioherbicide in Canada was BioMal®, a product containing spores of *Colletotrichum gloeosporioides* f.sp. *malvae* for control of round-leaved mallow (*Malva pusilla*) [31, 32]. The bioherbicide was registered for use in several crops in Canada, however, the small market size and technical difficulties associated with mass-production were two major considerations that deterred commercializaton of the product. More recently, however, Encore Technologies is pursuing U.S. EPA registration of the bioherbicide under the name Mallet WP. Further considerations include application of the product on small-flowered mallow (*M. parviflora*) in order to expand the market of the bioherbicide and field performance under cool temperatures and humid conditions (David Johnson, Encore Technologies, personal communication).

Dr. BioSedge® is an endemic rust, *Puccinia canaliculata*, registered for control of yellow nutsedge (*Cyperus esculentus*) in the U.S. [6, 33, 34]. Application of the rust fungus in early spring inhibits flowering and reduces tuber formation by as much as 46%. Because of the obligate parasitic nature of the pathogen, it is not economically feasible to mass-produce the fungus for commercial use. Another limitation of Dr. BioSedge is that it is very host specific and does not infect several biotypes of yellow nutsedge nor does it infect purple nutsedge (*C. rotundus*).

Chondrostereum purpureum is a saprophytic basidiomycetous fungus that, upon wounding or injury, will invade the cambium of hardwood tree species such as red alder (*Alnus rubra*), black cherry (*Prunus serotina*), white birch (*Betula papyrifera*), and aspen (*Populus* spp.) [35,36]. In the Netherlands, the fungus is sold by Koppert B.V. under the name BioChon® [6] and it is presently undergoing registration in Canada and the U.S. under the name CHONTROL (W.E. Hintz, MycoLogic, Inc., personal communication). Although the fungus has a very broad host range, including a number of broad-leaved trees and several rosaceous fruit trees, epidemiological studies provided evidence of the safe application and low risk of dispersal of the fungus onto nontarget plants [35].

Another wood-decaying fungus that has been developed as a mycoherbicide is *Cylindrobasidium laeve*, under the tradename Stumpout® [6, 9, 37]. It reduces regrowth of wattle (*Acacia mearnssi* and *A. pycnantha*) in the Cape Province of South Africa and is mass-produced on solid substrate and formulated as an oil-based paste.

Alternaria destruens is a fungal pathogen undergoing review by the U.S. EPA for registration as a bioherbicide, named Smolder®, for control of dodder (*Cuscuta* spp.) in cranberries in the U.S.A. [6, 9, 38]. Application of the pathogen using a pesta formulation in field plots resulted in control of dodder the following year. A new fungal species, *Phomopsis amaranthicola*, is currently being evaluated as a mycoherbicide for biological control of pigweeds and amaranths [9, 39]. Symptoms include leaf and stem blight and plant mortality occurs when spores are applied at concentrations of 1.5×10^6 to 1.5×10^7 conidia/ml at the two to four-leaf stage. Other pathogens that are currently in precommercial stages of evaluation include *Dactylaria higginsii* for control of yellow and purple nutsedges [40], *Exserohilum monoceras* for control of barnyard grass (*Echinochloa crus-galli*) [41], *Ascochyta caulina* for control of lambsquarters (*Chenopodium album*) [42, 43], and *Cercospora piaropi* (=*C. rodmanii*) for control of waterhyacinth (*Eichhornia crassipes*) [2].

2.2.2. Bacterial agents

Although the majority of microbial agents that are being developed as bioherbicides are fungal pathogens, there are examples of bacterial agents that are currently being explored for their biological control potential. An isolate of *Xanthomonas campestris* pv. *poae* that causes systemic wilting of annual bluegrass (*Poa annua*) has been developed and registered in Japan as the bioherbicide Camperico® [6, 9, 44]. Out of fourteen bacterial strains isolated from annual bluegrass in Japan, isolate JT-P482 was found to be the most effective for control the weed in turfgrass, providing more than 70% control [6]. Mowing of the turfgrass prior to application of the bacterium is a very effective means for enhancing efficacy and timing and rate of application are key factors for providing optimum levels of weed control in most golf course greens. A U.S.

strain of *Xanthomonas campestris* pv. *poae* is also under consideration for development as a bioherbicide by EcoSoil [1].

Pseudomonas syringae pv. *tagetis* (PST) is also being explored as a possible bioherbicide for control of Canada thistle (*Cirsium arvense*) and other Asteraceae weeds [45, 46]. The pathogen causes apical chlorosis and up to 78% reduction in Canada thistle stems/m^2 was observed when applied in the spring. In order for disease symptoms to occur, PST must be applied with the surfactant Silwet L-77 [46]. In addition, a sub-lethal rate (0.1 X) of the herbicide glyphosate causes greater apical chlorosis and significant reductions in fresh and dry weights of Canada thistle, when compared to the application of PST alone and PST with Silwet L-77 [47].

Rhizobacteria (root-colonizing bacteria) that have an inhibitory effect on plant growth are often referred to as deleterious rhizobacteria (DRB) [8, 48]. They are characterized as nonparasitic (exopathogenic) microbes and despite their ubiquitous nature in soil, this group of bacteria have often been ignored because their symptoms are subtle and not obvious in the field. Several bacteria that have been identified as DRB include strains of *Enterobacter, Klebsiella, Critrobacter, Arthrobacter* and fluorescent pseduomonads such as *Pseudomonas cichorii* and *P. fluorescens* [48, 49].

A number of projects in the US and Canada using DRB for biological control have been summarized in recent reviews [1, 8, 9]. Some of the first reports of root-colonizing *Pseudomonas* spp. on plants were described on wheat, potato, and sugarbeet [49-51]. The potential of DRB as biological weed control agents has been explored for 18 weed species including velvetleaf (*Abutilon theophrasti*), lambsquarters, pigweed (*Amaranthus retroflexus*), smartweed (*Amaranthus retroflexus*), cocklebur (*Xanthium* spp.), downy brome (*Bromus tectorum*), wild oat (*Avena fatua*), and green foxtail (*Setaria viridis*), and jointed goatgrass (*Aegilops cylindrica*) [1, 8, 49, 53-55]. The majority of the projects targeted annual weeds in cereal and row crops and several of these projects focused on preliminary screening of bacteria in laboratory experiments.

Kremer [56] found specific bacterial associations with five annual weed species where phytopathogenicity was observed in approximately 15% of all bacterial isolates evaluated. Seedling vigor and the nonviable portion of seeds in soil were reduced, while seedling emergence was decreased by 55% with the soil application of two bacterial isolates [57, 58]. Screening of several thousand bacterial strains against annual grass weeds such as downy brome, green foxtail and wild oat have been conducted in Canada and the U.S. [53, 55, 59]. In many instances, the weed suppressive activity of DRB has been attributed to the production of secondary metabolites and phytotoxins [1, 8]. Partial purification of a phytotoxin from the *P. fluorescens* D7 strain caused root discoloration and reduced root elongation [60] and this toxin was found to be complex, consisting of at least two polypeptides, fatty acid esters and lypopolysaccharides [61]. Attempts to further purify the compound resulted in near complete loss of phytotoxicity, suggesting that either the purification process damaged the phytotoxin or that more than one component in the complex is necessary for activity. Further investigation into the mode of action revealed that the active fraction from the D7 strain affected lipid synthesis and membrane integrity while there was no effect on cell division, respiration, or protein synthesis [62]. Other researchers have also demonstrated that the over production of auxin (indole-3-acetic acid [IAA]) of some bacterial strains, when L-tryptophan as the precursor is added to the media, results in greater root length inhibition of field bindweed [63-65].

Pre-emergent soil application of three DRB strains resulted in significant reductions in growth of two grass weeds [66]. Application of a bacterial strain in a granular formulation in the field

(1999 field season) resulted in reductions in aboveground biomass and weed emergence of wild oat by 57% and 64%, respectively. Improvements in formulations resulted in significant increases in weed suppression by the DRB (Figure 1). Patent applications were obtained for the screening and application of bacterial agents for biological control of downy brome and jointed goatgrass [67, 68]. Spray application of bacterial strains *Pseudomonas fluorescens* D7 and 2V19 to the soil surface at a rate of 10^8 cells/m^2 in 1 L water provided strong weed suppressive activity against natural populations of downy brome. The inhibition of the weed led to increases of winter wheat yields by 18 to 35% and the increases in wheat yield were attributed to reductions in the competition by downy brome [51]. Johnson *et al.* [69] showed that soil factors play a significant role on the efficacy of DRB. Temperatures of 10 to 15°C resulted in greater colonization and inhibition of downy brome roots and moist, cool soils were found to be more conducive to root colonization of *P. fluorescens*. When specific DRB strains were applied to soil in combination with reduced rates of herbicides (such as metribuzin and diclofop), significant increases in weed suppressive activity were observed [70, 71].

Many DRB have a wide spectrum of activity on various plant species and have often been criticized as being non-host specific. However, several researchers have discovered that there is a degree of host-specificity amongst bacterial strains [50, 54, 55]. Boyetchko and Holmström-Ruddick [72] reported that specific bacterial strains and their metabolites exhibiting suppressive properties to downy brome did not inhibit growth of spring and winter cereal crops. In some instances, plant-growth promotion was observed and was dependent on the bacterial strain and crop cultivar. However, Kremer and Kennedy [8] suggested that researchers should recognize the risk of utilizing DRB with a wide host-range because of the potential detrimental effects on crop plants.

2.3. Weed Seed Bank Management

Weed seed banks represent a major cause of weed infestations in agricultural production [73]. It has been suggested that weed seeds entering and surviving in the soil for at least 1 year comprise the persistent portion of the seed bank [74]. Weed-associated microbes can either indirectly attack seeds under specific agricultural cropping systems or can directly attack seeds as soil-applied biocontrol agents. Kremer [58] therefore proposed that soil microorganisms can be exploited to deplete the weed seed bank. Although quantitative evidence on the mortality of seeds by soil microbes is limited, the impact of microbial activity on seed bank mortality has been indirectly documented through buried seed studies. Kennedy and Smith [75] provided supporting evidence that suggested only a fraction (<1%) of the 1 million to ten billion bacteria contained in 1 g of soil are known or have only been partially characterized. Yet, the soil represents an excellent resource of genetic information and biological activity. Aldrich [76] was one of the first to suggest that weed seed numbers could be reduced in the soil by using microorganisms to remove part of the seed coat that provides protection against seed deterioration. Seedborne fungi associated with velvetleaf seeds were capable of attacking the seeds via seed coat openings [77]. If the seed coat of velvetleaf was punctured, more than 50% of the seeds and developing seedlings were attacked by the fungi. The isolation of several fungal and bacterial microbes from the seeds of giant foxtail and velvetleaf resulted in a reduction of root growth by up to 85% in lettuce seedling bioassays [7]. Production of phytotoxins by seed-borne bacteria may be responsible for affecting seed viability and seedling vigor [78]. On the other hand, fungi infesting wild oat seeds were not found to be pathogenic in seed bioassays, suggesting that poor seed viability could not be solely attributed to seedborne fungi [79].

2.4.2 The "Multiple-Pathogen Strategy"

Another approach that has been proposed is the use of multiple pathogens that are all applied using the bioherbicide or inundative application strategy. This approach combines multiple pathogens of the same weed or several pathogens for targeting several weed species. Boyette *et al.* [83] showed that a combination of *C. gloeosporioides* f.sp. *aeschynomene* and *C. gloeosporioides* f.sp. *jussiae* could be used to simultaneously control two target weeds, waterprimrose (*Jussiaea decurrens*) and northern jointvetch (*Aeschynomene virginica*). In this case, there was no synergistic effect found, but the study did demonstrate that it was feasible to use biological control agents to manage two weeds in the same cropping system.

The "multiple-pathogen strategy" is also being utilized by Chandramohan [84] in two different systems. A model system was first developed that uses a combination of *Phomopsis amaranthicola*, *Alternaria cassiae*, *Colletotrichum dematium* f.sp. *crotalaria*, and *Fusarium udum* f.sp. *crotalariae*. *Phomopsis amaranthicola* is a pathogen with a host range that is limited to the genus *Amaranthus*. *Senna obtusifolia* (sicklepod) and *Crotalaria spectabilis* (showy crotalaria) are major and alternative hosts respectively for *Alternaria cassiae* and *C. dematium* f.sp. *crotalaria* while *F. udum* f.sp. *crotalariae* is a pathogen of showy crotalaria. Each of these pathogens, with the exception of *F. udum* f.sp. *crotalariae*, provided excellent control when applied to the foliage of their target weed. When a mixed stand of pigweed, showy crotalaria, and sicklepod was inoculated with a mixture of the four pathogens, all weeds were controlled. Although there was no indication of a synergistic effect using the mixture, the lack of antagonism between these agents allowed for their application as a "cocktail" that demonstrated their utility for control of multiple weeds in this manner.

In the second system, the newly-isolated grass pathogens *Drechslera gigantea*, *Exserohilum longirostratum*, and *E. rostratum* for control of several grass weeds in citrus and sugarcane was studied [84]. The pathogens, when applied in a mixture, controlled crowfoot grass (*Dactyloctenium aegyptium*), Texas panicum (*Panicum texanum*), yellow foxtail (*Setaria glauca*), guineagrass (*Panicum maximum*), southern sandbur (*Cenchrus echinatus*), johnsongrass (*Sorghum halepense*), and large crabgrass (*Digitaria sanguinalis*). While all three pathogens caused at least 83% mortality of all of the weeds tested when applied singly, the mixture caused 100% mortality of all weeds tested. The use of the pathogen mixture allows for increased efficacy over a more broad range of grass weeds and could prevent the development of weed resistance to a single pathogen. Using a slightly different approach, Fernando *et al.* [85] found that phylloplane bacteria, such as *Pseudomonas* spp., in competing for nutrients or iron, caused the biological control agent *Colletotrichum coccodes* to move from a saprophytic phase to a parasitic phase. The fungus, a potential biological control agent to be used against velvetleaf, showed reduced germ tube formation and increased appressorium formation when *Pseudomonas* spp. were applied to leaf surfaces in conjunction with applications of the biological control agent. This research opens the door for evaluating many other combinations of phylloplane bacteria and fungi.

2.4.3. Potential combinations

There are several target weeds for which several pathogens are being developed in various laboratories. While providing some control to the target weed, applications of several pathogens together could be beneficial. Pigweeds and amaranths, for example, are being studied in laboratories around the world. Efforts to develop control programs for these weeds could benefit from combining pathogens, such as *Phomopsis amaranthicola* and *Microsphaeropsis amaranthi*. Preliminary work was conducted by Rosskopf (unpublished) who found no antagonism between these organisms, either in petri plate testing or in preliminary greenhouse inoculations and work in this area is continuing (R. Charudattan, personal communication). In addition, these fungi could be combined with those being developed under the European Research Program (COST-816), in which researchers are evaluating *Alternaria alternata* and *Trematophoma lignicola* [86].

Multiple pathogens have been under development for the control of *Mimosa pigra* var. *pigra* in Australia using two fungal plant pathogens, *Ploeospora mimosae-pigrae* and *Diabole cubensis* [87]. These pathogens were found to have very different environmental optima and could be used in a seasonal application approach. Rather than co-inoculating the weeds with these agents, they could be applied during the season when they are most effective, thereby providing a long-term strategy for management of this weed.

Many pathogens are also being developed as bioherbicides for control of nutsedges [88, 89], grass weeds [84, 90, 91], and kudzu [92]. The many potential benefits of combining fungal and bacterial agents with different modes of action that affect below and above-ground plant parts could be derived for these systems as well. The genetic variability found in weed populations within a country and throughout the world requires that multiple tactics be applied. These tactics can be highly specialized to accommodate population differences in various geographical areas, or a more general bioherbicide approach can be developed that incorporates multiple pathogens that have different pathogenicity determinants and will act over a more broad range of environmental conditions, similar to the approach taken for some post-harvest biological control programs (R. Guetsky, unpublished).

2.4.4. Combination of insects and pathogens as an alternative to the mycoherbicide/classical paradigm

The widespread success of biocontrol of *Opuntia stricta* by an insect/multiple pathogen synergism has indicated that the principle derived from that original study might have wider application. This seems especially applicable since weed biocontrol programs need to be increasingly concerned with adapting to greater regulatory scrutiny and concentrate on insects whose effects can produce synergistic effects. The potential utility of wilt pathogens has been foreseen [20] because of several traits that make such fungi suitable for biocontrol, such as their often very narrow host ranges. Deployment of fungal pathogens such as *Fusarium* spp. may straddle the boundary between the classical and mycoherbicide approaches. Because some *Fusarium* species, such as *F. oxysporum* have long-lived propagules (chlamydospores), this would render them self-perpetuating against a susceptible weed host. Furthermore, *F. oxysporum* as stated by one reviewer, "lacks any history of pathogenicity to graminaceous spp." [93], confirmed in one recent study with isolates of *F. oxysporum* and *F. solani* pathogenic to the perennial prairie weed *E. esula/virgata* [94]. The narrow host ranges of isolates of *F. oxysporum* and *F. solani* generally, along with high levels of virulence against some weeds that has been demonstrated in recent studies [94-97], indicates that such fungi are promising biocontrol agents of perennial weeds.

The capacity of insects to vector *Fusarium* spp. has been shown [98, 99]. The highly destructive nature attributed to the insect involved is likely due to its capacity to vector *F. solani* to act as a synergist in causing damage to coffee berries. Such associations may be hitherto unheralded but causal mechanisms for successful biocontrol largely attributed to certain insects. Recent research has provided new means of delivering beneficial microbes. Devices have been developed which by a specific chemical attractant draws insects into a chamber containing a microbial control agent [100]. The insect is able to reemerge from the chamber after acquiring the agent and deliver it to where it can exert effects at key sites of action [100, 101]. It has been determined in assays that were made of migrating active, adults of the leafy spurge biocontrol agent *Aphthona* spp., that 30% of adults are carrying one or more *Fusarium* spp. (Caesar, unpublished). Another approach that melds the mycoherbicide approach with the classical nature of *F. oxysporum* and potentially other *Fusarium* spp. is indicated by a recent study which showed that foliar and stem infection can lead to root infection of tomato and basil plants. Thus it is plausible that ensuring insect transmission of highly virulent isolates of *Fusarium* spp. pathogenic to leafy spurge and other perennial rangeland weeds can lead to root disease, essential for biological control of such weeds with aggressively expansive root systems. There are many other potential roles of plant pathogens in classical weed biocontrol not possible to cover here. In any case, given the demonstrated potential of plant pathogens for safe, effective impact on invasive weeds in natural areas and the continued under-emphasis in this area, there is a need for redoubling of effort by plant pathologists against long term target species that have not been impacted by years of effort using an exclusively insect-oriented approach.

3. SELECTION OF CANDIDATE PATHOGENS

There is a consensus that classical biological control of exotic weeds of rangeland, pastures, watersheds, riparian and other noncultivated areas is the most environmentally and economically sound weed management approach. Classical weed biocontrol in these areas has traditionally been based almost entirely on the release of insects. While current studies have begun to include plant pathogens as part of weed biocontrol research programs in some cases [102], unfortunately, others continue to be based on searches for insects exclusively [103, 104]. Such a situation runs counter to the principles elucidated in the very earliest successes in weed biocontrol, which stipulated that plant pathogens should be accounted for as part of the criteria for selecting insect agents and that they can play a role even when the individual research program was based primarily on insects as agents [105, 106].

One place in the scheme of weed biocontrol research that could benefit from the greater participation of plant pathologists is within the realm of a more thorough examination of potential direct and indirect effects of a candidate agent on the target weed that has been called for by close observers of weed biocontrol research [107]. Such information can help derive principles that lead to the optimal deployment of plant pathogens (and insects) as components of integrated control of weeds or can help determine whether or not release of a plant pathogenic agent is likely to provide impact. Laudably, some workers have attempted to investigate possible mechanisms that may lead to predicting the likely impact of a classical agent against the target weed [108] or standards by which to determine sampling procedures for new insect agents [109]. Currently there is a need to develop criteria and standards for selecting weed biocontrol agents based on their capacity to interact and synergize to produce long-term, sustained reductions in weed density. Furthermore, assessment of the potential of an agent for integration with cultural practices or reduced levels of herbicide application is also needed.

Two limiting factors that have often been ignored in the exploration for effective biological weed control agents are biological and ecological information of i) the target weed population and ii) the microbial agents (fungal and bacterial) that interact with various stages of life cycle of the weed [7]. Without such fundamental information, it becomes rather difficult to select an appropriate biological strategy into what is often a complex weed management system. Weed populations are genetically variable by nature, including the existence of several biotypes.

Careful selection of appropriate microbial strains as biological control agents must therefore be a major consideration in any weed biocontrol program.

3.1. Target Weed Population Structure

Understanding the population genetics of the target weed can greatly improve efforts to develop pathogens for biological control. Knowing the diversity of a target weed population can provide insight into the potential of a particular agent to have an impact on a total population. It was observed that there was differential susceptibility of nutsedge biotypes to the registered rust bioherbicide, *Puccinia canaliculata* [110, 111]. The rust fungus collected from a particular geographical location infected only yellow nutsedge from the same locations and did not infect purple nutsedge, although purple nutsedge was reported to be a host of the rust in other locations [89]. Studies conducted previously [112, 113] using isozyme analysis on limited populations of purple and yellow nutsedges indicated that the genetic diversity within these species was extremely low. Similarly, Okoli *et al.* [114], using the random amplified polymorphic DNA (RAPD) approach found that purple nutsedge populations from the United States and the Caribbean region showed very little genetic variation, representing a very large clonal population. Conversely, the RAPD technique revealed a high degree of interspecific variation between purple and yellow nutsedges and extensive variability within and between the yellow nutsedge populations sampled. This work contributes to the search for biological control agents for these weeds in that, as the authors suggest, different approaches should be sought for the two species. An obligate parasite, like *P. canaliculata*, would be more successfully applied to the homogeneous populations found in purple nutsedge and facultative parasites sought for control of the diverse populations of yellow nutsedge.

It has become increasingly more common to find molecular approaches for characterizing weed populations being used at the onset of biological control projects. Information gained by determining the genetic composition of a target weed population can assist in determining if it is advantageous to return to the centre of origin of a target weed or if the target population is genetically distinct from the plants at the origin, thereby preventing the need to survey for foreign pathogens and re-introduce them where the weed is a problem. Amsellem *et al.* [115] used the technique of Amplified Fragment Length Polymorphism (AFLP) to estimate genetic diversity of the highly aggressive weed *Rubus alcefolius* to be targeted on La Reunion island. The weed was believed to be introduced to the island directly from its native range in Asia or from Asia through Madagascar. The AFLP analysis revealed that the plants collected from the native range were grouped as one clade and had a high level of genetic diversity among individuals from the same location. The second clade was comprised of individuals from areas of introduction, including Madagascar. The populations from Madagascar showed levels of diversity that were comparable to the populations found in Asia, while those from La Reunion had reduced intra-locality variation.

Australian scientists are investigating biological control agents for the pasture pest Serrated tussock (*Nasella trichotoma*) [116]. Using RAPD analysis, populations of the weed from Argentina and Victoria and New South Wales in Australia were found to be genetically distinct. Morphological differences between Victoria populations and those found in new South Wales support the distinction of these two populations and biological control agents are now being sought that will control both populations. A similar project was undertaken to evaluate the genetic variability in *Hypericum androsaemum* populations in which the rust fungus *Melampsora hypericum* was applied. Some populations were readily infected by this pathogen and others were not. It was further discovered that populations of the rust and the weed were genetically distinct [117]. Work of this kind is progressing throughout the world and will contribute greatly to the search for effective biological control agents that will most effectively control the target weed populations and help to make the search effort as efficient as possible.

3.2. Identification and Characterization of Fungal Biocontrol Agents

Molecular tools that have been developed for use in the identification and phylogeny of fungi and bacteria are extremely helpful for characterization of pathogens. Combinations of molecular information along with morphological data can provide the most reliable identification of potential biological control agents. These tools can also assist in providing clear taxonomic classifications by contributing to background information necessary for regulatory decisions, patenting procedures, and the development of particular strains as commercial products. Use of sequence information from variable and conserved portions of the fungal genome has become increasingly common in biological control research. The internal transcribed spacer regions of the ribosomal DNA of fungi have been used in several cases to assist in species determinations and phylogenetic relationships. Rosskopf *et al.* [118] used this region to classify a fungal plant pathogen isolated from pigweed. The morphological characteristics of the fungus allowed for a tentative identification of the fungus as a new species belonging to the genus *Phomopsis*. Some characteristics were, however, uncommon or not reported for other species in the genus. These included: the presence of a third type of conidium, while two types are usually found; extremely large alpha-conidia; and multiple loci of conidiogenesis. These characteristics brought into question whether the isolate represented a new genus, however, comparison of the ITS region sequence with known species confirmed the identification of the pigweed isolate as a new species within the genus. These sequences are now being used to gain additional information concerning the phylogenetic relationships of these isolates to other identified members within the genus *Phomopsis* (Rossman, unpublished).

The water hyacinth pathogens *Cercospora rodmanii* and *C. piaropi* have been studied as potential biological control agents for this weed. These fungi were described as two distinct species. The second species described, *C. rodmanii*, was named as a new species based on minor differences in conidial morphology and disease development. Tessmann *et al.* [119] using partial DNA sequences from three protein-coding genes, elongation factor-1α, β-tubulin, and histone H3 found that separation of the species was not supported by the phylogenetic analysis. Based on combined datasets for 14 isolates from around the world, two clades were supported that provided differentiation based on geographical distribution rather than species separation. The description of *C. piaropi* was therefore amended to include *C. rodmanii* as a synonym.

Sands *et al.* [120], working with wilt of coca, isolated a *Fusarium oxysporum* and confirmed that it was the causal agent of the disease that was observed and previously reported in Hawaiian coca research plots. Through testing of the host range and vegetative compatibility of the Hawaiian isolate, it was determined that the fungus represented a new forma specialis. Using 10-base randomly amplifying polymorphic DNA primers, the Hawaiian isolate was found to be closely related to the isolates from South America, suggesting that the fungus was brought in on the transplants that were used to establish the Hawaiian research plantation. Knowledge of the origin of the fungus, in this case, would establish that a novel isolate for biological control of coca was not being introduced into South America.

4. BIOTECHNOLOGICAL APPLICATIONS

For the successful implementation of bioherbicides into a weed management system, practical approaches for their application requires an understanding of formulation and delivery and methods for assuring consistent field performance. Development of effective formulations can provide micro-environments that will ensure adequate propagule survival and facilitate the infection process into the target weed species. DNA technologies are also providing tools to enhance the pathogenicity or virulence of the pathogens and to also facilitate monitoring of the organisms in order to assess risk of introducing them into the environment.

4.1. Formulation

Although the literature is replete with examples of fungal and bacterial agents as potential bioherbicides, unreliable field performance resulting from environmental and technological constraints have been cited as major reasons for their lack of progression beyond the initial evaluation phase [9, 121-123]. Slow progress in formulation and application technology for delivery of bioherbicides and inadequate knowledge of the microbe-plant interface that affects delivery and deposition of the microbial agent have been significant hurdles to the development of bioherbicide products. Therefore, concerted efforts over the last couple of decades have focused on the evaluation and development of various formulations that are compatible with microbial-based technologies [124-126]. The major components comprising a formulation include the active ingredient (i.e. biological propagule), the carrier or solvent, and often an adjuvant to facilitate delivery to the target weed [9, 124-126]. Formulation can be used to overcome environmental constraints, particularly dew period requirements, reduce the amount of active ingredient required to achieve acceptable weed control, and facilitate spread and dispersal on the leaf surface while concomitantly preventing the propagules from washing off the plant surface and protecting against desiccation and UV irradiation. In some cases, the carrier may also serve as the delivery vehicle of the agent. Biocontrol products must also have long-term stability and shelf life for ease of transport from the manufacturer to the end-user [9]. There should be compatibility between the biological and formulation ingredients and compatibility with agricultural practices [127]. These formulated products are either applied as metabolically active or dormant propagules [128]. However, long shelf life is often achieved when the microbial agent is in a physiologically dormant state. Stabilization of microbial products by freezing to lower the metabolism or by the addition of membrane stabilizers such as skimmed milk, sugars, salts, glycerol, etc. have been used, but the recovery of viable propagules from frozen/refrigerated products can be significantly reduced and may often be dependent on the organism [126, 128].

With respect to foliar-applied fungal pathogens, factors such as temperature, free moisture and leaf wetness duration, and UV irradiation can significantly affect their ability to germinate, produce infection structures and ultimately cause plant disease [125]. Foliar bacterial agents often require wounds or natural openings to enter the plant [46, 92]. Unfortunately, the identification of promising foliar agents in a biocontrol project has resulted through evaluations by "spraying till runoff" and providing dew periods in excess of 12 hours for infection and penetration of the target weed [129-133]. These environmental conditions are often not achieved under natural field conditions and may overestimate the potential of a biocontrol agent. Although it has been proposed that timing of application provided by rain, irrigation and natural dew to take advantage of longer periods of free moisture and leaf wetness may help to address these environmental limitations in the field [121], it may be somewhat unrealistic to expect farmers to delay weed control efforts when weed competition that can result in significant yield losses may be the over-riding factor in determining when a farmer is willing to spray. Despite the success of Collego as a simple formulation (i.e. dried spores in a wettable powder) and its ability to cause disease epidemics under conditions that are conducive to high humidity (i.e. irrigated or paddy crops), Greaves *et al.* [130] has criticized researchers for continuing to evaluate bioherbicide candidates using simple formulations and inappropriate application methods to overcome the problems of extensive water volume and dew period requirements.

4.1.1. Liquid Formulations

One of the first bioherbicides was DeVine, a liquid formulation of chlamydospores, with a shelf life of 6 weeks when refrigerated [28, 29, 134]. Although not very stable, this product is marketable on a small scale and "made to order" by Abbott Laboratories (Abbott Park, Ill.). A simple adjuvant can be comprised of a water soluble nutrient base mixed with the water suspension containing a fungal propagule prior to application. For example, Collego is sold as a wettable powder that is gradually rehydrated by resuspending in water and mixing with a prepackaged amount of sugar that acts as a nutrient source and osmoticum [9]. Liquid formulations are considered appropriate for the application of foliar-applied bioherbicides for postemergence weed control. One of the simplest liquid formulations is the suspension of the biocontrol agent in water. Compounds that assist or modify the activity of the active ingredient are known as adjuvants, which include a variety of compounds including stickers, nutrients, inert carriers, humectants, sunscreening agents, antievaporation agents, and surfactants [9, 122, 125, 126, 135].

The waxy cuticular surface of the plant can prevent water-based formulations from becoming evenly distributed on the plant surface, therefore, adjuvants have been used to help in the dispersal of microbial biocontrol agents. Surfactants are the most commonly used adjuvants and are considered essential for facilitating the uniform distribution of the microbial propagule on the leaf surface, particularly if the surface is hydrophobic in nature. They can also serve as emulsifiers, stickers, and stabilizers and alter the leaf-wetting properties of the bioherbicide by modifying the surface tension of the spray droplets at the microbe-plant interface. Surfactants aid in the penetration of lipophilic barriers (i.e. leaf cuticle) which can improve the retention of the inocula, thereby increasing the probability of infection by the bioherbicide agent.

Nonionic surfactants such as Tween 20®, Tween 80®, Triton X-100®, and Tergitol® are commonly used in formulations. For example, using a concentration of 0.1-1% Tween 80® and 0.02M potassium phosphate buffer at pH 6.5, spore germination of *Alternaria cassiae* was induced while enhancement of infection of sicklepod seedlings was observed [136]. Conversely, neither Tween 20® nor glycerol improved mycoherbicidal activity of *C. orbiculare* on Bathurst burr compared to water alone as the liquid carrier, although higher spore doses were found to be associated with improved plant mortality [137]. The authors suggested that the failure of Tween 20® to enhance the bioactivity of the fungal pathogen may be more directly related to a faster rate of evaporation due to the reduced surface tension of the droplets applied to the leaf surface, while water droplets alone retained their surface tension resulting in run-off of the droplets into leaf axils and therefore greater moisture retention.

The addition of the adjuvant sorbitol to *Colletotrichum coccodes* for control of velvetleaf caused an increase of viable spores (20-fold) [138]. Furthermore, three 9-hour intermittent dew periods provided on consecutive nights were equally effective as one 18-hour dew period. Other water-retaining compounds used in a formulation to reduce moisture loss after application include skinning agents such as polyvinyl alcohol (PVA) and psyllium, a plant mucilloid [130]. For example, application of multiple pathogens (*Drechslera gigantea, Exserohilum rostratum*, and *E. longrostratum*) in psyllium (Metamucil®) provided superior control of green foxtail compared to the pathogens in water (Figure 2). A mixture of psyllium and PVA enhanced the efficacy of *Ascochyta caulina* on *Chenopodium album*. Also, the addition of Sylgard as a surfactant significantly reduced the dew period below 8 hours by overcoming the water repellent nature of the waxy leaf surfact of the weed. In some cases, surfactants such as Tween 20® and Tween 80® have been found to have fungitoxic properties to bioherbicides such as *Chondrostereum purpureum* [135]. In addition, off-the-shelf commercial formulations (e.g. adjuvants, sunscreen agents, pesticides) were tested for compatiblility with *C. purpureum* [139]. With the exception of Bond and Suntan Gel-2, the majority of products were found to be fungitoxic at concentrations of 0.1% or higher and it was concentration-dependent. It is unclear, however, if the main active ingredient in these commercial formulations was directly responsible for the incompatibility or whether one of the other components in these products caused the fungitoxic activity. This indicates that due care and diligence should be taken when testing off-the-shelf products as part of a formulation for a biological agent.

Organosilicone surfactants, such as Silwet L-77® and Silwet 408® represent a new generation of surfactants that enable the bioherbicide agent to penetrate the cuticular layer of the leaf and/or gain entry into natural plant openings such as stomata and hydathodes [1, 9, 92, 140]. For example, the penetration of *Ascochyta pteridis* for control of bracken (*Pteridum aquilinum*) was enhanced when formulated with Silwet L-77® [140, 141]. However, it was further observed that the phytotoxicity of the surfactant to the guard cells of the plant caused a hypersensitive response resulting in reduced disease development. Application of Silwet L-77® along with the bacterial agent *Pseudomonas syringae* pv. *tagetis* resulted in enhanced disease severity on Canada thistle, when compared to the application of the bacterial agent in water alone [46, 47]. Results were similar when *Pseudomonas syringae* pv. *phaseolicola* was applied with Silwet L-77® for control of kudzu (*Pueraria lobata*) [142]. Hot, dry, sunny conditions appeared to be conducive for entry of the formulated bacteria into the stomata, whereas environmental conditions that caused the stomata to close (i.e. low light levels experienced at dusk or dawn) did not permit entry of the bacteria into the plant [92]. However, some phytotoxicity to Silwet L-77® is exhibited in some plant species.

It has been suggested that one of the most critical factors in liquid formulations that needs to be addressed is the ability to retain moisture around the droplet containing the bioactive ingredient (e.g. microbial propagule) [130, 140, 143]. Invert emulsions are a type of formulation that is comprised of various compounds in a water-in-oil entrapment mixture in order to retard evaporation and enhance the germination and infection of microbial propagules. The formulation has moisture-retaining properties to overcome long dew period requirements by many fungal bioherbicides [121, 125, 126, 144]. For example, formulating *Alternaria cassiae* in a mixture of paraffin wax, mineral oil, soybean oil, and lecithin resulted in a significant decrease in water evaporation and enhanced spore germination on sicklepod, despite inadequate dew [145]. The amendment of the invert emulsion with nutrients and optimal pH conditions also enhanced spore germination [136, 146]. Similarly, formulating *Alternaria eichhorniae* with invert emulsions containing either 25 or 35% soybean oil resulted in higher spore germination [147].

spores applied in water alone. However, the results from the field were not consistent. Spore germination rates were not enhanced and very high water volumes (500-1000 L/ha) were required for adequate weed control. These high water volume requirements are not representative of the typical application rates (i.e. 100-200 L/ha) used in countries like Australia, Canada, and the U.S. for chemical herbicides. The authors suggested that the oil emulsions may have improved the leaf surface wetting properties at low concentrations (0.5–1%).

Retention of moisture using hydophylic polymers has also been investigated [158]. The ability of 8 polymers to enhance the pathogenicity, viability and germination of fungal spores was evaluated. Several gels, including Kelgin® HV, MV, LV, Kelzan® xanthan gum, Gellan gum, N-Gel™, Metamucil®, and Evergreen®500 were found to be equally capable of retaining moisture for up to 6 days , with Kelzan® xanthan gum being the most effective Pathogenicity was enhanced, especially with Kelzan® xanthan gum and Kelgin®-HV. In view of the lack of optimum dew periods needed for fungal pathogens to infect leaf surfaces, a common obstacle to several mycoherbicides, the ability of the hydrophylic polymers to enhance bioherbicidal activity and retain moisture is evidence of their potential as components in a microbial formulation.

Use of fruit pectin and plant filtrates as adjuvants led to the alteration of the host range of *A. crassa* [159]. A variety of crop and weed species previously exhibiting resistance to the pathogen alone exhibited various degrees of susceptibility when amended with either fruit pectin or water soluble filtrates from jimsonweed and hemp sesbania. The authors suggested that the weed control spectrum could be expanded with judicious timing of application using these novel adjuvants. Other researchers demonstrated that naturally occurring adjuvants, such as the extracellular conidial matrix of *C. orbiculare* [160, 161] or extracellular polysaccharides from bacteria [162, 163] could be used to enhance disease development of foliar fungal bioherbicides. Disease development and visual symptoms in spiny cocklebur increased by the conidial matrix of *C. orbiculare* [160]. Enzymes present in the matrix were known to be involved in the penetration of fungal spores on the plant surface. Fernando *et al.* [162] showed that disease on velvetleaf occurred four to five days earlier when co-inoculated with phylloplane *Pseudomonas* spp. and *C. coccodes* compared to *C. coccodes* alone. It was discovered that viscosin, a peptidolipid biosurfactant and mediator of phytopathogenicity is produced by a pectolytic strain of *Pseudomonas fluorescens* [163].

4.1.2. Solid or granular formulations

While liquid formulations are more suitable for application of postemergence bioherbicides, solid-based formulations have been explored for preemergence bioherbicides that infect the weeds at or below the soil surface [1, 9, 125, 126]. Several advantages of soil-based formulations include: i) they have a buffering capacity against environmental extremes; ii) carriers contained within the granule can act as a substrate for the microbe; iii) they are less likely to be washed away from the target site; iv) shelf life may be longer than with liquid formulations; and v) granules can be versatile in their size and standard weight, thereby facilitating field application [1, 126].

Fungal spores entrapped in calcium alginate as a formulation of mycoherbicides was first reported by Walker and Connick [164]. Homogeneous mixtures of microbial propagules and sodium alginate, along with other carriers such as kaolin clay, ground oatmeal, soy flour, and cornmeal are added dropwise into a solution of 0.25 M $CaCl_2$ [165-169]. For example, the amendment of sodium alginate with nutrients (e.g. oatmeal, cornmeal, or soy flour) improved conidial production and field efficacy of *Fusarium solani* f.sp. *cucurbitae* for control of Texas gourd [166]. Daigle and Cotty [146] found that spore yield of *A. cassiae* was optimal when fermentation media were amended with 2.4% dehydrated potato dextrose broth and 14% V-8 vegetable juice and when entrapped in alginate pellets containing corn cob grits as the filler. Other researchers also demonstrated that a rice alginate prill formulation could enhance populations of *Fusarium oxysporum* f.sp. *erythroxyli* for biological control of coca [170].

Another type of granular formulation that has been explored for fungal bioherbicides is the pesta formulation, based on the entrapment of fungal propagules in a wheat-gluten matrix consisting of semolina flour, kaolin and fungal biomass [171]. The pasta-like dough is kneaded, passed through a pasta maker to produce a homogeneous sheet which is subsequently dried and crumbled. This formulation has been used to deliver *Fusarium oxysporum* for control of narcotic plants such as coca and opium poppy [172]. The moisture content (water activity) and sucrose concentration can also be modified in order to optimize shelf life of the product [172, 173]. Processing the *F. oxysporum* in pesta granules using twin-screw extrusion and fluid-bed drying at a water activity of 0.12 a_w and concentration of 5×10^6 cfu/g chlamydospores provided a shelf life of at least one year [172].

4.2. Applied Biotechnology

Molecular tools utilizing recombinant DNA technology are providing exciting opportunities in bioherbicide research to explore ways of enhancing efficacy and virulence, constructing fungal and bacterial agents with novel traits, and modifying host-specificity. Moreover, many of these tools can also be used to monitor the introduction and dispersal of the introduced microbial agents in order to establish the level of risk to the environment.

4.2.1. Genetic manipulation for increased pathogenicity or host range modification

The use of mycoherbicides on a large scale has been limited by the host-specificity of most of the fungi and bacteria that are being developed. While it has been shown that the effects of bioherbicides can be enhanced if multiple pathogens or combinations of pathogens and chemical herbicides are used, there is also the possibility that fungal and bacterial weed control agents could be genetically modified to expand the range of target weeds that are affected. These agents could also be modified or isolates selected that have tolerance to chemical pesticides. In this manner, modified biological control agents could be used in tank mixes with the chemical products to which resistance has been conferred. Microbial agents could also be transformed with genes that code for virulence factors, such as those for production of non-selective toxins. This approach is limited with respect to fungal biological control agents because only large gene clusters have been identified for toxin production in fungi and transformation methodologies for gene clusters of this size have not yet been successful.

Although it is not yet possible to create fungal transformants using toxin or pathogenicity gene clusters, several other approaches have been used to create model systems for altering pathogenicity and host range. Charudattan *et al.* [174] explored the possibility of altering virulence and host range of a plant pathogenic bacterium by inserting herbicide production genes from a non-pathogenic organism. The "bialaphos production genes" that are found in the saprophytes *Streptomyces hygroscopicus* and *S. viridochromogenes* result in the production of the herbicidally active phosphinothricin. This chemical has been semi-synthetically produced and sold in Japan as a broad-spectrum herbicide. The goal of the study was to insert these genes into the bacterium *Xanthomonas campestris* pv. *campestris* (XCC) and determine the effects of their expression. XCC was chosen for this experiment because it has a relatively simple genome and the protocols for genetic manipulation of this pathogen are well established. In addition, the organism has natural resistance to bialaphos in concentrations up to 1000 ppm. After successful transfer of a 33.5-kb fragment from a 41-kb plasmid containing the reassembled bialaphos genes into XCC, which was confirmed using Southern blot, the transformed bacteria were evaluated for phenotypic characteristics. No changes were reported in the morphological or fatty acid characteristics of the transformed XCC (XCC/pIL-1). The effect of XCC/pIL-1 on non-host plants was different from that of the wild-type XCC. The wild-type XCC causes a black rot of the host plants cabbage and broccoli and a hypersensitive reaction in non-hosts, such as bean, pepper, sunflower, and tobacco. The transformed XCC caused typical blackrot symptoms on cabbage and broccoli, but no hypersensitive response in the non-hosts. Rather, these plants became chlorotic at the inoculation sites. The results from inoculation of the non-hosts with the cell-free broth of the XCC/pIL-1 were inconclusive, therefore it is difficult to discern if the non-host response was due to the production of bialaphos by the bacterium or if the insertion of the genes altered the host response. In either case, the transformation resulted in an altered host response to the pathogen and provides a model system for development of other modified biological control agents.

A second approach involving the use of bialaphos-related genes was implemented by Brooker *et al.* [175]. *Colletotrichum gloeosporioides* f.sp. *aeschynomene* (*C.g.a.*) is an excellent pathogen of northern jointvetch (*Aeschynomene virginica*), but has only a limited effect on Indian jointvetch (*A. indica*), a weed often found in mixed stands with *A. virginica*. Therefore, an approach using a bialaphos-resistant transformant of *C.g.a.* applied with reduced rates of the herbicide was proposed [175]. The *bar* gene encoding for acetyltransferase that catalyses the acetylation of phosphinothricin, rendering it inactive was inserted into *C.g.a.* by protoplast manipulation. A transformant was further tested based on maintenance of morphological and growth characteristics when compared to the wild-type. Northern blot analysis confirmed expression of the gene in the *C.g.a.* transformant that was chosen and no differences in the host ranges of the wild-type and transformed *C.g.a.* over a range of spore concentrations were found. There was also no effect on northern jointvetch of bialaphos at low concentrations (0.56kg/ha). When the wild-type and the transformant were applied to northern jointvetch with and without bialaphos, there was a detrimental impact on the *C.g.a.*, resulting in lower disease severity on the wild-type when low concentrations of spores (2.5×10^6-2.5×10^8 spores/ml) were used. This impact was overcome at higher spore concentrations (2.5×10^9 spores/ml or higher). There was no impact of bialaphos on disease severity caused by the transformant regardless of spore concentration used. Using Indian jointvetch, only low levels of disease resulted with the application of the wild-type with and without the addition of bialaphos and with the transformant without bialaphos. The combination of the transformant with bialaphos resulted in an 86% increase in disease severity over application of the transformant alone. This study demonstrates that the addition of a gene for resistance to a nonselective toxin can enhance the level of disease produced with lowered levels of inoculum and the combination can be used to broaden the range of plants on which the agent is effective. This type of transformation has many potential applications, including the addition of resistance to other types of herbicides, as well as fungicides.

Using *C.g.a.*, TeBeest [176] generated benomyl-resistant strains through ethyl methanesulfonate mutagenesis. While these mutants exhibited resistance to benomyl and could theoretically be used for tank mixing with the fungicide, additional tests with the mutants revealed that these isolates were highly variable with respect to several infection components [177]. In general, the benomyl-resistant strains were less competitive than wild-type strains as a result of producing fewer sporulating lesions, having slower production of sporulating lesions, having extended latent period and slower lesion expansion rate, and producing fewer lesions on inoculated plants than the better of the two wild-type strains tested.

In a similar study, actively growing mycelium of the pathogen *Colletotrichum gloeosporioides* f.sp. *malvae* (*C.g.m.)*, the causal agent of anthracnose of round-leaved mallow, was UV-irradiated to produce benomyl-resistant strains [178]. A single mutant strain was found to be as virulent as the wild-type when applied in the absence of benomyl. There was an interaction between timing of the benomyl application, strain used, and dew period. In the absence of benomyl, disease severity was similar between the benomyl-resistant strain and the wild-type regardless of whether the dew period was 24 or 48 hours. When benomyl was applied 24 hours before the strains, no disease developed with the wild-type under either dew treatment, but the benomyl-resistant strain caused 17% and 58% plant wilting with 24 and 48 hours of dew, respectively. Concurrent applications of benomyl with the fungus resulted in the wild-type strain causing no plant wilting and the benomyl-resistant strain causing 4% and 67% wilting with 24 and 48 hours of dew, respectively. When benomyl was applied 24 hours after the fungus, the wild-type caused 33% and 79% wilt after 24 and 48 hours of dew respectively and the resistant *C.g.m.* caused 68% and 100% respectively. Although the results of this test showed that tolerance of the mutant strain to benomyl was dependent upon the timing of application and that the benomyl-mutant could not be applied at the same time as a fungicide treatment, it does demonstrate that the use of a mutant could permit this agent to be part of an integrated pest management program as long as the recommendations concerning timing of pesticide applications are adhered to.

Sands and Miller [179] have argued that few plant pathogenic fungi that are limited to specific hosts are sufficiently lethal to be used as biological control agents. Many plant pathogenic fungi that show a high level of efficacy for biological control of the target weed are often extremely limited in the environment in which they are able to perform. The fungus *Sclerotinia sclerotiorum* is a broad-host range pathogen attacking many different plant species. Infected plants are killed rapidly and under a variety of environmental conditions. Sands *et al.* [180] demonstrated that the fungus, through ultraviolet or chemical mutagenesis, can be genetically altered to restrict its host range, spread, and production of survival structures. In examining the fungus for control of Canada thistle (*Cirsium arvense*) and spotted knapweed (*Centaurea maculosa*), high levels of plant mortality were achieved using wild-type strains. Generation of mutants yielded pyrimidine-dependent auxotrophs, nonsclerotia-forming mutants, and reduced host range/reduced virulence mutants. The auxotrophic and nonsclerotia-forming mutants are extremely limited in their ability to spread from the point of infection, do not survive adverse environmental conditions, and in the case of the auxotrophic mutants, are dependent upon an exogenous nutrient source. These characteristics could make it possible to use this typically broad host range fungus for the control of weeds without endangering susceptible crops.

4.2.2. Molecular markers for biocontrol agents

The application of molecular markers for biological control agents has faciliated the tracking of organisms that have been released into the environment and has allowed for strain differentiation that was not previously possible when based on morphological characteristics. The rust pathogen *Puccinia carduorum*, for example, was originally collected in Turkey [89], and was subsequently released in Virginia for the control of musk thistle (*Carduus thoermeri*). The spread of the pathogen throughout the United States was confirmed through the use of a specific DNA sequence identified within the Internal Transcribed Spacer 2 of the ribosomal DNA. The sequence information has allowed for the released Turkish isolate to be differentiated from an indigenous isolate of the same species from California that is pathogenic to the slenderflower thistle (*C. tenuiflorus*) [181] .

The fungus *Chondrostereum purpureum,* which is found throughout many temperate zones and has a broad host range, has been under development as a biological control agent for many deciduous forest weeds [36]. The biological control isolates of this pathogen are applied as a mycelial "paste" to cut stems of forest weed trees. In order to effectively determine the infection frequency of a specific isolate that is intentionally applied versus infection by other opportunistic fungi colonizing the cut stump surface, a two-part PCR-based system was developed that could first, quickly identify *C. purpureum* isolated from wounded trees, and second, distinguish amongst genetic individuals within populations of *C. purpureum* [182, 183]. Identification of *C. purpureum* through the use of PCR was accomplished by the development of a selective marker based on the large intergenic spacer (IGS or nontranscribed spacer region (NTS-L)) region of the ribosomal DNA. The second task was accomplished through the development of the APM22D13 sequence characterized amplified region (SCAR) primer set. After field inoculations with two formulations *of C. purpureum,* stumps were screened for the presence of the fungus, first by amplification of the IGS primers to identify all isolations of *C. purpureum.* Positive isolates were further screened using the isolate-specific APM22D13 primers and results of that analysis revealed that only two of the fungi isolated were endemic isolates of *C. purpureum.* In addition, the biocontrol fungus was only isolated from stumps to which it was applied. There was no cross-contamination of nontarget stumps and it was possible to determine which target species the biological control agent most effectively colonized. This work provides an excellent example of how a PCR- based system can allow large numbers of samples to be analyzed quickly and accurately and can provide important information on the safety and spread of a released biological control agent.

Past studies have relied upon the use of antibiotic resistant mutants to monitor populations of bacteria, but often these techniques are not very sensitive or reliable. *Lux* genes that encode for enzymes involved in bioluminescence have been used as markers to track pseudomonad bacteria in soil [184]. Use of the green fluorescent protein (*gfp*), that forms functional products with other proteins, have also been developed for visual inspection for fluorescence activity [185]. However, a criticism of these two systems is that the *lux* and *gfp* genes are carried on a transposable element thereby increasing the risk of horizontal transfer to another organism resulting in loss of specificity in the tracking system. A system for monitoring *Xanthomonas* bacteria was developed by tagging the bacterium with the phaseolotoxin gene cluster [186]. This marker system appears to be highly specific and unique, has no effect on other functions of the tagged bacterial strain other than colony color, and it can be used for other *Xanthomonas* species and pathovars.

5. IMPLEMENTATION OF BIOHERBICIDES INTO AGROECOSYSTEMS

While a great majority of the classical biological control and bioherbicide examples have clearly demonstrated their "potential" to biologically control weeds in controlled environment and small field plot studies, in reality, it is the end-user (i.e. the farmer) who will utilize such tools to manage weeds in a cropping situation. With this in mind, researchers must consider how biological control agents can be successfully implemented into crop production systems, given the complex crop rotations systems, pest management systems (including use of chemical insecticides, fungicides, and herbicides), and diverse weed populations within a particular field. Biological weed control strategies represent an excellent opportunity for farmers to practise integrated pest management, thereby mitigating some of the problems that have developed through traditional crop production practices.

5.1. Fitting Biologicals into Production Agriculture

Changes in availability of herbicides and other pesticides necessitate the search for other weed control alternatives. Current efforts to develop integrated weed management programs that include modifications to cultural practices, additions of cover crops, green manure, and use of soil solarization are systems that contribute to weed control, but would benefit from combinations with biological control agents. In addition, application of biological control agents compatible with chemical herbicides at reduced rates could be explored. In order to use biological control agents in the crop production systems, it is also imperative that the agents be tested for sensitivity to the other agricultural chemicals used for pest control. This is only one example in which biological control could provide the basis for an integrated approach to weed management in a cropping system. While a great deal of benefit can be derived from basic research involved in the development of biological control agents, such as understanding of host-pathogen interactions, pathogenicity determinants, basic biology and ecology of weeds and biology and epidemiology of pathogens, the ultimate goal of most biological control programs is to provide a weed management strategy that can be utilized in a land management or crop production system. This was accomplished with the introduction of DeVine into the citrus production system and with the utilization of Collego to manage northern jointvetch in rice production. Fundamental research evaluating the application of bioherbicides such as Collego in combination with other agricultural pesticides is essential if they are to be successfully combined with other crop production practices. These studies can help i) predict if the effects of the biological control agent can be enhanced if applied in combination with chemical herbicides, tillage, and crop rotations, or ii) identify unexpected benefits or drawbacks to exposure of the agents to chemical pesticides.

As the concept of integrated pest management moves or evolves from an abstract to a practically viable concept, the character of work within specialized fields must evolve to meet this development. Thus, the partitioning of the biological control of weeds into two mutually exclusive areas of work that has developed since its earliest stages must end in favor of an altered approach. Work on classical biological control of weeds has been compartmentalized into insect-centered approaches and pathogen-centered approaches [187]. The bulk of the efforts in the past have been devoted to the mycoherbicide approach, while the great potential for contributions to classical weed biocontrol in general remain unfulfilled. An impetus for the need to develop new approaches to how classical weed biocontrol is conducted is the high cost of research programs. As much as \$400,000-600,000 in expenditures per agent can be incurred for testing, which includes the search for and initial testing overseas and the continued or final testing [188]. Wide scale release, redistribution and monitoring of (mainly insect) releases easily accounts for an even greater eventual cost per agent. Usually, successful programs are determined by only a fraction of all agents initially tested and introduced [189], and often no more than 1-2 insects were causal for success. This perhaps indicates that there were key mechanisms that led to control of the individual weeds in each case. This concept would contrast with the notion that multiple insects per target weed should lead to an incrementally or cumulatively greater impact. Such a notion, uncritically applied, could be at least in part responsible for the introduction of a great many ineffective insects in classical weed biocontrol, each at great cost. Such key mechanisms discussed above may, as in the case of successful biocontrol of leafy spurge, necessarily include plant pathogens as synergists [187] for success. The demonstrated failure of insect agents alone to reduce biomass of the perennial weed *Centaurea maculosa* Lam. in controlled studies [190] along with the finding that one or more plant pathogens such as *Fusarium* spp. are associated with unthrifty, insect-damaged plants in foreign [97] and domestic stands of *C. maculosa* (Caesar, unpublished) further illuminates this point. Lessons learned in both insect-centered approaches and pathogen-based programs provide cautionary ideas and at the same time illustrate the need for basing future work on a more thorough examination of the ecology of the factors that apparently limit the weediness of a target species as called for recently [107]. Among the ecological factors that should be considered are microbes and especially plant pathogens. Thus, the greatest potential for increased success of plant pathogens in classical weed biocontrol is to use knowledge about their effects as participants with insects in weed suppression to reducing the time, effort and therefore cost of programs against weeds of noncultivated areas.

There are several types of interactions that can be evaluated in the area of biological control in production agriculture: i) detrimental effects of crop production pesticides on the biological control agent; ii) timing of application required if there is a detrimental effect; iii) possible enhanced activity of the biological control agent through combination with chemical herbicides; iv) unexpected benefits by combining the agent with other pesticides; v) possible alerations of non-target hosts by chemical pesticides that affect the activity of the biological control agent; vi) improved weed control by combining with cropping practices to minimize the need for chemical inputs; vii) possible reductions in the competitive interaction between weeds and the crop with application of a bioherbicide; and viii) possible combined effects of allelopathic crop cultivars with biological control to minimize chemical inputs.

5.2. Biological Control Agents as Components in Integrated Weed Management Systems

Combining biological control agents with other pest control measures requires an extensive amount of research if the agent is to be used in an integrated pest management system. Application of an agent, such as the use of DeVine®, in a production system can be relatively straight forward, or it may be significantly more complicated, as is the case with the use of Collego®. DeVine® can be applied as a liquid formulation and once applied, provides long-term control of milkweed vine (*Morrenia odorata*) in citrus in Florida and it required little development in terms of fit in the production system. According to the DeVine® label, the microbial agent should not be mixed with wetting agents, fertilizers, or other pesticides. This approach simply requires that a three-week interval be allowed before any applications of these materials.

Collego® has a more complicated history of research and development [5]. Rice and soybeans, the annual crops in which Collego® was developed for use, have a substantial number of weed, pathogen, and insect problems that limit production if not controlled. It was necessary to determine the crop protection and production practices that would be detrimental to the efficacy of bioherbicide. Smith [191] has summarized the specific tests and recommendations that have been conducted for integration of Collego® with other pesticides including sequential timings of Collego® with a variety of pesticides. For example, Collego® can be mixed with the herbicides acifluorfen and bentazon, but cannot be mixed with 2,4-D. Combining the microbial herbicide with the chemical herbicides allows for a broad-spectrum of weeds to be controlled. Fungicide applications (triphenyltin hydroxide) to control rice sheath blight can be applied one week after the biological control agent, but benomyl applications for controlling rice blast must not be used for three weeks after Collego®. Collego® could also be combined with insecticide treatments necessary for managing rice water weevil, fall armyworm, and rice stink bug without any loss in efficacy. Determining these constraints are not only necessary for labeling of bioherbicides, but are also crucial for enabling growers to adopt these practices successfully. In the case of Collego®, there was no enhanced control of the target weed through the integration of multiple pest control tactics, but the system approach allows the biological control agent to fit into an existing crop production system.

In some cases, the effect of the pathogen used for weed control can be enhanced when combined with chemicals. This is the case with control of Texas gourd (*Cucurbita texana*) using *Fusarium solani* f.sp. *cucurbitae* (FSC), in which the combination of FSC and trifluralin preplant incorporated significantly decreased seedling emergence over that achieved using FSC alone. Increased control of Texas gourd and a more broad range of weeds controlled was achieved. A helpful review of studies evaluating the effects of herbicides on plant root systems is provided by Altman *et al.* [192]. This information can contribute significantly to formulating ideas concerning combinations of biologicals and chemicals. Understanding the herbicide effects on plant morphological changes can help us to pair biological control agents with pesticides that may enhance their activity. Such is the case of the combination of *Colletotrichum coccodes* and the plant growth regulator thidiasuron (N-phenyl-N"-1,2,3-thiadiazol-5-yl-urea) for the control of velvetleaf. Application of thiadiazol, which has been used as a cotton defoliant, caused curling and cupping of leaves, stimulated axillary bud development and reduced several plant growth parameters. Co-application of the chemical with *C. coccodes* resulted in increased weed mortality compared to the application of the biological control agent alone.

Phatak *et al.* [34] reported that the combination of the rust fungus *Puccinia canaliculata* and paraquat (1,1"-dimethyl-4,4"-bipyridinium ion) provided 99% control of yellow nutsedge (*Cyperus esculentus*) compared to 60% control with the rust alone and 10% control with paraquat alone. This approach has also been used with the fungal pathogen in combination with imazaquin [2-[4,5-dihydro-4-methyl-4-(1-methylethyl)-5-oxo-1H-imidazol-2-yl]-3-quinoline carboxylic acid], bentazon [3-(1-methyl-ethyl-(1H)-2,1,3-benzothiadiazin-4(3H)-one 2,2-dioxide] [110], and metribuzin [4-amino-6-(1,1-dimethyl-ethyl)-3-(methylthio)-1,2,4-triazin-5(4H)-one]. These combinations can provide enhanced control of the biological control target and increase the spectrum of weeds controlled by the integrated management practices.

A great deal of information is available concerning the nontarget effects of chemical herbicides on various weeds and crops. This information can be utilized to determine the specific approach that might be employed for particular target weeds in crops. In addition to adding to the efficacy of the applied bioherbicides, this information can be used to make predictions concerning the potential non-target effects on crops. At this time, there are few generalizations that can be drawn that are based on the chemistries of pesticides concerning their compatibility with biological control agents. Therefore, the recommendation is to test any combinations that are applicable in the crop production system toward which the development of the biological control agent is geared. Reviews of some aspects of this approach have been provided [5, 192-194].

Several programs throughout the world are directing research to integrate biological control agents into production agriculture. While many of these are in their infancy, many programs are now geared in this direction from the start. Several programs are now searching for combinations of pathogens and insects that can be integrated from the start. In addition, several studies have been conducted that combine production practices with applications of biological control agents.

A "system management approach" proposed by Müller-Schärer and Frantzen [195], while in a preliminary stage of development, provides a useful model for the development of new projects or redirection of more mature ones. Using the *Senecio vulgaris*:*Puccinia lagenopharae* weed-pathosystem as a model, the authors proposed that in addition to the common isolation, screening, epidemiology, and genetic studies that have been routinely used in the development of a biological weed control agent, weed stress and cropping system aspects should be a component of the initial research program. These authors also propose that a change in the philosophy of weed control is essential for the success of biological control in that it is more consistent with retaining biological diversity to provide weed management tactics that are capable of keeping weed populations suppressed below economic thresholds, rather than using the "clean crop" approach. This approach allows us to study, particularly in cropping systems where a single weed problem is the main focus, how the biological control agent affects the competitive ability of the target weed. This may allow for the efficacy of biological control agents to be assessed based on crop yields, rather than on weed mortality. An example of this approach was investigated by Kadir *et al.* [88] in which the biological control agent *Dactylaria higginsii* was used to eliminate competition of purple nutsedge (*Cyperus rotundus*) with tomato. This system is directly applicable to the tomato production system, as other pre-plant incorporated herbicides are available in tomato that control most other weeds, but provide only limited or no control of nutsedge.

This approach has also been investigated for suppression of *Viola arvensis* in spring and winter cereals with the pathogen *Mycocentrospora acerina*. With high levels of inoculum, high mortality of the weed resulted and limited competition of the weed with the wheat crop. At lower levels of inoculum, when the weed was not killed, the application of the pathogen still resulted in a 28% reduction in competition from the weed [196]. Interestingly, a similar approach using the velvetleaf:*Colletotrichum coccodes* system demonstrated that the effects of the biological control agent when the weed was grown in competition with a soybean crop gave much greater weed suppression than when the biological control agent was applied to a pure stand of velvetleaf [197]. Although soybean yields increased in only one trial, this example provides the basis for developing a systems approach that includes competition from the crop as a component.

Application of the biological control agent *Ascochyta caulina* to common lambsquarters (*Chenopodium album*) in sugarbeet and corn resulted in a reduction of competition from the weed in the corn cropping system, but did not have a positive effect when applied in the sugar beet system. This example supports the idea that the successful development of bioherbicides can be enhanced if the crop characteristics are taken into account. While using the biological control agent alone may be a feasible approach for management of the weed in corn, a more integrated approach may need to be addressed if it is to be successful in sugarbeet production. Evaluating any given biological control agent for use in an agro-ecosystem presents an almost endless number of combinations that can be evaluated. The goal of the programs should be to find a system in which the biological control agent can have the greatest impact while adhering as closely as possible to the production system with which the growers are most familiar. It is unlikely that biological control will be adopted if the crop production practices have to be changed in a manner that is not economically viable for the grower or requires the grower to totally modify the system.

An excellent example of a bioherbicide for use in an annual cropping system is *Colletrichum truncatum* for the control of hemp sesbania (*Sesbania exaltata*) in soybean and other crops in the Mississippi Delta region [198, 199]. A great deal of applied research has been conducted with this agent in order to facilitate its implementation in the cropping system. Various formulations and applications methods have been tested, all of which could easily be adopted by growers. In some cases, practices may already have been adopted by growers that provide some measure of weed suppression into which a bioherbicide can be easily integrated. For example, in a maize cropping system in which the use of red clover as a living cover has been adopted, the weed hedge bindweed (*Calystegia sepium*) remains a problem. Combining the biological control agent *Stagnospora convolvuli* with the ground cover showed significantly better weed suppression than other approach used alone [200]. There are numerous examples in which cultural practices, such as the use of an allelopathic or "weed suppressive" cultivar, row spacing, tillage practices, use of green manures or cover crops, and soil solarization, as examples, provide some measure of weed suppression [201, 202]. These practices could very easily be coupled with the use of biological control agents and, in many cases, research has been proposed or is underway in this area, although there are few examples of these combinations being used in production agriculture. Changing the way that the research is approached, keeping these potential combinations in mind during the development process, can greatly enhance the level of success for biological control in specific crops and for biological control as a whole.

6. CONCLUSIONS

Changes in availability of herbicides and other pesticides as well as the development of herbicide-resistant weed populations have necessitated the search for weed control alternatives. Over the last three decades, eight bioherbicide products have been registered and one key question that has often been asked is why more of these products are not commercially available. When considering the level of investment by chemical companies for the development of chemical pesticides, with a success rate of less than 1% compared to a success ratio of 20:1 for pathogens as bioherbicides [4], bioherbicides are progressing extremely well. Utkhede [203] estimated that it costs as much as $80 million to develop a chemical pesticide compared to approximately $0.8 to 1.6 million for biopesticides. More importantly, it has been argued that lack of resources, infrastructure and numbers of scientists dedicated to biological control research have not been as generous when assessed against the chemical pesticide industry. Although research on bioherbicides has enjoyed more support than research on classical biological control, investment in long-term research coupled with the lack of support by adminstrators and funding agencies in general have been major impediments to the introduction of new classical biological control agents.

This review has provided several excellent examples of successful introductions of classical biological agents and bioherbicides that are either registered or are in precommercial evaluations. A change in philosophy in weed management is required, however, in order for biological weed control to enjoy greater success. As the concept of integrated weed management evolves, weed managers will have a better understanding and acceptance of the idea that complete eradication of weeds is not necessarily the goal, but that reducing weed competition below economic-thresholds will contribute to a more environmentally sustainable crop production system. It would be naive to suggest that biological control will replace chemical herbicides, rather this strategy should be considered one more tool in a weed management package for farmers to use. Current efforts to develop integrated weed management programs that include modifications to cultural practices, additions of cover crops, green manure, and use of soil solarization are systems that contribute to weed control, but would benefit from combinations with biological control agents. In addition, application of biological control agents compatible with chemical herbicides at reduced rates could be explored. In order to use biological control agents in crop production systems, it is imperative that the agents be tested for sensitivity to the other agricultural chemicals used for pest control. This is only one example in which biological control could provide the basis for an integrated approach to weed management. Finally, biological control requires the involvement of a team of researchers in several disciplines including plant pathology, entomology, weed biology and ecology, biochemistry, and agronomy if it is to progress from the early proof of concept phase to practical implementation in the field.

7. REFERENCES

1. Boyetchko, S.M. (1999). Innovative applications of microbial agents for biological weed control. *In:* K.G. Mukerji, B.P. Chamola, and K. Upadhyay (eds.),Biotechnological Approaches in Biocontrol of Plant Pathogens, Kluwer Academic/Plenum Publishers, London, pp. 73-97.
2. Charudattan, R. (1990). Biological control by means of fungi. *In:* K.J. Murphy and A. Pieterse (eds.), Aquatic Weeds, Oxford Univ. Press, Oxford, UK., pp. 186-201.
3. Charudattan, R. (1990). Pathogens with potential for weed control. *In:* R.E. Hoagland (ed.), Microbes and Microbial Products as Herbicides, ACS Symposium Series 439, Washington, D.C., pp. 132-154.
4. Charudattan, R. (1991). The mycoherbicide approach with plant pathogens. *In:* D.O. TeBeest (ed.), Microbial Control of Weeds, Chapman and Hall, Inc., New York, pp. 24-57.
5. Charudattan, R. (1993). The role of pesticides in altering biocontrol efficacy. *In:* J. Altman (ed.) Pesticide interactions in crop production: Beneficial and deleterious effects, CRC Press, Boca Raton, FL. , pp. 421-432.
6. Charudattan, R. (2000). Current status of biological control of weeds. *In:* C.G. Kennedy and T.B. Sutton (eds.), Emerging Technologies for Integrated Pest Management: Concepts, Research, and Implementation, APS Press, St. Paul, MN., pp. 269-288.
7. Kremer, R.J. (1998). Microbial interactions with weed seeds and seedlings and its potential for weed management. *In:* Integrated Weed and Soil Management, J.L. Hatfield, D.D. Buhler, and B.A. Stwart (Eds.), Ann Arbor Press, Chelsea, MI, pp. 161-179.
8. Kremer, R.J. and Kennedy, A.C. (1996). Rhizobacteria as biocontrol agents of weeds. Weed Technol. Weed Technol, 10 (1996) 601-609.
9. Rosskopf, E.N., Charudattan, R., and Kadir, J.B. (1999). Use of plant pathogens in weed control. *In:* T.S. Bellows and T.W. Fisher (eds.), Handbook of Biological Control, Academic Press, pp. 891-918.
10. TeBeest, D.O., Yang, X.B. and Cisar, C.R. (1992 The status of biological control of weeds with fungal pathogens. Annu. Rev. Phytopathol. 30:637-657.
11. Watson, A.K. (1991). The classical approach with plant pathogens. *In:* D.O. TeBeest (ed.), Microbial Control of Weeds, Chapman & Hall, New York, pp. 3-23.
12. Powell, K.A. and Jutsum, A.R. (1993). Technicial and commercial aspects of biocontrol products. Pesticide Sci. 37:315-321.
13. Aspelin, A.L. (1994). Pesticide Industry Sales and Usage-1992 and 1993 Market Estimates. U.S. EPA. Washington, D.C.
14. Hamill, A.S., Surgeoner, G.A., and Roberts, W.P. (1994). Herbicide reduction in North America: In Canada, an opportunity for motivation and growth in weed management. Weed Technol. 8:366-371.
15. Bridges, D.C. (1994). Impact of weeds on human endeavors. Weed Technol. 8:392-395.
16. Peeples, K.A. (1994). Agriculture's challenge to develop a vision for the future. Weed Technol. 8:372-375.
17. Bellinder, R.R., Gummesson, G., and Karlsson, C. (1994). Percentage-driven government mandates for pesticide reduction: The Swedish model. Weed Technol. 8:350-359.

18. Swanton, C.J., Harker, K.N., and Anderson, R.L. (1993). Crop losses due to weeds in Canada. Weed Technol. 7:537-542.
19. Holt, J.S. (1992 History of identification of herbicide resistant weeds. Weed Technol. 6:615-620.
20. Hasan, S. and Ayres, P.G. (1990). The control of weeds through fungi: Principles and prospects. New Phytol. 115-222.
21. Hasan, S. and Wapshere, A.J. (1973). The biology of *Puccinia chondrillina* a potential biological control agent of skeleton weed. Ann. Appl. Biol. 74 (1973) 325.
22. Supkoff, D. M., Joley, D.B., and Marois, J.J. (1988). Effect of introduced biological control organisms on the density of *Chondrilla juncea* in California. J. Appl. Ecology 25:1089-1095.
23. Kropp, B. R., Hansen, D., Flint, K.M., and Thomson, S.V. (1996). Artificial Inoculation and Colonization of Dyer's Woad (*Isatis tinctoria*) by the Systemic Rust Fungus *Puccinia thlaspeos*. Phytopathology. 86: 891-896.
24. Fröhlich, J., Fowler, S., Gianotti, A., Hill, R., Killgore, E., Morin, L., Sugiyama, L., Winks, C. (2000). Biological control of mist flower (*Ageratina riparia*, Asteraceae) : transferring a successful program from Hawaii to New Zealand. *In:* N.R. Spencer (ed.), Proc. X International Symposium on Biological Control of Weeds, 4-14 July, 1999, Montana State University, Bozeman, Montana, USA. pp. 37-43.
25. Hasan, S., Sobhian, R., and Knutson, L. (1999). Preliminary studies on *Ramularia crupinae* sp. nov. as a potential biological control agent for common *crupina (Crupina vulgaris)* in the USA. Ann. Appl. Biol. 135:489-494.
26. Hildebrand, P. D. and Jensen, K.I.N. (1991). Potential for the biological control of St. John's-wort (*Hypericum perforatum*) with an endemic strain *of Colletotrichum gloeosporioides*. Can. J. Plant Pathol. 13:60-70.
27. Littlefield, L. J. (1985). Biocontrol of leafy spurge with pathogenic fungi. *In*: E.S. Delfosse (ed.), Proc. VI International Symposium on Biological Control of Weeds, 19-25 August 1984, Vancouver, Canada. p. 633
28. Burnett, H.C., Tucker, D.P.H., and Ridings, W.H. (1974). *Phytophthora* root and stem rot of milkweed vine. Plant Dis. Rep. 58:355-357.
29. TeBeest, D.O. and Templeton, G.E. (1985). Mycoherbicides: Progress in the biological control of weeds. Plant Dis. 69:6-10.
30. Bowers, R.C. (1986). Commercialization of Collego - An industrialist's view. Weed Sci. 34(Suppl. 1):24-25.
31. Mortensen, K. (1988). The potential of an endemic fungus *Colletotrichum gloeosporioides*, for biological control of round-leaved mallow (*Malva pusilla*) and velvetleaf (*Abutilon theophrasti*). Weed Sci. 36:473-478.
32. Mortensen, K. and Makowski, R.M.D. (1989). Field efficacy at different rates of *Colletotrichum gloeosporioides* f.sp. *malvae* as a bioherbicide for round-leaved mallow (*Malva pusilla* Sm.). *In:* E.S. Delfosse (ed.) Proc. VIIth Int. Symp. Biol. Contr. Weeds, Rome, 1988. pp. 523-530.
33. Greaves, M.P. and MacQueen, M.D. (1992). Bioherbicides: their role in tomorrow's agriculture. *In:* Resistance: Achievements and Developments in Combating Pesticide Resistance, Sci. Symposium, Harpenden, England, UK. pp. 295-306.
34. Phatak, S.C., Wells, H.D., Sumner, D.R., Bell, D.K., and Glaze, N.C., (1983). Biological control of yellow nutsedge with the indigenous rust fungus *Puccinia canaliculata*. Science 219:1446-1447.
35. De Jong, M.D., Scheepens, P.C., and Zadoks, J.C. (1990). Risk analysis for biological control: a Dutch case study in biocontrol of *Prunus serotina* by the fungus *Chondrostereum purpureum*. Plant Dis. 74:189-194.
36. Dumas, M.T., Wood, J.E., Mitchell, E.G., and Boyonoski, N.W. (1997). Control of stump sprouting of *Populus tremuloides* and *P. grandidentata* by inoculation with *Chondrostereum purpureum*. Biol. Control 10:37-41.
37. Morris, M.J., Wood, A.R., and Den Breeyen, A. (1998). Development and registration of a fungal inoculant to prevent regrowth of cut wattle tree stump in South Africa, and a brief overview of other bioherbicide projects currently in progress. *In:* IV International Bioherbicide Workshop Programme and Abstracts, 6-7 August, 1998, Univ. of Strathclyde, Glasgow, Scotland. p. 15
38. Bewick, T.A., Binning, L.K., Stevenson, W.R., and Stewart, J. (1986). Development of biological control for swamp dodder. Proc. North Central Weed Control Conference 41:24.
39. Rosskopf, E.N. (1997). Evaluation of *Phomopsis amaranthicola* sp. nov. as a biological control agent for *Amaranthus* spp. Ph.D. Thesis, University of Florida, Gainesville, Florida.
40. Kadir, J. and Charudattan, R. (1996). *Dactylaria higginsii* (Lutrell) M.B. Ellis: A potential bioherbicide for nutsedge (*Cyperus* spp.). WSSA Abstracts 36:49.
41. Zhang, W.M., Moody, K. and Watson, A.K. (1996). Responses of *Echinochloa* species in rice (*Oryza sativa*) to indigenous pathogenic fungi. Plant Dis. 80:1053-1058.
42. Kempenaar, C., Horsten, P.F.M., and Scheepens, P.C. (1996). Growth and competitiveness of common lambsquarters (*Chenopodium album*) after foliar application of *Ascochyta caulina* as a mycoherbicide. Weed Sci. 44:609-614.

43. Kempenaar, C., Wanningen, R., and Scheepens, P.C. (1996). Control of *Chenopodium album* by soil application of *Ascochyta caulina* under greenhouse conditions. Ann. Appl. Biol. 129:343-354.
44. Imaizumi, S., Nishino, T., Miyabe, K., Fujimori, T., and Yamada, M. (1997). Biological control of annual bluegrass (*Poa annua* L.) with a Japanese isolate of *Xanthomonas campestris* pv. *poae* (JT-P482). Biol. Control 8:7-14.
45. Johnson, D.R., Wyse, D.L., and Jones, K.J. (1995). Efficacy of spring and fall applications of *Pseudomonas syringae* pv. *tagetis* for Canada thistle (*Cirsium arvense* L.) control in soybean. WSSA Abstracts 35:61.
46. Johnson, D.R., Wyse, D.L., and Jones, K.J. (1996). Controlling weeds with phytopathogenic bacteria. Weed Technol. 10:621-624.
47. Bailey, K.L., Boyetchko, S.M., Derby, J., Hall, W., Sawchyn, K., Nelson, T., and Johnson, D.R. (2000). Evaluation of fungal and bacterial agents for biological control of Canada thistle. *In:* N.R. Spencer (ed.), Proc. X International Symposium on Biological Control of Weeds, 4-14 July, 1999, Montana State University, Bozeman, Montana, USA. pp. 203-208.
48. Suslow, T.V. and Schroth, M.N. (1982). Role of deleterious rhizobacteria as minor pathogens in reducing crop growth. Phytopathology 72:111-115.
49. Kremer, R.J., Begonia, M.F.T., Stanley, L., and Lanham, E.T. (1990). Characterization of rhizobacteria associated with weed seedlings. Appl. Environ. Microbiol. 56:1649-1655.
50. Elliott, L.F. and Lynch, J.M. (1985). Plant growth-inhibiting pseudomonads colonizing winter wheat (*Triticum aestivum* L.) roots. Plant and Soil 84:57-65.
51. Fredrickson, J.K. and Elliott, L.F. (1985). Colonization of winter wheat roots by inhibitory rhizobacteria. Soil Sci. Soc. Am. J. 49:1172-1177.
52. Begonia, M.F.T., Kremer, R.J., Stanley, L., and Jamshedi, A. (1990). Association of bacteria with velvetleaf roots. Trans. Missouri Acad. Sci. 24:17-26.
53. Boyetchko, S.M. and Mortensen, K. (1993). Use of rhizobacteria as biological control agents of downy brome. *In:* Proc. Soils and Crops Workshop '93, Saskatoon, Saskatchewan, February 25-26, 1993. pp. 443-448.
54. Cherrington, C.A. and Elliott, L.F. (1987). Incidence of inhibitory pseudomonads in the Pacific Northwest. Plant Soil. 101:159-165.
55. Kennedy, A.C., Elliott, L.F., Young, F.L., and Douglas, C.L. (1991). Rhizobacteria suppressive to the weed downy brome. Soil Sci. Soc. Am. J. 55:722-727.
56. Kremer, R.J. (1987). Identity and properties of bacteria inhabiting seeds of selected broadleaf weed species. Microb. Ecol. 14:29-37.
57. Begonia, M.F.T. (1989). Characterization of attraction of rhizobacteria to weed seeds and seedlings. Ph.D. Dissertation, University of Missouri, Columbia, Missouri, USA, 165 pp.
58. Kremer, R.J. (1993). Management of weed seed banks with microorganisms. Ecol. Appl. 3:42-52.
59. Boyetchko, S.M. (1997). Efficacy of rhizobacteria as biological control agents of grassy weeds. *In:* Proc. Soils and Crops Workshop '97, Saskatoon, Saskatchewan, February 20-21, 1997. pp. 460-465
60. Tranel, P.J., Gealy, D.R. and Kennedy, A.C. (1993). Inhibition of downy brome (*Bromus tectorum*) root growth by a phytotoxin from *Pseudomonas fluorescens* strain D7. Weed Technol. 7:134-139.
61. Gurusiddaiah, S., Gealy, D.R., Kennedy, A.C. and Ogg, A.G., Jr. (1992). Production, isolation, and characterization of phytotoxic and fungistatic compounds for biocontrol of downy brome (*Bromus tectorum* L.) and plant pathogenic fungi. Weed Sci. Soc. Am. Abstr. 32:84.
62. Tranel, P.J., Gealy, D.R., and Irzyk, G.P. (1993) Physiological responses of downy brome (*Bromus tectorum*) roots to *Pseudomonas fluorescens* strain D7 phytotoxin. Weed Sci. 41:483-489.
63. Kremer, R.J. and Sarwar, M. (1995). Microbial metabolites with potential applications in weed management. Proc. Plant Growth Reg. Soc. Am. 21:48-51.
64. Sarwar, M. and Kremer, R.J. (1995). Enhanced suppression of plant growth through production of L-tryptophan-derived compounds by deleterious rhizobacteria. Plant Soil 172:261-269.
65. Sarwar, M. and Kremer, R.J. (1995). Determination of bacterially derived auxins using a microplate method. Letters Appl. Microbiol. 20:282-285.
66. Sawchyn, K., Nelson, T., Gibson, L., Leung, S., and Boyetchko, S.M. (2000). Development of rhizobacteria for biological control of grass weeds. *In:* Proc. Third International Weed Science Congress, Foz do Iguassu, Brazil, June 6-11, 2000. pp. 183
67. Elliott, L.F. and Kennedy, A.C. (1991). Method for screening bacteria and applicatin thereof for field control of the weed downy brome. U.S. Patent no. 5,030,562.
68. Kennedy, A.C., Ogg, A.G., Jr., and Young, F.L. (1992). Biocontrol of jointed goatgrass. U.S. Patent no. 5,163,991.

69. Johnson, B.N., Kennedy, A.C. and Ogg, A.G., Jr., (1993). Suppression of downy brome growth by a rhizobacterium in controlled environments. Soil Sci. Soc. Am. J. 57:73-77.
70. Harris, P.A. and Stahlman, P.W. (1992). Biological weed control in wheat using deleterious rhizobacteria. Weed Sci. Soc. Am. Abstr. 32:50.
71. Stubbs, T.L. and Kennedy, A.C. (1993). Effect of bacterial and chemical stresses in biological weed control systems. Agron. Abstr. 57:261.
72. Boyetchko, S.M. and Holmström-Ruddick, B. (1996). Host-range of rhizobacteria effective as biocontrol agents of downy brome. Can. J. Plant Pathol. 18:86-87.
73. Cavers, P.B. and Benoit, D.L. (1989). Seed banks in arable land. *In:* Ecology of Soil Seed Banks, M.A. Leck, V.T. Parker, and R.L. Simpson, (Eds.), Academic Press, San Diego, CA, pp. 309-328.
74. Thompson, K. and Grime, J.P. (1979). Seasonal variation in the seed banks of herbaceous species in ten contrasting habitats. J. Applied Ecol. 37:893-921.
75. Kennedy, A.C. and Smith, K.L. (1995). Soil microbial diversity and the sustainability of agricultural soils. Plant Soil 170:75-86.
76. Aldrich, R.J. (1984). Weed-crop ecology: principles in weed management. Breton, North Scituate, Massachusetts, USA.
77. Kremer, R.J., Hughes, L.B., Jr., and Aldrich, R.J. (1984). Examination of microorganisms and deterioration resistance mechanisms associatd with velvetleaf seed. Agronomy J. 76:745-749.
78. Frederickson, J.K. and Elliott, L.F. (1985). Effect on winter wheat seedling growth by toxin-producing rhizobacteria. Plant Soil 83:399-409.
79. Mortensen, K. and Hsiao, A.I. (1987). Fungal infestation of seeds from seven populations of wild oats (*Avena fatua* L.) with different dormancy and viability characteristics. Weed Res. 27:297-304.
80. Morin, L., Auld, B.A. and Brown, J.F. (1993). Synergy between *Puccinia xanthii* and *Colletotrichum orbiculare* on *Xanthium occidentale*. Biol. Control 3:296-310.
81. Hallett, S.G., Paul, N.D., and Ayres, P.G. (1995). A dual pathogen strategy for the biological control of groundsel (*Senecio vulgaris*). In: *VII International Symposium on Biological Control of Weeds,* DSIR/CSIRO, Canterbury, New Zealand. pp. 533.
82. Paul, N.D., Ayres, P.G. and Hallett, S.G. (1993). Mycoherbicides and other biocontrol agents for *Senecio* spp. Pesticide Sci. 37:323-329.
83. Boyette, C.D., Templeton, G.E. and Smith, R.J. , Jr. (1979). Control of winged waterprimrose (*Jussiaea decurrens*) and northern jointvetch (*Aeschynomene virginica*) with fungal pathogens. Weed Sci. 27:497-501.
84. Chandramohan, S. (1999). Multiple-pathogen strategy for bioherbicidal control of several weeds. Ph.D. dissertation, University of Florida, Gainesville, FL. 190 p.
85. Fernando, W.G.D., Watson, A.K., and Paulitz, T.C. (1996). The role of *Pseudomonas* spp. and competition for carbon, nitrogen, and iron in the enhancement of appressorium formation by *Colletotrichum coccodes* on velvetleaf. Eur. J. of Plant Pathol. 102:1-7.
86. Müller-Schärer, H. (2000). Biological control of weeds in European crops: recent achievements and future work. Weed Res. 40:83-98.
87. Seier, M.K. and Evans, H.C. (1996). Two fungal pathogens of *Mimosa pigra* var. *pigra* from Mexico: The finishing touch for biological control of this weed in Australia? *In:* V.C. Moran and J.H. Hoffmann (eds.), Proc. IX International Symposium on Biological Control of Weeds, January 19-26, 1996, University of Cape Town, Stellenbosch, South Africa. pp. 87-92
88. Kadir, J.B., Charudattan, R., Stall, W.M., and Bewick, T.A. (1999). Effect of *Dactylaria higginsii* on interference of *Cyperus rotundus* with *L. esculentum*. Weed Sci. 47:682-686.
89. Phatak, S.C., Callaway, M.B. and Vavrina, C.S. (1987). Biological control and its integration in weed management systems for purple and yellow nutsedge (*Cyperus rotundus* and *C. esculentus*). Weed Technol. 1:84-91.
90. Yang, Y-K., Kim, S-O., Chung, H-S. and Lee, Y-H. (2000). Use of *Colletotrichum graminicola* KA001 to control barnyard grass. Plant Dis. 84:55-59.
91. Zhang, W. and Watson, A.K. (1997). Efficacy of *Exserohilum monoceras* for the control of *Echinochloa* species in rice (*Oryza sativa*). Weed Sci. 45:144-150.
92. Zidack, N.K., Backman, P.A. and Shaw, J.J. (1992). Promotion of bacterial infection of leaves by an organosilicone surfactant: Implications for biological weed control. Biol. Control 2:111-117.
93. Evans, H. C. (1995). Fungi as biocontrol agents of weeds: a tropical perspective. Can J. Bot. 73(Suppl. 1):S58-S64.

94. Caesar, A. J., Campobasso, G. and Terragitti, G. (1999). Effects of European and U.S. strains of *Fusarium* spp. pathogenic to leafy spurge on North American grasses and pathogenicity and cultivated species. Biol. Control 14:130-136.
95. Caesar, A. J. (1996). Identification, pathogenicity and comparative virulence of *Fusarium* spp. associated with stand declines of leafy spurge (*Euphorbia esula*) in the Northern Plains. Plant Dis. 80:1395-1398.
96. Caesar, A. J. , Campobasso, G., and Terragitti, G. (1998). Identification, pathogenicity and comparative virulence of *Fusarium* spp. associated with diseased *Euphorbia* spp. in Eur. Biocontrol Sci. Technol. 8:313-319
97. Caesar, A. J., Campobasso, G., and Terragitti, G. (2001). Identification, pathogenicity and comparative virulence of *Fusarium* spp. associated with insect-damaged, diseased *Centaurea* spp. in Europe. BioControl: In Press
98. Morales-Ramos, J. A., Rojas, M.G., and Sittertz-Bhatkar, H. (2000). Symbiotic relationship between *Hypthenemus hampei* (Coleoptera: Scolytidae) and *Fusarium solani* (Moniliales: Tuberculariaceae). Ann. Entomol. Soc. Am. 93:541-547.
99. Rojas, M. G., Morales-Ramos, J.A., and Harrington, T.C. (1999). Association between *Hypthenemus hampei* (Coleoptera: Scolytidae) and *Fusarium solani* (Moniliales: Tuberculariaceae). Ann. Entomol. Soc. Am. 92:98-100.
100. Vega, F. E., Dowd, P.F., and Bartelt, R.J. (1995). Dissemination of microbial agents using an autoinoculating device and several insect species as vectors. Biol. Control 5:545-552.
101. Dowd, P. F., Vega, F.E., Nelsen, T.C., and Richard, J.L. (1998). Dusky sap beetle mediated dispersal of *Bacillus subtilus* to inhibit *Aspergillus flavus* and aflatoxin production in maize *Zea mays* L. Biocontrol Sci. Technol. 8:221-235.
102. Briese, D. T., McLaren, D.A., Pettit, W.J., Zapater, M., Anderson, F., Delhey, R., and Distel, R. (2000). New biological control initiatives against weeds of South American origin in Australia: nasella tussock grasses and blue heliotrope. *In:* N.R. Spencer (ed.), Proc. X International Symposium on Biological Control of Weeds, 4-14 July, 1999, Montana State University, Bozeman, Montana, USA. pp. 215-223
103. Ding, J., Fu, W., Wu, Y., and Reardon, R. (2000). Insects associated with mile-a-minute weed (*Polygonium perfoliatum* L.) in China: a three year survey report. Pages 25-31 *In:* N.R. Spencer (ed.), Proc. X International Symposium on Biological Control of Weeds, 4-14 July, 1999, Montana State University, Bozeman, Montana, USA. pp. 25-31
104. Watts, J. D. and Piper, G.L. (2000). The phytophagous insect fauna of scotch thistle, *Onopordum acanthium*, in southeastern Washington and northeastern Idaho. *In:* N.R. Spencer (ed.), Proc. X International Symposium on Biological Control of Weeds, 4-14 July, 1999, Montana State University, Bozeman, Montana, USA. pp. 33-39.
105. Dodd, A. P. (1940). The biological campaign against prickly pear. Commonwealth Prickly Pear Board, Brisbane, Australia, 177 pp.
106. Wilson, F. (1943). The entomological control of St. John's wort (*Hypericum perforatum* L.) with special reference to the weed in southern France. Australia Council of Science and Industrial Res. Bull. 169. 87 pp.
107. Louda, S. M. (2000). *Rhinocyllus conicus*-Insights to improve predictability and minimize risk of biological control of weeds. *In:* N.R. Spencer (ed.), Proc. X International Symposium on Biological Control of Weeds, 4-14 July, 1999, Montana State University, Bozeman, Montana, USA. pp. 39-45
108. Shishkoff, N. and Bruckart, W.L. (1996). Water stress and damage caused *by Puccinia jaceae* on two *Centaurea* species. Biol. Control 6:57-63.
109. Müller-Schärer, H., Lewinsohn, T.M., and Lawton, J.H. (1991). Searching for weed biocontrol agents-when to move on? Biocontrol Sci.Technol. 1:271-280.
110. Bruckart, W.L., Johnson, D.L., and Frank, J.R. (1988). Bentazon reduces rust-induced disease in yellow nutsedge (*Cyperus esculentus*). Weed Technol. 2:229-303.
111. Scheepens, P.C. and Hoogerbrugge, A. (1991). Host specificity of *Puccinia caniliculata*, a potential biological control agent for *Cyperus esculentus*. Netherlands J. Plant Pathol. 97:245-250.
112. Horak, M.J. and Holt, J.S. (1986). Isozyme variability and breeding systems in populations of yellow nutsedge (*Cyperus esculentus*). Weed Sci. 35:538-543.
113. Horak, M.J., Holt, J.S., and Ellstand, N.C. (1987). Genetic variation in yellow nutsedge (*Cyperus esculentus*). Weed Sci. 35:506-512.
114. Okoli, C.A.N., Shilling, D.G., Smith, R.L., and Bewick, T.A. (1997). Genetic diversity in purple nutsedge (*Cyperus rotundus* L.) and yellow nutsedge (*C. esculentus* L.) Biol. Control 8:111-118.
115. Amsellem, L., Le Bourgeois, T., Noyer, J.L., and Hossaert-McKey, M. (2000). Comparison of genetic diversity of the weed *Rubus alcefolius* in its introduced and native areas. *In:* Proc. X International Symposium

on Biological Control of Weeds, July 4-14, 1999, Montana State University, Bozeman, Montana, USA. pp. 253-260
116. Hussaini, I.P., Lawrie, A.C., and McLaren, D.A. (2000). Pathogens on and variation in *Nasella trichotoma* in Australia. *In:* Proc. X International Symposium on Biological Control of Weeds, July 4-14, 1999, Montana State University, Bozeman, Montana, USA. pp. 269-280.
117. Casonato, S.G., Lawrie, A.C., and McLaren, D.A. (2000). Biological control of *Hypericum androsaumum* with *Melampsora hypericorum* S.G. *In:* Proc. X International Symposium on Biological Control of Weeds, July 4-14, 1999, Montana State University, Bozeman, Montana, USA. pp. 133-134
118. Rosskopf, E.N., Charudattan, R., Shabana, Y.M., and Benny, G.L. (2000). *Phomopsis amaranthicola*, a new species from Amaranthus spp. Mycologia 92:114-122.
119. Tessmann, D.J., Charudattan, R., Kistler, H.C., and Rosskopf, E.N. (2001). A molecular characterization of *Cercospora* species pathogenic to water hyacinth and emendation of *C. piaropi*. Mycologia 99:1108-1112.
120. Sands, D.C., Ford, E.J., Miller, R.V., Sally, B.K., McCarthy, M.K., Anderson, T.W., Weaver, M.B., Morgan, C.T., and Pilgeram, A.L. (1997). Characterization of a vascular wilt of *Erythroxylum coca* caused by *Fusarium oxysporum* f.sp. *erythroxyli* forma specialis nova. Plant Dis. 81:501-504.
121. Auld, B.A. and Morin, L. (1995). Constraints in the development of bioherbicides. Weed Technol. 9:638-652.
122. Greaves, M.P., Holloway, P.J., and Auld, B.A. (1998). Formulation of microbial herbicides. *In:* H.D. Burges (ed.), Formulation of Microbial Biopesticides: Beneficial Microorganisms, Nematodes, and Seed Treatments, Kluwer Academic Publishers, Dordrecht. pp. 203-233.
123. Lumsden, R.D., Lewis, J.A., and Fravel, D.R. (1995). Formulation and delivery of biocontrol agents for use against soilborne plant pathogens. *In:* F.R. Hall and J.W. Barry (eds.), Biorational Pest Control Agents, Formulation and Delivery, ACS Symposium Series 595, Washington. pp. 166-182.
124. Boyetchko, S., Pedersen, E., Punja, Z., and Reddy, M. (1999). Formulations of biopesticides. *In:* F.R. Hall and J.J. Menn (eds.), Methods in Biotechnology, vol. 5: Biopesticides: Use and Delivery. Humana Press, Totowa, NJ. pp. 487-508.
125. Boyette, C.D., Quimby, P.C., Jr., Caesar, A.J., Birdsall, J.L., Connick, W.J., Jr., Daigle, D.J., Jackson, M.A., Egley, G.H., and Abbas, H.K. (1996). Adjuvants, formulations, and spraying systems for improvement of mycoherbicides. Weed Technol. 10:637-644.
126. Green, S., Stewart-Wade, S.M., Boland, G.J., Teshler, M.P., and Liu, S.H. (1998). Formulating microorganisms for biological control of weeds. *In:* G.J. Boland and L.D. Kuykendall (eds.), Plant-Microbe Interactions and Biological Control, Marcel Dekker, Inc., New York. pp. 249-281.
127. Greaves, M.P. and MacQueen, M.D. (1990). The use of mycoherbicides in the field. Aspects Appl. Biol. 24:163-168.
128. Paau, A.S. (1998). Formulation of beneficial organisms applied to soil. *In:* H.D. Burges (ed.), Formulation of Microbial Biopesticides: Beneficial Microorganisms, Nematodes, and Seed Treatments, Kluwer Academic Publishers, Dordrecht. Formulation of beneficial organisms applied to soil. pp. 235-254.
129. Lawrie, J., Greaves, M.P., Western, N.M., and Down, V. (1999). Application and formulation: Forgotten factors in the development of microbial herbicides. *In:* Proc. 1999 Brighton Conference - Weeds, Brighton, U.K. pp. 289-296.
130. Greaves, M.P., Dutton, L., and Lawrie, J. (2000). Formulation of microbial herbicides. Aspects Appl. Biol. 57:171-178.
131. Makowski, R.M.D. (1993). Effect of inoculum concentration, temperature, dew period, and plant growth stage on disease of round-leaved mallow and velvetleaf by *Colletotrichum gloeosporioides* f.sp. *malvae*. Phytopathology 83:1229-1234.
132. McRae, C.F. and Auld, B.A. (1988). The influence of environmental factors on anthracnose of *Xanthium spinosum*. Phytopathology 78:1182-1186.
133. Morin, L., Watson, A.K., and Reeleder, R.D. (1990). Effect of dew, inoculum density, and spray additives on infection of field bindweed by *Phomopsis convolvulus*. Can. J. Plant Pathol. 12:48-56.
134. Kenney, D.S. (1986). DeVine: The way it was developed - An industrialist's view. Weed Sci. 34(Suppl. 1):15-16.
135. Prasad, R. (1993). Role of adjuvants in modifying the efficacy of a bioherbicide on forest species: Compatibility studies under laboratory conditions. Pesticide Sci. 37:427-433.
136. Daigle, D.J., and Cotty, P.J. (1991). Factors that influence germination and mycoherbicidal activity of *Alternaria cassiae*. Weed Technol. 5:82-86.
137. Klein, T.A. and Auld, B.A. (1995). Evaluation of Tween 20 and glycerol as additives to mycoherbicide suspensions applied to bathurst burr. Plant Prot. Quart. 10:14-16.

138. Wymore, L.A. and Watson, A.K. (1986). An adjuvant increases survival and efficacy of *Colletotrichum coccodes*, a mycoherbicide for control of velvetleaf (*Abutilon theophrasti*). Phytopathology 76:1115.1116.
139. Prasad, R. (1994). Influence of several pesticides and adjuvants on *Chondrostereum purpureum* - A bioherbicide agent for control of forest weeds. Weed Technol. 8:445-449.
140. Womack, J.G. and Burge, M.N. (1993). Mycoherbicide formulation and the potential for bracken control. Pestic. Sci. 37:337-341.
141. McElwee, M., Irvine, J.I.M., and Burge, M.N. (1990). A mycoherbicidal approach to bracken control. *In:* C. Bassett, L.J. Whitehouse, and J.A. Zabkiewicz (eds.), Alternatives to the Chemical Control of Weeds, Proceedings of an International Conference, Rotorua, New Zealand, July 1989, Ministry of Forestry, FRI Bulletin 155. pp. 74-79.
142. Zidack, N.K. and Backman, P.A. (1996). Biological control of kudzu (*Pueraria lobata*) with the plant pathogen *Pseudomonas syringae* pv. *phaseolicola*. Weed Sci. 44:645-649.
143. Daigle, D.J., and Connick, W.J., Jr. (1990). Formulation and application technology for microbial weed control. *In:* R.E. Hoagland (ed.), Microbes and Microbial Products as Herbicides, ACS Symposium Series 439, Washington, D.C. pp. 288-304.
144. Connick, W.J., Jr., Lewis, J.A., and Quimby, P.C., Jr. (1990). Formulation of biocontrol agents for use in plant pathology. UCLA Symp. Mol. Cell Biol. pp. 345-372.
145. Daigle, D.J., Connick, W.J., Jr., Quimby, P.C., Jr., Evans, J., Trask-Morrell, B., and Fulgham, F.E. (1990). Invert emulsions: Carrier and water source for the mycoherbicide, *Alternaria cassiae*. Weed Technol. 4:327-331.
146. Daigle, D.J., and Cotty, P.J. (1992). Production of conidia of *Alternaria cassiae* with alginate pellets. Biol. Control 2:278-281.
147. Shabana, Y.M. (1997). Formulation of *Alternaria eichhorniae*, a mycoherbicide for waterhyacinth, in invert emulsions averts dew dependence. Z. Pflanzenkr. Pflanzanshutz. 104:231-238.
148. Amsellem, Z., Sharon, A., Gressel, J., and Quimby, P.C., Jr. (1990). Complete abolition of high inoculum threshold of two mycoherbicides (*Alternaria cassiae* and and *A. crassa*) when applied in invert emulsion. Phytopathology 80:925-929.
149. Yang, S.M. and Jong, S.C. (1995). Host range determination of *Myrothecium verrucaria* isolated from leafy spurge. Plant Dis. 79:994-997.
150. Yang, S.M., Johnson, D.R., Dowler, W.M., and Connick, W.J., Jr., (1993). Infection of leafy spurge by *Alternaria alternata* and *A. angustiovoidea* in the absence of dew. Phytopathology 83:953-958.
151. Womack, J.G., Eccleston, G.M., and Burge, M.N. (1996). A vegetable oil-based invert emulsion for mycoherbicide delivery. Biol. Control 6:23-28.
152. Auld, B.A. (1993). Vegetable oil suspension emulsions reduce dew dependence of a mycoherbicide. Crop Prot. 12:477-479.
153. Egley, G.H., and Boyette, C.D. (1995). Water-corn oil emulsion enhances conidia germination and mycoherbicidal activity of *Colletotrichum truncatum*. Weed Sci. 43:312-317.
154. Shabana, Y.M. (1997). Vegetable oil suspension emulsions for formulating the weed pathogen (*Alternaria eichhorniae*) to bypass dew. Z. Pflanzenkr. Pflanzanshutz. 104:239-245.
155. Boyette, C. D. (1994). Unrefined corn oil improves the mycoherbicidal activity of *Colletotrichum truncatum* for hemp sesbania (*Sesbania exaltata*) control. Weed Technol. 8:526-529.
156. Abbas, H.K. and Egley, G.H. (1996). Influence of unrefined corn oil and surface-active agents on the germination and infectivity of *Alternaria helianthi*. Biocontrol Sci. Technol. 6:531-538.
157. Klein, T.A., Auld, B.A., and Fang, W. (1995). Evaluation of oil suspension emulsions of *Colletotrichum orbiculare* as a mycoherbicide in field trials. Crop Prot. 14:193-197.
158. Shabana, Y.M., Charudattan, R., DeValerio, J.T., and Elwakil, M.A. (1997). An evaluation of hydrophilic polymers for formulating the bioherbicide agents *Alternaria cassiae* and *A. eichhorniae*. Weed Technol. 11:212-220.
159. Boyette, C.D. and Abbas, H.K. (1994). Host range alteration of the bioherbicidal fungus *Alternarioa crassa* with fruit pectin and plant filtrates. Weed Sci. 42:487-491.
160. McRae, C.F. and Stevens, G.R. (1990). Role of conidial matrix of *Colletotrichum orbiculare* in pathogenesis of *Xanthium spinosum*. Mycol. Res. 94:890-896.
161. Sparace, S.A., Wymore, L.A., Menassa, R., and Watson, A.K. (1991). Effects of the *Phomopsis convolvulus* conidial matrix on conidia germination and the leaf anthracnose disease of field bindweed (*Convolvulus arvensis*). Plant Dis. 75:1175-1179.
162. Fernando, W.G.D., Watson, A.K., and Paulitz, T.C. (1994). Phylloplane *Pseudomonas* spp. enhance disease

caused by *Colletotrichum coccodes* on velvetleaf. Biol. Control 4:125-131.
163. Laycock, M.V., Hildebrand, P.D., Thibault, P., Walter, J.A., and Wright, J.L.C. (1991). Viscosin, a potent peptidolipid biosurfactant and phytopathogenic mediator produced by a pectolytic strain of *Pseudomonas fluorescens*. J. Agric. Food Chem. 39:483-489.
164. Walker, H.L. and Connick, W.J., Jr. (1983). Sodium alginate for production and formulation of mycoherbicides. Weed Sci. 31:333-338.
165. Boyette, C.D. and Walker, H.L. (1986). Evaluation of *Fusarium lateritium* as a biological herbicide for controlling velvetleaf (*Abutilon theophrasti*), and prickly sida (*Sida spinosa*). Weed Sci. 34:106-109.
166. Weidemann, G.J. (1988). Effects of nutritional amendments on conidial production of *Fusarium solani* f.sp. *cucurbitae* on sodium alginate granules and on control of Texas gourd. Plant Dis. 72:757-759.
167. Weidemann, G.J. and Templeton, G.E. (1988). Control of Texas gourd, *Cucurbita texana*, with *Fusarium solani* f.sp. *cucurbitae*. Weed Technol. 2:271-274.
168. Weidemann, G.J. and Templeton, G.E. (1988). Efficacy and soil persistence of *Fusarium solani* f.sp. *cucurbitae* for control of Texas gourd (*Cucurbita texana*). Plant Dis. 72:36-38.
169. Zidack, N.K. and Quimby, P.C., Jr. (1999). Formulation and application of plant pathogens for biological weed control. *In:* F.R. Hall and J.J. Menn (eds.), Methods in Biotechnology, vol. 5: Biopesticides: Use and Delivery. Humana Press, Totowa, NJ. pp. 371-381.
170. Bailey, B.A., Hebbar, K.P., Strem, M., Darlington, L.C., and Lumsden, R.D. (1997). An alginate prill formulation of *Fusarium oxysporum* Schlechtend:Fr. f.sp. *erythroxyli* for biocontrol of *Erythroxylum coca* var. *coca*. Biocontrol Sci. Technol. 7:423-435.
171. Connick, W.J., Jr., Boyette, C.D., and McAlpine, J.R. (1991). Formulation of mycoherbicides using a pasta-like process. Biol. Control 1 (1991) 281-287.
172. Connick, W.J., Jr., Daigle, D.J., Pepperman, A.B., Hebbar, K.P., Lumsden, R.D., Anderson, T.W., and Sands, D.C. (1998). Preparation of stable, granular formulations containing *Fusarium oxysporum* pathogenic to narcotic plants. Biol. Control, 13 (1998) 79-84.
173. Connick, W.J., Jr., Daigle, D.J., Boyette, C.D., and McAlpine, J.R. (1996). Water activity and other factors that affect the viability of *Colletotrichum truncatum* conidia in wheat flour-kaolin granules ('Pesta'). Biocontrol Sci. Technol. 1 (1996) 277.
174. Charudattan, R., Prange, V.J., and DeValerio, J.T. (1996). Exploration of the "bialaphos genes" for improving bioherbicide efficacy. Weed Technol. 10:625-636.
175. Brooker, N.L., Mischke, C.F., Patterson, C.L., Mischke, S., Bruckart, W.L., and Lydon, J. (1996). Pathogenicity of *bar*-transformed *Colletotrichum gloeosporioides* f.sp. *aeschynomene*. Biol. Control 7:159-166.
176. TeBeest, D.O. (1984). Induction of tolerance to benomyl in *Colletotrichum gloeosporioides* f.sp. *aeschynomene* by ethyl methanesulfonate. Phytopathology 74:864.
177. Luo, Y. and TeBeest, D.O. (1997). Infection components of wild-type and mutant strains of *Colletotrichum gloeosporioides* f.sp. *aeschynomene* on northern jointvetch. Plant Dis. 81:404-409.
178. Holmström-Ruddick, B. and Mortensen, K. (1995). Factors affecting pathogenicity of a benomyl-resistant strain of *Colletotrichum gloeosporioides* f.sp. *malvae*. Mycol. Res. 99:1108-1112.
179. Sands, D.C. and Miller, R.V. (1993). Altering the host range of mycoherbicides by genetic manipulation. *In:* S.O. Duke, J.J. Menn, and J.R. Plimmer (eds.), ACS Symposium Series 524, Washington, D.C. pp. 101-109.
180. Sands, D.C., Miller, R.V., and Ford, E.J. (1990). Biotechnological approaches to control of weeds with pathogens. *In:* R.E. Hoagland (ed.), Microbes and microbial products as herbicides, ACS Symposium Series 439. pp. 184-190.
181. Luster, D.G., Berthier, Y.T., Bruckart, W.L., and Hack, M.A. (2000). Post-release spread of musk thistle rust monitored from Virginia to California using DNA sequence information. *In:* Proc. X International Symposium on Biological Control of Weeds, July 4-14, 1999, Montana State University, Bozeman, Montana, USA. pp. 302.
182. Becker, E.M., de la Bastide, P., Hahn, R.L., Shamoun, S.F., and Hintz, W.E. (2000). Molecular markers for monitoring mycoherbicides. *In:* Proc. X International Symposium on Biological Control of Weeds, July 4-14, 1999, Montana State University, Bozeman, Montana, USA. pp. 301.
183. Ramsfield, T.D., Becker, E.M., Rathief, S.M., Tang, Y., Vrain, T.C., Shamoun, S.F., and Hintz, W.E. (1996). Geographic variation of *Chondrostereum purpureum* detected by polymorphisms in the ribosomal DNA. Can. J. Bot. 74:1919-1929.
184. Kennedy, A.C. (1996). Molecular biology of bacteria and fungi for biological control of weeds. *In:* M. Gunasekaran and D.J. Weber (eds.), Molecular biology of the biological control of pests and diseases of plants, CRC Press Inc., Boca Raton, Florida. Pages 155-172.

185. Poppenborg, L., Friehs, K., and Flaschel, E. (1997). The green fluorescent protein is a versatile reporter for bioprocess monitoring. J. Biotechnol. 58:79-88.
186. Song, W-Y., Hutcheson, S.W., Hatziloukas, E., and Schaad, N.W. (1999). A gene-tagging system for monitoring *Xanthomonas* species. Plant Pathol. J. 15:137-143.
187. Caesar, A. J. (2000). Insect Pathogen synergisms are the foundation of weed biocontrol. *In:* N.R. Spencer (ed.), Proc. X International Symposium on Biological Control of Weeds, 4-14 July, 1999, Montana State University, Bozeman, Montana, USA. pp. 793-798.
188. Harris, P. (1993). Effects, constraints and the future of weed biocontrol. Agric. Ecosystems Environ. 46:289-303.
189. MacFadyen, R. (2000). Successes in Biological control of weeds. *In:* N.R. Spencer (ed.), Proc. X International Symposium on Biological Control of Weeds, 4-14 July, 1999, Montana State University, Bozeman, Montana, USA. pp. 3-14.
190. Callaway, R. M., DeLuca, T.H., and Belleveau, W.M. (1999). Biological-control herbivores may increase competitive ability of the noxious weed *Centaurea maculosa*. Ecology 80:1196-1201.
191. Smith, R.J., Jr. (1991). Integration of biological control agents with chemical pesticides. *In:* D.O. TeBeest (ed.), Microbial Control of Weeds, Chapman and Hall, New York. pp. 189-208.
192. Altman, J., Neate, S., Rovira, A.D. (1990). Herbicide-pathogen interactions and mycoherbicides as alternative strategies for weed control. *In*: R.E. Hoagland (ed.) , Microbes and microbial products as herbicides, ACS Symposium Series 439. pp. 240-259.
193. Christy, A.L., Herbst, K.A., Kostka, S.J., Mullen, J.P., and Carlson, P.S. (1993). Synergizing weed biocontrol agents with chemical herbicides. *In:* S.O. Duke, J.J. Menn, and J.R. Plimmer, (eds.) Pest control with enhanced environmental safety. ACS Symposium Series 524. pp. 87-100.
194. Hoagland, R.E. (1996). Chemical interactions with bioherbicides to improve efficacy. Weed Technol. 10:651-674.
195. Müller-Schärer, H. and Frantzen, J. (1996). An emerging system management approach for biological weed control in crops: *Senecio vulgaris* as a research model. Weed Res. 36:483-491.
196. Lawrie, J., Greaves, M.P., Down, V.M., and Lewis, J.M. (1999). Effects of the plant-pathogenic fungus *Mycocentrospora acerina* (Hartig) Deighton on growth and competition of *Viola arvensis* (Murr.) in spring wheat. Biocontrol Sci. Technol. 9:105-112.
197. DiTommaso, A., Watson, A.K., and Hallett, S.G. (1996). Infection by the fungal pathogen *Colletotrichum coccodes* affects velvetleaf (*Abutilon theophrasti*)-soybean competition in the field. Weed Sci. 44:924-933.
198. Boyette, C.D. (1991). Control of hemp sesbania with a fungal pathogen *Colletotrichum truncatum*. USDA patent, #5,034,328. July 23, 1991.
199. Boyette, C.D. (1993). Bioogical control of hemp sesbania (*Sesbania exaltata*) under field conditions with *Colletotrichum truncatum* formulated in an invert emulsion. Weed Sci. 41:497-500.
200. Guntli, D., Burgos, S., Kump, I., Heeb, M., Pfirter, H.A., and Défago, G. (1999). Biological control of hedge bindweed (*Calystegia sepium*) with *Stagonospora convolvuli* strain LA39 in combination with competition from red clover (*Trifolium pratense*). Biol. Control 15:252-258.
201. Ammon, H.U. and Müller-Schärer, H. (1999). Prospects for combining biological weed control with integrated crop production systems, and with sensitive management of alpine pastures in Switzerland. Z. Pflanzenkr. Pflanzenschutz 106:213-220.
202. Wyse, D.L. (1994). New technologies and approaches for weed management in sustainable agricultural systems. Weed Technol. 8:403-407.
203. Utkhede, R.S. (1996). Potential and problems of developing bacterial biocontrol agents. Can. J. Plant Pathol. 18:455-462.

Applied Mycology and Biotechnology
Volume 2. Agriculture and Food Production

Biotechnology of Arbuscular Mycorrhizas

Manuela Giovannetti[a] and Luciano Avio[b]
[a]Dipartimento di Chimica e Biotecnologie Agrarie, Università di Pisa, Via del Borghetto 80, 56124 Pisa, Italy; [b]Centro di Studio per la Microbiologia del Suolo, C. N. R., Via del Borghetto 80, 56124 Pisa, Italy (E-Mail: mgiova@agr.unipi.it).

Mycorrhizas are symbiotic associations established between thousands of species of soil-borne fungi and the roots of most terrestrial plant species. Host plants and fungal symbionts interact producing mutual benefits: fungi colonise roots and obtain carbon compounds which they are unable to synthesise and plants receive mineral nutrients absorbed and translocated by the large extraradical hyphal network spreading from mycorrhizal roots into the surrounding soil. Arbuscular mycorrhizas have a worldwide distribution, and occur in about 80% of plant species, including the most important temperate and tropical crops. They play a fundamental role in soil fertility and plant nutrition and in the maintenance of stability and biodiversity within plant communities. AM fungal symbionts belong to a small group of Zygomycetes, the Glomales. They are obligate biotrophs and, after establishing functional symbioses with host plants, produce spores in the soil which are able to germinate and grow, but are unable to produce extensive mycelia and to complete their life cycle in the absence of their hosts. Many laboratory and field experiments demonstrated large plant growth responses to inoculation of AM fungal endophytes, showing that the performance of each species and isolate may differ depending on plant species and environmental soil conditions, which greatly influence their development. Although AM symbionts cannot be grown in axenic culture, and knowledge of their biology is limited, yet some of the fundamental factors affecting their development have been studied. Here we review and analyse important data on the main parameters affecting fungal infectivity, efficiency and ability to survive, multiply and spread in different environments, which may contribute to the biotechnological exploitation and utilization of AM fungi in sustainable agriculture and biodiversity conservation. The symbiotic performance of AM endophytes depends first and foremost on the availability of highly infective strains. The possibility of surviving in the soil ecosystem of any super-efficient strain would depend on the ability of its mycelium to infect host plants rapidly and to compete for infection sites with indigenous endophytes, which natural selection likely made very infective. Most studies have investigated spore-related factors such as dormancy, germination and pre-symbiotic growth of germlings, as well as parameters affecting appressorium formation and intraradical growth. When considering the mechanisms underlying fungal efficiency, the most important features to be taken into account are fungal ability to develop extensive hyphal networks which can explore the soil, absorb, translocate and transfer soil mineral nutrients to the host plants, the viability of extraradical mycelium, the rate of hyphal nutrient uptake, translocation and transfer to host cells. AM fungi cannot be produced in large scale reactors given their recalcitrance to pure culture. Despite this constraint, many methods have been developed for manipulating AM fungi, inoculating them on suitable host plants, and reproducing large amounts of inoculum. In this way, *in vivo* cultures of isolates and species from temperate, tropical, semi-arid, polar and Mediterranean regions are presently propagated in *ex-situ* collections. Conservation of genetic resources of AM fungi requires information about the origin, ecosystem, type of soil and symbiotic performance under diverse environmental conditions of the isolates maintained. Since most information stored in the different international databases is not available, the establishment of linkages between them is one of the major aims to be pursued by AM researchers, for obtaining a more complete understanding, and accordingly a better biotechnological exploitation, of AM symbionts in the different world ecosystems.

1. INTRODUCTION

Mycorrhizas are symbiotic associations established between thousands of species of soil-borne fungi and the roots of most terrestrial plant species. Host plants and fungal symbionts interact producing mutual benefits: fungi colonise roots and obtain carbon compounds which they are unable to synthesise and plants receive mineral nutrients absorbed and translocated by the large extraradical hyphal network spreading from mycorrhizal roots into the surrounding soil. Mycorrhizal symbioses differ morphologically and physiologically, and their diversity depends on the different fungal and plant species. The most important types are represented by ectomycorrhizas, ectoendomycorrhizas and endomycorrhizas [1]. In ectomycorrhizas the fungal symbiont forms a thick mantle of interwoven hyphae around feeder roots and grows intercellularly in the cortex, forming a network termed Hartig net, but it never penetrates root cells. Ectomycorrhizas are widespread in temperate and boreal forest trees such as *Castanea*, *Quercus*, *Pinus*, *Tsuga*, *Pseudotsuga*, *Corylus*, *Eucalyptus*, *Betula*, *Tilia*, *Ulmus*, *Larix*. Fungal symbionts are represented by more than 5000 different species mainly belonging to Basidiomycota, which produce mushrooms and toadstools, such as *Amanita*, *Boletus*, *Cortinarius*, *Hebeloma*, *Laccaria*, *Russula*, *Suillus*, *Tricholoma*, and to Ascomycota, such as the hypogeous genus *Tuber* [2].

Ectoendomycorrhizas are structurally different from ectomycorrhizas, since the fungus is able to grow intracellularly in the first layer of root cells. Depending on the fungal/plant association the fungal mantle may be more or less developed. These mycorrhizas occur in some plant genera belonging to Ericales, *i. e. Arbutus* and *Arctostaphylos*, and are known as arbutoid mycorrhizas. The fungi able to form arbutoid mycorrhizas generally give rise to ectomycorrhizal symbioses when colonising other host plants [3, 4]. Similarly, a limited number of fungi infecting dominant ectomycorrhizal trees in natural forest ecosystems are able to colonise a small group of achlorophyllous epiparasitic plants, monotropes, (Monotropoideae, Ericaceae), forming ectoendomycorrhizas of the monotropoid type. The achlorophyllous species depend on fungal symbionts for their carbon nutrition, since the extraradical mycelium acts as a link between the different plants, transferring photosyntates from the autotrophic to the heterotrophic hosts [5, 6]. In young plants growing in nurseries another type of ectoendomycorrhizas has been described, occurring in some species of Pinaceae, which is produced by ascomycetoid fungi of the genus *Wilcoxina* and by isolates of an heterogeneous group of fungi known as dark septate endophtyes [7, 8]. Within the group of endomycorrhizas the three major types are represented by ericoid, orchid and arbuscular mycorrhizas. Ericoid mycorrhizas occur in some Ericales such as *Erica*, *Calluna*, *Rhododendron*, *Vaccinium* and the most frequently isolated fungal symbionts are either ascomycetous forms belonging to *Hymenoscyphus ericae* or mitosporic forms of the genus *Oidiodendron* [9]. Orchid mycorrhizas occur in a very large and diverse group of economically important plants, whose mycorrhizal status has been first described to occur in the protocorm, the juvenile stage of their life cycle. Fungal symbionts supply carbon and vitamins to the developing embryo seedlings, which they colonise shortly after germination.

The symbionts of both adult plants and protocorms are represented by *Rhizoctonia* (anamorph of *Tulasnella*, *Thanatephorus* and *Ceratobasidium*) and other Basidiomycetes such as *Marasmius*, *Fomes*, *Armillaria* [10]. Arbuscular mycorrhizal (AM) symbioses have a worldwide distribution, and occur in about 80% of plant species. They are formed by soil fungi belonging to the Glomales (Zygomycetes), which play a fundamental role in plant nutrition and soil fertility. AM fungi are obligate biotrophs and, after establishing functional symbioses with host plants, produce spores in the soil which are able to germinate and grow, but are unable to produce extensive mycelia and to complete their life cycle in the absence of their hosts [11, 12]. AM fungi are widely distributed in natural and agricultural ecosystems, and produce mycorrhizas in nearly all the crops of agronomic importance.

Studies on the distribution of mycorrhizal associations in nature have shown close relationships between major terrestrial biomes and predominant mycorrhizal types and distinctive patterns of mycorrhizal symbioses have been identified, depending on climatic, latitudinal, altitudinal or nutritional gradients. In extreme environments ericoid mycorrhizas are the most frequently recorded, followed by ectomycorrhizas, while arbuscular mycorrhizas are more common where phosphorus represents the main limiting mineral nutrient, in temperate and tropical areas [13]. These findings suggest different functional roles of mycorrhizas in the different edaphic and environmental conditions, whose knowledge may be of fundamental importance to understand the complex plant/fungal interactions existing in natural ecosystems. Plant competition and community structure have been demonstrated to be affected by mycorrhizal fungi, and a higher floristic diversity was found in experimental microcosms and in the field in the presence of AM fungi [14, 15]. Transfer of carbon compounds between plants interconnected by a common mycorrhizal mycelium was detected, suggesting that mycorrhizal fungi may represent important factors of interplant nutrient transfer and resource redistribution within plant and fungal communities [1, 16-20].

AM symbioses play a fundamental role in soil fertility and plant nutrition and in the maintenance of stability and biodiversity within plant communities, both in natural and in agro-ecosystems. Thousands of experiments in laboratory and in the field contributed to demonstrate large plant growth responses to inoculation of AM fungal endophytes, showing that the performance of each species and isolate may differ depending on plant species and environmental soil conditions, which greatly influence their development. Although AM symbionts cannot be grown in axenic culture, and knowledge of their biology is limited, yet some of the fundamental factors affecting their development have been studied. The aim of this chapter is to review and analyse important data on the main parameters affecting fungal infectivity and efficiency and on fungal ability to survive, multiply and spread in different environments, which may contribute to the biotechnological exploitation and utilization of AM fungi in sustainable agriculture and biodiversity conservation.

2. ARBUSCULAR MYCORRHIZAS

2.1. Distribution

Arbuscular mycorrhizas are very successful in evolutionary terms, since both fossil records and DNA sequence data have shown that they originated early in the history of land plants. Their occurrence in fossil plant roots since the Lower Devonian period led some authors to develop a theory considering mycorrhizas as fundamental to land colonisation by plants [21-23]. Other findings confirmed the occurrence of AM structures as early as 410-360 million years ago [24, 25]. Recently, fossilised fungal hyphae and spores similar to the extant AM fungi have been found in Ordovician dolomites of Wisconsin, dated 460 Ma [26]. Such fossil recordsare consistent with the molecular dating based on nucleotide substitution numbers, which established the origin of the AM genus *Glomus* about 462-353 Ma [27], and with the estimate of the divergence time of Glomales from the progenitor of ascomycete and basidiomycete lineages, about 490 Ma ago [580-410 Ma] [28]. Arbuscular mycorrhizas are widespread and are distributed from arctic to sub-antarctic regions [29-32], in temperate and tropical grassland and forests [33-35], scrub and desert ecosystems [36-38] from sand dunes [39- 41] to alpine sites [42-44].

2.2. Host Plants and Fungal Symbionts

AM symbioses are widely distributed within all phyla of land plants and occur in most plant families, except for genera and species belonging to Brassicaceae, Chenopodiaceae and Cyperaceae, and plant species, genera and families which are exclusively host of ectomycorrhizas, ectoendomycorrhizas, ericoid and orchid mycorrhizas [1, 45, 46]. The most important temperate and tropical crop plants form arbuscular mycorrhizas: cereals including rice, corn, barley, wheat, most legumes, most fruit trees including citrus, peach, grapevine, olive, most vegetables like onion, strawberry, tomato, potato, and other economically important species such as sunflower, cassava, cotton, sugarcane, tobacco, coffee, tea, cocoa, rubber, oil palm, banana. AM fungal symbionts belong to a small group of Zygomycetes, the Glomales, including about 180 described species. They have been isolated from all the continents, and establish symbioses with most plant species, showing no preferential association with any particular host. Their low host specificity was interpreted as an efficient strategy which permitted the survival of individuals and populations and the co-evolution of these "living fossils" with their hosts for more than 400 million years [12, 47]. This interpretation is further supported by considering that AM endophytes are obligate biotrophs and cannot be cultivated on synthetic media in the absence of the host. Such inability represents the major constraint to the large biotechnological application of AM fungi.

2.3. Development and Structure and of the Symbiosis

AM fungi are obligate biotrophs, and their life cycle cannot be completed in the absence of host plants. Germinating spores originate a short-lived pre-symbiotic mycelium, which is able to recognize the host and differentiate infection structures, the appressoria, on the root surface. Appressoria produce infective hyphae which grow within the root cortex, both intercellularly along the longitudinal root axis and intracellularly forming haustoria - the arbuscules - where nutrient exchanges between plant and fungus are supposed to occur. After establishing the symbiosis, the fungus drains carbon from the host, essential for its further extraradical growth, produces large mycelial networks that explore the surrounding environment and absorb mineral nutrients from the soil. Eventually the fungus is able to complete its life cycle by the formation of new spores [48] (Fig. 1). The characteristic structures of the AM system are represented by arbuscules, vesicles, resting spores. Arbuscules are the key structure of AM symbiosis, since the order Glomales was established on the basis of their presence in root cells. They are intracellular haustoria, which originate from the dichotomous branching of hyphae invading root cells: during branching, the hyphae progressively reduce their diameter, and the distal branches are
no more detectable under the light microscope (Fig. 2). Electron microscope investigations showed that even the finest hypha is surrounded by the plasma membrane of the host, and revealed the complete life cycle of the arbuscular structure [49-51]. Time-course studies showed that arbuscules developed as early as 42 hours since the first contact between germinated spores and host plants [52]. Nevertheless their development and occurrence are affected by different factors, such as host nutrient status, light, host phenological stages [53]. Vesicles are spore-like structures, 50-100 μm diameter, formed inside the root cortex, usually intercellularly, containing lipid globules, and for this reason regarded as storage organs (Fig.

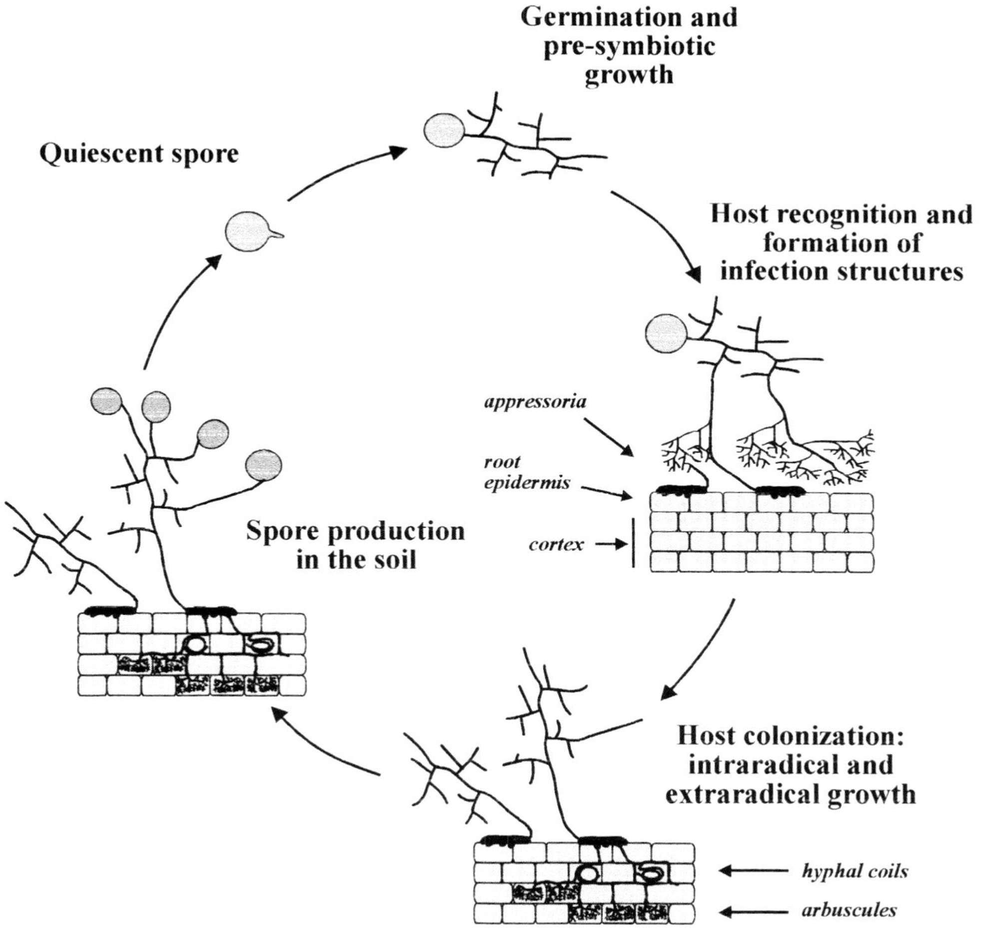

Figure 1. Flow diagram representing the life cycle of AM fungi and the fungal developmental switches occurring during the establishment of a functional symbiosis. Reprinted from: Giovannetti, Spore germination and pre-symbiotic mycelial growth, in "Arbuscular Mycorrhizae: Molecular Biology and Physiology, 2000. Courtesy of Kluwer Academic Publishers, The Netherlands.

3). They are formed only by species belonging to the suborder Glomineae within the Glomales. Some fungi produce spores in the roots, which are clearly detected under the dissecting microscope in unstained material, since they protrude from the roots, deforming their structure. In these cases, it is very difficult to distinguish between vesicles and intraradical spores, particularly in young infections.

Two different infection patterns have been described, the *Paris*-type and the *Arum*-type (Gallaud, 1905, quoted in 1). The *Arum*-type is the most studied and is characterised by the spread of AM fungi via the apoplastic space between cortical root cells. Vesicles are intercellular or intracellular and arbuscules are terminal on intracellular hyphal branches. The *Paris*-type is characterised by the absence of intercellular hyphae and by symplastic development of the fungus, which spreads from cell to cell within the cortex and forms many intracellular hyphal coils and intercalary arbuscules along the coils [54] (Fig. 4). The formation of *Arum*- or *Paris*-type AM infection appears to be under the genetic control of the host plant [55-58]. Extraradical mycorrhizal mycelium is the soil-based structure of AM symbiosis, and consists of a large network of hyphae extending from infected roots into the soil (Fig. 5). Its main functions are represented by the absorption of soil mineral nutrients, which is particularly efficient due to the high surface-to-volume ratio of hyphae, and by the aggregation of soil particles, although recent experiments stressed the important role played by mycorrhizal networks in the transfer of carbon among different plants [16, 14, 59]. The extraradical mycelium is dimorphic, as described in classical works [60, 61], and consists of coarse hyphae, 8-20 μm in diameter, and of highly branched, fine hyphae interpreted as absorbing structures [62]. Small vesicle-like structures called auxiliary cells are formed by some species of one suborder of Glomales, the Gigasporineae. Different factors affect the development of the extraradical hyphae, such as the amount of intraradical infection, host nutritional status and environmental conditions, which in turn may influence the production of resting spores.

2.4. Nutrient Uptake and Transport to Plant Hosts

Well-documented studies evidence that AM fungi have important effects on plant growth, increasing phosphoru content in mycorrhizal plants [1, 53, 63, 64]. Many experiments aimed at studying the mechanisms involved in the improved P nutrition demonstrated that mycorrhizal infection gave similar growth responses as added P, that growth improvements and higher P concentration in mycorrhizal plants were correlated with better P absorption by mycorrhizal roots, and that the enhanced P uptake varied depending on the amount of available P in the soil, on plant species and on fungal isolates [65-69]. The use of radioactive compounds allowed to determine that P was absorbed by AM hyphae from the available pool present in the soil solution, showing that mycorrhizal plants use the same P source as non-mycorrhizal ones [70, 71]. Since P is a poorly mobile element, in P deficient soils the zone at the root surface is rapidly depleted because roots absorb P faster than the soil can replenish it. Thus, the mechanism underlying the better P nutrition of mycorrhizal plants is the efficient soil exploration by hyphae extending beyond the depletion zone and absorbing P from the soil solution. Recent molecular works allowed to isolate a cDNA clone [GvPT] encoding a transmembrane phosphate transporter from *Glomus versiforme*, whose expression was localized in the external hyphae during mycorrhizal associations [72]. Other studies suggested that fungal membrane transporters may be differentially expressed in the extraradical hyphal network, confirming the important role played by AM fungi in P uptake from the soil [73]. Electron microscope studies and chemical analyses demonstrated that the transfer of P from the external mycelium is operated through the accumulation of granules of polyphosphate within hyphal vacuoles, which are probably translocated into the root-based hyphae by protoplasmic streaming [74-78]. The process of P release to the host cells appears

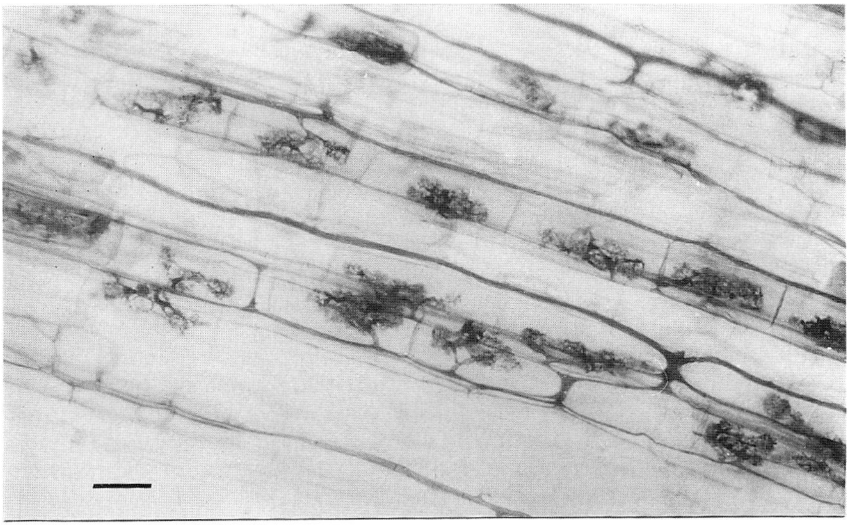

Figure 2. Light micrograph of typical *Arum*-type mycorrhizal infection, showing intracellular arbuscules originating from intercellular hyphae spreading along the longitudinal root axis in *Trifolium pratense*. Bar=40μm.

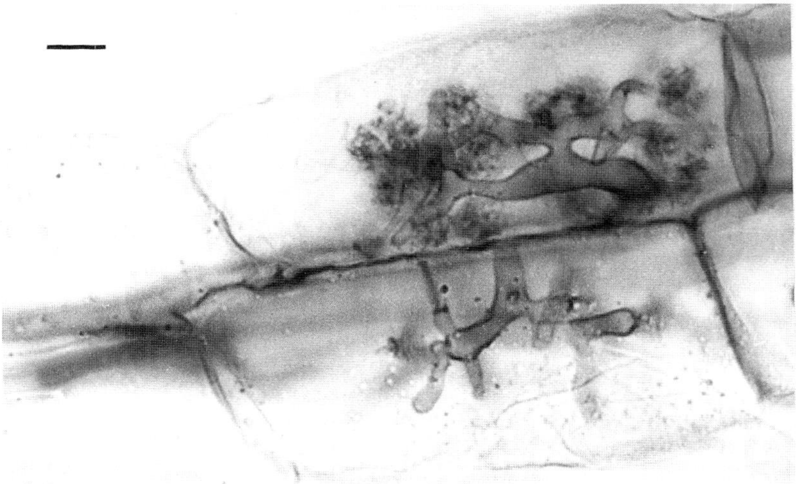

Figure 3. Light micrograph of two arbuscules formed within adjacent root cells, showing dichotomous branching of hyphae after cell penetration. Bar=10 μm.

to take place in the arbuscules and to be mediated by polyphosphatases and alkaline phosphatases [51, 79-84]. The fact that AM fungi are coenocytic organisms, which do not possess cross walls, is an important feature possibly facilitating high flow rates in the hyphae. Many studies have also demonstrated the major role played by AM fungi in the mobilization and transfer of soil nutrients such as N, Zn, Ca, S [68, 85-89]. Moreover, experimental works reported tranfer of carbon, nitrogen and phosphorus betweeen plant species interconnected by a common AM mycelium, highlighting the importance of the "mycorrhizal web" for the exploitation of soil resources [14, 16, 90, 91].

Many pot experiments with different host plants and soils of different P concentrations showed that there are strong differences among plant species in the extent to which they depend on mycorrhizas. For example, some plant species such as *Citrus* spp. are totally dependent on mycorrhizal AM inoculation during the seedling stage [92], and show no response to P addition, while other species, for example grasses, do not show any mycorrhizal dependency [93]. Other experiments showed that soybean and cowpea were less dependent on mycorrhizal inoculation than onion and cassava [94]. The different degrees of mycorrhizal dependency has been correlated with the amount and distribution of roots in soil and with the number and length of root hairs [95, 96]. The results of all the experimental works performed have shown that complex interactions exist between plant species and cultivar, soil P content, the amount of infection and the efficiency of the fungal species inoculated [97, 98].

2.5. Carbon Allocation to Fungal Symbionts

The basis of mutualism in AM symbiosis is the bidirectional nutrient transfer, mineral P and other poorly mobile ions from fungus to plant and organic C from plant to fungus. Carbon transfer from host root tissues is fundamental for the production of a large amount of fungal biomass, because AM fungi are obligate symbionts and their degree of dependence on plant-derived carbon compounds influences the outcome of the interaction. Since early experimental evidences of the transfer of photosynthetically incorporated ^{14}C from the host to external mycelium and spores [99-101], many data have been accumulated on the amounts and types of C compounds that are translocated to the fungus. Some authors showed that sucrose increased in mycorrhizal roots, whereas there is no direct evidence of sucrose hydrolysis to glucose and fructose [102]. Trehalose and polyols were detected in mycorrhizal roots, spores and extraradical mycelium of AM fungi [103, 104], and glucose was suggested as a likely substrate for conversion to trehalose [105]. However, transfer site has not been definitely assessed, although some evidences suggest that C compounds may be translocated to the fungus via intercellular hyphae rather than via arbuscules. Many observations showed that at the beginning of root colonization, fungal growth was boosted before the formation of any arbuscule [106]. In addition, cytochemical studies showed that fungal H+ATPases are distributed mostly on the fungal membranes of intercellular hyphae and coils, while they are erratically present in arbuscules [107]. These findings are very important because the spatial separation of the bidirectional transfer of mineral nutrients and carbon compounds could explain the different efficiency showed by diverse plant/fungus associations. Recently a sugar transporter gene has been cloned from *Medicago truncatula* mycorrhizal roots: the transcripts were upregulated following fungal colonization and were localized in the root sites where the hyphae were present, suggesting that the inducing signals are triggered by the occurrence of fungus in the cortex [108].

The portion of photosynthate drained by the fungal partner may be considerable. Different estimates showed that mycorrhizal plants obtain from 4 to 20% more C than non-mycorrhizal ones [109-111]. Variations of the estimates depend on experimental conditions and on

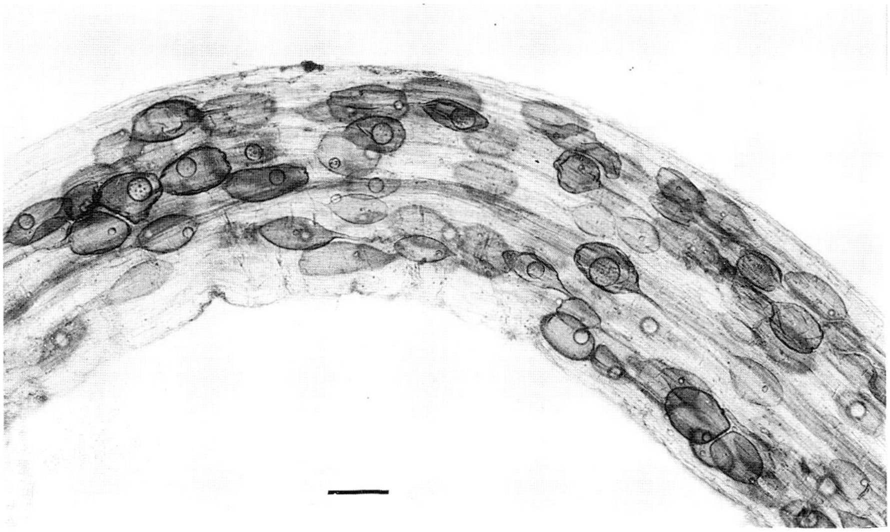

Figure 4. Light micrograph of vesicles in an old strawberry host root. Bar=100μm.

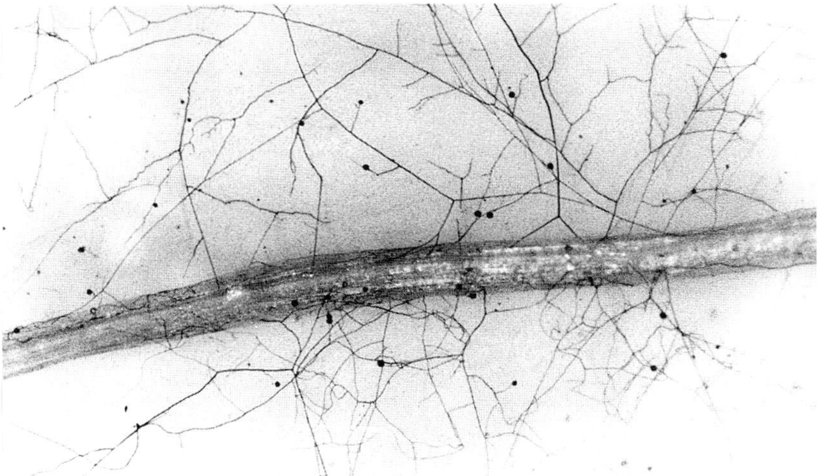

Figure 5. Light micrographs showing extraradical hyphae of *Glomus mosseae* spreading from mycorrhizal roots of *Allium porrum*. Bar=300μm.

differences between diverse plant-fungus associations, which affect fungal biomass, root and soil colonization, fungal and plant metabolic activity.

2.6. Non Nutritional Effects of AM

An important non nutritional effect of AM is related to their ability to enhance and maintain soil aggregation, which contribute to overall fertility, affecting soil tilth, water relations, root penetration and erosion potential. Extraradical hyphae of AM fungi were shown to be directly involved in the formation and stability of soil aggregates, by producing the mycelial network that holds together soil particles [112-114], and binds them through the production of a persistent glycoprotein termed glomalin, which was retrieved from the soil in amounts ranging from 2.8 to 14.8 mg/g [115-117]. Mycorrhizal plants are generally more tolerant to drought stress than non mycorrhizal plants, and this effect has been ascribed to improved plant nutrition [118], altered hormonal balances [119, 120], or to an increased water flow through the hyphae [121].

AM fungi may play a positive role in plant tolerance to soil borne plant pathogens. For example, *Glomus intraradices* and *Glomus mosseae* increased resistance to infection by root pathogens in tobacco and tomato plants [122, 123]. Other works reported increases in disease severity [124, 125] showing that the interactions between plants, pathogens, AM fungi and environment are complex and unpredictable. Plant protection effects were often attributed to the improved nutrient status of mycorrhizal plants [126, 127], although other mechanisms were proposed, since some diseases were limited only by the presence of AM fungi in the roots [128]. The activation of plant defences by mycorrhizal fungi has been supposed to elicit induced resistance in some plant-pathogen interactions [129]. However, recent experiments suggested that the protective role of AM fungi is mediated by a population of antagonistic microbes occurring on fungal spores and hyphae, the mycorrhizosphere [130].

3. THE BIOLOGY OF ARBUSCULAR MYCORRHIZAL FUNGI

3.1. Taxonomy

AM fungi produce asexual, multinucleate spores, on which their identification is based. Spores are produced on extramatrical hyphae, either single or aggregated, to form more complex structures, the sporocarps (Figs. 6, 7), and their diameters range from about 50 to 600 μm. They differ in morphological characteristics - namely shape, colour, size, spore wall and subtending hypha structure, sporocarp and peridium occurrence, modes of spore germination, spore ontogeny, infection patterns - which were used to describe many distinct species. The first taxonomical work placed AM fungi within the Endogonaceae family, with four genera, *Glomus*, *Gigaspora*, *Acaulospora* and *Sclerocystis*, and 29 species [131]. Since then more than 150 new species were described [132].

The phylogenetic interpretation of Morton and Benny [133] brought to the creation of the order Glomales, where all species forming arbuscular mycorrhizas were included, on the assumption that this is an original character, distinctive of the order. Three families, Glomaceae, Acaulosporaceae and Gigasporaceae, and six genera, *Glomus*, *Sclerocystis*, *Acaulospora*, *Entrophospora*, *Gigaspora*, *Scutellospora*, were founded on the basis of the mode of spore formation. This taxonomical structure has been largely supported by molecular systematic studies, which confirmed the monophyletic origin of the order [134] and the subdivision in three families [27]. The molecular approach has been used to delineate a new evolutionary pattern of these fungi, suggesting that two additional ancestral lineages may be recognised [135]. Recently, molecular analysis of the genus *Gigaspora* pointed out that isolates originally identified as five different species actually belonged to three subgeneric groups [136]. Accordingly, phylogenetic analyses of the only species left in the genus *Sclerocystis* [137] allowed to determine the definitive taxonomical position of all *Sclerocystis* species within the genus *Glomus* [138].

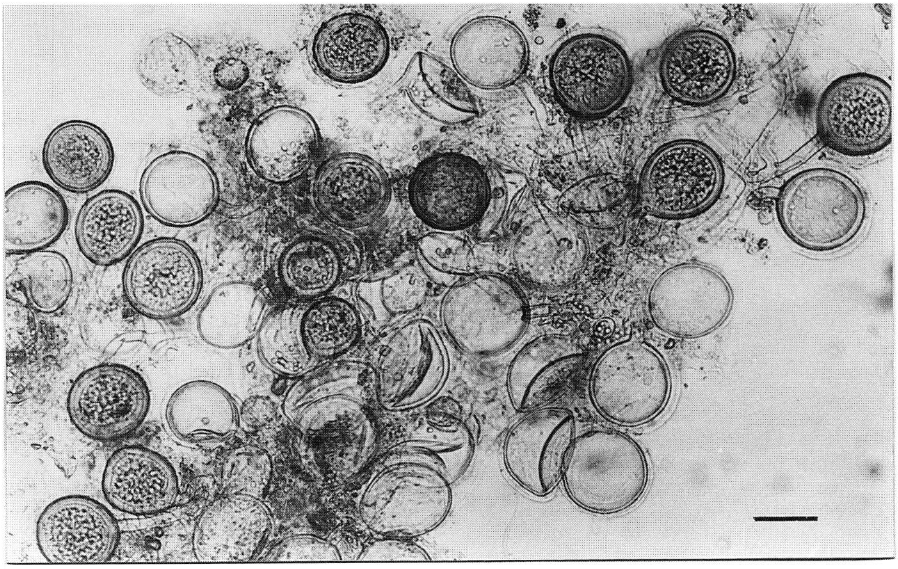

Figure 6. Light micrograph showing a cluster of spores produced by *Glomus intraradices*. Multilayered spore walls are evident. Bar=80μm.

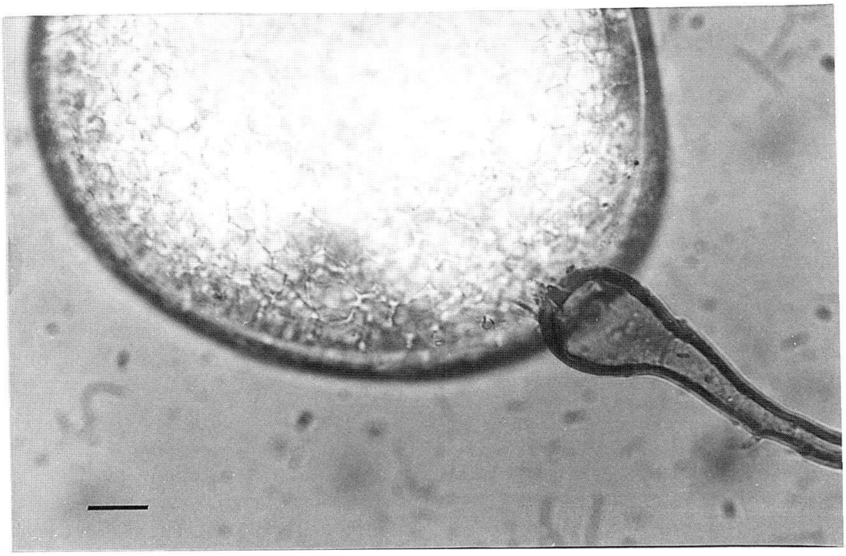

Figure 7. Light micrograph showing a spore of *Scutellospora* sp. formed singly at the tip of a bulbous hyphal attachment. Bar=20μm.

3.2. Biological, Biochemical and Molecular characterization

Morphological characterisation is fundamental for taxonomical species identification, though large biological diversity occurs within species, and isolates show different characteristics related to pH requirements, temperature tolerance, reactions to added fertilizers, tolerance to agricultural practices and other factors concerning the soil environment. The main biological characters of AM fungi affecting their life cycle and symbiotic performance have been studied in different genera, species and isolates, and concern spore dormancy, spore germination, pre-symbiotic growth, infectivity, efficiency, spore production, environmental tolerance. These studies provided a good indication of the large variability of AM symbionts, showing that the knowledge of biological characters influencing their occurrence and growth is of crucial importance for their biotechnological use in agriculture and for their management in environmental conservation. The biotechnologically relevant aspects of biological diversity will be considered in the next paragraph. Biochemical and molecular analyses improved our knowledge of AM fungi, supplied tools for the identification of different isolates directly in the roots and in the soil, and allowed the monitoring of their spread and activity. Many biochemical and molecular markers were proposed to study diversity at different taxonomic levels, the most successful being those concerning proteins, lipids and nucleic acids.

Enzyme mobility during polyacrylamide gel electrophoresis was used to compare spores of different isolates of morphologically similar species, leading to the detection of interspecific and intraspecific variations [139, 140], and it is actually used to support results of DNA analyses [136, 141] (Fig. 8). Identification of AM fungi in roots was performed at the species level by detecting mycorrhiza-specific isozymes [142, 143]. Besides the possibility to distinguish fungi producing morphologically similar root infection, isozyme analysis allows the detection of fungal metabolic activity, which is not possible with DNA-based techniques. Immunochemical methods were used to identify AM fungi by using both polyclonal and monoclonal antibodies. These tools showed different degrees of specificity, and allowed the detection of AM fungi in roots, although the presence of unspecific cross-reactions may limit the reliability of the results [144-148]. Electrophoretic analyses of spore soluble proteins have also been useful to investigate isolate diversity and species identification. For example, geographically different isolates of *G. intraradices* produced consistent profiles, showing 98% similarity, while isolates belonging to different species showed larger differences [141, 148, 149] (Fig. 9). The analysis of fatty acid methyl esters [FAME] was utilised to distinguish spores belonging to different species of AM fungi [150], yet it is not useful for the identification of species within the root or in the soil. The quantification of signature fatty acids in soil and roots was proposed to estimate arbuscular fungal biomass [151].

Molecular methods have been successfully used for identification of isolates and species since the introduction of the polymerase chain reaction (PCR). A sequence derived from different regions of DNA differs between species and isolates, and specific primers have been developed to identify AM fungi not only from spores but also from colonised roots and soil. Specific primers were produced from genes present in multiple copies such as the ribosomal RNA genes 18S, 5.8S, 25S, and ITS1 and ITS2 (internal transcribed spacers) containing sequences of different variability. Family-specific primers were obtained from 18S sequence used to detect AM fungi in roots [152, 153] and in soil [154]. Species-specific primers were obtained from more variable regions, namely ITS1, ITS2 and D1 and D2 regions of 25S, for ecological and phylogenetical studies [155, 156]. Other approaches to develop species-specific primers utilised fragments obtained from random amplified polymorphic DNA (RAPD) [157, 158], or highly repeated sequences [159]. Diversity of AM fungi was also

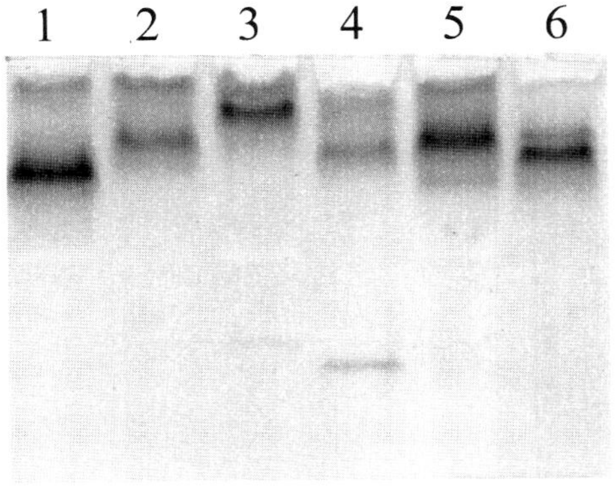

Figure 8. MDH profile from spores of *Glomus coronatum* isolates of distinct geographic origin: 1, Italy (BEG 28); 2, Italy (FO97); 3, Spain (BEG49); 4, Australia (Wum-2); 5, Abu Dhabi (BEG 139), and *Glomus caledonium*, 6 (BEG 20).

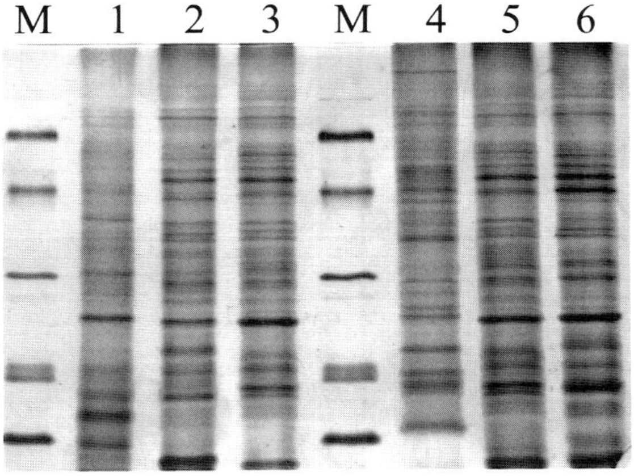

Figure 9. SDS-PAGE of soluble proteins extracted from spores of *Glomus mosseae* isolates of distinct geographic origin: 1, Italy (BI98); 2, England (BEG25); 3, England (BEG12); 4, Syria (INVAM-SY710); 5, USA (INVAM-IN101C); 6, USA (INVAM-AZ225C); M, standard molecular markers: kDa 97.4 - 66 - 45 - 31 - 21.5 - 14.5.

studied by using AFLP [160] and restriction analyses of amplified sequences PCR-RFLP [161] (Fig. 10). These molecular methods should be extended to study the whole diversity present within species of AM fungi, and could represent powerful tools for diagnostic purposes and for tracking introduced inocula in the field.

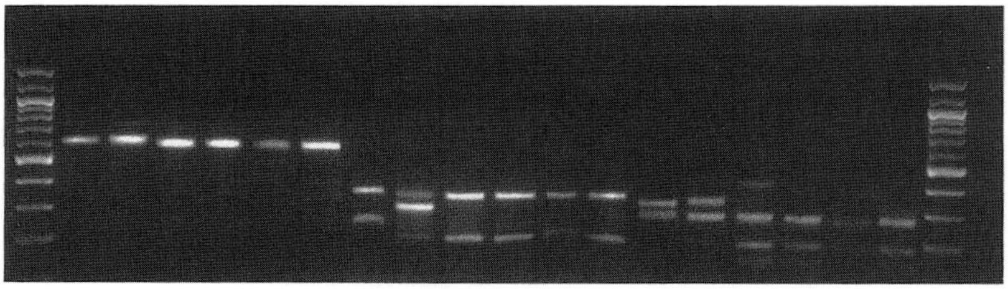

Figure. 10. Restriction fragment patterns of ITS fragments from geographically distinct isolates of *Glomus mosseae* and from *Glomus coronatum* (BEG 28). PCR products were digested with *Hae*III (lanes 1-6), *Hinf*I (lanes 7-12) and *Msp*I (lanes 13-18). Lanes 1-7-13: BEG 28; lanes 2-8-14: BI98; lanes 3-9-15: BEG25; lanes 4-10-16: BEG12; lanes 5-11-17: IN101C; lanes 6-12-18: AZ225C. (Courtesy of M. Agnolucci and C. Sbrana).

4. THE BIOTECHNOLOGY OF ARBUSCULAR MYCORRHIZAL FUNGI

4.1. Exploiting Fungal Biodiversity

Different AM fungi can differ greatly in the extent to which they improve host plant growth in various soil and environmental conditions. Because of the complexity of the interactions between fungal isolates and host/soil/environmental factors, no generalization is possible. Nevertheless, the criteria to be followed for the selection of good endophytes can be outlined. The symbiotic performance of AM endophytes is the result of the interaction of two fundamental fungal characters, infectivity and efficiency. Infectivity has been defined as the ability of a fungal isolate to establish rapidly an extensive mycorrhizal infection in the roots of a host plant and is correlated with complex soil-fungus-host factors affecting the pre-symbiotic steps of fungal life cycle, from germination and germling growth in the absence of the host to host recognition, differentiation of infection structures and root colonization [162]. The efficiency of different fungal endophytes cannot be measured simply in terms of host growth responses, since many factors contribute to determine their symbiotic performance. However, since the capacity of AM fungal hyphae to increase the uptake of P and other poorly mobile nutrients is considered one of the most important mechanisms of plant nutrition, as well as root developmental pattern and soil solution concentration [163], studies on the efficiency of different isolates of AM endophytes have mainly taken into account factors affecting the development of extramatrical fungal networks in the soil, the rate of P absorption, translocation and transfer from hyphae to host cells.

4.2. Selection of Infective Isolates

The biotechnological utilization of AM fungi depends first and foremost on the availability of highly infective strains. The possibility of surviving in the soil ecosystem of any super-efficient strain would depend on the ability of its mycelium to infect host plants rapidly and to compete for infection sites with indigenous endophytes, which natural selection likely made very infective. Although different kinds of fungal propagules may affect infectivity, most studies have investigated the ecology and physiology of spores, as they represent the most frequently used propagules in laboratory and field experiments. Therefore the parameters to be taken into account when selecting for infectivity mainly concern spore-related factors such as spore dormancy, spore germination and pre-symbiotic growth of spore germlings. Other important parameters concern appressorium formation and intraradical growth.

Spore dormancy. Spore dormancy represents an important character influencing the competitiveness of different species and isolates, since their symbiotic performance is dependent on infectivity, which is strictly related to rapid germination. Since the early experiments on spore germination, Godfrey [164] and Mosse [11] were aware of the existence of spore dormancy, which they successfully removed by storing spores on damp filter paper at 5°C for 6 weeks. In this way 80% spores belonging to *Endogone* sp. (presumably *G. mosseae*) germinated within 3-4 days. Only in 1983 Tommerup distinguished between dormancy and quiescence, defining dormant a spore failing to germinate although exposed to physical and chemical conditions supporting germination of apparently identical spores of the same species, which were defined quiescent spores [165]. Dormancy can be relieved by storage in many species: for example, freshly harvested spores of *G. mosseae* were able to germinate slowly compared to spores detached from sporocarps and stored at 6°C for 5 weeks [166]. Similar results were obtained with a North American isolate of *G. mosseae*, which showed that 10°C-stored spores germinated more readily than fresh ones [167]. Other species, such as *G. intraradices*, *Glomus clarum*, *Glomus caledonium*, *Glomus monosporum*, *Acaulospora laevis*, *Acaulospora longula,* showed spore dormancy [168-171]. On the contrary, *Gigaspora gigantea* germinated as early as one day after incubation [172] and *G. margarita* germ tubes emerged after 72 h incubation on water agar or within 3-5 days on agar media [173, 174]. Propagule dormancy should be further studied to understand whether the phenomenon is correlated with species, genera or isolates, and to investigate the molecular signals able to relieve spore dormancy and activate the cell cycle.

Spore germination. Germination is the first step in the life cycle of AM fungi and an important character to take into account when selecting endophytes for practical applications, since spores germinate from a quiescent-like state under different physical, chemical and microbiological conditions, which show a high degree of variation among isolates. Beyond spore dormancy, the environmental parameters influencing spore germination in different genera, species and isolates are represented by pH, temperature, moisture, mineral and organic nutrients, the presence of host/nonhost plants and microorganisms. Differences in spore germination were found to be correlated with the environment from which the different AM fungi were originally isolated. For example, spores of *A. laevis*, which is the predominant AM endophyte in low pH soils [175-177], were able to germinate well at pH between 4 and 5, and showed a poor germination at pH between 6.5 and 8 [178]. Similarly, *Gigaspora coralloidea* and *Gigaspora heterogama*, isolated from acidic soils, germinated best at pH from 4 to 6 [179], and different isolates of *G. mosseae*, collected from agricultural soils, showed a pH optimum for spore germination between 6 and 9, and failed to germinate at pH 4 and 5 [180, 181]. However, the pH values detected for optimum germination concern the isolate of the species used in each experiment, and it is conceivable that each geographical isolate represents an ecotype adapted to peculiar soil characteristics. AM fungal spore germination is greatly influenced by temperature, since fungal species isolated from the same site show different temperature limits for germination, ranging from 15-25°C in *A. laevis* to 10-30°C in *Gigaspora calospora* and 10-25°C in *G. caledonium* [182]. Also in the case of temperature, the differences between AM fungal species reflect the different environments from which they were isolated. As an example, *G. coralloidea* and *G. heterogama* isolated from Florida germinated best at 34°C, whereas *G. mosseae*, isolated from more northern latitudes, showed maximum germination at 20°C and failed to germinate at 34°C [180, 183]. Moisture was shown to influence spore germination of some AM fungal species, such as *G. mosseae*, *Glomus epigaeum*, *G. gigantea* and *A. longula* [170, 172, 180]. Laboratory experiments showed that AM fungal spore germination is not affected by nutrient content of soil. For example, *G. gigantea* germinated well at P concentrations ranging from 5 to 500 ppm [172], *G. mosseae* and *G. caledonium* were not affected by P concentrations in agar up to 30 mM [184], *G. margarita* germinated well up to 16 mM phosphate solution [185] and *G. epigaeum* was not influenced by levels of NH_4NO_3 and K_2SO_4 up to 200 ppm [180]. Inorganic ions such as Cu, Mn and Zn inhibit spore germination of *G. mosseae* and *G. caledonium* [166, 168]. Host-derived signals do not represent essential factors for spore germination of AM fungi, since their spores can germinate in axenic culture, although the presence of host roots and root exudates positively affects germination and/or germ tube growth [186-192]. On the contrary, root exudates of nonhost plant species do not generally affect the percentage of spore germination [193-195].

Soil microorganisms can stimulate germination in AM fungi [11, 196-198] and many bacterial species were identified affecting spore germination and hyphal extension. For example, *Streptomyces orientalis* stimulated the germination of *G. mosseae* spores [199], *Pseudomonas* and *Corynebacterium* were capable of enhancing germination of *G. versiforme* spores [200], *Klebsiella pneumoniae* increased hyphal extension of *Glomus deserticola* germlings [201], *Trichoderma* spp. enhanced the development of mycelium from germinating spores of *G. mosseae* [202]. Recent studies have demonstrated an intimate association of AM fungal spores with many bacteria which were localized on and within the outer spore wall or found to be embedded in the electron-dense spore wall in different *Glomus* species [203, 204, 205, 206].

Pre-symbiotic growth. Hyphae originating from a germinated spore generally grow forward with a strong apical dominance and regular, right-angled branches. Hyphae are thick-walled, aseptate and about 5-10 μm wide and contain many nuclei. Germlings of AM fungi are able to elongate and originate a coenocytic mycelial network generally extending no more than 50-200 mm (Fig. 11). For example, in axenic culture *G. caledonium* mycelium measured from 10 to 40 mm after 10-15 days growth, and its mean growth rate, during the early growth phase, was 1.97 μm/min [12]. Similarly, *Gigaspora margarita* germlings ranged from 18 mm to 25 mm after 9 days growth [188, 207] and in *G. clarum* new hyphae extended up to 8 mm after 10 days incubation [169] (Table 1). Host roots and root exudates promote germling growth of most AM fungi and induce fungal differential morphogenesis characterized by an increase in hyphal branching, probably functional to the production of infection structures [191, 208]. In the absence of the host young germlings cease growth within 15-20 days of germination [11, 12, 167, 172, 207, 209, 210]. The growth arrest is always correlated with an increase in the proportion of empty hyphae, which are originated by the progressive retraction of protoplasm from the tips, followed by the production of cross walls separating empty segments from viable ones. Recent experiments with *G. caledonium* showed that although hyphal tracts devoid of protoplasm were not metabolically active, small tracts proximal to the spore were viable, and capable of establishing root infection, even after 6 months [12], confirming the long-term ability of some AM fungi to retain infectivity in the absence of the

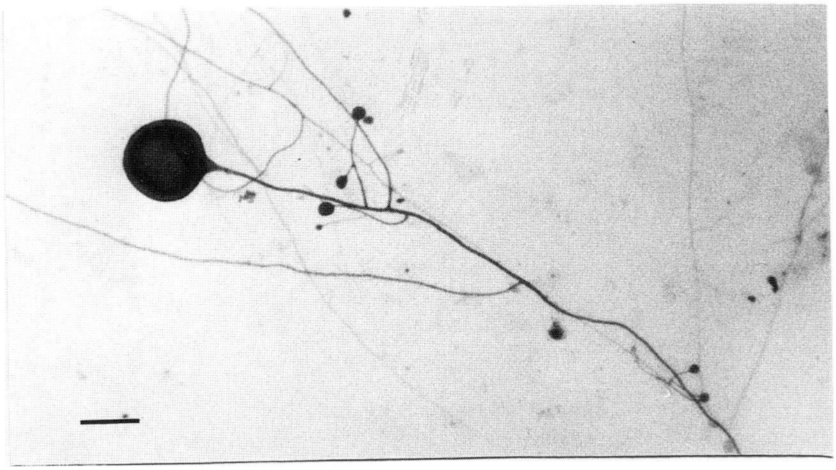

Figure 11. Light micrograph showing the poor growth ability of a spore of *Glomus coronatum* germinating in axenic culture in the absence of the host. Bar=120 μm.

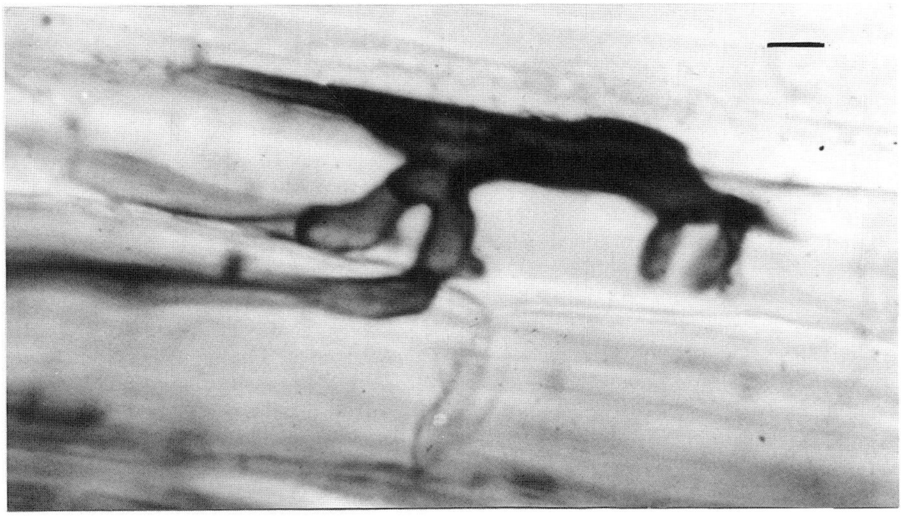

Figure 12. Light micrograph showing an appressorium produced by *Glomus mosseae* on the root surface of *Medicago sativa*. Bar=7 μm.

host [211]. Protoplasm withdrawal from hyphal tips towards the mother spore can be interpreted as a process of resource reallocation, functional to maintaining the limited energy resources of germlings, which are never totally depleted when germlings cease growth [12, 168, 172, 209] since they are able of multiple germination [11, 212].

An important character correlated with the survival capacity of pre-symbiotic mycelium is represented by its ability to form hyphal networks by means of anastomosis between contacting hyphae, since in natural conditions germlings growing in the absence of the host could plug into older and larger mycelia, extending their possibility to contact host roots [213]. Recent works showed that pre-symbiotic mycelia of some fungal species and isolates are characterized by a high frequency of hyphal fusions, ranging from 51-57% in *G. mosseae* to 35-54% in *G. caledonium* and 58-69% in *G. intraradices*, in hyphae belonging to the same germling, and from 40% in *G. mosseae* to 34% in *G. caledonium* and 98% in *G. intraradices*, in hyphae belonging to different individually germinated spores. On the contrary, other species such as *Gigaspora rosea* and *Scutellospora castanea* never formed anastomoses, even between hyphae belonging to the same germling [213], suggesting that inability to form large individual hyphal networks may be responsible for their disappearance from agricultural soils, which was recently observed [214].

Appressorium formation. Appressoria of AM fungi are inflated, multinucleate infection structures which are formed on the root surface of host plants (Fig. 12). They have been studied in a few fungal species, in connection with host factors affecting their production [216-217]. The time required by a mycorrhizal endophyte to form infection structures is an important factor of its infectivity and was investigated in time-course experiments, which showed that the first appressoria are formed as early as 36-48 hours since the beginning of the interaction between germinated spores and host plants, depending on plant species, and that their number vary in different host/symbiont combinations. For example, *G. mosseae* formed appressoria on basil roots after 36 h, *G. caledonium* after 48 h and *Glomus* A6 [*Glomus viscosum*] after 60 h. The number of appressoria was also variable, since after 72 h *G. mosseae* produced 22 appressoria, *G. caledonium* 2 and *G. viscosum* 11 per plant root system [52]. The number of appressoria ranged from 2.6-21.1 to 4.6-10.7 per mm root length in strawberry and apple, respectively, colonised by unidentified AM endophytes [61] and from 0.2 to 10 per mm root length in micropropagated plum rootstocks infected with four different *Glomus* species [218]. Rapid production of many appressoria represents a highly desirable character in the selection of infective AM symbionts for agricultural use, since it is conceivable that competition with indigenous endophytes is realized on the basis of infection performance.

Intraradical fungal growth. The capacity of AM fungi to develop within the root is important for nutrient transfer from the symbiont to the host. Thus infection rate might be considered an important parameter for comparisons between different endophytes. Yet, values of mycorrhizal colonisation show large variability, depending on complex interaction of host plants, fungal symbionts, soil and environmental factors. Thus, due to the complexity of the system, results of different combinations are unpredictable and it is difficult to extrapolate general principles from experimental data: in fact, sometimes only 10% root colonization may give large growth improvements, while the same percentage of root infection by different endophytes may differentially improve plant growth [219]. This is particularly important when considering that the outcome of the symbiosis depends on a delicate balance between mineral nutrients supplied by the fungus and carbon compounds allocated to the fungus [220, 221].

4.3. Selection of Efficient Isolates

Beyond infectivity, the other fundamental character to be considered, when selecting AM endophytes for biotechnological utilization, is fungal efficiency. It has already been mentioned that many factors contribute to determine the symbiotic performance of different AM fungi, yet most experimental works concern only the ability of different species and

Table 1. Lengths of pre-symbiotic (top) and symbiotic extramatrical (bottom) mycelium of arbuscular mycorrhizal fungi reported in different experimental works.

Fungal species	Experimental system, incubation times	Hyphal length per spore [mm]	References
Acaulospora laevis	in soil, 21d	98 ÷ 134	[211]
	in vitro, 21d	0.33 ÷ 17.6	[182]
Gigaspora calospora	*in vitro*, 21d	13.9 ÷ 16	[182]
Gigaspora gigantea	*in vitro*, 10 ÷ 18d	318 ÷ 544	[283]
Gigaspora rosea	*in vitro*, 8d	16.5 ÷ 72.8	[236]
Gigaspora margarita	*in vitro*, 9d	1.7 ÷ 17.7	[188]
	in vitro, 28d	20 ÷ 25	[207,210]
Glomus caledonium	in soil, 14d	90 ÷ 104	[211]
	in vitro, 15 ÷ 28d	10 ÷ 66	[168, 12]
	in vitro, 14d	11.4 ÷ 13.1	[182]
Glomus epigaeum	*in vitro*, 7d	0.4 ÷ 0.6	[186]
	in vitro, 6 months	70.8	[12]
Glomus etunicatum	*in vitro*, 21 ÷ 28d	0.14 ÷ 0.5	[190]
Glomus fasciculatum	*in vitro*, 14d	0.25 ÷ 0.28	[284]
Glomus mosseae	*in vitro*, 8d	10	[166]
	in grit, 20d	17.6	[191]
	in soil, 43d	7.1 ÷ 8.5	[285]

Fungal species	Experimental system, incubation times	Mycelial length	References
Acaulospora laevis	in soil, 7- 14d	1.1 ÷ 6.9 m/g soil	[224]
Gigaspora margarita	*in vitro*, 24d	1.1 m	[210]
Gigaspora rosea	in soil, 135d	4.8 m/g soil	[286]
Glomus caledonium	in soil, 49d	3 ÷ 5 m/g soil	[73]
Glomus etunicatum	in soil 135d	3.3 m/g soil	[286]
Glomus intraradices	*in vitro*, 120d	25 ÷ 100 cm/cm^2	[227]
Glomus mosseae	in microcosm, 7d	5.3 m/cm^3 soil	[195]
	in soil, 135d	8.1 m/g soil	[286]
Glomus spp.	in the field	7.8 m/g soil 49 m/m root	[113]
Scutellospora calospora	in soil, 7-14d	1.34 ÷ 9.22 m/g soil	[224]
	in soil, 49d	9 ÷ 10 m/g soil	[73]

isolates to increase plant nutrition and growth. When considering the mechanisms underlying fungal efficiency, in terms of plant production, the most important parameters to be taken into account are fungal ability to develop extensive hyphal networks which can explore the soil, absorb, translocate and transfer soil mineral nutrients to the host plants, the viability of extraradical mycelium, the rate of hyphal nutrient uptake, translocation and transfer to host cells. Investigations on the length of the extramatrical mycelium of AM fungi, performed after destructive extraction from the soil in laboratory conditions, reported length which ranged from 1 to 10 m/g soil (Table 1). Experiments aimed at comparing the relative amounts

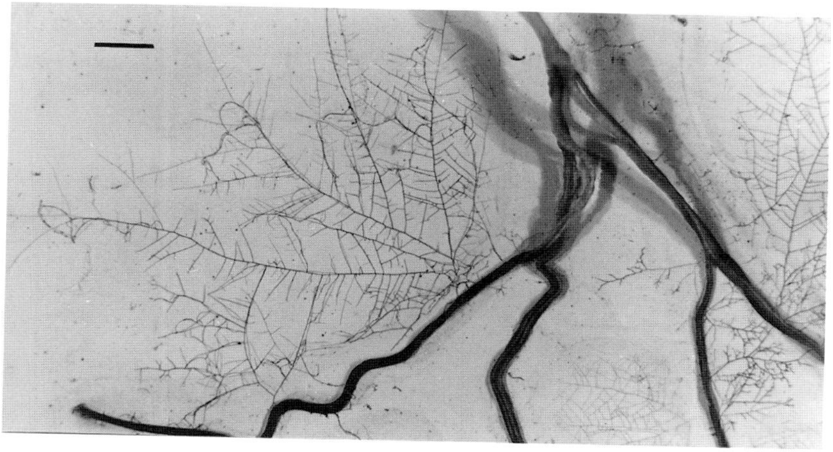

Figure 13. Visualization of the development of intact extraradical mycelium produced by *Glomus mosseae*, spreading from mycorrhizal roots of *Allium porrum* and showing densely branched and anastomosed hyphae. Bar=15 mm.

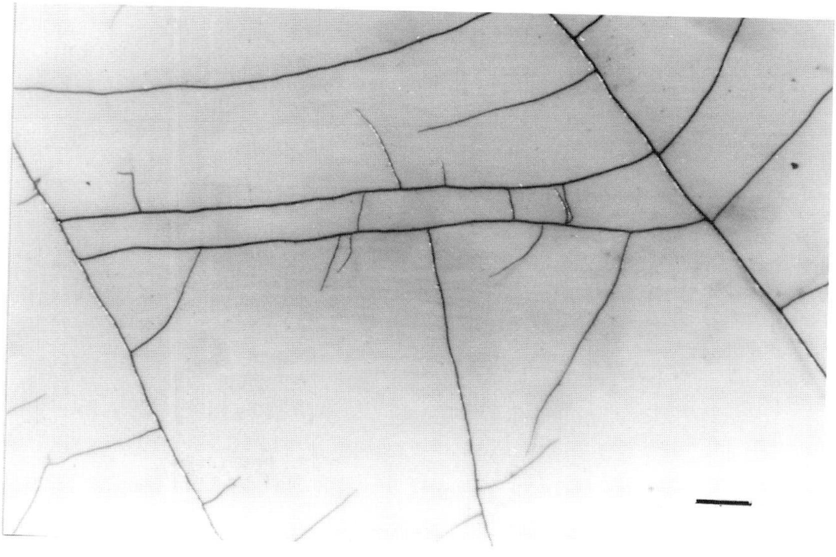

Figure 14. Light micrograph showing anastomosis formation among extraradical hyphae of *Glomus mosseae* growing in symbiosis with *Allium porrum*. Bar=180 μm.

of mycelium produced by AM fungi reported that *G. mosseae* produced twice as much extraradical hyphae as *Glomus geosporum* per unit length of mycorrhizal root [222], confirming previous results on the large variability shown by different isolates [223]. In addition, also the spread of extraradical hyphae from colonised roots shows large variations and the average rate of hyphal spread of *A. laevis* was reported to be 4 times as high as that of other two species belonging to different genera [224]. Recent findings showed that the spread of *G. mosseae* extraradical hyphae ranged from 738 to 1067 mm/day, after 7 days' growth in a two-dimensional membrane system [225]. It is interesting to note the high growth rate of AM fungi growing in symbiosis with host plants, when compared with the poor growth ability of the presymbiotic mycelium, 2.8 mm/day, after seven days [12].

Some nondestructive investigations on the structure of extramatrical mycelium of AM fungi provided qualitative information on the architecture of AM mycelium, and its development before and after symbiosis establishment [226, 227]. However, virtually nothing is known of the biological basis for the establishment of extensive extraradical mycelial networks through which nutrients are proposed to flow. One of these mechanisms is represented by the formation of anastomoses, which have been recently visualized and quantified in presymbiotic mycelium [213]. Similar studies on the structure of extraradical mycelium have shown that the frequency of anastomoses per hyphal contact ranged from 67% to 77% in *G. mosseae* hyphae growing in symbiosis with different plants [Figs. 13-14]. The establishment of protoplasmic flow between fusing hyphae was evidenced by histochemical localization of succinate-dehydrogenase-activity [SDH] and by the detection of nuclei in hyphal bridges [225].

Another important parameter of fungal efficiency, determining the actual fungal biomass actively functioning in the soil for nutrient uptake and translocation, is represented by the viability of extramatrical mycelium. Metabolic activity of hyphae growing in the soil, measured by using different vital stains, ranged from 63% in 6 weeks-old *G. intraradices* hyphae to 100% in 3-weeks-old *G. clarum* hyphae [228, 229]. Other data showed that the activity of extraradical mycelium ranged from 0 to 32% in hyphae extracted from soil, increasing greatly in hyphae attached to colonized roots - 96% in 6, 9 and 13 weeks-old *G. mosseae* and *G. intraradices* hyphae [230]. A recent work reported that the length of vital extraradical hyphae ranged from 20 to 40 m/m-colonized root in *Eucalyptus coccifera* seedlings inoculated with 3 different AM fungi [231]. Alkaline phosphatase used as a molecular marker of the efficiency of the symbiosis showed differences in phosphate metabolism between extraradical hyphae of *G. rosea* and *Glomus manihotis* [232].

The efficiency of different AM endophytes may also depend on the inflow rates of P in extraradical and intraradical hyphae. Experimental data on translocation flux of ^{32}P in *G. mosseae* extraradical hyphae ranged from 2 to 20 x 10^{-6} mol m^{-2} s^{-1} [68, 233, 234]. Other findings showed that *G. caledonium* external mycelium was the most effective of three AM fungi for uptake and transfer of ^{32}P, confirming that fungal efficiency is correlated with P transport to the host plant [235]. Recent time-lapse and video-enhanced microscopy studies allowed to measure, on the basis of particles movement [presumably vacuoles, nuclei, fat droplets, organelles, granules], bidirectional protoplasmic flow rate, which ranged from 2.98 to 4.27 μm/s in living pre-symbiotic hyphae of *G. caledonium* [236]. Such values are consistent with those obtained by Nielsen and Jakobsen in extramatrical hyphae of *G. intraradices*, 1-5 μm/s (personal communication). An elegant experiment found that *G. caledonium* was the most efficient endophyte in the exchange of ^{14}C and ^{32}P, compared with other two species, showing that variations in efficiency are also dependent on the below-ground C allocation in relation to P uptake and transfer [235]. One of the most important experimental data to be obtained in order to compare hyphal transport efficiency of different AM endophytes is represented by P flux through appressoria, which was calculated to correspond to $3.8x10^{-8}$ mol $cm^{-2}s^{-1}$ [66]. It is important to note that the production of a high number of appressoria on the root surface is an essential feature relevant not only to the infectivity, but also to the efficiency of isolates, since appressoria represent the only link connecting extraradical and intraradical mycelium and only through them mineral nutrients can be transferred from the hyphae developed in the soil to the root-based hyphae and then released to the host plants.

4.4. Inoculum Production and Inoculation Technology

AM fungi cannot be produced in large-scale reactors given their recalcitrance to pure culture. Despite this constraint, many methods have been developed for manipulating AM fungi, inoculating them on suitable host plants, and reproducing large amounts of inoculum. It has already been mentioned that AM fungi are maintained in growth chambers or in greenhouses, in pot cultures together with their hosts. The first step in any attempt to produce pure AM fungal cultures is the collection of spores from the soil. Different procedures for retrieving and handling AM fungal spores can be utilized, but the fundamental method for recovering spores from the soil is the wet-sieving and decanting technique [237]. This method is based on the fact that soil particles are heavier than spores, which, in water, tend to float. Generally spores are extracted from soil by using sieves down to a mesh size of 50 μm, all the sievings are transferred into petri dishes, examined under a dissecting microscope and spores are manually isolated by using capillary pipettes. Each spore type can be then mounted on microscope slides to be observed and measured for taxonomic identification under light microscope. The different spore types are then inoculated nearby the roots of a suitable host plant, and allowed to grow and establish a functional symbiosis. The plants can be cultivated in soil, sand, expanded clay, peat, or other substrates, after their sterilisation by steam, fumigation or gamma irradiation. Usually about three months after plant inoculation new spores are produced in the substrate of the pot cultures, on the extraradical mycelium surrounding the root system. Because arbuscular mycorrhizas are not visible at the naked eye (although yellow patches in the roots indicate the presence of the fungus in species like onion and maize), the assessment of mycorrhizal infection is carried out by clearing and staining root samples. The most widely used method to detect mycorrhizal infection is based on clearing the root tissue with KOH and staining fungal hyphae with Trypan blue or other chemicals [238, 239]. Detailed observations under dissecting and light microscopes of intraradical fungal structures are useful to estimate the percentage of mycorrhizal infection [240] and to check for the presence of fungal pathogens and eelworms. Such sanitary test is absolutely necessary before any further utilization of the entire content of the pot culture, termed "crude inoculum" and consisting of mycorrhizal roots, growing substrate infested with external mycelium, spores and/or sporocarps. If no contamination is present, 10-20 g of crude inoculum can be used for another cycle, so that a large quantity of inoculum may be produced in about six months. Other techniques have been developed to produce large quantities of soil-free inoculum, based on hydroponic and aeroponic cultivation systems [241].

Many attempts were performed to obtain pure dual cultures and, after the development of aseptic two-membered culture systems between AM fungi and whole plants [242, 243], techniques utilizing root organ cultures were proposed utilising red clover and tomato roots infected with *G. mosseae* and *G. margarita*, respectively [106, 244]. Nevertheless, since the two organisms showed different nutritional requirements, only the use of excised root transformed by *Agrobacterium rhizogenes* proved successful [245, 246]. The most common host plant used in this system is carrot, but the method works well with several hosts, for example tomato or lucerne, provided they can be transformed [247, 248, 249]. However, transformed mycorrhizal roots are generally used as experimental model systems for investigations on host/symbionts interactions, on the biology and physiology of AM fungi, on the effects of xenobiotic compounds [208, 248-251].

All the methods for producing AM inoculum deviced so far are highly expensive and the implementation of mycorrhizal inoculation technology in agriculture is strictly correlated to the high value of the products. One of the plant production systems which could benefit from AM biotechnology is plant micropropagation, since the large perturbations routinely applied to the microflora make its restoration, through AM fungi, an appropriate and necessary tool. Several works showed the feasibility and importance of AM inoculation in a large number of horticulture, fruit and ornamental micropropagated plants [252]. Since micropropagated plants are greatly affected by transplant stress during the transfer from *in vitro* to *in vivo* systems, the phase best suited for AM inoculation was proved to be the weaning stage [253, 254]. Moreover, the selection of suitable endophytes was demonstrated to play a fundamental role in reducing weaning time and increasing growth after transplant [218, 255, 256].

5. IMPACT OF AGRICULTURAL PRACTICES

When natural ecosystems are converted to agro-ecosystems, cultural practices such as tillage, crop rotation, fertilizer and pesticide treatments exert strong selective pressures on AM fungal communities, so that only the most tolerant strains and species are able to survive. Qualitative and quantitative changes in composition of AM fungal population in agricultural soil were found to be affected by anthropogenic disturbance and cultural stresses in arable fields [257-260]. Accordingly, reduction in biodiversity and in population densities of different AM fungal species was evidenced in cultivated sites compared to natural ecosystems [214, 261, 262]. Many reports demonstrated the predominance of spore types of the genus *Glomus*, both in number of species and population densities, in agricultural soils [214, 263, 264, 265]. By contrast, the genera *Scutellospora*, *Gigaspora* and *Acaulospora* generally occur more abundantly in natural and low-nutrient sandy soils [40, 266, 267]. This evidence may be directly correlated either with the capacity of abundant sporulation of most *Glomus* species, or with their ability to form wide "mycorrhizal webs" which may represent an important source of soil inoculum, contributing to enhancing soil infectivity [268, 269]. It is important to note that glomalean spore types found in many surveys of natural and agricultural soils do not probably represent the whole population of AM fungi, since some endophytes colonizing abundantly the roots of host plants may produce none or very low numbers of spores [270]. Nevertheless, the fact that spores of certain species and/or genera disappear depending on agronomic regimes suggests that spore quantity and quality may represent important parameters to be taken into account in any programme aimed at selecting suitable strains for agricultural application, since they reflect the differential capacity of sporulation of AM fungal species and their ability to tolerate different agronomic regimes, to survive and to produce new fungal propagules. Knowledge of the factors influencing the dynamics and diversity of the AM fungal communities, such as chemical inputs, continuous monoculture or soil management, is essential in any study aimed at identifying the soundest farming practices for preventing ecosystem degradation and in any attempt to use them in sustainable agriculture [271, 272]. The adaptation or reassessment of some agronomic practices such as the extensive use of pesticides and fertilizers, should also be pursued, for full exploitation of mycorrhizal potential.

6. CONSERVATION OF GENETIC RESOURCES

Resting spores of AM fungi can be easily retrieved from the soil, germinated and inoculated on host plant roots. After establishing the symbiosis they are able to produce thousands of new spores within a few weeks. Thus, despite the inability to be grown and reproduced on synthetic media in axenic culture, AM fungi can be routinely propagated on living host plants, in growth chambers or in greenhouses. In this way, *in vivo* cultures of isolates and species of AM fungi from temperate, tropical, semi-arid, polar and Mediterranean regions are presently propagated in *ex-situ* collections in many laboratories all over the world. Such germplasm collections usually maintain fungal isolates, originated from the most diverse environments, in growth conditions and soil-free substrates or soil mixtures, which may differ from those they were isolated from. This represents a strong selective pressure, which may lead to the conservation of the only strains able to survive and reproduce in the standardised culture conditions of the collections. Moreover many isolates are presently grown starting from single spores, for preserving the genetic integrity of single individuals, and also this practice may contribute to decrease the overall genomic diversity of isolates, since spores of the same culture and isolate show large molecular diversity [273-278]. Morphological, biochemical and molecular analyses of "pedigree" single spore cultures, whose relationships with each other are known, could reveal whether and how genetic diversity decreases from one generation of spores to the other [279].

In vitro collections of some species of AM fungi were established using transformed roots by *Agrobacterium rhizogenes* and a limited number of fungal cultures are now maintained, belonging to *Gigaspora* and *Glomus*. The monoxenic cultivation system was proposed as a tool for long-term maintenance of high quality cultures in germplasm collections [280], although differences in growth and sporulation ability may be found among different isolates [281]. As well as in *in vivo* collections, in this type of propagation system a selection is exerted towards genotypes best fitted to the standardized conditions of the monoxenic cultures, possibly leading not only to a reduction in biological diversity, but also to a decrease in infectivity and efficiency of the maintained endophytes [280]. Nevertheless, whatever the kind of *ex-situ* collection, concerns about the loss in genetic diversity remain. A long-term strategy for safeguarding AM fungal biodiversity is represented by *in situ* conservation, through the protection of the habitats or ecosystems where they show the highest variation, and characterised by the absence of anthropogenic disturbance, such as natural reserves [282, 162].

Conservation of genetic resources of AM fungi, both in *ex-situ* and *in-situ* collections, requires information about the origin, ecosystem, type of soil, symbiotic performance under diverse environmental conditions of the isolates maintained. Besides the large collection existing in the U. S. A. (INVAM, West Virginia), and the Stock Centres without walls operating in Europe (BEG), many other research laboratories throughout the world maintain these organisms in germplasm collections and most of them probably register taxonomical, physiological, biochemical and molecular data of their isolates. Unfortunatley most information stored in the different international databases is not available, and the establishment of linkages between them is one of the major aims to be pursued by AM researchers, for obtaining a more complete understanding, and accordingly a better biotechnological exploitation, of AM symbionts in the different world ecosystems.

7. CONCLUSIONS

AM fungi represent fundamental factors of plant productivity since they are capable, by means of extraradical hyphae, to mediate transfer of nutrients from the soil to host plants and between different plants linked by a common mycorrhizal network. Their management appears essential for their use as biofertilisers, for reducing the inputs of chemical fertilisers and pesticides, in the perspectives of sustainable agriculture and of conservation of natural resources. Since large variations in symbiotic performance have been found in the different host/fungus/soil combinations, a more systematic approach is necessary to detect and select infective and efficient strains to be used for inoculation in diverse host plants and soil conditions. Recent developments have demonstrated a large biochemical and molecular diversity between isolates of different geographical origin belonging to the same species complex, and also between different spores of the same isolate. These findings are very important and evidence our poor understanding of the range of diversity existing in AM fungi, since we do not yet know how large are the genetic and functional differences among different isolates of the same species, among individual spores of the same isolate, and probably also among individual nuclei of the same spore. Progress in knowledge has been hindered by the obligately biotrophic status of AM fungi and the absence of sexual reproduction. Recent successful application of molecular methods to the analysis of gene expression in AM fungal hyphae may open the way to the identification of molecular mechanisms regulating phosphate uptake and transport efficiency in different host/symbiont combinations. More information is required on the biological, genetic and molecular mechanisms involved in the development of the symbiosis and on the key molecular signals exchanged between plant host and fungal symbiont leading to reciprocal recognition before the establishment of a compatible interaction. While much information is available on the functioning of AM symbiosis, less is known of the mechanisms underlying fungal behaviour, its biology and genetics. For example, we do not yet know what are the signal molecules triggering the expression of fungal genes regulating the sequence of developmental switches involved in hyphal committment to the symbiotic status. These studies deserve further research and efforts because a sound scientific knowledge is necessary for the improvement of AM biotechnology aimed at selecting highly infective and efficient inoculants and predicting their success as biofertilizers in sustainable plant production systems.

8. REFERENCES

1. Smith, S. E. and Read, D. J. (1997). Mycorrhizal symbiosis. Academic Press, London, pp. 1-605.
2. Molina, R., Massicotte, H. and Trappe, J. M. (1992). Specificity phenomena in mycorrhizal symbioses: community-ecological consequences and practical implications. In: Mycorrhizal Functioning. (Ed. Allen M. F.). Chapman & Hall, New York, pp. 357-423.
3. Giovannetti, M. and Lioi, L. (1990). The mycorrhizal status of *Arbutus unedo* in relation to compatible and incompatible fungi. Can. J. Bot. 68: 1239-1244.
4. Massicotte, H. B., Melville, L. H., Molina, R. and Peterson, R. L. (1993). Structure and histochemistry of mycorrhizae synthesized betweeen *Arbutus menziesii* (Ericaceae) and two basidiomycetes, *Pisolithus tinctorius* (Pisolithaceae) and *Piloderma bicolor* (Corticiacieae). Mycorrhiza 3: 1-11.
5. Leake, J. R. (1994). The biology of myco-heterotrophic 'saprophytic' plants. New Phytol. 127: 171-216.
6. Cullings, K. W., Szaro, T. M. and Bruns, T. D. (1996). Evolution of extreme specialisation with a lineage of ectomycorrhizal parasites. Nature 379: 63-66.
7. Egger, K. N. (1996). Molecular systematics of E-strain mycorrhizal fungi: *Wilcoxina* and its relationship to *Tricharina* (Pezizales). Can. J. Bot. 74: 773-779.
8. Jumpponen, A. and Trappe, J. M. (1998). Dark septate endophytes: a review of facultative biotrophic root-colonizing fungi. New Phytol. 140: 295-310.
9. Straker, C. J. (1996). Ericoid mycorrhiza: Ecological and host specificity. Mycorrhiza 6: 215-225.
10. Peterson, R. L., Uetake, Y. and Zelmer, C. (1998). Fungal symbioses with orchid protocorms. Symbiosis 25: 29-55.
11. Mosse, B. (1959). The regular germination of resting spores and some observations on the growth requirements of an *Endogone* sp. causing vesicular- arbuscular mycorrhiza. Trans. Br. mycol. Soc. 42: 273-286.
12. Logi C., Sbrana, C. and Giovannetti, M. (1998). Cellular events involved in survival of individual arbuscular mycorrhizal symbionts growing in the absence of the host. Appl. Environm. Microbiol. 64: 3473-3479.
13. Read, D. J. (1991). Mycorrhizas in ecosystems. Experientia 47: 376-391.
14. Grime, J. P., Mackey, J. M. L., Hillier, S. H. and Read, D. J. (1987). Floristic diversity in a model system using experimental microcosms. Nature 328: 420-422.
15. van der Heijden, M. G. A., Boller, T., Wiemken, A. and Sanders, I. R. (1998). Different arbuscular mycorrhizal fungal species are potential determinants of plant community structure. Ecology 79: 2082-2091.

16. Francis, R. and Read, D. J. (1984). Direct transfer of carbon between plants connected by vesicular-arbuscular mycorrhizal mycelium. Nature 307: 53-56.
17. Finlay, R. D. and Read, D. J. (1986). The structure and function of the vegetative mycelium of ectomycorrhizal plants. I. Translocation of ^{14}C-labelled carbon between plants interconnected by a common mycelium. New Phytol. 103: 143-156.
18. Read, D. J. (1997). Mycorrhizal fungi - The ties that bind. Nature 388: 517-518.
19. Simard, S. W., Perry, D. A., Jones, M. D., Myrold, D. D., Durall, D. M. and Molina, R. (1997). Net transfer of carbon between ectomycorrhizal tree species in the field. Nature 388: 579-582.
20. Robinson, D. and Fitter, A. (1999). The magnitude and control of carbon transfer between plants linked by a common mycorrhizal network. J Exp. Bot. 50: 9-13.
21. Pirozynski, K. A. and Malloch, D. W. (1975). The origin of land plants: a matter of mycotrophism. BioSystems 6: 153-164.
22. Stubblefield, S. P., Taylor, T. N. and Trappe, J. M. (1987). Fossil mycorrizae: a case for symbiosis. Science 237: 59-60.
23. Pirozynski, K. A. and Dalpé, Y. (1989). Geological history of the Glomaceae, with particular reference to mycorrhizal symbiosis. Symbiosis 7: 1-36.
24. Phipps, C. J. and Taylor, T. N. (1996). Mixed arbuscular mycorrhizae from the Triassic of Antarctica. Mycologia 88: 707-714.
25. Remy, W., Taylor, T. N., Hass, H. and Kerp, H. (1994). Four hundred-million-year-old vesicular arbuscular mycorrhizae. Proc. Natl. Acad. Sci. USA 91: 11841-11843.
26. Redecker, D., Kodner, R. and Graham, L. E. (2000). Glomalean fungi from the Ordovician. Science 289: 1920-1921.
27. Simon, L., Bousquet, J., Levesque, R. C. and Lalonde, M. (1993). Origin and diversification of endomycorrhizal fungi and coincidence with vascular plants. Nature 363: 67-69.
28. Berbee, M. L. and Taylor, J. W. (1993). Dating the evolutionary radiations of the true fungi. Can. J. Bot. 71: 1114-1127.
29. Christie, P. and Nicolson, T. N. (1983). Are mycorrhizas absent from the Antarctic? Trans. Br. mycol. Soc. 80: 557-561.
30. Laursen, G. A., Treu, R., Seppelt, R. D. and Stephenson, S. L. (1997). Mycorrhizal assessment of vascular plants from subantarctic Macquarie Island. Arctic. Alp. Res. 29: 483-491.
31. Dalpé, Y. and Aiken, S. G. (1998). Arbuscular mycorrhizal fungi associated with *Festuca* species in the Canadian high arctic. Can. J. Bot. 76: 1930-1938.
32. Strullu, D. G., Frenot, Y., Mauric, D., Gloaguen, J. C. and Plenchett,e C. (1999). First study of mycorrhizae in the Kerguelen islands. Comptes Rendus Acad. Sci. Ser. III-Sci. Vie-Life Sci. 322: 771-777.
33. Brundrett, M. C., Ashwath, N. and Jasper, D. A. (1996). Mycorrhizas in the Kakadu region of tropical Australia .1. Propagules of mycorrhizal fungi and soil properties in natural habitats. Plant Soil 184: 159-171.
34. Moutoglis, P. and Widden, P. (1996). Vesicular-arbuscular mycorrhizal spore populations in sugar maple (*Acer saccharum* Marsh.) forests. Mycorrhiza 6: 91-97.
35. Moyersoen, B., Fitter, A. H. and Alexander, I. J. (1998). Spatial distribution of ectomycorrhizas and arbuscular mycorrhizas in Korup National Park rain forest, Cameroon, in relation to edaphic parameters. New Phytol. 139: 311-320.
36. Dhillion,S. S., Vidiella, P. E., Aquilera, L. E. (1995). Mycorrhizal plants and fungi in the fog-free pacific coastal desert of Chile. Mycorrhiza 5: 381-386.
37. Stutz, J. C. and Morton, J. B. (1996). Successive pot cultures reveal high species richness of arbuscular endomycorrhizal fungi in arid ecosystems. Can. J. Bot. 74: 1883-1889.
38. Jacobson, K. M. (1997). Moisture and substrate stability determine VA-mycorrhizal fungal community distribution and structure in an arid grassland. J. Arid Environ. 35: 59-75.
39. Koske, R. E. and Halvorson, W. L. (1981). Ecological studies of vesicular-arbuscular mycorrhizae in a barrier sand dune. Can. J. Bot. 59: 1413-1422.
40. Giovannetti, M. and Nicolson, T. H. (1983). Vesicular-arbuscular mycorrhizas in Italian sand dunes. Trans. Br. mycol. Soc. 80: 552-557.
41. Gemma, J. N. and Koske, R. E. (1997). Arbuscular mycorrhizae in sand dune plants of the north Atlantic Coast of the US: Field and greenhouse inoculation and presence of mycorrhizae in planting stock. J. Environ. Manage. 50: 251-264.
42. Read, D. J. and Haselwandter, K. (1981). Observations on the mycorrhizal status of some alpine plant communities. New Phytol. 88: 341-352.
43. Allen, E. B., Chambers, J. C., Conner K. F., Allen M. F. and Brown R. W. (1987). Natural reestablishment of mycorrhizae in disturbed alpine ecosystems. Arctic. Alp. Res. 19: 11-20.

44. Gardes, M. and Dahlberg, A. (1996). Mycorrhizal diversity in arctic and alpine tundra: An open question. New Phytol. 133: 147-157.
45. Newman, E. I. and Reddell, P. (1987). The distribution of mycorrhizas among families of vascular plants. New Phytol. 106: 745-751.
46. Trappe, J. M. (1987). Phylogenetic and ecological aspects of mycotrophy in the Angiosperms from an evolutionary standpoint. In: Ecophysiology of VA Mycorrhizal Plants. (Ed. Safir G. R.). CRC Press, Boca Raton, pp. 5-25.
47. Giovannetti, M. (2001). Survival strategies in arbuscular mycorrhizal symbionts. In: Symbiosis. (Ed. Seckbach J.). Kluwer Academic Publishers, Dordrecht
48. Giovannetti, M. (2000). Spore germination and pre-symbiotic mycelial growth. In: Arbuscular Mycorrhizae: Physiology and Function. (Eds. Kapulnik Y. and Douds D. D.). Kluwer Academic Publishers, Dordrecht, pp. 47-68
49. Kinden, D. A. and Brown, M. F. (1975). Electron microscopy of vesicular-arbuscular mycorrhizae of yellow poplar. III. Host endophyte interactions during arbuscular development. Can. J. Bot. 21: 1930-1939.
50. Kinden, D. A. and Brown, M. F. (1976). Electron microscopy of vesicular-arbuscular mycorrhizae of yellow poplar. IV. Host endophyte interactions during arbuscular deterioration. Can. J. Bot. 22: 64-75.
51. Cox, T. S. and Tinker, P. B. (1976). Traslocation and transfer of nutrients in vesicular-arbuscular mycorrhizas. I. The arbuscule and phosphorus transfer: a quantitative ultrastructural study. New Phytol. 77: 371-378.
52. Giovannetti, M. and Citernesi, A. S. (1993). Time-course of appressorium formation on host plants by arbuscular mycorrhizal fungi. Mycol. Res. 97: 1140-1142.
53. Smith, S. E. and Gianinazzi-Pearson, V. (1988). Physiological interactions between symbionts in vesicular- arbuscular mycorrhizal plants. Annu. Rev. Plant Physiol. Plant Mol. Biol. 39: 221-244.
54. Smith, F. A. and Smith, S. E. (1997). Structural diversity in (vesicular)-arbuscular mucorrhizal symbioses. New Phytol. 137: 373-388.
55. Gerdemann, J. W. (1965). Vesicular-arbuscular mycorrhizae formed on maize and tulip tree by *Endogone fasciculata*. Mycologia 57: 562-575.
56. Jacquelinet-Jeanmougin, S. and Gianinazzi-Pearson, V. (1983). Endomycorrhizas in the Gentianaceae. I. The fungi associated with *Gentiana lutea* L. New Phytol. 95: 663-666.
57. Brundrett, M. C. and Kendrick, B. (1990). The roots and mycorrhizas of herbaceous woodland plants. II. Structural aspects of morphology. New Phytol. 114: 469-479.
58. Bedini, S., Maremmani, A. and Giovannetti, M. (2000). Paris-type mycorrhizas in Smilax aspera L. growing in a Mediterranean sclerophyllous wood. Mycorrhiza 10: 9-13.
59. Martins, M. A. (1993). The role of the external mycelium of arbuscular mycorrhizal fungi in the carbon transfer process between plants. Mycol. Res. 97: 807-810.
60. Nicolson, T. H. (1959). Mycorrhiza in the Gramineae: I. Vesicular-arbuscular endophytes, with special reference to the external phase. Trans. Br. mycol. Soc. 42: 421-438.
61. Mosse, B. (1959). Observations on the extra-matrical mycelium of a vesicular-arbuscular endophyte. Trans. Br. mycol. Soc. 42: 439-448.
62. Bago, B., Azcon-Aguilar, C., Goulet, A. and Piche, Y. (1998). Branched absorbing structures (BAS): a feature of the extraradical mycelium of symbiotic arbuscular mycorrhizal fungi. New Phytol. 139: 375-388.
63. Harley, J. L. and Smith, S. E. (1983). Mycorrhizal symbiosis. Academic Press, London, pp. 1-483.
64. Abbott, L. K. and Robson, A. D. (1984). The effect of VA mycorrhizae on plant growth. In: VA mycorrhiza. (Eds. Powell C. L. and Bagyaraj D. J.). CRC Press, Boca Raton, pp. 113-130.
65. Sanders, F. E. and Tinker, P. B. (1971). Mechanism of absorption of phosphate from soil by *Endogone* mycorrhizas. Nature 233: 278-279.
66. Sanders, F. E. and Tinker, P. B. (1973). Phosphate flow into mycorrhizal roots. Pesticide Science 4: 385-395.
67. Rhodes, L. H. and Gerdemann, J. W. (1975). Phosphate uptake zones of mycorrhizal and non-mycorrhizal onions. New Phytol. 75: 555-561.
68. Cooper, K. M. and Tinker, P. B. (1978). Translocation and transfer of nutrients in vesicular-arbuscular mycorrhizas. II. Uptake and translocation of phosphorus, zinc and sulphur. New Phytol. 81: 43-52.
69. Smith, S. E. (1982). Inflow of phosphate into mycorrhizal and non-mycorrhizal *Trifolium subterraneum* at different levels of soil phosphate. New Phytol. 90: 293-303.
70. Mosse, B., Hayman, D. S. and Arnold, D. J. (1973). Plant growth responses to vesicular-arbuscular mycorrhiza. V. Phosphate uptake by three plants species from P-deficient soils labelled with ^{32}P. New Phytol. 72: 809-815.

71. Powell, C. L. (1975). Plant growth responses to vesicular-arbuscular mycorrhiza. VIII. Uptake of P by onion and clover infected with different *Endogone* spore types in ^{32}P labelled soil. New Phytol. 75: 563-566.
72. Harrison, M. J. and van Buuren, M. L. (1995). A phosphate transporter from the mycorrhizal fungus *Glomus versiforme* . Nature 378: 626-629.
73. Smith, F. A., Jakobsen, I. and Smith, S. E. (2000). Spatial differences in acquisition of soil phosphate between two arbuscular mycorrhizal fungi in symbiosis with Medicago truncatula. New Phytol. 147: 357-366.
74. Ling-Lee, M., Chilvers, G. A. and Ashford, A. E. (1975). Polyphosphate granules in three different kinds of tree mycorrhiza. New Phytol. 75: 551-554.
75. Callow, J. A., Capaccio, L. C. M., Parish, G. and Tinker, P. B. (1978). Detection and estimation of polyphosphate in vesicular-arbuscular mycorrhizas. New Phytol. 80: 125-134.
76. White, J. A. and Brown, M. R. (1979). Ultrastructure and X-ray analysis of phosphorus granules in a vesicular-arbuscular mycorrhizal fungus. Can. J. Bot. 57: 2812-2818.
77. Cox, G., Moran, K. J., Sanders, F. E., Nockolds, C. and Tinker, P. B. (1980). Translocation and transfer of nutrients in vesicular-arbuscular mycorrhizas. III. Polyphosphate granules and phosphorus translocation. New Phytol. 84: 649-659.
78. Strullu, D. G., Gourret, J. P., Garrec, J. P. and Fourcy, A. (1981). Ultrastructure and electron-probe microanalysis of the metachromatic vacuolar granules occurring in *Taxus* mycorrhizas. New Phytol. 87: 537
79. Cox, G. and Sanders, F. E. (1974). Ultrastructure of the host-fungus interface in a vesicular-arbuscular mycorrhiza. New Phytol. 73: 901-912.
80. Gianinazzi-Pearson, V. and Gianinazzi, S. (1978). Enzymatic studies on the metabolism of vesicular-arbuscular mycorrhiza. II. Soluble alkaline phosphatase specific to mycorrhizal infection in onion roots. Physiol. Plant Pathol. 12: 45-53.
81. Gianinazzi, S., Gianinazzi-Pearson, V. and Dexheimer, J. (1979). Enzymatic studies on the metabolism of vesicular-arbuscular mycorrhiza. III. Ultrastructural localisation of acid and alkaline phosphatase in onion roots infected with *Glomus mosseae* (Nicol. & Gerd.). New Phytol. 82: 127-132.
82. Capaccio, L. C. M. and Callow, J. A. (1982). The enzymes of polyphosphate metabolism in vesicular-arbuscular mycorrhizas. New Phytol. 91: 81-91.
83. Marx, C., Dexheimer, J., Gianinazzi-Pearson, V. and Gianinazzi, S. (1982). Enzymatic studies on the metabolism of vesicular-arbuscular mycorrhiza. IV. Ultracytoenzymological evidence (ATPase) for active transfer process in the host-arbuscular interface. New Phytol. 90: 37-43.
84. Tisserant, B., Gianinazzi-Pearson, V., Gianinazzi, S. and Gollotte A. (1993). *In planta* histochemical staining of fungal alkaline phosphatase activity for analysis of efficient arbuscular mycorrhizal infections. Mycol. Res. 97: 245-250.
85. Rhodes, L. H. and Gerdemann, J. W. (1978). Hyphal translocation and uptake of sulphur by vesicular-arbuscular mycorrhizae. Soil Biol. Biochem. 10: 355-360.
86. Rhodes, L. H. and Gerdemann, J. W. (1978). Translocation of calcium and phosphate by external hyphae of vesicular-arbuscular mycorrhizae. Soil Sci. 126: 125-126.
87. Smith, S. E., Smith, F. A. and Nicholas, D. J. D. (1981). Effects of endomycorrhizal infection on phosphate and cation uptake by *Trifolium subterraneum*. Plant Soil 63: 57-64.
88. Johansen, A., Jakobsen, I. and Jensen, E. S. (1993). Hyphal transport by a vesicular-arbuscular mycorrhizal fungus of N applied to the soil as ammonium or nitrate. Biol. Fert. Soils 16: 66-70.
89. Mader, P., Edenhofer, S., Boller, T., Wiemken, A. and Niggli, U. (2000). Arbuscular mycorrhizae in a long-term field trial comparing low-input (organic, biological) and high-input (conventional) farming systems in a crop rotation. Biol. Fert. Soils 31: 150-156.
90. Chiariello, N., Hickman, J. C. and Mooney, H. A. (1982). Endomycorrhizal role for interspecific transfer of phosphorus in a community of annual plants *Plantago erecta*. Science 217: 941-943.
91. Francis, R., Finlay, R. D. and Read, D. J. (1986). Vesicular-arbuscular mycorrhiza in natural vegetation systems. IV.Transfer of nutrients in inter- and intra-specific combinations of host plants. New Phytol. 102: 103-111.
92. Menge, J. A., Johnson, E. L. V. and Platt, R. G. (1978). Mycorrhizal dependency of several citrus cultivars under three nutrient regimes. New Phytol. 81: 553-559.
93. Sparling, G. P. and Tinker, P. B. (1978). Mycorrhizal infection in Pennine grassland. II. Effects of mycorrhizal infection on the growth of some upland grasses on g-irradiated soils. J. Appl. Ecol. 15: 951-958.
94. Yost, R. S. and Fox ,R. L. (1979). Contribution of mycorrhizae to P nutrition of crops growing on an oxisol. Agron. J. 71: 903-908.

95. Baylis, J. T. S. (1975). The magnolioid mycorrhiza and mycotrophy in root systems derived from it. In: Endomycorrhizas. (Eds. Sanders F. E., Mosse B. and Tinker P. B.). Academic Press, London, pp. 373-389.
96. Nye, P. and Tinker, P. B. (1977). Solute movement in the soil-root system. Blackwell Scientific, Oxford, pp. 1-322.
97. Koide, R. T., Goff, M. D. and Dickie, I. A. (2000). Component growth efficiencies of mycorrhizal and nonmycorrhizal plants. New Phytol. 148: 163-168.
98. Smith, F. A. (2000). Measuring the influence of mycorrhizas. New Phytol. 148: 4-6.
99. Ho, I. and Trappe, J. M. (1973). Translocation of ^{14}C from *Festuca* plants to their endomycorrhizal fungi. Nature 244: 30-31.
100. Bevege, D. I., Bowen, G. D. and Skinner, M. F. (1975). Comparative carbohydrate physiology of ecto and endomycorrhizas. In: Endomycorrhizas. (Eds. Sanders F. E., Mosse B. and Tinker P. B.). Academic Press, London, pp. 49-174.
101. Snellgrove, R. C., Splittstoesser, W. E., Stribley, D. P. and Tinker, P. B. (1982). The distribution of carbon and the demand of the fungal symbiont. New Phytol. 92: 75-87.
102. Amijee, F., Stribley, D. P. and Tinker, P. B. (1993). The development of endomycorrhizal root systems. VIII. Effects of soil phosphorus and fungal colonization on the concentration of soluble carbohydrates in roots. New Phytol. 123: 297-306.
103. Bécard, G., Doner, L. W., Rolin, D. B., Douds, D. D. and Pfeffer, P. E. (1991). Identification and quantification of trehalose in vesicular-arbuscular mycorrhizal fungi by in vivo ^{13}C NMR and HPLC analyses. New Phytol. 118: 547-552.
104. Schubert, A., Wyss, P. and Wiemken, A. (1992). Occurrence of trehalose in vesicular-arbuscular mycorrhizal fungi and in mycorrhizal roots. J. Plant Physiol. 140: 41-45.
105. Shachar-Hill, Y., Pfeffer, P. E., Douds, D., Osman, S. F., Doner, L. W. and Ratcliffe, R. G. (1995). Partitioning of intermediary carbon metabolism in vesicular- arbuscular mycorrhizal leek. Plant Physiol. 108: 7-15.
106. Mosse, B. and Hepper, C. M. (1975). Vesicular-arbuscular mycorrhizal infections in root organ cultures. Physiol. Plant Pathol. 5: 215-223.
107. Gianinazzi-Pearson, V., Smith, S. E., Gianinazzi, S. and Smith ,F. A. (1991). Enzymatic studies on the metabolism of vesicular-arbuscular mycorrhizas. V. Is H^+-ATPase a component of ATP-hydrolysing enzyme activities in plant-fungus interfaces? New Phytol. 117: 61-74.
108. Harrison, M. J. (1996). A sugar transporter from *Medicago truncatula*: altered expression pattern in roots during vesicular-arbuscular (VA) mycorrhizal associations. Plant J. 9: 491-503.
109. Koch, K. E. and Johnson, C. R. (1984). Photosynthate partitioning in split-root citrus seedlings with mycorrhizal and nonmycorrhizal root systems. Plant Physiol. 75: 26-30.
110. Douds, D. D., Johnson, C. R. and Koch, K. E. (1988). Carbon cost of the fungal symbiont relative to net leaf P accumulation in a split-root VA mycorrhizal symbiosis. Plant Physiol. 86: 491-496.
111. Jakobsen, I. and Rosendahl, S. (1990). Carbon flow into soil and external hyphae from roots of mycorrhizal cucumber plants. New Phytol. 115: 77-83.
112. Sutton, J. C. and Sheppard, B. R. (1976). Aggregation of sand dune soil by endomycorrhizal fungi. Can. J. Bot. 54: 326-333.
113. Tisdall , J. M. and Oades, J. M. (1979). Stabilization of soil aggregates by the roots systems of ryegrass. Aust. J. Soil Res. 17: 429-441.
114. Jastrow, J. D., Miller, R. M. and Lussenhop, J. (1998). Contributions of interacting biological mechanisms to soil aggregate stabilization in restored prairie. Soil Biol. Biochem. 30: 905-916.
115. Wright, S. F., FrankeSnyder, M., Morton, J. B. and Upadhyaya, A. (1996). Time-course study and partial characterization of a protein on hyphae of arbuscular mycorrhizal fungi during active colonization of roots. Plant Soil 181: 193-203.
116. Wright, S. F. and Upadhyaya, A. (1998). A survey of soils for aggregate stability and glomalin, a glycoprotein produced by hyphae of arbuscular mycorrhizal fungi. Plant Soil 198: 97-107.
117. Rillig, M. C., Wright, S. F., Allen, M. F. and Field, C. B. (1999). Rise in carbon dioxide changes in soil structure. Nature 400: 628
118. Safir, G. R., Boyer, J. S. and Gerdemann, J. W. (1972). Nutrient status and mycorrhizal enhancement of water transport in soybean. Plant Physiol. 49: 700-703.
119. Levy, Y. and Krikun, J. (1980). Effect of VA mycorrhizae on *Citrus jambhiri* water relations. New Phytol. 85: 25-31.
120. Allen, M. F., Moore, T. S. and Christensen, M. (1982). Phytormone changes in Bouteloua gracilis. II. Altered levels of gibberellin-like substances and abscisic acid in the host plant. Can. J. Bot. 60: 468-471.
121. Faber, B. A., Zasoski, R. J., Munns, D. N. and Shakel, K. (1991). A method for measuring hyphal nutrient and water uptake in mycorrhizal plants. Can. J. Bot. 69: 87-94.

122. Baltruschat, H. and Schönbeck, F. (1975). Untersuchungen über den Einfluss der endotrophen Mycorrhiza auf den Befall von Tabak mit *Thielaviosis basicola*. Phytopath. Z. 84: 172-188.
123. Dehne, H. W. and Schönbeck, F. (1975). Untersuchungen über den Einfluss der endotrophen Mycorrhiza auf die *Fusarium*-Welke der Tomate. Z. Pflanzenk. Pflanzens. 82: 630-632.
124. Ross, J. P. (1972). Influence of *Endogone* mycorrhiza on *Phytophthora* rot of soybean. Phytopathology 62: 896-897.
125. Davis, R. M. and Menge, J. A. (1980). Influence of *Glomus fasciculatus* and soil phosphorus on *Phytophthora* root rot of citrus. Phytopathology 70: 447-452.
126. Zambolin, L. and Schenck, N. C. (1983). Reduction of the effects of pathogenic, root-infecting fungi on soybean by the mycorrhizal fungus, *Glomus mosseae*. Phytopathology 73: 1402-1405.
127. Graham, J. H. and Egel, D. S. (1988). *Phytophthora* root rot development on mycorrhizal and phosphorus- fertilized nonmycorrhizal sweet orange seedlings. Plant dis. 72: 611-614.
128. Caron, M., Fortin, J. A. and Richard, C. (1986). Effect of phosphorus concentration and *Glomus intraradices* on *Fusarium* crown and root rot of tomatoes. Phytopathology 76: 942-946.
129. Cordier, C., Pozo, M. J., Barea, J. M., Gianinazzi, S. and Gianinazzi-Pearson, V. (1998). Cell defense responses associated with localized and systemic resistance to *Phytophthora parasitica* induced in tomato by an arbuscular mycorrhizal fungus. Mol. Plant-Microbe Interact. 11: 1017-1028.
130. Pfleger, F. L. and Linderman, R. G. (1994). Mycorrhizae and Plant Health. APS, St.Paul, pp. 1-344.
131. Gerdemann J. W. and Trappe J. M. (1974). Endogonaceae in the Pacific Northwest. Mycologia Mem. 5: 1-76.
132. Morton, J. B. (1988). Taxonomy of VA mycorrhizal fungi: classification, nomenclature, and identification. Mycotaxon 32: 267-324.
133. Morton, J. B. and Benny, G. L. (1990). Revised classification of arbuscular mycorrhizal fungi (Zygomycetes): a new order, Glomales, two new suborders, Glomineae and Gigasporinae, and two new families, Acaulosporaceae and Gigasporaceae, with an emendation of Glomaceae. Mycotaxon 37: 471-491.
134. Gehrig, H., Schüssler, A. and Kluge, M. (1996). *Geosiphon pyriforme*, a fungus forming endocytobiosis with *Nostoc* (Cyanobacteria) is an ancestral member of the Glomales: evidence by SSU rRNA analysis. J. Mol. Evol. 43: 71-81.
135. Redecker, D., Morton, J. B. and Bruns, T. D. (2000). Ancestral lineages of arbuscular mycorrhizal fungi (Glomales). Mol. Phylogenet. Evol. 14: 276-284.
136. Bago, B., Bentivenga, S. P., Brenac, V., Dodd, J. C., Piche, Y. and Simon, L. (1998). Molecular analysis of *Gigaspora* (Glomales, Gigasporaceae). New Phytol. 139: 581-588.
137. Almeida, R. T. and Schenck, N. C. (1990). A revision of the genus *Sclerocystis* (Glomaceae, Glomales). Mycologia 82: 703-714.
138. Redecker, D., Morton, J. B. and Bruns, T. D. (2000). Molecular phylogeny of the arbuscular mycorrhizal fungi Glomus sinuosum and Sclerocystis coremioides. Mycologia 92: 282-285.
139. Sen, R. and Hepper, C. M. (1986). Characterization of vesicular-arbuscular mycorrhizal fungi (*Glomus* spp.) by selective enzyme staining following polyacrylamide gel electrophoresis. Soil Biol. Biochem. 18: 29-34.
140. Hepper, C. M., Sen, R., Azcon-Aguilar, C. and Grac, C. (1988). Variation in certain isozymes amongst different geographical isolates of the vesicular-arbuscular mycorrhizal fungi *Glomus clarum*, *Glomus monosporum* and *Glomus mosseae*. Soil Biol. Biochem. 20: 51-59.
141. Dodd, J. C., Rosendahl, S., Giovannetti, M., Broome, A., Lanfranco, L. and Walker, C. (1996). Inter- and intraspecific variation within the morphologically similar arbuscular mycorrhizal fungi *Glomus mosseae* and *Glomus coronatum*. New Phytol. 133: 113-122.
142. Hepper, C. M., Sen, R. and Maskall, C. S. (1986). Identification of vesicular-arbuscular mycorrhizal fungi in roots of leek (*Allium porrum* L.) and maize (*Zea mays* L.) on the basis of enzyme mobility during polyacrylamide gel electrophoresis. New Phytol. 102: 529-539.
143. Tisserant, B., Brenac, V., Requena, N., Jeffries, P. and Dodd, J. C. (1998). The detection of *Glomus* spp. (arbuscular mycorrhizal fungi) forming mycorrhizas in three plants, at different stages of seedling development, using mycorrhiza-specific isozymes. New Phytol. 138: 225-239.
144. Aldwell, F. E. B., Hall, I. R. and Smith, J. M. B. (1983). Enzyme-linked immunosorbent assay (ELISA) to identify endomycorrhizal fungi. Soil Biol. Biochem. 15: 377-378.
145. Aldwell, F. E. B., Hall, I. R. and Smith, J. M. B. (1985). Enzyme-linked immunosorbent assay as an aid to taxonomy of the Endogonaceae. Trans. Br. mycol. Soc. 84: 399-402.
146. Wright, S. F., Morton, J. B. and Sworobuk, J. E. (1987). Identification of a vesicular-arbuscular mycorrhizal fungus by using monoclonal antibodies in an enzyme-linked immunosorbent assay. Appl. Environm. Microbiol. 53: 2222-2225.

147. Hahn, A., Bonfant,e P., Horn, K., Pausch, F. and Hock, B. (1993). Production of monoclonal antibodies against surface antigens of spores from arbuscular mycorrhizal fungi by an improved immunization and screening procedure. Mycorrhiza 4: 69-78.
148. Thingstrup, I., Rozycka, M., Jeffries, P., Rosendahl, S. and Dodd, J. C. (1995). Detection of the arbuscular mycorrhizal fungus *Scutellospora heterogama* within roots using polyclonal antisera. Mycol. Res. 99: 1225-1232.
149. Avio, L. and Giovannetti, M. (1998). The protein pattern of spores of arbuscular mycorrhizal fungi: comparison of species, isolates and physiological stages. Mycol. Res. 102:985-990: 985-990.
150. Graham, J. H., Hodge, N. C. and Morton, J. B. (1995). Fatty acid methyl ester profiles for characterization of glomalean fungi and their endomycorrhizae. Appl. Environm. Microbiol. 61: 58-64.
151. Olsson, P. A., Thingstrup, I., Jakobsen, I. and Baath, F. (1999). Estimation of the biomass of arbuscular mycorrhizal fungi in a linseed field. Soil Biol. Biochem. 31: 1879-1887.
152. Simon, L., Lalonde, M. and Bruns, T. D. (1992). Specific amplification of 18s fungal ribosomal genes from vesicular-arbuscular endomycorrhizal fungi colonizing roots. Appl. Environm. Microbiol. 58: 291-295.
153. Di Bonito, R. D., Elliott, M. L. and des Jardin, E. A. (1995). Detection of an arbuscular mycorrhizal fungus in roots of different plant species with the PCR. Appl. Environm. Microbiol. 61: 2809-2810.
154. Claassen, V. P., Zasoski, R. J. and Tyler, B. M. (1996). A method for direct soil extraction and PCR amplification of endomycorrhizal fungal DNA. Mycorrhiza 6: 447-450.
155. Millner, P. D., Mulbry, W. W., Reynolds, S. L. and Patterson, C. A. (1998). A taxon-specific oligonucleotide probe for temperate zone soil isolates of *Glomus mosseae*. Mycorrhiza 8: 19-27.
156. Van Tuinen, D., Jacquot, E., Zhao, B., Gollotte, A. and Gianinazzi-Pearson, V. (1998). Characterization of root colonization profiles by a microcosm community of arbuscular mycorrhizal fungi using 25S rDNA-targeted nested PCR. Mol. Ecol. 7: 879-887.
157. Lanfranco, L., Wyss P., Marzachí, C. and Bonfante, P. (1995). Generation of RAPD-PCR primers for the identification of isolates of *Glomus mosseae*, an arbuscular mycorrhizal fungus. Mol. Ecol. 4: 61-68.
158. Abbas, J. D., Hetrick, B. A. D. and Jurgenson, J. E. (1996). Isolate specific detection of mycorrhizal fungi using genome specific primer pairs. Mycologia 88: 939-946.
159. Zézé ,A., Hosny M., Gianinazzi-Pearso,n V. and Dulieu, H. (1996). Characterization of a highly repeated DNA sequence (SC1) from the arbuscular mycorrhizal fungus *Scutellospora castanea* and its detection in planta. Appl. Environm. Microbiol. 62: 2443-2448.
160. Rosendahl, S. and Taylor, J. W. (1997). Development of multiple genetic markers for studies of genetic variation in arbuscular mycorrhizal fungi using AFLP^{TM}. Mol. Ecol. 6: 821-829.
161. Redecker, D., Thierfelder, H., Walker, C. and Werner, D. (1997). Restriction analysis of PCR-amplified internal transcribed spacers of ribosomal DNA as a tool for species identification in different genera of the order glomales. Appl. Environm. Microbiol. 63: 1756-1761.
162. Giovannetti ,M. and Gianinazzi-Pearson, V. (1994). Biodiversity in arbuscular mycorrhizal fungi. Mycol. Res. 98: 705-715.
163. Clarkson, D. T. (1985). Factors affecting mineral nutrient acquisition by plants. Annu. Rev. Plant Physiol. 36: 77-115.
164. Godfrey, R. M. (1957). Studies on British species of *Endogone*. III. Germination of spores. Trans. Br. mycol. Soc. 40: 203-210.
165. Tommerup, I. C. (1983). Spore dormancy in vesicular-arbuscular mycorrhizal fungi. Trans. Br. mycol. Soc. 81: 37-45.
166. Hepper, C. M. and Smith, G. A. (1976). Observation on the germination of *Endogone* spores. Trans. Br. mycol. Soc. 66: 189-194.
167. Daniels, B. A. and Graham ,S. O. (1976). Effects of nutrition and soil extracts on germination of *Glomus mosseae* spores. Mycologia 68: 108-116.
168. Hepper ,C. M. (1979). Germination and growth of *Glomus caledonium* spores: the effects of inhibitors and nutrients. Soil Biol. Biochem. 11: 269-277.
169. Louis, I. and Lim ,G. (1988). Effect of storage of inoculum on spore germination of a tropical isolate of *Glomus clarum*. Mycologia 80: 157-161.
170. Douds, D. D. and Schenck, N. C. (1991). Germination and hyphal growth of VAM fungi during and after storage in soil at five matric potentials. Soil Biol. Biochem. 23: 177-183.
171. Gazey, C., Abbott, L. K. and Robson, A. D. (1993). VA mycorrhizal spores from three species of *Acaulospora*: germination, longevity and hyphal growth. Mycol. Res. 97: 785-790.
172. Koske, R. E. (1981). *Gigaspora gigantea*: observations on spore germination of a VA-mycorrhizal fungus. Mycologia 73: 288-300.
173. Sward, R. J. (1981). The structure of the spores of *Gigaspora margarita*. III. Germ tube emergence and growth. New Phytol. 88: 667-673.

174. Siqueira, J. O., Hubbell D. H. and Schenck N. C. (1982). Spore germination and germ tube growth of a vesicular-arbuscular mycorrhizal fungus in vitro. Mycologia 74: 952-959.
175. Abbott, L. K. and Robson, A. D. (1977). The distribution and abundance of vesicular-arbuscular endophytes in some Western Australian soils. Aust. J. Bot. 25: 515-522.
176. Nicolson, T. H. and Schenck, N. C. (1979). Endogonaceous mycorrhizal endophytes in Florida. Mycologia 71: 178-198.
177. Morton, J. B., Bentivenga, S. P. and Wheeler, W. W. (1993). Germ plasm in the International collection of arbuscular and vesicular-arbuscular mycorrhizal fungi (INVAM) and procedures for culture development, documentation and storage. Mycotaxon 48: 491-528.
178. Hepper, C. M. (1984). Regulation of spore germination of the vesicular-arbuscular mycorrhizal fungus *Acaulospora laevis* by soil pH. Trans. Br. mycol. Soc. 83: 154-156.
179. Green, N. E., Graham, J. H. and Schenck, N. C. (1976). The influence of pH on the germination of vesicular-arbuscular mycorrhizal spores. Mycologia 68: 929-934.
180. Daniels, B. A. and Trappe, J. M. (1980). Factors affecting spore germination of the vesicular-arbuscular mycorrhizal fungus, *Glomus epigaeus*. Mycologia 72: 457-471.
181. Douds, D. D. (1997). A procedure for the establishment of Glomus mosseae in dual culture with Ri T-DNA-transformed carrot roots. Mycorrhiza 7: 57-61.
182. Tommerup, I. C. (1983). Temperature relations of spore germination and hyphal growth of vesicular-arbuscular mycorrhizal fungi in soil. Trans. Br. mycol. Soc. 81: 381-387.
183. Schenck, N. C., Graham, S. O. and Green N. E. (1975). Temperature and light effect on contamination and spore germination of vesicular-arbuscular mycorrhizal fungi. Mycologia 67: 1189-1192.
184. Hepper, C. M. (1983). Effect of phosphate on germination and growth of vesicular- arbuscular mycorrhizal fungi. Trans. Br. mycol. Soc. 80: 487-490.
185. Tawaraya, K., Saito M., Morioka, M. and Wagatsuma, T. (1996). Effect of concentration of phosphate on spore germination and hyphal growth of arbuscular mycorrhizal fungus, *Gigaspora margarita* Becker & Hall. Soil Sci. Plant Nutr. 42: 667-671.
186. Graham, J. H. (1982). Effect of citrus exudates on germination of chlamydospores of the vesicular-arbuscular mycorrhizal fungus, *Glomus epigaeum*. Mycologia 74: 831-835.
187. Bécard, G. and Piché, Y. (1990). Physiological factors determining vesicular-arbuscular mycorrhizal formation in host and non-host Ri T-DNA transformed roots. Can. J. Bot. 68: 1260-1264.
188. Gianinazzi-Pearson, V., Branzanti, B. and Gianinazzi, S. (1989). In vitro enhancement of spore germination and early hyphal growth of a vesicular-arbuscular mycorrhizal fungus by host root exudates and plant flavonoids. Symbiosis 7: 243-255.
189. Nair,M. G., Safir, G. R. and Siqueira, J. O. (1991). Isolation and identification of vesicular-arbuscular mycorrhiza- stimulatory compounds from clover (*Trifolium repens*) roots. Appl. Environm. Microbiol. 57: 434-439.
190. Tsai, S. M. and Phillips, D. A. (1991). Flavonoids released naturally from alfalfa promote development of symbiotic *Glomus* spores in vitro. Appl. Environm. Microbiol. 57: 1485-1488.
191. Giovannetti, M., Sbrana, C., Avio, L., Citernesi, A. S. and Logi, C. (1993). Differential hyphal morphogenesis in arbuscular mycorrhizal fungi during pre-infection stages. New Phytol. 125: 587-593.
192. Tawaraya, K., Watanabe, S., Yoshida, E. and Wagatsuma, T. (1996). Effect of onion (*Allium cepa*) root exudates on the hyphal growth of *Gigaspora margarita*. Mycorrhiza 6: 57-59.
193. El-Atrach, F., Vierheilig, H. and Ocampo, J. A. (1989). Influence of non-host plants on vesicular-arbuscular mycorrhizal infection of host plants and on spore germination. Soil Biol. Biochem. 21: 161-163.
194. Giovannetti, M., Sbrana, C. and Logi, C. (1994). Early processes involved in host recognition by arbuscular mycorrhizal fungi. New Phytol. 127: 703-709.
195. Giovannetti, M. and Sbrana, C. (1998). Meeting a non-host: the behaviour of AM fungi. Mycorrhiza 8: 123-130.
196. Azcòn, R. (1987). Germination and hyphal growth of *Glomus mosseae* in vitro : effects of rhizosphere bacteria and cell-free culture media. Soil Biol. Biochem. 19: 417-419.
197. Azcòn, R. (1989). Selective interaction between free-living rhizosphere bacteria and vesicular-arbuscular mycorrhizal fungus *Glomus mosseae*. Soil Biol. Biochem. 21: 639-644.
198. Azcon-Aguilar, C., Diaz-Rodriguez R. M. and Barea J. M. (1986). Effect of soil micro-organisms on spore germination and growth of the vesicular-arbuscular mycorrhizal fungus *Glomus mosseae*. Trans. Br. mycol. Soc. 86: 337-340.
199. Mugnier, J. and Mosse,B. (1987). Spore germination and viability of a vesicular arbuscular mycorrhizal fungus, *Glomus mosseae*. Trans. Br. mycol. Soc. 88: 411-413.
200. Mayo, K., Davis ,R. E. and Motta, J. (1986). Stimulation of germination of spores of *Glomus versiforme* by spore-associated bacteria. Mycologia 78: 426-431.

201. Will, M. E. and Sylvia, D. M. (1990). Interaction of rhizosphere bacteria, fertilizer, and vesicular-arbuscular mycorrhizal fungi with sea oats. Appl. Environm. Microbiol. 56: 2073-2079.

202. Calvet, C., Barea, J. M. and Pera, J. (1992). In vitro interactions between the vesicular-arbuscular mycorrhizal fungus *Glomus mosseae* and some saprophytic fungi isolated from organic substrates. Soil Biol. Biochem. 24: 775-780.

203. Ames, R. N., Mihara, K. L. and Bayne, H. G. (1989). Chitin-decomposing actinomycetes associated with a vesicular-arbuscular mycorrhizal fungus from a calcareous soil. New Phytol. 111: 67-71.

204. Walley, F. L. and Germida ,J. J. (1996). Failure to decontaminate *Glomus clarum* NT4 spores is due to spore wall-associated bacteria. Mycorrhiza 6: 43-49.

205. Filippi, C., Bagnol,i G., Citernesi, A. S. and Giovannetti, M. (1998). Ultrastructural spatial distribution of bacteria associated with sporocarps of *Glomus mosseae*. Symbiosis. 24: 1-12.

206. Maia, L. C. and Kimbrough, J. W. (1998). Ultrastructural studies of spores and hyphae of a *Glomus* species. Int. J. Plant Sci. 159: 581-589.

207. Bécard, G. and Piché, Y. (1989). New aspects on the acquisition of biotrophic status by a vesicular-arbuscular mycorrhizal fungus, *Gigaspora margarita*. New Phytol. 112: 77-83.

208. Bue,e M., Rossignol, M., Jauneau, A., Ranjeva, R. and Becard, G. (2000). The pre-symbiotic growth of arbuscular mycorrhizal fungi is induced by a branching factor partially purified from plant root exudates. Mol. Plant-Microbe Interact. 13: 693-698.

209. Beilby, J. P. and Kidby, D. K. (1980). Biochemistry of ungerminated and germinated spores of the vesicular-arbuscular mycorrhizal fungus, *Glomus caledonius*: changes in neutral and polar lipids. J. Lipid Res. 21: 739-750.

210. Bécard, G. and Piché, Y. (1989). Fungal growth stimulation by CO_2 and root exudates in vesicular-arbuscular mycorrhizal symbiosis. Appl. Environm. Microbiol. 55: 2320-2325.

211. Tommerup, I. C. (1984). Persistence of infectivity by germinated spores of vesicular-arbuscular mycorrhizal fungi in soil. Trans. Br. mycol. Soc. 82: 275-282.

212. Koske, R. E. (1981). Multiple germination by spores of *Gigaspora gigantea*. Trans. Br. mycol. Soc. 76: 328-330.

213. Giovannetti, M., Azzolini, D. and Citernesi, A. S. (1999). Anastomosis formation and nuclear and protoplasmic exchange in arbuscular mycorrhizal fungi. Appl. Environm. Microbiol. 65: 5571-5575.

214. Helgason, T., Daniell, T. J., Husband, R., Fitter, A. H. and Young, J. P. W. (1998). Ploughing up the wood-wide web? Nature 374: 431-431.

215. Lackie, S. M., Garriock, M. L., Peterson, R. L. and Bowley, S. R. (1987). Influence of host plant on the morphology of the vesicular-arbuscular fungus, *Glomus versiforme* (Daniels and Trappe) Berch. Symbiosis 3: 147-158.

216. Garriock, M. L., Peterson, R. L. and Ackerley, C. A. (1989). Early stages in colonization of *Allium porrum* (leek) roots by the vesicular-arbuscular mycorrhizal fungus, *Glomus versiforme*. New Phytol. 112: 85-92.

217. Giovannetti M., Avio L., Sbrana C. and Citernesi A. S. (1993). Factors affecting appressorium development in the vesicular-arbuscular mycorrhizal fungus *Glomus mosseae* (Nicol. & Gerd.) Gerd. & Trappe. New Phytol. 123: 114-122.

218. Fortuna, P., Citernesi, A. S., Morini, S., Giovannetti, M. and Loreti F. (1992). Infectivity and effectiveness of different species of arbuscular mycorrhizal fungi in micropropagated plants of Mr.S. 2/5 plum rootstocks. Agronomie 12: 825-829.

219. Graham, J. H., Linderman, R. G. and Menge, J. A. (1982). Development of external hyphae by different isolates of mycorrhizal *Glomus* spp. in relation to root colonisation and growth of Troyer citrange. New Phytol. 91: 183-189.

220. Kucey, R. M. N. and Pau,l E. A. (1982). Biomass of mycorrhizal fungi associated with bean roots. Soil Biol. Biochem. 14: 413-414.

221. Pearson, J. N. and Jakobsen, I. (1993). Symbiotic exchange of carbon and phosphorus between cucumber and three arbuscular mycorrhizal fungi. New Phytol. 124: 481-488.

222. Green, D. C., Vilarino, A., Newsam, R., Jeffries, P. and Dodd, J. C. (1994). Quantification of mycelial development of arbuscular mycorrhizal fungi using image analysis. Mycorrhiza 5: 105-113.

223. Abbott, L. K. and Robson, A. D. (1985). Formation of external hyphae in soil by four species of vesicular-arbuscular mycorrhizal fungi. New Phytol. 99: 245-255.

224. Jakobsen, I., Abbott, L. K. and Robson, A. D. (1992). External hyphae of vesicular-arbuscular mycorrhizal fungi associated with *Trifolium* subterraneum L. New Phytol. 120: 371-380.

225. Giovannetti, M., Fortuna, P., Citernesi, A. S., Morini, S. and Nuti, M. P. (2001). The occurrence of anastomosis formation and nuclear exchange in intact arbuscular mycorrhizal networks. New Phytol. 151: 717-724.

226. Friese, C. F. and Allen, M. F. (1991). The spread of VA mycorrhizal fungal hyphae in the soil: inoculum types and external hyphal architecture. Mycologia 83: 409-418.

227. Bago, B., Azcon-Aguilar, C. and Pich, Y. (1998). Architecture and developmental dynamics of the external mycelium of the arbuscular mycorrhizal fungus *Glomus intraradices* grown under monoxenic conditions. Mycologia 90: 52-62.
228. Schubert, A., Marzachí, C., Mazzitelli, M., Cravero M. C. and Bonfante-Fasolo P. (1987). Development of total and viable extraradical mycelium in the vesicular-arbuscular mycorrhizal fungus *Glomus clarum* Nicol.& Schenck. New Phytol. 107: 183-190.
229. Hamel, C., Fyles, H. and Smith, D. L. (1990). Measurement of development of endomycorrhizal mycelium using three vital stains. New Phytol. 115: 297-302.
230. Sylvia, D. M. (1988). Activity of external hyphae of vesicular-arbuscular mycorrhizal fungi. Soil Biol. Biochem. 20: 39-43.
231. Jones, M. D., Durall, D. M. and Tinker, P. B. (1998). Comparison of arbuscular and ectomycorrhizal *Eucalyptus coccifera*: growth response, phosphorus uptake efficiency and external hyphal production. New Phytol. 140: 125-134.
232. Boddington, C. L. and Dodd, J. C. (1999). Evidence that differences in phosphate metabolism in mycorrhizas formed by species of *Glomus* and *Gigaspora* might be related to their life-cycle strategies. New Phytol. 142: 531-538.
233. Cooper, K. M. and Tinker, P. B. (1981). Translocation and transfer of nutrients in vesicular-arbuscular mycorrhizas. IV. Effect of environmental variables on movement of phosphorous. New Phytol. 88: 327-339.
234. Pearson, V. and Tinker, P. B. (1975). Measurements of phosphorus fluxes in the external hyphae of endomycorrhizas. In: Endomycorrhizas. (Eds. Sanders F. E., Mosse B. and Tinker P. B.). Academic Press, London, pp. 277-287.
235. Pearson, J. N. and Jakobsen, I. (1993). The relative contribution of hyphae and roots to phosphorus uptake by arbuscular mycorrhizal plants, measured by dual labelling with ^{32}P2 and ^{33}P. New Phytol. 124: 489-494.
236. Giovannetti ,M., Sbrana, C. and Logi, C. (2000). Microchambers and video-enhanced light microscopy for monitoring cellular events in living hyphae of arbuscular mycorrhizal fungi. Plant·Soil 226: 153-159.
237. Gerdemann, J. W. and Nicolson, T. H. (1963). Spores of mycorrhizal *Endogone* species extracted from soil by wet sieving and decanting. Trans. Br. mycol. Soc. 46: 235-246.
238. Phillips, J. M. and Hayman, D. S. (1970). Improved procedure for clearing roots and staining parasites and vesicular-arbuscular mycorrhizal fungi for rapid assessment of infection. Trans. Br. mycol. Soc. 55: 158-161.
239. Vierheilig, H., Coughlan, A. P., Wyss, U. and Piche, Y. (1998). Ink and vinegar, a simple staining technique for arbuscular-mycorrhizal fungi. Appl. Environm. Microbiol. 64: 5004-5007.
240. Giovannetti,M. and Moss,e B. (1980). An evaluation of techniques for measuring vesicular arbuscular mycorrhizal infection in roots. New Phytol. 84: 489-500.
241. Jarstfer, A. G. and Sylvia, D. M. (1995). Aeroponic culture of VAM fungi. In: Mycorrhiza. (Eds. Varma A. and Hock B.). Springer-Verlag, Berlin, pp. 427-441.
242. Mosse, B. and Phillips, J. M. (1971). The influence of phosphate and other nutrients on the development of vesicular-arbuscular mycorrhiza in culture. J. Gen. Microbiol. 69: 157-166.
243. Macdonald, R. M. (1981). Routine production of axenic vesicular-arbuscular mycorrhizas. New Phytol. 89: 87-93.
244. Miller-Wideman, M. A. and Watrud, L. S. (1984). Sporulation of *Gigaspora margarita* on root cultures of tomato. Can. J. Microbiol. 30: 642-646.
245. Mugnier, J. and Mosse B. (1987). Vesicular-arbuscular mycorrhizal infection in transformed root-inducing T-DNA root grown axenically. Phytopathology 77: 1045-1050.
246. Bécard, G. and Fortin J. A. (1988). Early events of vesicular-arbuscular mycorriza formation on Ri T-DNA transformed roots. New Phytol. 108: 211-218.
247. Simoneau, P., Louisy-Louis N., Plenchette C. and Strullu D. G. (1994). Accumulation of new polypeptides in Ri T-DNA transformed roots of tomato (*Lycopersicon esculentum*) during the development of vesicular-arbuscular mycorrhizae. Appl. Environm. Microbiol. 60: 1810-1813.
248. Bago, B., Vierheilig H., Piche, Y. and Azcon-Aguilar, C. (1996). Nitrate depletion and pH changes induced by the extraradical mycelium of the arbuscular mycorrhizal fungus *Glomus intraradices* grown in monoxenic culture. New Phytol. 133: 273-280.
249. Douds, D. D., Galvez, L., Bécard, G. and Kapulnik Y. (1998). Regulation of arbuscular mycorrhizal development by plant host and fungus species in alfalfa. New Phytol. 138: 27-35.
250. Rhlid, R. B., Chabo,t S., Piche Y. and Chenevert R. (1993). Isolation and identification of flavonoids from Ri T-DNA- transformed roots (*Daucus carota*) and their significance in vesicular-arbuscular mycorrhiza. Phytochemistry 33: 1369-1371.

251. Wan, M. T., Rahe, J. E. and Watts, R. G. (1998). A new technique for determining the sublethal toxicity of pesticides to the vesicular-arbuscular mycorrhizal fungus *Glomus intraradices*. Environ. Toxicol. Chem. 17: 1421-1428.
252. Lovato P. E., Gianinazzi-Pearson V., Trouvelot A. and Gianinazzi S. (1996). The state of art mycorrhizas micropropagation. Adv. Hort. Sci. 10: 46-52.
253. Pons,S. F., Gianinazzi-Pearson,V., Gianinazzi,S. and Navatel,J. C. (1983). Studies of VA mycorrhizae in vitro: mycorrhizal synthesis of axenically propagated wild cherry (*Prunus avium*) plants. Plant Soil 71: 217-221.
254. Ravolanirina,F., Blal B., Gianinazzi,S. and Gianinazzi-Pearso, V. (1989). Mise au point d'une méthode rapide d'endomycorhization de vitroplant. Fruits 44 (3): 165-170.
255. Guillemin,J. P., Gianinazzi,S. and Trouvelot,A. (1992). Screening of arbuscular endomycorrhizal fungi for establishment of micropropagated pineapple plants. Agronomie 12: 831-836.
256. Azcon-Aguilar,C., Padilla,I. G., Encina,C. L., Azcòn,R. and Barea,J. M. (1996). Arbuscular mycorrhizal inoculation enhances plant growth and changes root system morphology in micropropagated *Annona cherimola* Mill. Agronomie 16: 647-652.
257. Hayman,D. S. (1975). The occurrence of mycorrhiza in crops as affected by soil fertility. In: Endomycorrhizas. (Eds. Sanders F. E., Mosse B. and Tinker P. B.). Academic Press, London, pp. 495-509.
258. Kruckelmann,H. W. (1975). Effects of fertilizers, soils, soil tillage, and plant species on the frequency of Endogone chlamydospores and mycorrhizal infection in arable soils. In: Endomycorrhizas. (Eds. Sanders F. E., Mosse B. and Tinker P. B.). Academic Press, London, pp. 511-525.
259. Johnson,N. C. and Pfleger,F. L. (1992). Vesicular-arbuscular mycorrhizae and cultural stresses. In: Mycorrhizae in sustainable agriculture. (Eds. Bethlenfalvay G. J. and Linderman R. G.). ASA, Madison, pp. 71-99.
260. Land,S., von Alten, nd Schönbeck,F. (1993). The influence of host plant, nitrogen fertilization and fungicide application on the abundance and seasonal dynamics of vesicular-arbuscular mycorrhizal fungi in arable soils of northern Germany. Mycorrhiza 2: 157-166.
261. Schenck,N. C. and Kinloch,R. A. (1980). Incidence of mycorrhizal fungi on six field crops in monoculture on newly cleared woodland site. Mycologia 72: 445-456.
262. Sieverding,E. (1991). Vesicular-arbuscular mycorrhiza management in tropical agrosystems. Deutsche Gesellschaft Technische Zusammenarbeit (GTZ) GmbH, Eschborn, pp. 1-371.
263. Hayman,D. S. and Stovold G. E. (1979). Spore population and infectivity of vesicular-arbuscular mycorrhizal fungi in New South Wales. Aust. J. Bot. 27: 227-233.
264. Jensen,A. and Jakobsen,I. (1980). The occurrence of vesicular-arbuscular mycorrhiza in barley and wheat grown in some danish soils with different fertilizer treatments. Plant Soil 55: 403-414.
265. Land,. and Schönbeck,F. (1991). Influence of different soil types on abundance and seasonal dynamics of vesicular arbuscular mycorrhizal fungi in arable soils of North Germany. Mycorrhiza 1: 39-44.
266. Koske, E. (1975). *Endogone* spores in Australian sand dunes. Can. J. Bot. 53: 668
267. Gemma, N., Koske, R. E. and Carreiro, M. (1989). Seasonal dynamics of selected species of VA-mycorrhizal fungi in a sand dune. Mycol. Res. 92: 317-321.
268. Addy, H. D., Miller, M. H. and Peterson, R. L. (1997). Infectivity of the propagules associated with extraradical mycelia of two AM fungi following winter freezing. New Phytol. 135: 745-753.
269. McGee, P. A., Pattinson, G. S., Heath, R. A., Newman, C. A. and Allen, S. J. (1997). Survival of propagules of arbuscular mycorrhizal fungi in soils in eastern Australia used to grow cotton. New Phytol. 135: 773-780.
270. Clapp, J. P., Young, J. P. W., Merryweather, J. and Fitter, A. H. (1995). Diversity of fungal symbionts in arbuscular mycorrhizas from a natural community. New Phytol. 130: 259-265.
271. Allen, M. F. (1991). The ecology of mycorrhizae. Cambridge University Press, Cambridge, pp. 1-184.
272. Bethlenfalvay,G. J. and Linderman, R. G. (1992). Mycorrhizae in sustainable agriculture. ASA, Madison, pp. 1-124.
273. Sanders, I. R., Alt M., Groppe, K., Boller, T. and Wiemken, A. (1995). Identification of ribosomal DNA polymorphisms among and within spores of the Glomales: application to studies on the genetic diversity of arbuscular mycorrhizal fungal communities. New Phytol. 130: 419-427.
274. LloydMacgilp, S. A., Chambers, S. M., Dodd J. C., Fitter, A. H., Walker, C. and Young, J. P. W. (1996). Diversity of the ribosomal internal transcribed spacers within and among isolates of *Glomus mosseae* and related mycorrhizal fungi. New Phytol. 133: 103-111.
275. Zézé, A., Sulistyowati, E., Ophel-Keller, K., Barker, S. and Smith, S. (1997). Intersporal genetic variation of *Gigaspora margarita*, a vesicular arbuscular mycorrhizal fungus, revealed by M13 minisatellite-primed PCR. Appl. Environm. Microbiol. 63: 676-678.
276. Clapp, J. P., Fitter, A. H. and Young,, J. P. W. (1999). Ribosomal small subunit sequence variation within spores of an arbuscular mycorrhizal fungus, *Scutellospora* sp. Mol. Ecol. 8: 915-921.

277. Lanfranco L., Delpero, M. and Bonfante, P. (1999). Intrasporal variability of ribosomal sequences in the endomycorrhizal fungus *Gigaspora margarita*. Mol. Ecol. 8: 37-45.

278. Antoniolli, Z. I., Schachtman, D. P., Ophel, K. K. and Smith, S. E. (2000). Variation in rDNA ITS sequences in Glomus mosseae and Gigaspora margarita spores from a permanent pasture. Mycol. Res. 104 Part 6: 708-715.

279. Bever, J. D. and Morton, J. (1999). Heritable variation and mechanisms of inheritance of spore shape within a population of *Scutellospora pellucida*, an arbuscular mycorrhizal fungus. Amer. J. Bot. 86: 1209-1216.

280. Declerck, S., Strullu, D. G. and Plenchette, C. (1998). Monoxenic culture of the intraradical forms of *Glomus* sp. isolated from a tropical ecosystem: a proposed methodology for germplasm collection. Mycologia 90: 579-585.

281. Pawlowska, T. E., Douds, D. D. and Charvat I. (1999). In vitro propagation and life cycle of the arbuscular mycorrhizal fungus *Glomus etunicatum*. Mycol. Res. 103: 1549-1556.

282. Hawksworth, D. L. (1991). The fungal dimension of biodiversity: magnitude, significance, and conservation. Mycol. Res. 95: 641-655.

283. Douds, D. D., Nagahashi, G. and Abney, G. D. (1996). The differential effects of cell wall-associated phenolics, cell walls, and cytosolic phenolics of host and non-host roots on the growth of two species of AM fungi. New Phytol. 133: 289-294.

284. Elias, K. S. and Safir, G. R. (1987). Hyphal elongation of *Glomus fasciculatus* in response to root exudates. Appl. Environm. Microbiol. 53: 1928-1933.

285. Powell, C. L. (1976). Development of mycorrhizal infections from *Endogone* spores and infected root fragments. Trans. Br. mycol. Soc. 66: 439-445.

286. Schreiner, R. P., Mihara, K. L., McDaniel, H. and Bethlenfalvay G. J. (1997). Mycorrhizal fungi influence plant and soil functions and interactions. Plant Soil 188: 199-209.

Applied Mycology and Biotechnology
Volume 2. Agriculture and Food Production

Arbuscular Mycorrhizal Fungi as Biostimulants and Bioprotectants of Crops

L. J. C. Xavier and S. M. Boyetchko
Agriculture and Agri-Food Canada, Saskatoon Research Centre, 107, Science Place, Saskatoon SK S7N 0X2, Canada

Arbuscular mycorrhizal fungi (AMF) are ubiquitous soil fungi that form obligate associations with >80% of plants. These associations can alter plant productivity, because AMF can act as biofertilizers, bioprotectors or biodegraders. Mycorrhizal fungi modify the quality and abundance of rhizosphere microfloral assemblage leading to an alteration in the overall rhizosphere microbial activity. Benefits arising from synergistic interactions between AMF and other rhizosphere microflora such as rhizobia, non-symbiotic diazotrophs, phytohormone producers and phosphate solubilizers are well recognized. Equally significant is the role of AMF as bioprotectors of plants against phytopathogenic bacteria and fungi. Despite these benefits, interactions between AMF and rhizosphere microflora are underexploited in agriculture, horticulture and forestry, in part due to constraints posed by their obligately biotrophic status and the lack of appropriate technology for mass production. Nonetheless, the use of AMF-rhizosphere microflora combinations appears to be a promising strategy to enhance plant growth or protect plants against pathogens and other environmental stresses, that warrants further attention.

1. INTRODUCTION

The world is facing an enormous challenge to meet the consumption needs of an increasing human population, with limited new arable land [1,2]. In order to meet the consumption demands of the world's population, intensive agriculture using advanced technologies has become a single-element solution for many [3,4]. In a majority of cases, these technologies have been the result of research from the local, national or international scientific community. However, sometimes these solutions do not remedy or rectify the problem. For example, the indiscriminate use of chemical pesticides decades ago are the serious problems of the present day, in that they have caused the development of pesticide resistance and potentially altered the natural flora and fauna in the ecosystem [4]. Furthermore, the development of pesticide resistance in some cases has led to the application of larger than recommended quantities or a mixture of pesticides. This not only presents problems for the ultimate quality of the product but also results in the loss of beneficial members of the soil or rhizosphere community and a loss in the quality of food. In order to circumvent this problem, researchers have resorted to alternate ways such as the use of soil microorganisms to augment or substitute chemicals [5-7].

The high density and types of soil microorganisms make the soil-root interface a very complex area. The complexity of the soil-root interface and the association between plant roots and soil microorganisms such as diazotrophs, phytopathogenic fungi and ectomycorrhizal fungi have been studied in detail [8-12] and have prompted their use and exploitation for plant growth enhancement. Today, microorganisms are not only used as biofertilizers and bioprotectors, but also as biodegraders of pollutants. The addition of microorganisms to soil or other substrates for various purposes is collectively termed "Rhizosphere Technologies" [13,14]. These technologies have long past the realm of academia and have also become resources of large business corporations and other institutions such as government research institutions. On the other hand, these rhizosphere technologies are one of the sole means available to millions of people in the developing world who live on sustenance farming, (i.e., depend on their land as the sole means for survival), due to the lack of resources for procurement of high-priced farm inputs. As research continues to reveal, these rhizosphere microorganisms can not only be augmented to soil, but also manipulated in the rhizosphere by the implementation of simple and old agronomic strategies [15]. Obviously, the choice of manipulation strategies depends upon (i) resources, (ii) infrastructure, and (iii) the demand for and use of such technologies including genetically modified traits [16-18].

2.0 ARBUSCULAR MYCORRHIZAL FUNGI AND THE MYCORRHIZOSPHERE

2.1. Arbuscular Mycorrhizal Fungi

The soil ecosystem is a complex environment that contains several different groups of microorganisms that are beneficial or detrimental to each other and the host plant. The AMF are a group of fungi that are ubiquitous, and that form an intimate association with > 80% of land plants which typically benefits the host [19]. Mycorrhizal propagules have been found in all the climatic zones, including the Arctic and the Antarctic. The host plant regulates the life, growth and multiplication of AMF [20,21]. Using a molecular clock, researchers have estimated that AMF are 360 - 420 million years old [22-24]. Interestingly, this time period coincides with the colonization of land by plants, and it is widely proposed that AMF aided plants to colonize land [22]. This co-evolution between plants and AMF has led to the speculation that the AMF lost their capacity for the synthesis of essential biological compounds for independent survival [19,25]. This theory does not appear to be inconceivable, as horizontal gene transfer from a symbiotic mycorrhizal fungus to an angiosperm has been reported [26]. Despite or as a result of their obligately biotrophic status, AMF have been shown to determine plant diversity, ecosystem variability and productivity [27]. From spore germination until the hyphae comes into contact with plant roots, AMF interact with many groups of microorganisms in soil. These interactions can be beneficial or detrimental to the partners involved and to the plant, and sometimes, may be a preferred activity for enhanced plant growth and productivity.

2.2. The Mycorrhizosphere

The zone of soil that loosely covers the roots of plants constitutes the "rhizosphere" in which complex microbiological and biochemical activities occur [28,29]. Sloughed-off root cells, mucilage and root exudates serve as a steady source of nutrients for large populations of different microbial species that have an impact on soil and plant health [30]. The AMF induce changes in the host root exudation pattern following host root colonization that results in an alteration of the microbial equilibrium in the rhizosphere [31,32]. These changes are collectively referred to as the "mycorrhizosphere effect" which are simply, the changes in the rhizosphere of plants following mycorrhizal colonization of plant roots [33]. The mycorrhizosphere effect assumes significance because of the transient or permanent fluctuations in the quality and abundance of rhizosphere microbial populations subsequent to AMF colonization of plant roots [15,34-38], that may promote the elimination or proliferation of pathogens or the selection of a group of beneficial microflora [37,38]. Paulitz and Linderman [36] reported that the germination of *Glomus etunicatum* Becker & Gerdemann in raw soil was temporarily delayed by two strains of *Pseudomonas fluorescens*. Shalaby and Hanna [39] noted that the rhizosphere of soybean (*Glycine max* L.) plants colonized by *G. mosseae* (Nicol. &Gerd.) Gerdemann and Trappe had lower levels of phytopathogenic *Pseudomonas syringae* compared to uninoculated soybean plants. These modifications are generally mediated by alterations in host root membrane permeability resulting in modifications to root exudate composition [31,32].

3. APPLICATIONS OF AMF-MICROFLORA INTERACTIONS

3.1. Improved Host Nutrition and Plant Yield

Augmentation of soil with different beneficial microorganisms including AMF species that results in plant growth promotion has been demonstrated by several workers over four decades [8,40-43]. A great majority of these reports focus on the interactions between the most obvious biofertilizers (e.g., diazotrophs) and AMF, while some reports have addressed interactions between AMF and unusual, nonetheless beneficial microorganisms [44-47]. The most commonly studied organisms for interactions with AMF are bacteria, which are known for their capacity for nitrogen fixation, phytohormone production and phosphate solubilization. However, synergistic interactions between AMF and some saprophytic fungi such as *Penicillium bilaiae* Chalab. that are involved in P-solubilization are also documented [48].

3.1.1. Interactions between AMF and symbiotic organisms

Diazotrophs, using their nitrogenase enzyme complex, fix atmospheric nitrogen that is then transferred to the host. However, the site of nitrogen fixation varies with different diazotrophs and hence the direct involvement of the host in the nitrogen fixation process. The nature of interactions between the diazotroph and the host also indicates the significance of the association for both partners involved in the symbiosis. Dual inoculation of legumes with AMF and suitable *Rhizobium* sp. strains can enhance legume productivity [40,49]. In some cases, dual inoculation of legumes has been included as one of the many revegetation strategies for the reclamation and rehabilitation of desertified and disturbed ecosystems [50-52]. Whatever the purpose, co-inoculation of legumes with AMF and suitable *Rhizobium* sp. strains has been mostly beneficial to the host plant. The growth, yield and nutrition of legumes co-inoculated with AMF and rhizobia were generally superior to that of plants singly inoculated with the *Rhizobium* sp. strain or the AMF species [49,50,53-56]. Increases in legume productivity as a result of co-inoculation with AMF and *Rhizobium* strains were in excess of 50% compared to plants not receiving AMF.

The selection of appropriate AMF species for enhanced legume productivity may be as important as the selection of *Rhizobium* sp. strain for inoculation purposes. This is because nodulated plants benefit from the uptake of P by the AMF extraradical mycelium, which is required for the process of nitrogen fixation. In general, the selection of AMF species for co-inoculation purposes has been based on the availability of AMF species for experimentation. The inclusion of several AMF species offers the potential for differences in the growth promoting ability between AMF species in the manner in which they interact with the host and the *Rhizobium* symbiont [41,57,58]. Ianson and Linderman [41] assessed the response of pigeonpea (*Cajanus cajan* L.) to an effective *Rhizobium* sp. strain and seven different AMF isolates of *G. etunicatum*, *G. aggregatum* Schenck & Smith, *G. deserticola* Trappe, Bloss &

Menge, *G. mosseae*, *Gigaspora margarita* Becker & Hall and *G. intraradices* Schenck & Smith. They found that the AMF species exerted very different effects on nodulation and nitrogen fixation. They also suggested that these differences were probably mediated by a specific inter-endophyte interaction between the *Rhizobium* sp. strain and the different AMF species, which became evident under N-limiting conditions. In general, legume response to AMF-*Rhizobium* combinations varies significantly depending on the AMF species in combination. It is an interesting fact that different AMF species can alter the outcome of legume-*Rhizobium* sp. combination, indicating their significance in the tripartite symbiosis.

The success of the tripartite symbiosis depends on all partners. In a compatible and effective tripartite symbiosis, the AMF species mobilizes adequate amounts of P, which is utilized by the *Rhizobium* sp. strain for nitrogen fixation. Some workers have examined the response of mycorrhizal leguminous hosts to different strains of rhizobia, in an attempt to understand the role of the *Rhizobium* sp. strain in the tripartite symbiosis [59,60]. However, no difference was noted between the *Rhizobium* sp. strains (n=2 or 3) tested. In these cases, the lack of an adequate number of *Rhizobium* sp. strains may have resulted in insufficient information with regard to the role of the *Rhizobium* sp. strains in the tripartite symbiosis. Alternatively, it could imply that there were no differences in the efficacy of the *Rhizobium* sp. strains included. Recently, Xavier and Germida [43] reported that an incompatible AMF partner could reduce the efficacy of an effective *Rhizobium* sp. strain on a legume host. Furthermore, they found that the growth and yield of lentil (*Lens culinaris* L.) co-inoculated with AMF species and a *Rhizobium* sp. strain was highest when the AMF-*Rhizobium* combination included an effective *Rhizobium* sp. strain. This suggests that the *Rhizobium* symbiont plays a significant role in the overall health of the dual symbiotic legume host. On the other hand, some legumes do not respond to inoculation with diazotrophs in the absence of AMF, indicating the importance of AMF in the tripartite symbiosis [61,62].

It is important to co-select compatible rhizobia and AMF which will result in desired plant effects. Various workers have attempted to develop superior combinations of AMF and rhizobia for different legume species [40,44,48,50,63,64]. These reports suggest that the symbiotic efficiency of a tripartite symbiosis depends on the particular AMF-*Rhizobium* sp. strain combination. This important finding signifies the high level of interactions between the three partners involved in the symbiosis.

3.1.2. Interactions between AMF and non-symbiotic organisms

3.1.2.1. Diazotrophs

Microorganisms in the rhizosphere live in close proximity because of the steady supply of nutrients around the roots. Due to the high density of bacteria in the rhizosphere and the ubiquitous nature of AMF, interactions occur. Some of these interactions may be potentially beneficial to plant growth. Many workers have assessed the co-inoculation response of mycorrhizal hosts to various rhizobacteria [34,35,44,62,65-70]. For example, *Acetobacter diazotrophicus* is a diazotroph that colonizes the aerial parts of sugar cane (*Saccharum officinarum* L.) and sweet potato (*Ipomoea batatus* L.) plant [8]. However, unlike other AMF-associative diazotroph associations, *A. diazotrophicus* is beneficial to sugarcane only when co-inoculated with AMF [8]. Paula and co-workers [62] assessed the synergistic effects of co-inoculating sweet potato with *G. clarum* Nicolson & Schenck and *A. diazotrophicus*. They found that *A. diazotrophicus* infected aerial plant parts and enhanced tuber production only when co-inoculated with *G. clarum* spores. In addition, AMF intraradical growth and sporulation was enhanced by the bacterium. The microaerophilic diazotroph, *Azospirillum* spp. has been combined with AMF for the purpose of enhancing plant yield in cereals. For example, Barea and co-workers [66] found that dual inoculation of maize (*Zea mays* L.) with *A. brasilense* and a yellow-vacuolate AMF spore type most closely resembling *G. mosseae* produced plants of a similar size and N content but a higher P content than those fertilized with N and P. Pacovsky *et al.* [67] examined the influence of edaphic factors on the interactions between *Azospirillum* species and *G. fasciculatum* (Thaxter sensu Gerdemann) Gerdemann & Trappe. They found that plants inoculated with AMF and *Azospirillum* had higher plant dry weight, shoot-root-ratios and N content than those inoculated with either AMF or the bacterium alone. They also found that inoculation with *Azospirillum* stimulated AMF colonization of roots. Although there are general interactions between AMF and *Azospirillum* in the rhizosphere, not all of these interactions result in a synergistic response [68]. For example, Rao and co-workers [68] assessed the response of barley (*Hordeum vulgare* L.) to combinations of an *Azospirillum* inoculant and/or five different AMF species grown in potted field soil. They found that while some combinations of AMF and bacteria increased seed and shoot yield of barley, others did not have an effect.

Diazotrophic *Azotobacter* spp. are free living bacteria that enhance plant growth through fixation of atmospheric nitrogen [71] or the production of plant growth promoting substances [72,73]. Azcon [69] and El-Raheem *et al.* [70] examined the effects of interactions between *Azotobacter* and AMF on the growth and nutrient content of tomato (*Lycopersicum esculentum* L.) plants. For example, Azcon [69] examined the response of tomato to co-inoculation with *A. vinelandii* strain ATCC 12837 and *G. mosseae*, *G. fasciculatum* or *Glomus* sp. type E3 in a sand-vermiculite medium. They found that the response of tomato plants to the different *Azotobacter*-AMF associations was not the same. *Azotobacter vinelandii* did not alter the response of tomato plants to the *G. mosseae* or the E3 inoculant, but enhanced the growth response of plants inoculated with *G. fasciculatum*, indicating the specificity of AMF-*Azotobacter* interactions. Although the *Azotobacter* strain influenced the growth of mycorrhizal tomato plants, the AMF colonization of tomato roots was not altered by the *Azotobacter* strain. Similarly, El-Raheem and co-workers [70] studied the effects of *A. chroococcum* -*G. fasciculatum* interactions on tomato growth and nutrition in a steam-sterilized sandy soil. They reported that dual inoculation of tomato with *A. chroococcum* and *G. fasciculatum* resulted in enhanced AMF root colonization, and improved shoot growth and N, Ca, Mg and K contents. Therefore, it appears that the simultaneous application of diazotrophs and AMF improves the N and P nutrition of the host, resulting in enhanced plant growth compared to uninoculated plants.

3.1.2.2. Phosphate solubilizers

The ability of AMF to solubilize sparingly soluble phosphorus has not been a subject for great consideration, as compared to the use of AMF in transporting P after release by a P-solubilizing microorganism. Interactions between AMF and other rhizosphere bacteria such as phosphate solubilizing bacteria (PSB) have been reported [65,74-79]. The rationale for co-inoculating plants with AMF and PSB is to exploit the ability of the PSB to release P from bound or organic P sources in soil, which may then be absorbed by AMF hyphae and effectively translocated to the host plant. For example, Raj *et al.* [65] assessed the response of finger millet (*Eleusine coracana* L.) co-inoculated with *G. fasciculatum* and a PSB *Bacillus circulans* and grown in a 1:1 sterilized soil: sand mixture amended with either radio-labeled superphosphate or tricalcium phosphate. They found that plants inoculated with the AMF removed more P from soil, and plants co-inoculated with *G. fasciculatum* and *B. circulans* yielded more dry matter and

P. Similarly, Azcon-Aguilar and co-workers [74] examined the interactions between soybean, two unidentified PSB and two AMF species, *G. mosseae* and a *Glomus* sp. E3 in gamma-irradiated soil amended with three levels of tricalcium phosphate. They found that only inoculation with *Glomus* sp. E3 improved the aboveground dry matter production and utilization of tricalcium phosphate. However, they also found that the proportion of P from the radio-labeled tricalcium phosphate in the shoot material was lower in mycorrhizal plants, indicating that most of the available P in soil was probably derived from other sources. Recently, Toro and co-workers [78] examined the response of alfalfa (*Medicago sativa* L.) to co-inoculation with a PSB *Enterobacter* sp. and *G. mosseae* in soil containing indigenous populations of PSB and amended with rock phosphate. They found that mycorrhizal plants absorbed more P from soil compared to non-mycorrhizal plants, and that the *Enterobacter* inoculant improved the use of rock phosphate in all plants. This suggests that P released by the PSB was effectively transported to the host by the AMF, resulting in enhanced host P nutrition. Kim *et al.* [79] reported synergistic interactions between the PSB *Enterobacter agglomerans* and *G. etunicatum* that resulted in enhanced P uptake in tomato plants. They also noted that co-inoculation treatment with the highest oxalic acid concentration at 35 days after planting was associated with the highest P content at 55 days after planting.

Reports on the interactions between AMF and a phosphorus-solubilizing fungus, *P. bilaiae* constitute the only AMF-non-mycorrhizal fungal interaction that generally leads to an improvement in host nutrition and growth [48,80]. Although Kucey [48] reported synergistic interactions between AMF and *P. bilaiae* that resulted in enhanced wheat and field bean yield, Iqbal and Waterer [80] did not observe a marked improvement in the growth and yield of co-inoculated pepper (*Capsicum annuum* L.) plants. It appears that, in some cases, host P nutrition significantly improved following an increase in P availability as a result of P solubilization by *P. bilaiae* and movement of P to host roots by the AMF species.

3.1.2.3. Phytohormone producing rhizobacteria

Arbuscular mycorrhizal fungi interact specifically with phytohormone producing rhizobacteria, specifically *Pseudomonas* spp. [35,81,82]. For example, Meyer and Linderman [35] examined the response of subterranean clover (*Trifolium subterraneum* L.) to dual inoculation with an indigenous AMF mixed culture and *P. putida* grown in non-sterile soil. The shoot dry weight and mycorrhizal colonization of co-inoculated plants was significantly greater than plants left uninoculated or inoculated with the AMF mixture or bacteria alone. Furthermore, nodulation of these plants by the indigenous soil rhizobia was enhanced, and the concentrations of Fe, Cu, Al, Zn, Co and Ni were greater in co-inoculated plants. Similarly, Gryndler and Vosatka [81] studied the response of maize to *G. fistulosum* Skou & Jakobsen and various culture fractions of *P. putida*. They found that living cells of *P. putida* and *G. fistulosum* increased the leaf area and shoot dry weight of plants more than by *G. fistulosum* alone. They also reported that mycorrhizal colonization of maize roots was higher in plants co-inoculated with living cells or a dialyzed cell extract of *P. putida*. They also found that root infection by extraneous filamentous fungi was reduced when plants were co-inoculated with live *P. putida* cells, indicating the protective effect of the bacterium on plants. Walley and Germida [82] investigated the response of spring wheat (*Triticum aestivum* L.) to interactions between *Pseudomonas* spp. strains and *G. clarum* NT4 under gnotobiotic and field soil conditions. They found that there were specific interactions between *G. clarum* NT4 and the bacterial strains, but that the effect on plant growth was dependent on the bacterial strain. They also found that *G.*

clarum NT4 spores did not germinate *in vitro* in the presence of some *Pseudomonas* spp. strains and suggested that a non-volatile substance might have inhibited spore germination.

3.1.2.4. Interactions with other bacteria

Interactions between AMF and various other bacteria have also been examined, but results appear to vary depending on the bacterial species or strain employed [69,44]. For example, Azcon [69] inoculated tomato seedlings with an indigenous Enterobacteriaceae isolate and *G. mosseae*, *G. fasciculatum* or *Glomus* sp. E3 isolate. They found that mycorrhizal tomato plants exhibited enhanced vigor and growth in the presence of this bacterial isolate. They reported, however, that the extent of AMF root colonization was not related to the effect of this bacterium on plant growth, indicating that both organisms did not act together, but exerted their individual effects on plant growth. In an attempt to establish sea oats (*Uniola paniculata* L.) on replenished beaches quickly, Will and Sylvia [44] inoculated sea oats with a mixture of *Glomus* sp. S238 and *G. deserticola* and *Klebsiella pneumoniae*, *Bacillus polymyxa* and *Alcaligenes denitrificans*. However, they found no consistent evidence supporting a synergistic effect on plant growth. Andrade and co-workers [45] reported that the response of pea (*Pisum sativum* L.) co-inoculated with *G. mosseae* and a *Bacillus* sp. previously isolated from surface-sterilized spores of *G. mosseae* was not positive. Recently, Xavier [47] isolated several bacterial species from intact, surface sterilized *G. clarum* spores and assessed the response of pea to co-inoculation with *G. clarum* and some of these bacterial isolates. It was noted that some of the bacterial isolates stimulated AMF colonization and mycorrhizal pea growth and nutrition, whereas other bacterial isolates not only inhibited spore germination, but also reduced plant growth.

In most cases, interactions between AMF and plant growth-promoting microorganisms resulted in enhanced host growth and nutrition. Thus, AMF may interact synergistically with these microbes which modify host plant growth and nutrition through a number of mechanisms: atmospheric nitrogen fixation, P release from less readily available sources, phytohormone production, siderophore production, production of antibiotics, and/or aggressive colonization of the endorhizosphere. Interactions between these beneficial bacteria and AMF may directly influence plant growth or indirectly alter microbial activities in the rhizosphere. However, careful consideration must be given to the selection of these organisms and their effect on each other and on the host when in combination.

3.2. Plant Disease Control

The AMF control plant diseases in at least three ways, (i) preventing pathogen infection by eliciting resistance mechanisms such as systemic induced resistance, (ii) reducing pathogen inoculum in the rhizosphere, thereby reducing the incidence of infection, and (iii) limiting the spread of disease by competition. In some cases, mycorrhizal plants tend to tolerate pathogen infection better than non-mycorrhizal plants.

3.2.1. Interactions between phytopathogenic bacteria and AMF

Mycorrhizal plants appear to influence the activities of phytopathogenic bacteria, resulting in a change in pathogen activity. Shalaby and Hanna [39] found that *P. syringae* infection of soybean plants was reduced as a result of inoculation with *G. mosseae*. In addition, they found that the population density of the bacterial pathogen was reduced in soybean rhizosphere. Similarly, Sharma *et al.* [83] reported that mulberry (*Morus alba* L.) plants fertilized with 60 to

90 kg of P per hectare per year and inoculated with *G. fasciculatum* or *G. mosseae* reduced the incidence of bacterial blight caused by *P. syringae* pv. *mori*. Waschkies *et al.* [84] reported that inoculation of grapevines with AMF reduced the incidence of grapevine (*Vitis vinifera* L.) replant disease. They found that the number of fluorescent pseudomonads on the grapevine root surface was significantly reduced. Eggplant (*Solanum melongena* L.) and cucumber (*Cucumis sativus* L.) infection by *P. lacrymans* was reduced when plants were inoculated with *G. macrocarpum* Tul. & Tul. [85]. Otto and Winkler [86] found a significant negative correlation between AMF root colonization and the colonization of apple (*Malus sylvestris* L.) seedling rootlets by actinomycetes causing replant disease.

Suresh and Rai [87] reported that the root extract of tomato plants inoculated with *G. fasciculatum* was detrimental to *P. solanacearum* in nutrient broth, and reduced *P. solanacearum* populations. This inhibitory effect of *G. fasciculatum* on *P. solanacearum* may be the result of alterations in the plant such as pH changes unfavorable for the phytopathogenic bacterium, or induced phytochemical alterations that are directly inhibitory to the pathogen. The authors also noted that these inhibitory effects were specific towards the pathogenic bacterium and not towards beneficial or other rhizosphere bacteria such as *Azotobacter* spp., *P. putida or P. fluorescens* [87]. Plant pathogenic bacteria can be effectively managed by employing management strategies such as selection of highly mycotrophic hosts that will induce changes in the microbial equilibrium that may result in the elimination of pathogens or reduce pathogen activity.

3.2.2. Interactions between phytopathogenic fungi and AMF

The AMF or mycorrhizal plants have been used with considerable success for the control of phytopathogenic fungi [36,88-94]. Success of AMF-mediated control of phytopathogenic fungi ranges from disease suppression or elimination of disease to disease resistance by the host plant [88,90,94-96]. Few workers have investigated the stage of AMF differentiation at which biological control of pathogens is initiated [97]. Furthermore, there are fewer reports that include all possible facets of mycorrhizae-mediated protection of plants against pathogens. Kapoor *et al.* [91] assessed microbial interactions in the mycorrhizosphere of dill (*Anethum graveolens* L.) and found that several soil-borne pathogens were suppressed following colonization of roots by *G. macrocarpum*. Similarly, Kumar *et al.* [98] reported that the incidence of diseases caused by phytopathogenic fungi in cashew (*Anacardium occidentale* L.) nurseries was reduced significantly in mycorrhizal cashew rootstocks compared to non-mycorrhizal controls.

Several studies investigating the potential of AMF for biological control of *Fusarium oxysporum* Schlecht. f. sp. *lycopersici* Jarvis & Shoemaker have been reported [99-106]. Significant reductions in wilting by the pathogen were attributed, in part, to increased lignin deposition in the plant cell walls as a result of mycorrhizal colonization, which may have restricted the spread of the pathogen [101]. Other studies demonstrated that severity of *Fusarium* crown and root rot of tomato and the number of pathogen propagules were significantly reduced when tomato was colonized by *G. intraradices* [103-105]. The interaction between the AMF species and the pathogen was influenced by various factors such as the growth medium used [102] and the time at which disease symptoms were observed [103]. Improved P nutrition did not influence the severity of *Fusarium* root rot, shoot growth of the plant and final concentration of pathogen inoculum [104]. Pre-infestation of soil with AMF prior to introduction of the pathogen also resulted in significant reductions in disease severity and final number of propagules of *F. oxysporum* f. sp. *lycopersici* [105]. The authors suggested that induced plant resistance, rather than improved P nutrition may have caused a reduction in the disease.

A number of researchers have reported on the AMF-mediated reduction of root rot disease in cereal crops [107-110]. Boyetchko and Tewari [108] reported that AMF suppressed common root rot, caused by *Bipolaris sorokiniana* (Sacc.) Shoem. *Glomus dimorphicum* Boyetchko & Tewari, *G. mosseae*, and *G. intraradices* reduced the severity of root rot in barley, but *G. dimorphicum* was not as effective as the other AMF species tested [107,111]. Improved P nutrition by the AMF played a role in suppressing barley root rot disease, but the authors concluded that improved P uptake alone was not solely responsible for the reduced disease severity, and that disease suppression might involve more than one mechanism.

Sreeramulu *et al.* [112] reported that the combined application of *G. fasciculatum and Trichoderma harzianum* Rifai was very effective at controlling damping-off caused by *Pythium aphanidermatum* Hesse and black shank disease caused by *Phytophthora parasitica* f. sp. *nicotianae* Van Breda de Haan in tobacco seedlings. Kulkarni *et al.* [95] reported that the addition of *G. fasciculatum*, *Gi. margarita*, *Acaulospora laevis* Gerdemann & Trappe and *Sclerocystis dussii* (Pat.) von Höhn significantly nullified the destructive effect of the *Sclerotium rolfsii* Sacc. on ground nut (*Arachis hypogaea* L.). Feldmann and Boyle [113] noted an inverse relationship between *G. etunicatum* colonization of three begonia cultivars and the intensity of powdery mildew caused by *Oidium begoniae* Link. Matsubara *et al.* [114] assessed the response of 12 week-old asparagus (*Asparagus officinalis* L.) inoculated with *G. fasciculatum*, *G. mosseae*, *Glomus* sp. R10 or *Gi. margarita* to the violet root rot pathogen *Helicobasidium mompa* Tanaka. They found that asparagus plants inoculated with *Gi. margarita* and *Glomus* sp. R10 were more effective at tolerating the pathogen than *G. fasciculatum* and *G. mosseae*. Recently, Norman and Hooker [94] reported that root extracts of mycorrhizal strawberry (*Fragaria vesca* L.) plants were able to reduce the spore numbers of *Phytophthora fragariae* Hickman compared to non-mycorrhizal plants. This suggests that AMF alter the phytochemistry of plants in a manner that affects pathogens.

Management practices in combination with mycorrhizal fungi effectively control diseases caused by phytopathogens [115,116]. For example, Prashanthi *et al.* [115] have shown that a combination of wheat straw, carbendazim, *G. fasciculatum*, and *T. viride* Persoon: Fries protected safflower (*Carthamus tinctorius* L.) seedlings from the root rot pathogen *Macrophomina phaseolina* (Tassi) Goid., and achieved 100% seedling survival. Sharma *et al.* [116] effectively managed ginger yellows caused by *F. oxysporum* f. sp. *zingiberi* Trujillo by using a combination of organic amendments (pine needles), fungicide treatments and biological agents. They found that the combination of pine needle amendment, *T. harzianum* and *Gi. margarita* delivered the best control for ginger yellows.

Using an *in vitro* system, Filion *et al.* [97] demonstrated that extracts from the extraradical mycelium of *G. intraradices* reduced the *conidial germination of F. oxysporum f. sp. chrysanthemi* Litrell, G.M. Armstr. & J.K. Armstr. They also noted that this biocontrol effect was directly proportional to the extract concentration, indicating that compounds released from the external mycelium can control pathogen survival and activity. Alternatively, this indicates that extracts from the extraradical mycelium altered the equilibrium in the mycorrhizosphere in a manner that resulted in inhibition of the pathogen. Researchers proposed that the invasion of pathogenic fungi is preventable by using an aggressively root colonizing mycorrhizal fungal species [113]. For example, a low degree of *G. etunicatum* root colonization was associated with high susceptibility to powdery mildew, whereas a high degree of mycorrhizal root colonization decreased pathogen infection [113]. However, it is not clear whether mycorrhizal root colonization preceded infection by the pathogen and whether the physical exclusion was accompanied by other mechanisms of biocontrol. Kjoller and Rosendahl [117] found that pea plants pre-inoculated with *G. intraradices* did not exhibit symptoms of root rot caused by *Aphanomyces euteiches* Drechsl. even though the pathogen was present in the roots. They further noted that *A. euteiches* enzymatic activity in mycorrhizal pea was different from that of non-mycorrhizal pea plants. Similar cases of tolerance or physical exclusion by AMF leading to the control of diseases caused by pathogens such as *Phytophthora* spp., *Pythium* spp., *Fusarium* spp., *Rhizoctonia* spp. and *Cylindrocarpon* spp. have been reported by Sierota [118] for damping-off in forest nurseries.

There are conflicting reports on the target specificity of the biocontrol activities of mycorrhizal fungi. Jaizme-Vega *et al.* [119] found that *G. intraradices* and a *Glomus* spp. isolate reduced rhizome necrosis and external disease symptoms caused by *F. oxysporum* f. sp. *cubense* (E.F. Smith) Wollenw. on micropropagated banana (*Musa paradisiaca* L.) plants, but differences between the efficacy of *G. intraradices* or the *Glomus* spp. isolate were not noted. This indicates that the bioprotection offered to mycorrhizal plants by AMF is non-specific and is not AMF species-dependent. In contrast, Li *et al.* [92] transplanted eggplant and cucumber seedlings into soils inoculated with *G. mosseae*, *G. versiforme (*Karsten) Berch, *Glomus* spp.-1 and *Glomus* spp.-2 and subsequently inoculated seedlings with *Verticillium dahliae* Kleb. and *P. lacrymans*. They found that the effects of eggplant and cucumber wilt caused by *V. dahliae* were alleviated only by *G. versiforme*, indicating species-specific antagonistic symbiont-pathogen interactions. Using differences in the qualitative and quantitative expression of glucanases (phytoalexin elicitor-releasing factor), Pozo *et al.* [120] demonstrated that there were significant differences between *G. mosseae* and *G. intraradices* in the elicitation of basic glucanase isoforms in tomato root extracts against the pathogen *P. parasitica f. sp. nicotianae*. Plants colonized by *G. mosseae* and challenged with the fungal pathogen contained two new basic glucanase isoforms in the root extracts that were not detected in plants colonized by *G. intraradices* and the pathogen. This indicates that AMF do exhibit specificity in the elicitation of host response to plant pathogens.

Phytopathogenic fungi have also been a target for organisms associated with AMF spores, as recently demonstrated by Budi *et al.* [121]. A *Paenibacillus* sp. strain isolated from the mycorrhizosphere of *G. mosseae*-inoculated sorghum (*Sorghum bicolor* (L.) Moench) had antifungal activity against *P. parasitica* Dastur, *F. oxysporum*, *F. culmorum* (W.G. Smith) Sacc., *A. euteiches*, *Chalara elegans* Nag Raj & Kendr., *Pythium* sp., and *R. solani* Kühn, while concomitantly enhancing mycorrhizal development. Adequate host nutrition ensured by mycorrhizal colonization can result in a healthy plant that is able to ward off or tolerate pathogens effectively. Rabie [122] studied the effect of co-inoculation of plants with AMF and rhizobia on disease resistance. This was accomplished by co-inoculating *Vicia faba* L. with *G. mosseae* and *R. leguminosarum* challenged with the plant pathogen *Botrytis fabae* Sard. The author found that plants co-inoculated with AMF and rhizobia had higher levels of total nitrogen, superior growth and number of nodules. It was also noted that an inverse relationship existed between the levels of phenolic substances, Ca, Mg and Zn concentrations in nodulated plants and disease severity. This suggests that a healthy plant is able to withstand, tolerate or prevent diseases. It is possible that, as suggested by Graham *et al.* [32], host membrane permeability following mycorrhization and nodulation of *V. faba* restricted the entry of this pathogen. This effect is usually accompanied by the restoration of overall plant health by adequate host nutrition. Alternatively, it has been reported that some mycorrhizal plants tolerate pathogens [123]. For example, Guillemin *et al.* [123] found that *Glomus* sp. stimulated the growth of micropropagated pineapple (*Ananas comosus* L.), enhanced mineral nutrition and helped tolerate the effect of the pathogen *P. cinnamoni* Rands. In this case, it is possible that improved general vigor of the mycorrhizal pineapple in the presence of the pathogen did not severely alter plant productivity. Similar cases of reduction in host disease severity following mycorrhizae formation have been reported for citronella (*Java citronella* L.) using *G. aggregatum* by Ratti *et al.* [124].

3.2.3. Interactions between phytopathogenic viruses and AMF

Viruses remain the least studied amongst all the plant disease-causing target organisms for mycorrhizae-mediated biocontrol. Nevertheless, it does not diminish the fact that viruses can cause serious crop losses, and that their effects can be altered depending on the appropriate control measures. Based on previous reports, the general response of mycorrhizal plants to the presence of viral pathogens is as follows: (1) mycorrhizal plants apparently enhance the rate of multiplication of viruses in some plants [125,126]; (2) more leaf lesions are found on mycorrhizal plants than on non-mycorrhizal plants [127,128]; and (3) the density of AMF spores is reduced considerably in the rhizosphere [126,129]. Enhanced viral activity may be due to the high P levels in mycorrhizal plants compared to non-mycorrhizal plants, because non-mycorrhizal plants fertilized with P were also found to enhance viral multiplication and activity [125]. Early studies using electron microscopy revealed that virus particles were absent in the AMF hyphae and around arbuscules, suggesting that mycorrhizae were not viral vectors, and that mycorrhizal fungi did not appear to interact with viruses [130]. Thus far, observed responses of mycorrhizal plants to infection by pathogenic viruses have not been positive.

3.3. Enhanced Environmental Stress Tolerance

3.3.1. Drought

Mycorrhizal plants are known to tolerate drought more than non-mycorrhizal plants [131-133]. For example, Goicoechea *et al.* [132,133] reported that alfalfa plants co-inoculated with *G. fasciculatum* and *R. meliloti* exhibited delay in leaf senescence and maintained or enhanced the number of stems under drought-stressed conditions, as compared with uninoculated plants. Uninoculated plants had the largest drop in cytokinin levels, which was associated with lower number of stems and early leaf senescence. In contrast, the cytokinin levels in the co-inoculated plants were maintained during drought. Furthermore, they suggested that while water stress can lead to inhibition of nodule protein synthesis, nodule respiration, nodule metabolism and nitrogenase activity, plants co-inoculated with AMF and rhizobia were able to maintain nodule activity. Similar positive effects of co-inoculation with bradyrhizobia and AMF on drought-stressed lupine were noted by Lynd and Ansman [134].

3.3.2. Salinity

Fungi are generally known to tolerate osmotic stress because of their low water activity. The osmotic stress tolerance of mycorrhizal plants has been demonstrated under controlled conditions [135-138], however, reports on the osmotic stress tolerance of AMF under natural conditions, particularly that of salt stress are scarce. Nonetheless, the salt stress response of mycorrhizal plants has been characterized well. Reports such as those of Hirrel and Gerdemann

[139] established that AMF could enhance growth of onion (*Allium cepa* L.) and pepper (*Piper nigrum* L.) in saline soils. They attributed this response to enhanced P levels in mycorrhizal plants. This response was further investigated in a more recent study conducted by Azcon and El-Atrash [137] using AMF and rhizobia. They established that no amount of plant-available P was able to protect nodulated alfalfa plants more effectively than *G. mosseae* in soils with saturation extract electrical conductivities in excess of 43 dS m^{-1}. They also speculated that nodulated and mycorrhizal legumes meet their nutritional requirements by employing alternative mechanisms and maintain their physiological status under high salt stress conditions. Furthermore, the authors speculated that the bioprotective mechanism of *G. mosseae* was different from that involved in improved P uptake by mycorrhizal plants. Tsang and Maun [140] determined the salt tolerance of a nodulated and mycorrhizal trailing wild bean plant (*Strophostyles helvola* L.) commonly found in coastal foredunes under greenhouse conditions. They found that increasing salt concentrations did not reduce the number of nodules or AMF hyphal colonization, but reduced the number of arbuscules and vesicles in mycorrhizal plants, and concluded that the negative effects of increasing salt concentrations can be somewhat mitigated by AMF.

3.3.3. Soil Disturbance

Soil disturbance results in loss of key soil properties such as soil structure, plant nutrient availability, organic matter content, and a loss in microbial activity [141]. The first indicator of a disturbed ecosystem often includes loss of or disturbance in the natural plant communities. Arguably, the first step to rehabilitation of disturbed ecosystems is re-establishment of the disturbed site with indigenous plant communities or native plant species that would thrive under disturbed soil conditions. Several researchers have determined the value of AMF in the reclamation of disturbed ecosystems [27,44,51,142-146]. Will and Sylvia [44] evaluated the co-inoculation response of transplanted sea oats on a replenishment project in northeastern Florida to a mixture of *Glomus* sp. S238 and *G. deserticola* and *K. pneumoniae*, *B. polymyxa* and *A. denitrificans*. They found that after two growing seasons, the level of AMF colonization was low and sporulation was significantly lower than those found in the established dunes. On the other hand, Herrera *et al.* [51] reported success in the use of AMF and rhizobia to restore indigenous nitrogen-fixing woody legumes to rehabilitate disturbed Mediterranean ecosystems. Recently, in two long-term experiments, Requena *et al.* [53] demonstrated that co-inoculation of legume plant species, which constitute the key plant species in a disturbed Mediterranean ecosystem, with AMF and suitable rhizobia enhanced soil fertility and quality. They reported that as a result of the tripartite symbiosis, soil N content, organic matter, water stable aggregates and N transfer from legume to non-legume species were significantly enhanced within the natural succession of plants.

3.3.4. Toxic pollutants

Toxic pollutants include heavy metals and other recalcitrant compounds such as xenobiotics. Few studies have addressed the role of AMF in the remediation process despite the potential benefits of AMF in remediation [147-152]. Bioremediation of toxic pollutants such as xenobiotics has also been accomplished using rhizobacteria such as *Pseudomonas* spp. [153-155]. In a majority of these cases, the source of the bacterium and the AMF species used for phytoremediation was a contaminated site. Therefore, it is possible that the diversity of these organisms has been modified by the existing soil conditions, and that the microbial community is adapted to certain levels of xenobiotics [149]. However, most researchers have overlooked the most obvious research strategy of using phytoeffective combinations of AMF and rhizobacteria that may be synergistic, resulting in the enhanced removal or degradation of xenobiotics. El-Kherbawy *et al.* [152] evaluated the growth and heavy metal uptake of AMF and *Rhizobium*-inoculated alfalfa using industrially polluted soil contaminated with high levels of Zn and Cd at different levels of soil pH. They concluded that co-inoculation aided in the uptake of heavy metal by alfalfa, however, only at neutral soil pH. Recently, Stommel *et al.* [156] found a metallothionein-encoding gene in *Gi. rosea* Nicolson & Schenck spores that probably is involved in heavy-metal binding which further strengthens the case for the potential for AMF in the bioremediation process.

4. FACTORS INFLUENCING AMF - RHIZOSPHERE MICROFLORA INTERACTIONS

4.1. Plant factors

Host plant mycotrophicity is dependent upon host plant susceptibility to mycorrhizal root colonization [157-159]. All plant species generally fall into three broad categories: highly mycotrophic, facultatively mycotrophic and non-mycotrophic. Highly mycotrophic host plants such as those belonging to the families Leguminosae, Rosaceae and Alliaceae possess thick root hairs that are not suitable for efficient nutrient absorption [19]. It is in these hosts that AMF render the most benefits. Facultatively mycotrophic hosts tend to be less mycorrhizal compared to the highly mycotrophic hosts because they possess fine root hairs that are well suited for nutrient absorption. However, this condition tends to be more of an exception than a rule, as facultatively mycorrhizal host plants vary in their ability to become susceptible to mycorrhizal colonization. Furthermore, the ability of different genotypes of the same host to associate with AMF varies tremendously [158,160]. The third group includes plants that are resistant to or do not form mycorrhizae, and are referred to as non-mycotrophic. Members of the plant families Caryophyllaceae, Brassicaceae, Cruciferae, Chenopodiaceae and Cyperaceae are generally non-mycotrophic. Several workers have attempted to deduce the mechanism of resistance to mycorrhizae formation in non-mycotrophic plants [159,161-163]. The proposed cause for non-mycotrophicity is the presence of glucosinolates in plant roots [159,161,162]. Recently, Vierheilig *et al.* [159] indicated that the presence of a particular glucosinolate, 2-phenylethylglucosinolate, was apparently the inhibitory factor in non-mycotrophic plants (e.g., *Barbarea* spp., *Brassica* spp., *Carica papaya* L., *Lepidium* spp., *Nasturtium* sp., *Reseda* app., *Sinapis alba* L., and *Tropaeolum* spp.) they tested.

Studies have shown that in the nodulation and mycorrhizal symbioses, host defence responses are not activated by the invading microorganisms [20,164], whereas the host defence system is highly active, when the invader is a plant pathogen [164]. Blilou *et al.* [165] studied plant resistance to AMF colonization using pea mutants, in an effort to understand the mechanism of resistance to mycorrhizae formation [165]. They found that endogenous salicylic acid levels were high in plant mutants that were resistant to AMF colonization, but not in the wild type plants. Future studies will help explain whether a relationship exists between salicylic acid levels and differences in AMF susceptibility to different genotypes of the same host.

Plants and AMF have a very intimate and specific association that is influenced by genotypic differences in the host, which can alter the mycorrhizal dependency of the host [166,167]. The tripartite symbiosis between AMF, rhizobia and legumes can be influenced by differences in the genotypes of a legume host [49,57,167,168]. Some of these reports indicate that the symbiotic efficiency of the host depend on the particular combination of a cultivar, *Rhizobium* sp. strain and AMF species [49,57]. Such specificity in the interactions between the three partners indicates that the symbiosis is judiciously regulated and that this may be an effective strategy for maximizing legume productivity.

4.2. Soil factors

Most soils can rarely sustain high levels of plant productivity without the addition of fertilizers. On the other hand, larger than optimal amounts of N and P fertilizers, can severely impair the nodulation and mycorrhizal symbioses. High levels of N repress nitrogenase activity and the potential N benefits from rhizobia to plants, whereas high P levels reduce benefits from AMF and can cause severe growth depression [169]. Therefore, legume response to co-inoculation with AMF and rhizobia is generally assessed in mineral or marginal soils with little or no N and P fertilizer.

The role of P in the biological nitrogen fixation (BNF) process and the mycorrhizal symbiosis places phosphorus at a central place in the tripartite symbiosis. The reduction of one molecule of nitrogen to ammonia through the rhizobia-mediated BNF process requires 16 ATP molecules [170]. Therefore, an increase in the availability of P may directly increase the nitrogenase activity of the *Rhizobium* sp. strain. In contrast to rhizobia, the effect of P on the AMF species depends on the tolerance of the AMF species to P and the P requirement of the host. Generally, high levels of P are detrimental to the germination of spores and/or the growth and differentiation of intraradical and extraradical hyphae [171]. Schubert and Hayman [172] evaluated the response of onion plants to *G. mosseae*, *G. epigaeum* Daniels & Trappe, *G. macrocarpum*, *G. caledonium* (Nicol. & Gerd.) Trappe & Gerdemann, *Glomus* sp. E3, *G. clarum* and *Gi. margarita* in sterilized soil amended with different levels of P. They found that plants responded positively to *G. mosseae* and *G. epigaeum* at low to moderate P levels. However, *G. clarum* was ineffective at low P, whereas *G. macrocarpum* and *Gi. margarita* enhanced plant growth at the medium P level. In contrast, *G. caledonium* and *Glomus* sp. E3 were effective at low, medium and high P levels.

The efficacy of rhizobia in the tripartite symbiosis is only ensured when critical factors such as macro and micronutrient levels are maintained. Therefore, the selection of an appropriate concentration of P is crucial for the proper functioning of the tripartite symbiosis. Various workers have studied the effect of P on the tripartite symbiosis [47,74,173-176]. There is a consensus among all these reports that high P levels were not beneficial in terms of enhancing plant productivity and nutrition if not inhibitory, whereas low to moderate P levels were highly beneficial for the association.

Although some nutrients assume greater importance than others because of the special roles they play, all nutrients essential for plant growth must be present at optimal levels for the maximum growth of plants. For example, trace elements such as Fe and Mo are essential for the proper functioning of the nitrogenase enzyme complex consisting of the dinitrogenase and dinitrogenase reductase [177,178]. The effect of less than optimal levels of these trace elements on the tripartite symbiosis has not been evaluated yet, but it can be safely assumed that these elements may have a significant effect on nitrogen fixation and therefore, on the tripartite symbiosis. Soil characteristics other than nutrient levels can alter the efficacy of interactions between AMF and other beneficial soil microflora. For example, extreme soil pH, soil temperature and soil moisture levels can lead to a reduction in the efficacy of the microbes in action [179]. In general, the overall performance of AMF and rhizobacteria depends upon appropriate plant-soil conditions.

4.3. Compatibility between AMF and microflora

Legume response to various AMF and rhizobia depends on the particular combination of AMF and rhizobia [40,43,49,63,64], which suggests that inter-endophyte compatibility is required for the establishment and functioning of the tripartite symbiosis. Researchers have demonstrated the apparent closeness in the signal transduction pathways for the nodulation and mycorrhizal symbioses [180-182]. Some workers have speculated that, given the proposed time of existence of AMF, some plant processes leading to nodulation probably evolved from pathways established for mycorrhizal symbiosis [180]. Others suggest that AMF signals activate the signal transduction pathway for the nodulation symbiosis [182]. To a large extent, these studies have been accomplished using pea mutants with Nod$^-$ and Myc$^-$ phenotypes, i.e., plants that do not form nodules and are resistant to mycorrhizal colonization [183-186]. In Myc$^-$ pea mutants, host-AMF recognition occurred but fungal penetration of epidermal cells and fungal differentiation were restricted, whereas in the wild type pea, typical fungal entry and differentiation were observed [20]. These Myc$^-$ pea mutants were obtained as a result of mutation at a single locus that pertains to three complementation groups, i.e., *p*, *c*, and *a* that correspond to the sym8, sym19 and sym30 nodulation genes [20]. Furthermore, it was noted that the Myc$^-$ trait does not separate from the Nod$^-$ trait in the F_2 progeny, and therefore, it was speculated that the same gene controls both traits. The Nod$^-$ Myc$^-$ pea mutants were not colonized by field AMF populations or laboratory isolates, under any growth conditions [20,187], suggesting that gene products that define the Nod- and Myc- mutations must play a central role in the establishment of the mycorrhizal and rhizobial symbioses. Recently, Morandi *et al.* [188] assessed the influence of genes regulating supernodulation on the colonization of *G. mosseae* in pea mutants. The two supernodulating pea mutants were affected at the sym28 and sym29 genes [189], and upon colonization by *G. mosseae* exhibited high levels of colonization and significantly lower root mass. This was also associated with a high number of arbuscules. The authors speculated that supernodulation and intensive arbuscular colonization may be associated with an alteration in the endogenous gibberellic acid levels, and that the observed effects were probably the result of common factors involved in the regulation of the nodulation and mycorrhizal symbiosis. However, there were generally no significant differences in the shoot mass of plants.

Results from Nod$^-$ Myc$^-$ soybean mutant studies contradict the results of the Nod$^-$ Myc$^-$ plant mutant studies. For example, Duc *et al.* [183], Balaji *et al.* [185] and Sagan *et al.* [186] report that Nod$^-$ pea mutants are also Myc$^-$. In contrast, Wyss *et al.* [190] found that Nod$^-$ soybean mutants and the wild type plants were colonized equally well by AMF. They also found that the translational products of polyadenylated RNA obtained from nodule-free and wild type plants cross reacted with antiserum against soluble nodulins and membrane-bound nodulins. Similarly, Xie *et al.* [191] observed that *G. mosseae* not only colonized non-nodulating mutants inoculated with an effective *Bradyrhizobium japonicum* strain but also wild type soybean plants inoculated with ineffective rhizobia. They further noted that a *Rhizobium* sp. mutant (NGR□*nod*ABC) did not have an effect on mycorrhization of soybean, and that the addition of 10^{-7} to 10^{-9} M concentrations of highly purified nodulation (Nod) factors (acetylated NodNGR-V and sulfated NodNGR-V) obtained from strain NGR234 had different effects on the colonization of soybean roots by *G. mosseae*. It is unclear from these studies what aspects of the nodulation process or what *nod* genes are involved in the stimulation of the mycorrhiza symbiosis of the same legume host.

As a rule, it appears that all partners have to be compatible for the effective functioning of the tripartite symbiosis. It is relatively easier to assess the effectiveness of a *Rhizobium* strain compared to an AMF species. This is because of the uniform conditions under which a *Rhizobium* strain functions. For example, a *R. leguminosarum* bv. *viceae* strain has a limited host range and is usually repressed by high levels of soil N [192]. Furthermore, in the absence of other competitors and adverse soil conditions such as suboptimal pH and salinity, the *Rhizobium* strain can fix nitrogen and enhance its host growth. However, it is unclear whether a functional rhizobial symbiosis is critical for mycorrhizae formation. This issue has not been addressed in detail so far. Workers attempting to enhance legume productivity through the use of microbial inoculants have selected rhizobia because of their effectiveness on the host or because of strain availability or both [40,49]. However, in attempting to identify the importance of 'Nod' factors on the elicitation of AMF, Xie *et al.* [191] inoculated soybean plants with ineffective *Rhizobium* sp. strains and found that mycorrhizal colonization was not diminished in the soybean plants, and suggested that a functional rhizobial symbiosis was not critical for mycorrhization of the host. It is difficult to make inferences on host growth or productivity from this study, as host growth parameters were not included. On the other hand, the effectiveness of an AMF partner is not well defined with regard to soil and plant species. Depending on the soil nutrient levels (P, N, and micronutrients) and host species, the AMF-host association may be mutualistic or parasitic. Therefore, it is important to select for AMF with appropriate hosts and their rhizobia to maximize the benefits from the tripartite symbiosis.

5. AMF-RHIZOSPHERE MICROFLORA BIOTECHNOLOGY

The wealth of information available on mycorrhizal associations with various plants and their impact on plant productivity testifies to the importance of this symbiosis. The AMF have been reported to function as biofertilizers, biostimulators and bioprotectors, and therefore, biotechnology with regard to AMF can benefit agriculture, horticulture and sylviculture immensely [193-195]. However, the success in commercializing some organisms such as *Rhizobium* spp. and *P. bilaiae* has not been achieved with AMF. The greatest impediment to AMF commercialization is the obligate biotrophicity of the fungi, which necessitates the use of a living host for sustained survival and propagation. Therefore, the cost of inoculum production can be relatively high as compared to other organisms. Furthermore, the lack of consistency in product efficacy and poor market demands have contributed to the insignificant and slow progress in this area.

5.1. Plant biotechnology

Biotechnology refers to the combination of science and technology that uses our knowledge of biological systems for practical applications. The term 'Plant biotechnology' generally refers to and involves drawing principles from various biological sciences to integrate and create technologies that can genetically improve plants, create new methods for pest control and enhance plant nutrition. Enhanced plant nutrition and pest control have traditionally been areas wherein the AMF were naturally involved. However, the application of AMF and other rhizobacteria or fungi into non-traditional systems such as transplanted crops for the purposes of biological control, enhanced competitiveness and improved plant growth and yield have made them a target for plant biotechnology [193,196-200].

Few researchers have determined the response of field-grown crops to AMF and rhizobia [56,201-208]. In all these cases, plants have been inoculated separately with AMF and the diazotrophs. Although most of these reports indicate a synergistic effect leading to enhanced plant productivity, the technology has not been implemented widely in agriculture. The constraints associated with mass production of AMF inocula and the lack of appropriate application technology for AMF have further made the use of AMF in traditional agriculture almost impossible.

Microbial inoculants are steered towards horticultural crops mainly because of the relatively smaller market for fruits and vegetables and the small land size. Furthermore, since most horticultural crops are transplanted, AMF inoculation is made even simpler, by inoculating pre-transplant stage plants or seedlings in artificial substrates, potting mixtures or disinfected nursery beds [197,200]. The major advantages of this technology include early vigor, improved plant growth and productivity at reduced soil nutrient levels, early flowering and fruit set, early competitiveness against weeds, improved crop uniformity, and disease resistance [197,209,210]. In addition, it requires relatively smaller amounts of inoculum compared to field application. Xavier *et al.* [197] reported that co-inoculating pre-transplant bell pepper seedlings with AMF isolates (*G. clarum* and *G. mosseae*) and plant growth promoting rhizobacteria such as *Bacillus* sp. and *Xanthomonas maltophilia* increased pepper yield at medium levels of applied P fertilizer. Gianinazzi *et al.* [211] demonstrated the AMF inoculation enhanced flower production in roses (*Rosa* spp.). Arias and Cargeeg [212] reported that growth promotion by AMF was the basis of commercial application of an AMF inoculant, Vaminoc□□ to cyclamen (*Cyclamen persicum* L.) and chrysanthemum (*Chrysanthemum leucanthemum* L.) grown in Japan. Inoculation of plants with AMF and other bacterial inoculants is particularly feasible for micropropagated plants. Problems such as survival and poor development of shoot and root systems usually associated with micropropagated plants can be effectively eliminated by inoculation with AMF [193]. Vosatka and Gryndler [200] found that co-inoculation of micropropagated potatoes (*Solanum tuberosum* L.) with selected AMF species and bacteria such as *B. subtilis* and *Agrobacterium radiobacter* resulted in higher total weight of minitubers per plant and higher weight per minituber compared to uninoculated plants. This approach not only reduces high fertilizer inputs but also provides other benefits to the developing micropropagated plants.

Systems other than the conventional agriculture and horticulture such as golf greens and herb production can also benefit from AMF and other bacterial inoculations. For example, Gange [213, 214] reported that the abundance of AMF was associated with a reduction in annual meadow grass (*Poa annua* L.) of golf greens. This association was further investigated under controlled conditions. Results show that AMF can cause a growth reduction in *P. annua*, and therefore suggest the possibility of AMF inoculation as a means for biological weed control on golf greens. The co-inoculation of weeds using selective AMF and rhizobacterial inoculants with bioherbicidal properties proven to reduce weeds in horticultural and specialized systems such as golf greens or other important medicinal plants can be an effective alternative to chemical weed control.

The types of mycorrhizal associations generally discussed with tree species are the ectomycorrhizas. Woody leguminous and non-leguminous species such as *Acacia nilotica L.*, *Leucaena leucocephala* (Lam.) de Wit, *Prosopis juliflora* (Mol.) Stuntz. and *Parasponia* spp. derive a great benefit from rhizobia. Co-inoculation of tree species with suitable combinations of rhizobia and AMF at the nursery stage may tremendously aid plants in establishing themselves, especially for those that will be introduced into less than ideal environmental conditions such as reclamation sites. Another non-mycorrhizal species that benefits plants is *Frankia*. Diem and Gauthier [215] co-inoculated *Casuarina equisetifolia* L. with crushed nodules of *Frankia* and *G. mosseae* and found significant improvements in nodulation and plant growth, indicating synergistic interactions between *Frankia* and *G. mosseae*. As Linderman [216] and Torrey [217] suggest, the time has come to address issues of mycorrhizosphere management by streamlining the recognition and development of microbial associations with plants, the characterization, investigational re-establishment of the various associations, expression of growth and developmental parameters, archiving, and application protocols. Indeed, given the advancement in the understanding of AMF biology and genetics at the present time, the success of the co-inoculation of plants with the obligately biotrophic AMF and other rhizosphere microflora rests upon the success of development of efficient culture systems and effective delivery systems.

5.2. State of the Art of AMF Culture Systems and their Limitations

Extensive sporulation of AMF is still limited to the presence of a living host, because of a cell cycle check point which arrests AMF growth in the absence of a living host [218]. Therefore, the routine use of AMF can be achieved by mass producing them under a living host. It is obvious then that culture systems for the large-scale propagation of AMF is an area that has not seen much progress, although some attempts have been made. The traditional AMF culture system (i.e., pot cultures), poses problems of lack of purity, high production cost and lack of uniformity. Advanced systems such as hydroponics [219] and aeroponics [220] not only require specialized equipment for production of inoculum but may also pose problems of impurity in the final product because of the long-term growth conditions.

The development of root organ cultures has revolutionized AMF research in two different ways, (i) understanding the biology of AMF [221-224], and (ii) for the production of plantlets [223] and AMF inoculum [225,226]. St-Arnaud *et al.* [225] described a two compartment *in vitro* system for the production of *G. intraradices* spores. Using this method, they obtained in excess of 34,000 clean spores (mean of 15,000 viable spores) in the distal compartment that did not contain roots. Although root organ cultures are free of external contamination, these systems are not very cost-effective. Jolicoeur *et al.* [226] proposed a second-generation bioprocess for AMF propagation using bioreactors which follows growth in a Petri dish. In all these cases, the high cost of producing mycorrhizal inoculants may generally dissuade the consumer from procuring the high-priced final product. The most ideal way to remove hurdles associated with mass-production of AMF spores is to thoroughly understand the biology and molecular genetics of AMF so that manipulations can be carried out to "rectify" the obligate biotrophicity status.

5.3. Co-delivery Systems and Multiagent Inoculants

Inoculation of plants with a co-culture of various microorganisms for inoculation purposes appears to be an attractive and efficient alternative to separate inoculations [227,228]. For example, Rice *et al.* [228] reported that a common production and delivery system was used for co-culturing *R. meliloti* and *P. bilaiae*. While such a production system would not be feasible for AMF because of the need for a living host, the delivery system can be, such as in the form of a multiagent inoculant. Co-delivery of AMF propagules such as in a sheared form [220] along with *Rhizobium* cells in a liquid formulation that would preserve the viability of the *Rhizobium* cells without loss of viability of the AMF propagules would be ideal for legume seed treatment. However, such products will require extensive research in order to determine feasibility of production, storage conditions, duration of effectiveness, compatibility between the different agents in the formulation, the cost of production and the demand for such products. In the same context, a co-delivery system for selective combinations of *P. bilaiae* and AMF may enhance the availability of P for various crops particularly legumes and horticultural crops. With particular reference to AMF, the private sector has made a great deal of investment in the production of multiagent AMF inoculants containing different AMF species with various beneficial characteristics. The augmentation of soil with multiagent AMF inoculants is aimed at offering a choice of AMF species to the host from which it may select. In field soils interestingly, this may be achieved by adopting appropriate cropping practices, resulting in enhanced biodiversity! The utility of multiagent inoculants may be particularly beneficial in floriculture, herb and spice production and in weed control.

5.4. Gene Transfer in AMF

Gene manipulations carried out with other filamentous fungi are rare or have never been carried out with AMF. The complexity of the symbiosis, the source of specialized structures such as arbuscules, the timing of different fungal and fungal-plant activities are some of the issues that need to be further resolved before manipulations can be carried out. However, great success with AMF can be achieved only after the discovery of the factors that contribute to the cell cycle arrest following spore germination. A recent report by Requena *et al.* [218] shows that a homologue of the yeast (*Saccharomyces cerevisiae* Meyen ex Hansen) TOR2 protein which encodes a cell cycle check point arresting further growth has been found in AMF. Further analysis of the genes encoding this protein and genetic manipulations may determine whether or not this protein synthesis can be repressed leading to normal and unlimited growth in AMF under axenic conditions. Despite many hurdles, many research groups around the world are currently involved in cracking the code of several AMF genes, a step that is critical to the complete understanding and effective use of AMF.

Gene transfer in AMF occurs naturally by hyphal anastamosis [229], thus sustaining and creating genetic diversity. Reports on the genetic transformation of AMF are scarce. Forbes *et al.* [230] used biolistics to transform *Gi. rosea* for the expression of the GUS reporter gene. The success of this transformation presents the potential for genetically manipulation of AMF for various beneficial characteristics, such as enhanced glomalin production [146], improved salinity or drought tolerance, or enhanced heavy metal uptake [156], along with a reporter gene such as GFP or GUS for monitoring purposes. Another important trait that needs to be addressed is the difference in the dependency of plant genotypes of the same host on AMF. It has been proposed that in wheat cultivars, "mycorrhizal responsiveness" genes of newer cultivars have been inadvertently eliminated as a result of breeding for other desired plant characteristics [158]. Understanding the function of the "mycorrhizal responsiveness" genes in terms of the proteins they encode may enable us to verify whether restoration of the "mycorrhizal responsiveness" genes in newer cultivars may confer mycorrhizal dependency. It may be worthwhile to note that given the current enthusiasm among researchers for the molecular genetic analyses of AMF genes, although currently there are no suggestions of transformation of an AMF species for the expression of the nitrogenase enzyme complex, the suggestion and execution of similar strategies may not seem impossible in the future. It appears that there are already reports of AMF naturally harboring intracellular bacteria that contain *nifHDK* genes [231] that may prompt such gene transfer strategies.

6. CONCLUSIONS

Research on the mycorrhizosphere and its potential and the role of AMF as growth promoters, bioprotectors and providers of special benefits such as salinity and drought tolerance have been demonstrated beyond doubt. It is, however, important to realize that the AMF are one of the many groups of microorganisms occupying the rhizosphere. Therefore, interactions between various groups of microbes and AMF do occur, some of which assume significance because of their effect on plant yield. It may be very beneficial to utilize these various associations in order to enhance productivity without resorting to the application of excess of fertilizers or pesticides. In fact, given the nature of AMF it may be ideal to exploit these naturally occurring associations under reduced input levels.

7. REFERENCES

1. Barbier, E.B. (2000). The economic linkages between rural poverty and land degradation: some evidence from Africa. Agric. Ecosys. Environ. 82: 355-370.
2. Bennett, A.J. (2000). Environmental consequences of increasing production: some current perspectives. Agric. Ecosys. Environ. 82: 89-95.
3. Carpentier, C.L., Stephen, V.A., and Witcover, J. (2000). Intensified production systems on western Brazilian Amazon settlement farms: could they save the forest ? Agric. Ecosys. Environ. 82: 73-88.
4. Singh, R.B. (2000). Environmental consequences of agricultural development: a case study from the Green Revolution state of Haryana, India. Agric. Ecosys. Environ. 82: 97-103.
5. Kucheryava, N., Fiss, M., Auling, G., and Kroppenstedt, R.M. (1999). Isolation and characterization of epiphytic bacteria from the phyllosphere of apple, antagonistic *in vitro* to *Venturia inaequalis*, the causal agent of apple scab. Syst. Appl. Microbiol. 22: 472-478.
6. Thrane, C., Olsson, S., Nielsen, T.H., and Sorensen, J. (1999). Vital fluorescent stains for detection of stress in *Pythium ultimum* and *Rhizoctonia solani* challenged with viscosinamide from *Pseudomonas fluorescens* DR54. FEMS Microbiol. Ecol. 30: 11-23.
7. Rosskopf, E.N., Charudattan, R., Shabana, Y.M., and Benny, G.L. (2000). *Phomopsis amaranthicola*, a new species from *Amaranthus* sp. Mycologia 92: 114-122.
8. Paula, M. A., Reis, V. M., and Dobereiner, J. (1991). Interactions of *Glomus clarum* with *Acetobacter diazotrophicus* in infection of sweet potato (*Ipomoea batatus*), sugarcane (*Saccharum* spp.) and sweet sorghum (*Sorghum vulgare*). Biol. Fertil. Soils. 11: 111-115.
9. Bledsoe, C.S. (1992). Physiological ecology of ectomycorrhizae: Implications for field application. *In*: Mycorrhizal Functioning: An Integral Plant-Fungal Process (M.F. Allen ed.), Chapman and Hall, New York. pp. 424-437.
10. Dobbelaere, S., Croonenborghs, A., Thys, A., Broek, A., and Vanderleyden, J. (1999). Phytostimulatory effect of *Azospirillum brasilense* wild type and mutant strains altered in IAA production on wheat. Plant Soil. 212 :155-164.
11. Rodelas, B., Gonzalez-Lopez, J., Pozo, C., Salmeron, V., and Martinez-Toledo, M.V. (1999). Response of Faba bean (*Vicia faba* L.) to combined inoculation with *Azotobacter* and *Rhizobium leguminosarum* bv. *viceae*. Agric. Ecosyst. Environ. 12: 51-59.
12. Sanginga, N., Carsky, R.J., and Dashiell, K. (1999). Arbuscular mycorrhizal fungi respond to rhizobial inoculation and cropping systems in farmers' fields in the Guinea savanna. Biol. Fertil. Soils 30: 179-186.
13. Anderson, T.A., Kruger, E.L., and Coats, J.R. (1994). Enhanced degradation of a mixture of three herbicides in the rhizosphere of a herbicide-tolerant plant. Chemosphere 28: 1551-1557.
14. Donnelly, P.K. and Fletcher, J.S. (1994). Potential use of mycorrhizal fungi as bioremediation agents. *In*: Bioremediation through rhizosphere technology. American Chemical Society, Washington, DC. pp. 93-99.

15. Edwards, S. G., Young, J. P. W., and Fitter, A. H. (1998). Interactions between *Pseudomonas fluorescens* biocontrol agents and *Glomus mosseae*, an arbuscular mycorrhizal fungus, within the rhizosphere. FEMS Microbiol. Lett. 166: 297-303.
16. Paau, A.S. (1991). Improvement of *Rhizobium* inoculants by mutation, genetic engineering and formulation. Biotechnol. Adv. 9: 173-184.
17. Kuykendall, L.D., Abdel-Wahab, S.M., Hashem, F.M., and Berkum, P. van. (1994). Symbiotic competence and genetic diversity of *Rhizobium* strains used as inoculants for alfalfa and berseem clover. Lett. Appl. Microbiol. 19: 477-482.
18. Robleto, E. A., Kmiecik, K., Oplinger, E. S., Nienhuis, J., and Triplett, E. W. (1998). Trifolitoxin production increases nodulation competitiveness of *Rhizobium etli* CE3 under agricultural conditions. Appl. Environ. Microbiol. 64: 2630-2633.
19. S.E. Smith and D.J. Read (eds.) (1997). Mycorrhizal Symbiosis, Second Edition, Academic Press, Toronto, Canada, 605 pp.
20. Gianinazzi-Pearson, V. (1996). Plant cell responses to arbuscular mycorrhizal fungi: Getting to the roots of the symbiosis. Plant Cell. 8: 1871-1883.
21. Hirsch, A.M. and Kapulnik, Y. (1998). Signal transduction pathways in mycorrhizal associations: comparisons with the *Rhizobium*-legume symbiosis. Fungal Genet. Biol. 23: 205-212.
22. Simon, L., Bousquet, J., Levesque, R. C., and Lalonde, M. (1993). Origin and diversification of endomycorrhizal fungi and coincidence with vascular land plants. Nature 363: 67-69.
23. Remy, W., Taylor, T.N., Hass, H., and Kerp, H. (1994). Four hundred-million-year-old vesicular arbuscular mycorrhizae. Proc. Natl. Acad. Sci. USA. 91: 11841-11843.
24. Phipps, C.J. and Taylor, T.N. (1996). Mixed arbuscular mycorrhizae from the Triassic of Antarctica. Mycologia 88: 707-714.
25. Giovannetti, M., Sbrana, C., Citernesi, A.S., and Avio, L. (1996). Analysis of factors involved in fungal recognition responses to host-derived signals by arbuscular mycorrhizal fungi. New Phytol. 133: 65-71.
26. Vaughn, J.C., Mason, M.T., Sper-whitis, G.L., Kuhlman, P., and Palmer, J.D. (1995). Fungal origin in horizonal transfer of a plant mitochondrial group I intron in the chimeric coxI gene of *Peperomia*. J. Mol. Evol. 41: 563-572.
27. Van der Heijden, M.G.A., Klironomos, J.N., Ursic, M., Moutoglis, P., Streitwolf-Engel., R., Boller, T., Wiemken, A., and Sanders, I.R. (1998). Mycorrhizal fungal diversity determines plant biodiversity, ecosystem variability and productivity. Nature 396: 69-72.
28. Hornby, D., Bateman, G.L., Gutteridge, R.J., Lucas, P., Montfort, F., and Cavelier, A. (1990). Experiments in England and France on fertilisers, fungicides and agronomic practices to decrease take-all. Brighton Crop Prot. Conf. Pests Dis. Surrey : British Crop Protection Council 2: 771-776.
29. Lynch, J. M. (1990). Introduction: Some consequences of microbial rhizosphere competence for plant and soil. *In*: The Rhizosphere (J. M. Lynch ed.), John Wiley and Sons, West Sussex, England, p. 1-10.
30. Zheng, J., Sutton, J.C., and Yu, H. (2000). Interactions among *Pythium aphanidermatum*, roots, root mucilage, and microbial agents in hydroponic cucumbers. Can. J. Plant Pathol. 22: 368-379.
31. Ratnayake, M., Leonard, R. T., and Menge, J. A. (1978). Root exudation in relation to supply of phosphorus and its possible relevance to mycorrhiza formation. New Phytol. 81: 543-552.
32. Graham, J. H., Leonard, R. T., and Menge, J. A. (1981). Membrane-mediated decrease in root exudation responsible for inhibition of vesicular-arbuscular mycorrhiza formation. Plant Physiol. 68: 548-552.
33. Linderman, R. G. (1988). Mycorrhizal interactions with the rhizosphere microflora. The mycorrhizosphere effect. Phytopathol. 78: 366-371.
34. Bagyaraj, D. J. and Menge, J. A. (1978). Interaction between vesicular-arbuscular mycorrhizal fungi and *Azotobacter* and their effects on rhizosphere microflora and plant growth. New Phytol. 80: 567-573.
35. Meyer, J. R. and Linderman, R. G. (1986). Response of subterranean clover to dual inoculation with vesicular-arbuscular mycorrhizal fungi and a plant growth promoting bacterium *Pseudomonas putida*. Soil Biol. Biochem. 18: 185-190.
36. Paulitz, T. C. and Linderman, R. G. (1989). Interactions between fluorescent pseudomonads and VA mycorrhizal fungi. New Phytol. 113: 37-45.
37. Nemec, S. (1994). Soil microflora associated with pot cultures of *Glomus intraradix*-infected *Citrus reticulata*. Agric. Ecosyst. Environ. 1: 299-306.
38. Timonen, S., Jorgensen, K. S., Haahtela, K., and Sen, R. (1998). Bacterial community structure at defined locations of *Pinus sylvestris-Suillus bovinus* and *Pinus sylvestris-Paxillus involutus* mycorrhizospheres in dry pine forest humus and nursery peat. Can. J. Microbiol. 44: 499-513.

39. Shalaby, A. M. and Hanna, M. M. (1998). Preliminary studies on interactions between VA mycorrhizal fungus *Glomus mosseae*, *Bradyrhizobium japonicum* and *Pseudomonas syringae* in soybean plants. Acta Microb. Polonica 47: 385-391.
40. Azcon, R., Rubio, R., and Barea, J. M. (1991). Selective interactions between different species of mycorrhizal fungi and *Rhizobium meliloti* strains, and their effects on growth, N_2-fixation (^{15}N) and nutrition of *Medicago sativa* L. New Phytol. 117: 399-404.
41. Ianson, D. C. and Linderman, R. G. (1993). Variation in the response of nodulating pigeonpea (*Cajanus cajan*) to different isolates of mycorrhizal fungi. Symbiosis 15: 105-119.
42. Biro, B., Koves-Pechy, K., Voros, I., Takacs, T., Eggenberger, P., and Strasser, R. J. (2000). Interrelations between *Azospirillum* and *Rhizobium* nitrogen-fixers and arbuscular mycorrhizal fungi in the rhizosphere of alfalfa in sterile, AMF-free or normal soil conditions. Appl. Soil Ecol. 15: 159-168.
43. Xavier, L.J.C. and Germida, J.J. (2001). Response of lentil under controlled conditions to co-inoculation with arbuscular mycorrhizal fungi and rhizobia varying in efficacy. Soil Biol. Biochem. 33 (In Press)
44. Will, M. E. and Sylvia, D. M. (1990). Interaction of rhizosphere bacteria, fertilizer, and vesicular-arbuscular mycorrhizal fungi with sea oats. Appl. Environ. Microbiol. 56: 2073-2079.
45. Andrade, G., Azcon, R., and Bethlenfalvay, G. J. (1995). A rhizobacterium modifies plant and soil responses to the mycorrhizal fungus *Glomus mosseae*. Appl. Soil Ecol. 2: 195-202.
46. Bethlenfalvay, G. J., Andrade, G., and Azcon-Aguilar, C. (1997). Plant and soil responses to mycorrhizal fungi and rhizobacteria in nodulated and nitrate-fertilized peas (*Pisum sativum* L.). Biol. Fertil. Soils 24: 164-168.
47. Xavier, L.J.C. (L. Johnny) (1999). Effects of interactions between arbuscular mycorrhizal fungi and *Rhizobium leguminosarum* on pea and lentil. Ph.D. Thesis, The University of Saskatchewan, Saskatoon, SK, Canada. 263 pp.
48. Kucey, R. M. N. (1987). Increased phosphorus uptake by wheat and field beans inoculated with a phosphorus-solubilizing *Penicillium bilaii* strain and with vesicular-arbuscular mycorrhizal fungi. Appl. Environ. Microbiol. 53: 2699-2703.
49. Ahmad, M.H. (1995). Compatibility and co-selection of vesicular-arbuscular mycorrhizal fungi and rhizobia for tropical legumes. Crit. Rev. Biotech. 15: 229-239.
50. Redente, E. F. and Reeves, F. B. (1981). Interactions between vesicular-arbuscular mycorrhiza and *Rhizobium* and their effect on sweetvetch growth. Soil Sci. 132: 410-415.
51. Herrera, M. A., Salamanca, C. P., and Barea, J. M. (1993). Inoculation of woody legumes with selected arbuscular mycorrhizal fungi and rhizobia to recover desertified mediterranean ecosystems. Appl. Environ. Microbiol. 59: 129-133.
52. Requena, N., Perez-Solis, E., Azcon-Aguilar, C., Jeffries, P., and Barea, J.M. (2001). Management of indigenous plant-microbe symbioses aids restoration of desertified ecosystems. Appl. Environ. Microbiol. 67: 495-498.
53. Manjunath, A., Bagyaraj, D. J., and Gowda, H. S. G. (1984). Dual inoculation with VA mycorrhiza and *Rhizobium* is beneficial to *Leucaena*. Plant Soil. 78: 445-448.
54. Pacovsky, R. S., Fuller, G., and Stafford, A. E. (1986). Nutrient and growth interactions in soybeans colonized with *Glomus fasciculatum* and *Rhizobium japonicum*. Plant Soil. 92: 37-45.
55. Purcino, A. A. C., Lurlarp, C., and Lynd, J. Q. (1986). Mycorrhiza and soil fertility effects with growth, nodulation and nitrogen fixation of *Leucaena* grown on a typic Eutrustox. Commun. Soil Plant Anal. 17: 473-489.
56. Kucey, R. M. N. and Bonetti, R. (1988). Effect of vesicular-arbuscular mycorrhizal fungi and captan on growth and N_2 fixation by *Rhizobium*-inoculated field beans. Can. J. Soil Sci. 68: 143-149.
57. Ibijbijen, J., Urquiaga, S., Ismail, M., Alves, B. J. R., and Boddey, R. M. (1996). Effect of arbuscular mycorrhizal fungi on growth, mineral nutrition and nitrogen fixation of three varieties of common beans (*Phaseolus vulgaris*). New Phytol. 134: 353-360.
58. Saxena, A. K., Rathi, S. K., and Tilak, K. V. B. R. (1997). Differential effect of various endomycorrhizal fungi on nodulating ability of green gram by *Bradyrhizobium* sp. (Vigna) strain S24. Biol. Fertil. Soils 24: 175-178.
59. Vejsadova, H., Siblikova, D., Hrselova, H., and Vancura, V. (1992). Effect of the VAM fungus *Glomus* sp. on the growth and yield of soybean inoculated with *Bradyrhizobium japonicum*. Plant Soil. 140: 121-125.
60. Thiagarajan, T. R. and Ahmad, M.H. (1993). Influence of a vesicular-arbuscular mycorrhizal fungus on the competitive ability of *Bradyrhizobium* spp. for nodulation of cowpea *Vigna unguiculata* (L.) Walp in non-sterilized soil. Biol. Fertil. Soils 15: 294-296.

61. Azcon-Aguilar, C., Barea, J. M., Azcon. R., and Olivares, J. (1982). Effectiveness of *Rhizobium* and VA mycorrhiza in the introduction of *Hedysarum coronarium* in a new habitat. Agric. Ecosys. Environ. 7: 199-206.
62. Paula, M. A., Urquiaga, S., Siqueira, J. O., and Dobereiner, J. (1992). Synergistic effects of vesicular-arbuscular mycorrhizal fungi and diazotrophic bacteria on nutrition and growth of sweet potato (*Ipomoea batatus*). Biol. Fertil. Soils 14: 61-66.
63. Ruiz-Lozano, J. M. and Azcon, R. (1993). Specificity and functional compatibility of VA mycorrhizal endophytes in association with *Bradyrhizobium* strains in *Cicer arietinum*. Symbiosis 15: 217-226.
64. Redecker, D., von Berswordt-Wallrabe, P., Beck, D. P., and Werner, D. (1997). Influence of inoculation with arbuscular mycorrhizal fungi on stable isotopes of nitrogen in *Phaseolus vulgaris*. Biol. Fertil. Soils 24: 344-346.
65. Raj, J., Bagyaraj, D. J., and Manjunath, A. (1981). Influence of soil inoculation with vesicular-arbuscular mycorrhiza and a phosphate-dissolving bacterium on plant growth and ^{32}P-uptake. Soil Biol. Biochem. 13: 105-108.
66. Barea, J. M., Bonis, A. F., and Olivares, J. (1983). Interaction between *Azospirillum* and vesicular-arbuscular mycorrhizae and their effects on growth and nutrition of maize and ryegrass. Soil Biol. Biochem. 1: 705-710.
67. Pacovsky, R. S., Fuller, G., and Paul, E. A. (1985). Influence of soil on the interactions between endomycorrhizae and *Azospirillum* in sorghum. Soil Biol. Biochem. 17: 523-531.
68. Rao, N. S. S., Tilak, K. V. B. R., and Singh, C. S. (1985). Synergistic effect of vesicular-arbuscular mycorrhizae and *Azospirillum brasilei e* on the growth of barley in pots. Soil Biol. Biochem. 1: 121-129.
69. Azcon, R. (1989). Selective interaction between free-living rhizosphere bacteria and vesicular-arbuscular mycorrhizal fungi. Soil Biol. Biochem. 21: 639-644.
70. El-Raheem, A., El-Shanshoury, R., Hassan, M. A., and Abdel-Ghaffar, B.A. (1989). Synergistic effect of vesicular-arbuscular mycorrhizas and *Azotobacter chroococcum* on the growth and the nutrient contents of tomato plants. Phyton (Austria) 29: 203-212.
71 Azcon, R., Barea, J. M., and Callao, V. (1973). Selection of phosphate-solubilizing and nitrogen-fixing bacteria for using as biological fertilizers. Cuad Circ. Biol. 2: 23-30.
72. Barea, J. M. and Brown, M. G. (1974). Effect on plant growth produced by *Azotobacter paspali* related to synthesis of plant growth regulating substances. Appl. Bacteriol. 37: 583-593.
73. Azcon, R. and Barea, J. M. (1975). Synthesis of auxins, gibberellins and cytokinins by *Azotobacter vinelandii* and *Azotobacter beijerinckii* related to effects produced on tomato plants. Plant Soil 43: 609-619.
74. Azcon-Aguilar, C., Gianinazzi-Pearson, V., Fardeau, J. C., and Gianinazzi, S. (1986). Effect of vesicular-arbuscular mycorrhizal fungi and phosphate-solubilizing bacteria on growth and nutrition of soybean in a neutral-calcareous soil amended with ^{32}P-^{45}Ca-tricalcium phosphate. Plant Soil 96: 3-15.
75. Vejsadova, H., Catska, V., Hrselova, H., and Gryndler, M. (1993). Influence of bacteria on growth and phosphorus nutrition of mycorrhizal corn. J. Plant Nutr. 16: 1857-1866.
76. Toro, M., Azcon, R., and Herrera, R. (1996). Effects on yield and nutrition of mycorrhizal and nodulated *Pueraria phaseoloides* exerted by P-solubilizing rhizobacteria. Biol. Fertil. Soils 21: 23-29.
77. Toro, M., Azcon, R., and Barea, J.M. (1997). Improvement of arbuscular mycorrhiza development by inoculation of soil with phosphate-solubilizing rhizobacteria to improve rock phosphate bioavailability (^{32}P) and nutrient cycling. Appl. Environ. Microbiol. 63: 4408-4412.
78. Toro, M., Azcon, R., and Barea, J. M. (1998). The use of isotopic dilution techniques to evaluate the interactive effects of *Rhizobium* genotype, mycorrhizal fungi, phosphate-solubilizing rhizobacteria and rock phosphate on nitrogen and phosphorus acquisition by *Medicago sativa*. New Phytol. 138: 265-273.
79. Kim, K.Y., Jordan, D., and McDonald, G.A. (1998). Effect of phosphate-solubilizing bacteria and vesicular-arbuscular mycorrhizae on tomato growth and soil microbial activity. Biol. Fertil. Soils 26: 79-87.
80. Iqbal, M. and Waterer, D.R. (1993). TITLE Proceedings of the 9th NACOM, Guelph, Ontario, Canada, 1993, p. 94.
81. Gryndler, M. and Vosatka, M. (1996). The response of *Glomus fistulosum*-maize mycorrhiza to treatments with culture fractions from *Pseudomonas putida*. Mycorrhiza 6: 207-211.
82. Walley, F. L. and Germida, J. J. (1997). Response of spring wheat (*Triticum aestivum*) to interactions between *Pseudomonas* species and *Glomus clarum* NT4. Biol. Fertil. Soils 24: 365-371.
83. Sharma, D. D., Govindaiah, Katiyar, R. S., Das, P. K., Janardhan, L., Bajpai, A. K., Choudhry, P. C., and Janardhan, L. (1995). Effect of VA-mycorrhizal fungi on the incidence of major mulberry diseases. Indian J. Seric. 34: 34-37.

84. Waschkies, C., Schropp, A., and Marschner, H. (1994). Relations between grapevine replant disease and root colonization of grapevine (*Vitis* sp.) by fluorescent pseudomonads and endomycorrhizal fungi. Plant Soil 162: 219-227.
85. Li, S. L., Zhao, S. J., Zhao, L. Z., Li, S. L., Zhao, S. J., and Zhao, L. Z. (1997). Effects of VA mycorrhizae on the growth of eggplant and cucumber and control of diseases. Acta Phytophylac. Sinica 24: 117-120.
86. Otto, G. and Winkler, H. (1995). Colonization of rootlets of some species of Rosaceae by actinomycetes, endotrophic mycorrhiza, and endophytic nematodes in a soil conducive to specific cherry replant disease. Z. Pflanzenkr. Pflanzenschutz. 102: 63-68.
87. Suresh, C. K. and Rai, P. V. (1991). Interaction of *Pseudomonas solanacearum* with antagonistic bacteria and VA mycorrhiza. Curr. Res. 20: 36-37.
88. Krishna, K. R. and Bagyaraj, D. J. (1983). Interaction between *Glomus fasciculatum* and *Sclerotium rolfsii* in peanut. Can. J. Bot. 61: 2349-2351.
89. Duchesne, L. C., Peterson, R. L., and Ellis, B. E. (1989). The time-course of disease suppression and antibiosis by the ectomycorrhizal fungus *Paxillus involutus*. New Phytol. 111: 693-698.
90. Boyetchko, S. M. and Tewari, J. P. (1996). Use of VA mycorrhizal fungi in soil-borne disease management. *In*: Management of Soil Borne Diseases (R.S. Utkhede and V.K. Gupta eds.), Kalyani Publishers, New Delhi. pp. 146-163.
91. Kapoor, A. R., Mukherji, K. G., and Kapoor, R. (1998). Microbial interactions in mycorrhizosphere of *Anethum graveolens* L. Phytomorph. 48: 383-389.
92. Kegler, H. and Gottwald, J. (1998). Influence of mycorrhizas on the growth and resistance of asparagus. Arch. Phytopath. Plant Protect. 31: 435-438.
93. Becker, D. M., Bagley, S. T., and Podila, G. K. (1999). Effects of mycorrhizal-associated streptomycetes on growth of *Laccaria bicolor*, *Cenococcum geophilum*, and *Armillaria* species and on gene expression in *Laccaria bicolor*. Mycologia 91: 33-40.
94. Norman, J.R. and Hooker, J.E. (2000). Sporulation of *Phytophthora fragariae* shows greater stimulation by exudates of non-mycorrhizal than by strawberry roots. Mycol. Res. 104: 1069-1073.
95. Kulkarni, S. A., Kulkarni, S., Sreenivas, M. N., and Kulkarni, S. (1997). Interaction between vesicular-arbuscular (VA) mycorrhizae and *Sclerotium rolfsii* Sacc. in groundnut. Karnat. J. Agric. Sci. 10: 919-921.
96. Cordier, C., Pozo, M. J., Barea, J. M., Gianinazzi, S., and Gianinazzi-Pearson, V. (1998). Cell defense responses associated with localized and systemic resistance to *Phytophthora* induced in tomato by an arbuscular mycorrhizal fungus. Mol. Plant-Microbe Interact. 11: 1017-1028.
97. Filion, M., St. Arnaud, M., and Fortin, J. A. (1999). Direct interaction between the arbuscular mycorrhizal fungus *Glomus intraradices* and different rhizosphere microorganisms. New Phytol. 141: 525-533.
98. Kumar, D. P., Hedge, M., Bagyaraj, D. J., and Rao, A. R. M. (1998). Influence of biofertilizers on the growth of cashew (*Anacardium occidentale* L.) rootstocks. Cashew 12: 3-9.
99. Dehne, H. W., and Schönbeck, F. (1975). The influence of the endotrophic mycorrhiza on the fusarial wilt of tomato. Pflanzendrankh. Pflanzenpathol. Pflanzenschutz. 82:630-632.
100. Dehne, H. W., and Schönbeck, F. (1979a). Investigations on the influence of endotrophic mycorrhiza on plant diseases. I. Colonization of tomato plants by *Fusarium oxysporum* f.sp. *lycopersici*. Phytopathol. Z. 95:105-110.
101. Dehne, H. W., and Schönbeck, F. (1979b). Investigations on the influence of endotrophic mycorrhiza on plant diseases. II. Phenol metabolism and lignification. Phytopathol. Z. 95:210-216.
102. Caron, M., Fortin, J. A., and Richard, C. (1985). Influence of substrate on the interaction of *Glomus intraradices* and *Fusarium oxysporum* f.sp. *radicis-lycopersici* on tomatoes. Plant Soil. 87:233-239.
103. Caron, M., Fortin, J. A., and Richard, C. (1986a). Effect of *Glomus intraradices* on the infection by *Fusarium oxysporum* f.sp. *radicis-lycopersici* in tomatoes over a 12-week period. Can. J. Bot. 64:552-556.
104. Caron, M., Fortin, J. A., and Richard, C. (1986b). Effect of phosphorus concentration and *Glomus intraradices* on *Fusarium* crown and root rot of tomatoes. Phytopathol. 76:942-946.
105. Caron, M., Fortin, J. A., and Richard, C. (1986c). Effect of preinfestation of the soil by a vesicular-arbuscular mycorrhizal fungus, *Glomus intraradices*, on *Fusarium* crown and root rot of tomatoes. Phytoprotect. 67:15-19.
106. Caron, M., Fortin, J. A., and Richard, C. (1986d). Effect of inoculation sequence on the interaction between *Glomus intraradices* and *Fusarium oxysporum* f.sp. *radicis-lycopersici* in tomatoes. Can. J. Plant Pathol. 8:12-16.
107. Boyetchko, S. M. (1991). Biological control of the common root rot of barley through the use of vesicular-arbuscular mycorrhizal fungi. Ph.D. Thesis, The University of Alberta, Edmonton, Alberta, Canada, 192 pp.
108. Boyetchko, S. M. and Tewari, J. P. (1988). The effect of VA mycorrhizal fungi on infection by *Bipolaris sorokiniana* in barley. Can. J. Plant Pathol. 10:361. [Abstr]

109. Grey, W. E., van Leur, J. A. G., Kashour, G., and El-Naimi, M. (1989). The interaction of vesicular-arbuscular mycorrhizae and common root rot (*Cochliobolus sativus*) in barley. Rachis 8:18-20.
110. Rempel, C. B. and Bernier, C. C. (1990). *Glomus intraradices* and *Cochliobolus sativus* interactions in wheat grown under two moisture regimes. Can. J. Plant Pathol. 12:338 [Abstr]
111. Boyetchko, S. M. and Tewari, J. P. (1992). Interaction of VA mycorrhizal fungi with the common root rot of barley. *In:* Proc. International Common Root Rot Workshop, Saskatoon, Saskatchewan, August 11-14, 1991. pp. 166-169.
112. Sreeramulu, K. R., Onkarappa, T., and Swamy, H. N. (1998). Biocontrol of damping off and black shank disease in tobacco nursery. Tobac. Res. 24: 1-4.
113. Feldmann, F. and Boyle, C. (1998). Concurrent development of arbuscular mycorrhizal colonization and powdery mildew infection on three *Begonia hiemalis* cultivars. Z. Pflanzenkr. Pflanzenschutz. 105: 121-129.
114. Matsubara, Y., Kayukawa, Y., Yano, M., and Fukui, H. (2000). Tolerance of asparagus seedlings infected with arbuscular mycorrhizal fungus to violet root rot caused by *Helicobasidium mompa*. J. Japan. Soc. Hortic. Sci. 69: 552-556.
115. Prashanthi, S. K., Kulkarni, S., Sreenivasa, M. N., and Kulkarni, S. (1997). Integrated management of root rot disease of safflower caused by *Rhizoctonia bataticola*. Environ. Ecol. 15: 800-802.
116. Sharma,, S., Dohroo, N. P., Sharma, S. (1997). Management of ginger yellows through organic amendment, fungicide seed treatment and biological methods. Ind. Cocoa Arecanut Spice J. 21: 29-30.
117. Kjoller, R. and Rosendahl, S. (1996). The presence of the arbuscular mycorrhizal fungus *Glomus intraradices* influences enzymatic activities of the root pathogen *Aphanomyces euteiches* in pea. Mycorrhiza 6: 487-491.
118. Sierota, Z. (1997). Protection of nurseries against parasitic fungi. Sylwan 141: 5-13.
119. Jaizme-Vega, M. C., Sosa-Hernandez, B., Hernandez-Hernandez, J. M., and Galan-Sauco, V. (1998). Interaction of arbuscular mycorrhizal fungi and the soil pathogen *Fusarium oxysporum* f. sp. *cubense* on the first stages of micropropagated Grande Naine banana. Acta-Hortic. 490: 285-295.
120. Pozo, M. J., Azcon-Aguilar, C., Dumas-Gaudot, E., and Barea, J. M. (1999). β-1, 3-glucanase activities in tomato roots inoculated with arbuscular mycorrhizal fungi and/or *Phytophthora parasitica* and their possible involvement in bioprotection. Plant Sci. 141: 149-157.
121. Budi, S.W., van Tuinen, D.,, Martinotti, G., and Gianinazzi, S. (1999). Isolation from the *Sorghum bicolor* mycorrhizosphere of a bacterium compatible with arbuscular mycorrhiza development and antagonistic towards soilborne fungal pathogens. Appl. Environ. Microbiol. 65: 5148-5150.
122. Rabie, G. H. (1998). Induction of fungal disease resistance in *Vicia faba* by dual inoculation with *Rhizobium leguminosarum* and vesicular-arbuscular mycorrhizal fungi. Mycopathologia 141: 159-166.
123. Guillemin, J. P., Gianinazzi, S., Gianinazzi-Pearson, V., and Martin-Prevel, P. (1997). Endomycorrhiza biotechnology and micropropagated pineapple (*Ananas comosus* L. (Merr.). Acta Hortic. 425: 267-275.
124. Ratti, N., Alam, M., Sharma, S., and Janardhanan, K. K. (1998). Effects of *Glomus aggregatum* on lethal yellowing disease of *Java citronella* caused by *Pythium aphanidermatum*. Symbiosis 24: 115-126.
125. Daft, M. J. and Okusanya, B. O. (1973). Effect of *Endogone* mycorrhiza on plant growth. V. Influence of infection on the multiplication of viruses in tomato, petunia and strawberry. New Phytol. 72: 975-983.
126. Nemec, S. and Myhre, D. (1984). Virus - *Glomus etunicatum* interactions in citrus rootstocks [Sour orange, *Citrus macrophylla*, Duncan grapefruit, potential of mycorrhizal citrus rootstock seedlings to protect against growth suppression by viruses]. Plant Dis. 68: 311-314.
127. Schönbeck, F. (1978). Influence of the endotrophic mycorrhiza on disease resistance of higher plants. Z. Pflanzenkr. Pflanzenschutz. 85: 191-196.
128. Schönbeck, F. and Dehne, H. W. (1979). Investigations on the influence of endotrophic mycorrhiza on plant diseases. 4. Fungal parasites on shoots, *Olpidium brassicae*, TMV. Z. Pflanzenkr. Pflanzenschutz. 86: 103-112.
129. Jayaraman, J., Kumar, D., Jayaraman, J., and Kumar, D. (1995). Influence of mungbean yellow mosaic virus on mycorrhizal fungi associated with *Vigna radiata* var. PS16. Ind. Phytopathol. 48: 108-110.
130. Jabaji-Hare, S. H. and Stobbs, L. W. (1984). Electron microscopic examination of tomato roots coinfected with *Glomus* sp. and tobacco mosaic virus. Phytopathol. 74: 277-279.
131. El-Saidi, M.T. (1997). Salinity and its effect on growth, yield and some physiological processes of crop plants. *In*: Strategies for improving salt tolerance in higher plants (N.H. Enfield, ed.), Science Publications, pp. 111-127.
132. Giocoechea, N., Dolezal, K., Antolin, M.C., Strnad, M., and Sanchez-Diaz, M. (1995). Influence of mycorrhizae and *Rhizobium* on cytokinin content in drought-stressed alfalfa. J. Exp. Bot. 46: 1543-1549.

133. Giocoechea, N., Antolin, M.C., Strnad, M., and Sanchez-Diaz, M. (1996). Root cytokinins, acid phosphatase and nodule activity in drought-stressed mycorrhizal or nitrogen-fixing alfalfa plants. J. Exp. Bot. 47: 683-686.
134. Lynd, J.Q. and Ansman, T.R. (1995). Mycorrhizal etiology of favorable proteoid rhizogenesis, nodulation, and nitrogenase of lupines. J. Plant Nutr. 18: 2365-2377.
135. Dixon, R. K., Rao, M. V., and Garg, V.K. (1993). Inoculation of *Leucaena* and *Prosopis* seedlings with *Glomus* and *Rhizobium* species in saline soil: rhizosphere relations and seedling growth. Arid Soil Res. Rehab. 7: 133-144.
136. Baker, A., Sprent, J.I., and Wilson, J. (1995). Effects of sodium chloride and mycorrhizal infection on the growth and nitrogen fixation of *Prosopis juliflora*. Symbiosis 19: 39-51.
137. Azcon, R. and El-Atrach, F. (1997). Influence of arbuscular mycorrhizae and phosphorus fertilization on growth, nodulation and N_2 fixation (^{15}N) in *Medicago sativa* at four salinity levels. Biol. Fertil. Soils 24: 81-86.
138. McMillen, B.G., Juniper, S., and Abbott, L.K. (1998). Inhibition of hyphal growth of a vesicular-arbuscular mycorrhizal fungus in soil containing sodium chloride limits the spread of infection from spores. Soil Biol. Biochem. 30: 1639-1646.
139. Hirrel, M. C. and Gerdemann, J. W. (1980). Improved growth of onion and bell pepper in saline soils by two vesicular-arbuscular mycorrhizal fungi. Soil Sci. Soc. Am. J. 44: 654-655.
140. Tsang, A. and Maun, M.A. (1999). Mycorrhizal fungi increase salt tolerance of *Strophostyles helvola* in coastal foredunes. Plant Ecol. 144: 159-166.
141. Skujins, J. and Allen, M.F. (1986). Use of mycorrhizae for land rehabilitation. J. Appl. Microbiol. Biotechnol. 2: 161-176.
142. Allen, M.F. (1989). Mycorrhizae and rehabilitation of disturbed arid soils: processes and practices. Arid Soil Res. Rehab. 3: 229-241.
143. Azcon, R. and Barea, J.M. (1997). Mycorrhizal dependency of a representative plant species in mediterranean shrublands (*Lavandula spica* L.) as a key factor to its use for revegetation strategies desertification-threatened areas. Agric,-ecosyst-environ. 7: 83-92.
144. Bethlenfalvay, G.J., Cantrell, I.C., Mihara, K.L., and Schreiner, R.P. (1999). Relationships between soil aggregation and mycorrhizae as influenced by soil biota and nitrogen nutrition. Biol. Fertil. Soils 28: 356-363.
145. Estaun, V., Save, R., and Biel, C. (1997). AM inoculation as a biological tool to improve plant revegetation of a disturbed soil with *Rosmarinus officinalis* under semi-arid conditions. Appl. Soil Ecol. 6: 223-229.
146. Wright, S.F. and Upadhyaya, A. (1998). A survey of soils for aggregate stability and glomalin, a glycoprotein produced by hyphae of arbuscular mycorrhizal fungi. Plant Soil. 198: 97-107.
147. Entry, J.A., Vance, N.C., Hamilton, M.A., Zabowski, D., Watrud, L.S., and Adriano, D.C. (1996). Phytoremediation of soil contaminated with low concentrations of radionuclides. Water, Air Soil Pollut. 88: 167-176.
148. Leyval, C. and Binet, P. (1998). Effect of polyaromatic hydrocarbons in soil on arbuscular mycorrhizal plants. J. Environ. Qual. 27: 402-407.
149. Del Val, C., Barea, J.M., and Azcon-Aguilar, C. (1999). Diversity of arbuscular mycorrhizal fungus populations in heavy-metal contaminated soils. Appl. Environ. Microbiol. 65: 718-723.
150. Khan, A.G., Kuek., C., Chaudhry, T.M., Khoo, C.S., and Hayes, W.J. (2000). Role of plants, mycorrhizae and phytochelators in heavy metal contaminated land remediation. Chemosphere 41: 197-207.
151. Kelly, J.J., Haggblom, M., and Tate, R.L.III. (1999). Changes in soil microbial communities over time resulting from one time application of zinc: a laboratory microcosm study. Soil Biol. Biochem. 31: 1455-1465.
152. El-Kherbawy, M., Angle, J.S., Heggo, A., and Chaney, R.L. (1989). Soil pH, rhizobia, and vesicular-arbuscular mycorrhizae inoculation effects on growth and heavy metal uptake of alfalfa (*Medicago sativa* L.). Biol. Fertil. Soils 8: 61-65.
153. Siciliano, S.D. and Germida, J.J. (1998). Biolog analysis and fatty acid methyl ester profiles indicate that pseudomonad inoculants that promote phytoremediation alter the root-associated microbial community of *Bromus biebersteinii*. Soil Biol. Biochem. 30: 1717-1723.
154. Bizily, S.P., Rugh, C.L., Summers, A.O., and Meagher, R.B. (1999). Phytoremediation of methylmercury pollution: merB expression in *Arabidopsis thaliana* confers resistance to organomercurials. Proc. Natl. Acad. Sci. USA. 96: 6808-6813.

155. Zablotowicz, R.M., Locke, M.A., and Hoagland, R.E. (1997). Aromatic nitroreduction of acifluorfen in soils, rhizospheres, and pure cultures of rhizobacteria. *In*: Phytoremediation of soil and water contaminants, American Chemical Society, pp. 38-53.
156. Stommel, M., Mann, P., and Franken, P. (2001). EST-library construction using spore RNA of the arbuscular mycorrhizal fungus *Gigaspora rosea*. Mycorrhiza 10: 281-285.
157. Baylis, G. T. S. (1972). Fungi, phosphorus and the evolution of root systems. Search 3: 257-258.
158. Hetrick, B. A. D., Wilson, G. W. T., Gill, B. S., and Cox, T. S. (1995). Chromosome location of mycorrhizal responsive genes in wheat. Can. J. Bot. 73: 891-897.
159. Vierheilig, H., Bennett, R., Kiddle, G., Kaldorf, M., and Ludwig-Miller, J. (2000). Differences in glucosinolate patterns arbuscular mycorrhizal status of glucosinolate-containing plant species. New Phytol. 146: 343-352.
160. Xavier, L. J. C. and Germida, J. J. (1998). Response of spring wheat cultivars to *Glomus clarum* NT4 in a P-deficient soil containing arbuscular mycorrhizal fungi. Can. J. Soil Sci. 78: 481-484.
161. Hirrel, M. C., Mehavaran, H., and Gerdemann, J. W. (1976). Vesicular-arbuscular mycorrhizae in the Chenopodiaceae and Cruciferae: do they occur? Can. J. Bot. 56: 2813-2817.
162. Schreiner, R. P. and Koide, R. T. (1993). Antifungal compounds from the roots of mycotrophic and non-mycotrophic plant species. New Phytol. 123: 99-105.
163. DeMars, B. G. and Boerner. R. E. (1996). Vesicular arbuscular mycorrhizal development in the Brassicaceae in relation to plant life span. Flora 191: 179-189.
164. Mohr, U., Lange, J., Boller, T., Wiemken, A., and Vogeli-Lange, R. (1998). Plant defence genes are induced in the pathogenic interaction between bean roots and *Fusarium solani*, but not in the symbiotic interaction with the arbuscular mycorrhizal fungus *Glomus mosseae*. New Phytol. 138: 589-598.
165. Blilou, I., Ocampo, J.A., and Garcia-Garrido, J.M. (1999). Resistance of pea roots to endomycorrhizal fungus or *Rhizobium* correlates with enhanced levels of endogenous salicylic acid. J. Exp. Bot. 50: 1663-1668.
166. Azcon, R. and Ocampo, J. A. (1981). Factors affecting the vesicular-arbuscular infection and mycorrhizal dependency of thirteen wheat cultivars inoculated with *Glomus mosseae*. New Phytol. 87: 677-685.
167. Ibrahim, K. K., Arunachalam, V., Rao, P. S. K., and Tilak, K. V. B. R. (1995). Seasonal response of groundnut genotypes to arbuscular mycorrhiza-*Bradyrhizobium* inoculation. Microbiol. Res. 150: 218-224.
168. Thakur, A. K. and Panwar, J. D. S. (1997). Response of *Rhizobium*-vesicular arbuscular mycorrhizal symbionts on photosynthesis, nitrogen metabolism and sucrose translocation in greengram (*Phaseolus radiatus*). Ind. J. Agric. Sci. 67: 245-248.
169. Peng, S., Eissenstat, D.M., Graham, J.H., Williams, K., and Hodge, N.C. (1993). Growth depression in mycorrhizal citrus at high-phosphorus supply. Analysis of carbon costs. Plant Physiol. 101: 1063-1071.
170. Atkins, C. A. and Rainbird, R. M. (1982). Physiology and biochemistry in biological nitrogen fixation in legumes. *In*: Advances in Agricultural Microbiology (N. S. Subba Rao ed.), Butterworth, London, pp. 26-52.
171. Miranda, J. C. C. and Harris, P. J. (1994). Effects of soil phosphorus on spore germination and hyphal growth of arbuscular mycorrhizal fungi. New Phytol. 128: 103-108.
172. Schubert, A. and Hayman, D. S. (1986). Plant growth responses to vesicular-arbuscular mycorrhiza. XVI. Effectiveness of different endophytes at different levels of soil phosphate. New Phytol. 103: 79-90.
173. Barea, J. M., Escudero, J. L., and Azcon-Aguilar, C. (1980). Effects of introduced and indigenous VA mycorrhizal fungi on nodulation, growth and nutrition of *Medicago sativa* in phosphate-fixing soils as affected by P fertilizers. Plant Soil. 54: 283-296.
174. Morton, J. B., Yarger, J. E., and Wright, S. F. (1990). Soil solution P concentrations necessary for nodulation and nitrogen fixation in mycorrhizal and non-mycorrhizal red clover (*Trifolium pratense* L.). Soil Biol. Biochem. 22: 127-129.
175. Rahman, M. K. and Parsons, J. W. (1997). Effects of inoculation with *Glomus mosseae*, *Azorhizobium caulinodans* and rock phosphate on the growth of and nitrogen and phosphorus accumulation in *Sesbania rostrata*. Biol. Fertil. Soils 25: 47-52.
176. Stamford, N. P., Ortega, A. D., Temprano, F., and Santos, D. R. (1997). Effects of phosphorus fertilization and inoculation of *Bradyrhizobium* and mycorrhizal fungi on growth of *Mimosa caesalpiniaefolia* in an acid soil. Soil Biol. Biochem. 29: 959-964.
177. Munns, D. N. (1977). Mineral nutrition and the legume symbiosis. *In*: A Treatise on Dinitrogen Fixation (R. W. F. Hardy and A. H. Gibson eds.), John Wiley and Sons, New York, pp. 353-391.
178. Fischer, H. M. (1994). Genetic regulation of nitrogen fixation in rhizobia. Microbiol. Rev. 58: 352-386.

179. Klopatek, C.C., DeBano, L.F., and Klopatek, J.M. (1990). Impact of fire on the microbial processes in pinyon-juniper woodlands: management im‸lications. Gen-Tech-Rep-RM-Rocky-Mt-For-Range-Exp-Stn-U-S-Dep-Agric-For-Serv. Fort Collins, Colo. 191: 197-205.
180. La Rue, T.A. and Weeden, N.F. (1994). The symbiosis genes of the host. *In*: Proceedings of the First European Nitrogen Fixation Conference (G.B. Kiss and G. Endre, eds.) Szeged, Hungary Officina Press, pp. 147-151.
181. Van Rhijn, P., Fang, Y., and Galili, S. (1997). Expression of early nodulin genes in alfalfa mycorrhizae indicates that signal transduction pathways used in forming arbuscular mycorrhizae and *Rhizobium*-induced nodules may be conserved. Proc. Natl. Acad. Sci. USA 94: 5467-5472.
182. Albrecht, C., Geurts, R., Lapeyrie, F., and Bisseling, T. (1998). Endomycorrhizae and rhizobial Nod factors both require SYM8 to induce the expression of the early nodulin genes *PsENOD5* and *PsENOD12A*. Plant J. 15: 605-614.
183. Duc, G., Trouvelot, A., Gianinazzi-Pearson, V., and Gianinazzi, S. (1989). First report of non-mycorrhizal plant mutants (Myc-) obtained in pea (*Pisum sativum* L.) and fababean (*Vicia faba* L.). Plant Sci. 60: 215-222.
184. Gianinazzi-Pearson, V., Gianinazzi, S., Guillemin, J. P., Trouvelot, A., and Duc, G. (1991). Genetic and cellular analysis of resistance to vesicular-arbuscular (VA) mycorrhizal fungi in pea mutants. *In*: Advances in Molecular Genetics of Plant-Microbe Interactions (H. Hennecke and D. P. S. Verma eds.), Kluwer Academic Publishers, London. pp. 336-342.
185. Balaji, B., Ba, A. M., LaRue, T. A., Tepfer, D., and Piche, Y. (1994). *Pisum sativum* mutants insensitive to nodulation are also insensitive to invasion *in vitro* by the mycorrhizal fungus, *Gigaspora margarita*. Plant Sci. 102: 195-203.
186. Sagan, M., Morandi, D., Tarenghi, E., and Duc, G. (1995). Selection of nodulation and mycorrhizal mutants in the model plant *Medicago truncatula* (Gaertn.) after alpha-ray mutagenesis. Plant Sci. 111: 63-71.
187. Sagan, M., Messager, A., and Duc, G. (1993). Specificity of the *Rhizobium*-legume symbiosis obtained after mutagenesis in pea (*Pisum sativum* L.). New Phytol. 125: 757-761.
188. Morandi, D., Sagan, M., Prado-Vivant, E., and Duc, G. (2000). Influence of genes determining supernodulation on root colonization by the mycorrhizal fungus *Glomus mosseae* in *Pisum sativum* and *Medicago truncatula* mutants. Mycorrhiza 10: 37-42.
189. Sagan, M. and Duc, G. (1996). Sym28 and Sym29, two new genes involved in regulation of nodulation in pea (*Pisum sativum* L.). Symbiosis 20: 229-245.
190. Wyss, P., Mellor, R. B., and Wiemken, A. (1990). Vesicular-arbuscular mycorrhizas of wild-type soybean and non-nodulating mutants with *Glomus mosseae* contain symbiosis-specific polypeptides (mycorrhizins), immunologically cross-reactive with nodulins. Planta 182: 22-26.
191. Xie, Z. P., Staehelin, C., Vierheilig, H., Wiemken, A., Jabbouri, S., Broughton, W. J., Vogeli-Lange, R., and Boller, T. (1995). Rhizobial nodulation factors stimulate mycorrhizal colonization of nodulating and nonnodulating soybeans. Plant Physiol. 108: 1519-1525.
192. Novak, K., Skrdleta, V., Nemcova, M., and Lisa, L. (1993). Behaviour of pea nodulation mutants as affected by increasing nitrate level. Symbiosis 15: 195-206.
193. Budi, S.W. Cordier, C., Trouvelot, A., Gianinazzi-Pearson, V., Gianinazzi, S., Lemoine, M.C., and Blal, B. (1998). Arbuscular mycorrhiza as a way of promoting sustainable growth of micropropagated plants. Acta Hortic. 457: 71-77.
194. Kendrick, B. and Berch, S. (1985). Mycorrhizae: Applications in agriculture and forestry. *In*: Comprehensive biotechnology : the principles, applications, and regulations of biotechnology in industry, agriculture, and medicine (M. M. Young) Pergamon, Oxford. pp. 109-152.
195. Okon, Y., Bloemberg, G.V., and Lugtenberg, B.J.J. (1998). Biotechnology of biofertilization and phytostimulation. *In*: Agricultural Biotechnology (A. Altman ed.), Marcel Dekker, New York, pp. 327-349.
196. Varma, A. and Schuepp, H. (1995). Mycorrhization of the commercially important micropropagated plants. Crit. Rev. Biotechnol. 15: 313-328.
197. Xavier, L.J.C., de Freitas, J. R., and Germida, J.J. (1996) Response of pepper (*Capsicum annuum* L.) at different levels of P to co-inoculation with arbuscular mycorrhizal fungi and plant growth promoting rhizobacteria. Abstracts of the 46th Annual Meeting of the Canadian Society of Microbiology, AMp2, p49, Charlottetown, Prince Edward Island.
198. Azcon-Aguilar, C. and Barea, J.M. (1997). Applying mycorrhiza biotechnology to horticulture: significance and potentials. Scientia Hortic. 68: 1-24.

199. Vosatka, M., Jansa, J., Regvar, M., Sramek, F., and Malcova, R. (1999). Inoculation with mycorrhizal fungi--a feasible biotechnology for horticulture. Phyton. 39: 219-244.
200. Vosatka, M. and Gryndler, M. (2000). Response of micropropagated potatoes transplanted to peat media to post-vitro inoculation with arbuscular mycorrhizal fungi and soil bacteria. Appl. Soil Ecol. 15: 145-152.
201. Azcon-Aguilar, C. and Barea, J.M. (1981). Field inoculation of *Medicago sativa*, alfalfa with V-A vesicular-arbuscular mycorrhiza and *Rhizobium* in phosphate-fixing agricultural soil. Soil Biol. Biochem. 13: 19-22.
202. Barea, J.M. and Azcon-Aguilar, C. (1983). Mycorrhizas and their significance in nodulating nitrogen-fixing plants [Mycorrhizal physiology, soil ecosystems]. Adv. Agron. 36: 1-54.
203. Puthru, J.T., Prasad, K.V.S.K., Sharmila, P., and Pardha-saradhi, P. (1998). Vesicular arbuscular mycorrhizal fungi improves establishment of micropropagated *Leucaena leucocephala* plantlets. Plant Cell 53: 41-47.
204. Franco, A.A. and Faria, S.M. de. (1997). The contribution of N_2-fixing tree legumes to land reclamation and sustainability in the tropics. Soil Biol. Biochem. 29: 897-903.
205. Tilak, K.V.B.R. (1987). Recent developments in the field of *Azospirillum* and its interaction with vesicular-arbuscular mycorrhizal (VAM) fungi in increasing the crop yields. *In*: Crop Productivity (H.C. Srinivastasa, S. Bhaskaran, K.K.G. Menon), Oxford & IBH Publishing Co. pp. 287-305.
206. Barea, J. M., Azcon-Aguilar, C., and Azcon, R. (1987). Vesicular-arbuscular mycorrhiza improve both symbiotic N_2-fixation and N uptake from soil as assessed with a ^{15}N technique under field conditions. New Phytol. 106: 717-725.
207. Barea, J. M., El-Atrach, F., and Azcon, R. (1989). Mycorrhiza and phosphate interactions as affecting plant development, N_2-fixation , N-transfer and N-uptake from soil in legume-grass mixtures by using a ^{15}N dilution technique. Soil Biol. Biochem. 21: 581-589.
208. Vejsadova, H., Catska, V., Hrselova, H., and Gryndler, M. (1993). Influence of bacteria on growth and phosphorus nutrition of mycorrhizal corn. J. Plant Nutr. 16: 1857-1866.
209. Barea, J.M., Azcon, R., and Azcon-Aguilar, C. (1993). Mycorrhiza and crops. *In*: Advances in Plant Pathology Vol. 9. Mycorrhiza: A synthesis. (I. Tommerup ed.), Academic Press, London. pp. 167-189.
210. Chang, D.C.N. (1994). What is the potential for management of vesicular-arbuscular mycorrhizae in horticulture ? *In*: Management of Mycorrhizas in Agriculture, Horticulture and Forestry (A.D. Robson, L.K.Abbott and N. Malajchuk eds.), Kluwer, Dordrecht. pp. 187-190.
211. Gianinazzi, S., Trouvelot, A., and Gianinazzi-Pearson, V. (1990). Role of mycorrhizas in horticultural crop production. XXII International Horticultural Congress, Florence, 1990, p. 25-30.
212. Arias, I. And Cargeeg, R.E.P. (1992). Vaminoc □ a commercial VA mycorrhizal inoculant. International Symposium on Management of Mycorrhizas in Agriculture, Horticulture and Forestry, Perth, 105-109.
213. Gange, A.C. (1998). A potential microbiological method for the reduction of *Poa annua* L. in golf greens. J. Turfgrass Sci. 74: 40-45.
214. Gange, A.C., Lindsay, D. E., and Ellis, L.S. (1999). Can arbuscular mycorrhizal fungi be used to control the undesirable grass *Poa annua* on golf courses. J. Appl. Ecol. 36: 909-919.
215. Diem, H.G. and Gauthier, D. (1982). Effect of endomycorrhizal infection (*Glomus mosseae*) [fungi] on the modulation and growth of *Casuarina equisetifolia*. C. R. Seances. Acad. Sci. Ser. III Sci. Vie. 294: 215-218.
216. Linderman, R.G. (1991). Mycorrhizal interactions in the rhizosphere. *In*: The rhizosphere and plant growth (D.L. Keister and P.B. Cregan eds.), Kluwer Academic Publishers, Dordrecht, 1991, p. 343-348.
217. Torrey, J.G. (1992). Can plant productivity be increased by inoculation of tree roots with soil microorganisms. Can. J. For. Res. 22: 1815-1823.
218. Requena, N., Mann, P., and Franken, P. (2000). A homologue of the cell cycle check point TOR2 from *Saccharomyces cerevisiae* exists in the arbuscular mycorrrhizal fungus *Glomus mosseae*. Protoplasma 212: 89-98.
219. George, E., Haussler, K.U.,Vetterlein, K.U., Gorgus, E., and Marschner, H. (1992). Water and nutrient translocation by hyphae of *Glomus mosseae*. Can. J. Bot. 70: 2130-2137.
220. Sylvia, D.M. and Jarstfer, A.G. (1992). Sheared-root inocula of vesicular-arbuscular mycorrhizal fungi. Appl. Environ. Microbiol. 58: 229-232.
221. Chabot, S., Becard, G., and Piche, Y. (1992). Life cycle of *Glomus intraradix* in root organ culture. Mycologia 84: 315-321.
222. Benhamou, N., Fortin, J. A., Hamel, C., St-Arnaud, M., and Shatilla, A. (1994). Resistance responses to mycorrhizal Ri T-DNA-transformed carrot roots to infection by *Fusarium oxysporum* f. sp. *chrysanthemi*. Phytopathol. 84: 958-968.

223. Mathur, N. and Vyas, A. (1995). In vitro production of *Glomus deserticola* in association with *Ziziphus nummularia*. Plant Cell Rep. 14: 735-737.
224. Rousseau, A., Benhamou, N., Chet, I., and Piché, Y. (1996). Mycoparasitism of the extramatrical phase of *Glomus intraradices* by *Trichoderma harzianum*. Phytopathol. 86: 434-443.
225. St-Arnaud, M., Hamel, C., Vimard, B., Caron, M., and Fortin, J.A. (1996). Enhanced hyphal growth and spore production of the arbuscular mycorrhizal fungus *Glomus intraradices* in an *in vitro* system in the absence of host roots. Mycol. Res. 100: 328-332.
226. Jolicoeur, M., Williams, R.D., Chavarie, C. Fortin, J.A., and Archambault, J. (1999). Production of *Glomus intraradices* propagules, an arbuscular mycorrhizal fungus, in an airlift bioreactor. Biotechnol. Bioeng. 63: 224-232.
227. Wood, T. and Cummings, B. (1992). Biotechnology and the future of VAM commercialization. *In*: Mycorrhizal Functioning: An Integral Plant-Fungal Process (M.F.Allen ed.), Chapman and Hall, New York. pp. 468-487.
228. Rice, W.A., Olsen, P.E., and Leggett, M.E. (1994). Co-culture of *Rhizobium meliloti* and a phosphorus-solubilizing fungus (*Penicillium bilaii*) in sterile peat. Soil Biol. Biochem. 27: 703-705.
229. Giovannetti, M., Azzolini, D., and Citernesi, A.S. (1999). Anastomosis formation and nuclear and protoplasmic exchange in arbuscular mycorrhizal fungi. Appl. Environ. Microbiol. 65: 5571-5575.
230. Forbes, P. J., Millam, S., Hooker, J. E., and Harrier, L. A. (1998). Transformation of the arbuscular mycorrhiza *Gigaspora rosea* by particle bombardment. Mycol. Res. 102: 497-501.
231. Bianciotto, V., Bandi, C., Minerdi, D., Sironi, M., Tichy, H.V., and Bonfante, P. (1996). An obligately endosymbiotic mycorrhizal fungus itself harbors obligately intracellular bacteria. Appl. Environ. Microbiol. 62: 3005-3010.

Applied Mycology & Biotechnology

Volume 2: Agriculture and Food Production
Edited By G.G. Khachatourians and D.K. Arora
Keyword index